Technological Advancements and Applications in Mobile Ad-Hoc Networks:

Research Trends

Kamaljit I. Lakhtaria
Sir Padampat Singhania University, India

Information Science
REFERENCE

Managing Director:	Lindsay Johnston
Senior Editorial Director:	Heather Probst
Book Production Manager:	Sean Woznicki
Development Manager:	Joel Gamon
Development Editor:	Hannah Abelbeck
Acquisitions Editor:	Erika Gallagher
Typesetter:	Lisandro Gonzalez
Cover Design:	Nick Newcomer, Lisandro Gonzalez

Published in the United States of America by
Information Science Reference (an imprint of IGI Global)
701 E. Chocolate Avenue
Hershey PA 17033
Tel: 717-533-8845
Fax: 717-533-8661
E-mail: cust@igi-global.com
Web site: http://www.igi-global.com

Library of Congress Cataloging-in-Publication Data

Technological advancements and applications in mobile ad-hoc networks: research trends / Kamaljit I. Lakhtaria, editor.
p. cm.
Includes bibliographical references and index.
Summary: “This book offers a current look into the latest research in the ad-hoc network field, including frameworks for development and future directions of a wide range of applications, theories, and use across industrial, commercial, and domestic settings”--Provided by publisher.
ISBN 978-1-4666-0321-9 (hardcover) -- ISBN 978-1-4666-0322-6 (ebook) -- ISBN 978-1-4666-0323-3 (print & perpetual access) 1. Ad hoc networks (Computer networks) I. Lakhtaria, Kamaljit I., 1982-
TK5105.77.T43 2012
621.39'81--dc23

2011049482

British Cataloguing in Publication Data
A Cataloguing in Publication record for this book is available from the British Library.

All work contributed to this book is new, previously-unpublished material. The views expressed in this book are those of the authors, but not necessarily of the publisher.

To my Father & Mother:
Shri Ishwarlal J. Lakhtaria (Doctor), Shrimati Prabhaben I. Lakhtaria
& Wife Panna

Editorial Advisory Board

List of Reviewers

Table of Contents

Detailed Table of Contents

When designing MANET, several challenges arise due to the shared nature of the wireless medium, limited transmission range of wireless devices, and node mobility. Node mobility poses a greater challenge as it makes routes more prone to frequent ruptures, which in turn degrades the overall network performance. If the future movement of the node can be predicted in a precise way, the impact of mobility can be decreased and the overall performance of the network can be improved. This chapter provides an introduction and a critical review on some of the existing mobility prediction models and their effect on the performance of existing routing protocols.

Multicast routing has attracted a lot of attention in group oriented computing due to supporting data transmission from a single source node to multiple destinations concurrently. The advantage of multicast routing lies in its capability of reducing the communication cost and saving the network resources by sending only one copy of the message over the shared link leading to different destinations. Generally speaking, this survey classifies the multicast routing protocols into four categories based on the underlying routing structure: tree-based, mesh-based, hybrid and stateless multicast routing protocols. This survey summarizes the well-known most recent protocols of each category and compares their objectives, performances, advantages and disadvantages. This survey study helps to choose the best multicasting protocol for each network application with respect to the requirements of application.

network describing node's height and status of all connected links. The importance of these heights in TORA routing protocol are that a node in network may only forward data packets downstream instead of upstream. TORA are more appropriate for QOS in ad hoc networks with a higher mobility and data transfer rate.

Section 2
MANET Protocols

Chapter 7

Due to the inherent broadcast nature of the wireless medium, one-to-many multicast communications are characteristic of most of the MANET applications. MANET multicast routing protocols are primarily classified as tree-based and mesh-based protocols depending on the underlying topology used for communication. Tree-based protocols are further classified to source-tree based and shared-tree based depending on whether the tree is rooted at the multicast source or a common node. Further, there are several sub-categories of source-tree based and shared-tree based multicast routing protocols. The mesh-based multicast protocols are primarily categorized to source-initiated and receiver-initiated protocols depending on which entity (the source node or the receiver nodes) initiates the mesh formation. The book chapter explains in detail the working of eleven different multicast routing protocols, covering all of the sub-categories of tree-based and mesh-based routing protocols.

Chapter 8

Due to limited transmission range of wireless network more than single hops may be required to transfer data across the network. In order to reliable communication within the network, a routing protocol is used which are call MANET routing protocol. The major function of such an MANET routing protocol is to establishment short and real route between a pair of nodes so that messages may be delivered in a timely manner. This chapter examines two routing protocols for mobile ad hoc networks– the Destination Sequenced Distance Vector (DSDV), the pro-active routing protocol and the Ad hoc On- Demand Distance Vector routing (AODV), and re-active routing protocol. The chapter compares both protocols based on packet delivery fraction and average delay with changing number of source nodes and pause time.

Chapter 9

The mobility of nodes in mobile ad hoc networks and absence of any centralized control cause unpredictable changes in the network topologies. This makes routing a challenging task. Several routing protocols for mobile ad-hoc networks came into the existence. The protocols are classified in mainly in three categories: proactive, reactive, and hybrid. In this chapter a study of one of each proactive and reactive protocol respectively Destination Sequence Distance Vector routing (DSDV) and Dynamic Source Routing (DSR) is presented. The performance of above said protocols has been measured under

This chapter brings in the idea of deterministic approach to key generation. There is no need to assume a probability for any two nodes to be neighbors. Combinatorial based approach for key generation using a block design technique called symmetric Balanced Incomplete Block Design (BIBD). The main contribution to this work is the use of multiple key spaces to increase resilience and decrease memory utilization and still retain connectivity. It eliminates the dependency between the number of keys in the key-chain and number of nodes in the network. OMNeT++ with Mobility Framework is used for simulating the sensor network. The simulation shows each node broadcasting its key-chain, comparing its key-chain with that of its neighbor and then disconnecting the connection with those nodes that do not have a common key with its key-chain. It also generates an optimal number of key-chains to support birth of nodes after the network is deployed.

This chapter covers issues related to security in mobile ad hoc networks. It acts as a comprehensive survey material covering the cryptographic schemes and trust modeling techniques traditionally found in mobile ad hoc network (MANET) survey articles. The need for hybrid security techniques, involving both cryptographic approach and trust based model, in a resource constrained ad hoc network is also emphasized in this work. Additionally, the lack of realism in the research works related to ad hoc network security is also pointed out. The state of the art in ad hoc security should employ hybrid techniques which can be easily implemented in an ad hoc network. This will lead to the large scale deployment of mobile ad hoc networks in various context-aware applications.

The security of multimedia data in Wireless Ad hoc Networks (WANet) is commonly provided by encryption, which consists in transforming a plaintext message into an unintelligible ciphertext. Nevertheless, the classical and modern ciphers have all been developed without taking into consideration the intrinsic characteristics of multimedia flows. In this chapter, we propose to reduce the computational requirements for the multimedia flow's encryption when the energy is a limited resource, as it is the case of WANets. This research based on defining a new profile of Secure Real Time Protocol (SRTP) named EE_SRTP. EE_SRTP provides energy efficiency to SRTP which provides essentially confidentiality, message authentication, and replay protection to the RTP traffic. The performance evaluation demonstrates clearly that using this new scheme in the context of wireless ad hoc nodes allows saving energy while ensuring a high level of content confidentiality, without adding an overhead to the wireless network.

The unique characteristics of MANETs present a new set of non trivial challenges to security design. These challenges include open network architecture, shared wireless medium, stringent resource constraints, and highly dynamic network topology. Consequently, the existing security solutions for wired networks do not directly apply to the MANET domain. The ultimate goal of the security solution for MANET is to provide security services, such as authentication, confidentiality, integrity, anonymity and availabil-

modelling tools (ns-2, OPNET and OMNeT++) is explored in this chapter, concluding that OMNeT++ is worthy of study as an alternative to the other two more established tools. As an illustration of the use of OMNeT++, two modelling schemes are simulated and compared against the theory to determine both bit-level correctness, but also to demonstrate ease of modelling and analysis.

Many strategies have been devised over years for improving performance of wireless sensor networks with special consideration to energy efficiency, accuracy of the sensed data, increasing network lifetime, providing QoS, et cetera. Such techniques include design of various MAC protocols, routing algorithms, aggregation techniques, and many more. Data fusion has been developed to be one of such techniques which has a span of benefits across all the above mentioned performance criteria. The basic concept of data fusion is to apply various mathematical concepts on the raw sensor data to achieve an output with better fidelity and accuracy. Various mathematical, signal processing and probabilistic techniques can be implemented to fuse the data. This might be done for decreasing the volume of the data, improving the precision of the data, deriving an inference from the raw data which was not present earlier, and many other similar. The motive of this chapter is to introduce data fusion and why it is important in case of wireless sensor networks. Further, it classifies data fusion, explains the major techniques, models as well as the implementation aspects of data fusion, and concludes with the challenges and limitations posed for data fusion to be useful.

Cognitive radio is a paradigm for wireless communication in which either network or wireless node itself changes particular transmission or reception parameters to execute its tasks efficiently. This parameter alteration is based on observations of several factors from external and internal cognitive radio environment, such as radio frequency spectrum, user behavior, and network state. The chapter presents the relation between cognitive radio and ant colony. The bio-inspired cognitive radio algorithm is presented.

Research in MANET is like playing in a vast playground with only few known rules and large number of unknown rules applied to its different areas. Some areas are still unknown and some areas are at its early stage. Some of the research challenges that MANET present are dynamic topologies, battery lifetime, disconnected operations, security, et cetera. In general there are three types of MANET: vehicular ad hoc networks, intelligent vehicular ad hoc network, and inteRnet based mobile ad hoc networks. Each of these three types has its own research areas.

Foreword

With the emergence of numerous interfaces and standards in the domain of wireless communication systems, the world has witnessed a substantial amount of research towards developing high-end and compatible devices that are lighter and quicker in respect of seamless connectivity as well as user end requirement. While initial development was on infrastructure-based networks, which represented an integration of wireless and wired networks, the latest research has focused on purely wireless networks without any fixed backbone called as ad-hoc networks in which wireless entities form a network among themselves as and when required. Various classes of ad hoc networks have been developed, namely Mobile Ad hoc Networks (MANET), VANET, et cetera. The underlying principle that differentiates these variations of networks is in the entities involved in formation of the ad hoc network. While a VANET is formed between fixed devices mounted on high speed vehicles, a MANET is formed between low speed portable devices with low end computing needs. Interests in MANETs arose due to its ability to provide distributed and instant wireless networking solution in situations where infrastructure based networks are infeasible and/or expensive to be deployed. Moreover, they have greater flexibility towards congestion and robustness against single-point failures.

Mobile Ad hoc Networks (MANETs) have special resource requirements and topology features, which make them different from classic computer networks in resource management, routing, media access control, and QoS provisioning. Some of our day-to-day situations directly relate to ad hoc wireless network applications, such as self-organization, mobility management, and energy efficient design.

With more than twenty self-contained chapters, this volume provides a complete survey of the state-of-the-art research that encompasses all areas Mobile Ad hoc. Written by distinguished researchers in the field, these chapters focus on the theoretical and experimental study of advanced research topics involving security and trust, broadcasting and multicasting, power control and energy efficiency, and QoS provisioning.

This book is a great reference tool for graduate students, researchers, and mathematicians interested in studying mobile ad hoc and sensor networks.

This is the best book to give you a quick overview of all important aspect of Mobile Ad Hoc Networks, as this book covers technical study and real time performances in the area of ad hoc networks.

I feel proud fro Dr. Kamaljit and his authors, for working on such project and gather all utmost emerging information on MANET technology in a single book.

I recommend this book for four reasons; first, this book's logic and chapter flow make sense about this technology. Second, this book includes such basic concept for each aspect and raises this concept to advance most level of experiment and usage for MANET technology. Third, this book represents almost all real-world implementation, and four, the information presented in this book is definitely in current demand for researcher and scientists.

A.C.M. Fong
Auckland University of Technology, New Zealand

A.C.M. Fong *is author of "Multimedia Engineering: A Practical Guide for Internet Implementation." Prof. Dr. Fong is Professor of Computing Science at the School of Computing and Mathematical Sciences at Auckland University of Technology, New Zealand.*

Preface

As the importance of staying connected in ubiquitous environment in our daily life increases, it also sets new demands for connectivity. Wired solutions have been around for a long time, but there is increasing demand on working wireless solutions for connecting to the Internet, reading and sending E-mail messages, changing information in a meeting, and so on.

This book on Mobile Ad Hoc Networks concerns both time-intensive and real time implementation of wireless technologies looked for in research. In the last few years there has been a communication transformation to adopt MANET. MANETs have introduced many challenges and many opportunities to academic and industrial research as well. A comprehensive and unified concept of MANETs in a book is required. The goal of this book is to explore the developments and current/future challenges in the area of Mobile Ad Hoc Networks.

This book aimed to serve as a powerful reference and background in the area of MANETs by tackling the prevalent research challenges that hinder a fully deployable Mobile Ad hoc Network. This book compensates relevance areas of MANET, real time implementation problem, solution, findings, and future directions of the research as well.

The book is divided into four sections, wherein each section groups several MANET-related research topics:

Section 1: MANET Protocols
Section 2: Emerging Trends of Research in MANET
Section 3: MANET Threat and Security
Section 4: Mobility Issue in MANET

Kamaljit I. Lakhtaria
Sir Padampat Singhania University, India

Acknowledgment

This Book Project, is my brain child since my Doctorate research, but transformed into a form of book. I am thankful to many persons.

First I am thankful to Kristin M. Klinger, Erika Carter, and Hannah Abelbeck from IGI Global for their constant support and guidance.

I thank my Editorial Advisory Board Members for discussing this project with me and I am happy with everything. I specially extend thanks to Dr. Ashraf Darwish, Dr. Ajith Abraham, Dr. Natarajan Meghanathan, and Dr. Shyama Subramaniam for their prompt reply during initial stage of Book. Thanks to all the contributing authors and reviewers for spending their valuable time and effort for this Project.

I would like to thank my wife, Panna, for her love and inspiration to me.

Section 1
Mobility Research Models, Issues, and Findings in MANET

Chapter 1
Mobility Prediction and Mobile-aware Routing Protocols in MANETs

Lamiaa Khalid
WINCORE Laboratory, Ryerson University, Canada

Muhammad Jaseemmuddin
WINCORE Laboratory, Ryerson University, Canada

Alagan Anpalagan
WINCORE Laboratory, Ryerson University, Canada

ABSTRACT

Ad hoc wireless networks comprise of mobile devices that use wireless transmission for communication without the need for a pre-existing infrastructure. One of the main characteristics of ad hoc networks is that they are self-organizing, which necessitates that all networking functions be performed by the nodes themselves. When designing mobile ad hoc networks, several challenges arise due to the shared nature of the wireless medium, limited transmission range of wireless devices, and node mobility. Node mobility poses a greater challenge as it makes routes more prone to frequent ruptures, which in turn degrades the overall network performance. If the future movement of the node can be predicted in a precise way, the impact of mobility can be decreased, and the overall performance of the network can be improved. This chapter provides an introduction and a critical review on some of the existing mobility prediction models and their effect on the performance of existing routing protocols.

1. INTRODUCTION

A mobile ad hoc network (MANET) is a self-configurable network of mobile nodes that are connected through wireless links without a pre-defined infrastructure. Every node in the network is capable of functioning as a mobile router, which facilitates the multi-hop forwarding of packets from a source node to a destination node. The objective of this network architecture is to achieve increased flexibility, mobility and ease of management compared to wireless networks with fixed infrastructure (McDonald & Znati,

DOI: 10.4018/978-1-4666-0321-9.ch001

1999). However, the multi-hop nature and the lack of infrastructure in MANETs pose a number of challenges on the design of routing protocols and the supported applications. Since all the nodes can move freely anytime, the network topology is highly dynamic and random. As a result, traditional routing protocols designed to operate in networks with fixed infrastructure exhibit poor behavior and incur large overhead when operating in MANETs. The main reason for mobility to impact the performance of routing protocols is due to its impact on the average link duration between two nodes which in turn affects the path duration. Vehicular ad hoc networks (VANETs) are emerging within the realm of intelligent transportation systems and many of the concepts and discussions related to MANETs can be readily applied to the VANETs, however with appropriate considerations for relatively higher speed, existing smart control and communication entities and relatively predictable movement directions in VANET infrastructure. In this chapter, some of the mobility prediction technologies are reviewed and discussed.

BACKGROUND

Most of the research in MANETs is simulation-based with performance evaluation done under varying conditions. These simulations have several parameters such as user traffic and radio communication pattern, mobility model, propagation model, etc. The mobility model has a major impact on the link and route lifetime distribution which in turn impacts the protocol and application performance. Several mobility models have been proposed in the literature to provide better understanding and quantify mathematically the mobility of the ad hoc network nodes (Sadagopan, Bai, Krishnamachari & Helmy, 2003). One of the motivations behind the mobility models is the potential of mobility prediction. Knowing deterministically or probabilistically helps make an informed decision. Various stochastic mobility models, such as the random waypoint or the random reference group mobility model, are used to study the impact of mobility on the link and route lifetime (Camp, Boleng & Davies, 2002).

Mobility prediction has been extensively used in cellular networks in handoff management and resource reservation. In MANETs, mobility prediction techniques can mitigate the problems emerging from their mobile nature and can be applied to design efficient routing protocols (Su, Lee & Gerla, 2001). The quest of mobility prediction in MANETs is to predict the link quality as a measure of the mobility state of a node (Farkas, Hossmann, Ruf & Plattner, 2006). A viable mobility prediction scheme should offer a high level of prediction accuracy with minimal control overhead.

MOBILITY PREDICTION PROBLEM

Mobility prediction in general is the problem of estimating the trajectory of future positions of the nodes in mobile networks. Mobility prediction in MANETs is used in location-aided routing and mobility-aware topology control protocols. Various mobility metrics have been proposed as a measure of the topological change in MANETs. Mobility prediction can also be used to improve routing protocol performance, to reduce routing overhead and to aid in proactive route reconstruction. The general assumptions made in the problem of mobility prediction are (Farkas et al. 2006):

- It is possible to observe the current mobility states of the node.
- The behavior of the nodes shows some mobility pattern.

However, there is an inherent difference between mobility prediction in cellular networks and MANETs due to the difference in the network structure and the behavior of the nodes. In cellular networks, the base station which is one end of the communication link between two nodes is fixed and resource reservation is done through the base station normally in a centralized manner. On the other hand, both ends of the communication link in MANETs are mobile and each node should be able to predict its own future and convey this information to other nodes. This makes the problem of mobility prediction in MANETs more difficult partly due to the distributed nature of implementation.

Organization

The remainder of this chapter is organized as follows. In section 2, we introduce the taxonomy used for the different mobility metrics and discuss the advantages and disadvantages of some of the mobility prediction models proposed for MANETs. In section 3, we show how mobility prediction can be used to enhance the performance of existing routing protocols or aid in designing new mobility-aware routing protocols. Section 4 concludes with a discussion of possible future work.

2. TAXIMONY OF MOBILITY METRICS

Let us define some of the mobility measures for links and paths used in the literature (Xu, Blackmore & Jones, 2007). In many of these definitions, the links are assumed to be ON or OFF at any point in time. If the link exists then it is assumed to be ON, if not, it is assumed to be OFF. The metrics that characterize the links can be:

- *Link (Path) Availability:* It is the probability that a link (path) is active at a certain time in the future, given that it is currently active. The link is considered available even if it experienced failure during the intervening period. Therefore, the term link (path) availability describes a non-continuous case where the link can disappear and re-appear during the wait time (McDonald & Znati, 1999).
- *Link (Path) Persistence:* It is the probability that a link (path) continuously lasts until a future time T given that it exists at time to. Therefore, link (path) persistence describes the continuous availability of a certain link for a time $to + T$ (Jiang, He & Rao, 2005).
- *Link (Path) Residual Time:* It is the length of time the link (path) will continue to exist until it is broken. It can be measured from any point in the life of the link (path) (Su et al. 2001).
- *Link (path) Duration:* It is the average time that a link (path) lasts without breaking (Sadagopan et al. 2003) and it is measured from the time the link (path) is first ON until the next time it is OFF. Therefore, it is a measure of the link (path) stability (a link is considered stable if it exists for certain duration of time (Gerharz, de Waal, Martini & James, 2003) and stable paths are established along stable links).

The different mobility metrics defined above have been studied by various authors in the literature. In the following sections we will give a brief overview of some of the work that studied those metrics.

Link Availability

One of the earliest analyses in this area was done in (MacDonald & Znati, 1999). We present and discuss this analysis as it represents a fair number of models available in the literature. The authors used a slightly modified random walk mobility model and derived analytical expressions for the probability of link and path availability. Based on this mobility model, each node's movement consists of a sequence of random length intervals called mobility epochs during which a node moves in a constant direction with a constant speed. The speed and direction of each node vary randomly from epoch to epoch. Their assumptions reflect a network in which a large number of nodes operate in a truly ad hoc fashion. The link and path availability depend on the node speed, mobility model and transmission radius. The following assumptions are made:

- The epoch lengths are independent and identically distributed (i.i.d.) exponential random variables.
- The speed during each epoch is an i.i.d. random variable (e.g. normal, uniform distribution) and is constant only for the duration of the epoch.
- The direction of the mobile node (MN) during each epoch is i.i.d. uniform random variable over $[0; 2\pi)$ and is constant only for the duration of the epoch.
- Speed, direction and epoch length are uncorrelated.
- Node mobility is uncorrelated and links fail independently.

To deal with the joint node mobility, the authors used a transform method which was accomplished by logically treating one of the nodes (*n*) to be stationary, while the other node (*m*) moves with an equivalent random mobility vector $\vec{R}_{m,n}(t)$

They assumed the magnitude of $\vec{R}_{m,n}(t)$ at time *t* to be Rayleigh distributed while the direction is uniformly distributed over $[0; 2\pi)$. Also, they adopted the widely used, albeit optimistic, model (Hong & Rappaport, 1986) assumption that the transmission range of node *n* can be approximated by a circle of radius *Req*. Node *m* is assumed to correctly receive node *n*'s transmission if it is located within this circle. The link availability between nodes *n* and *m*, *Am,n*(*t*), is therefore given by:

$$Am,n(t) \equiv \Pr(Rm,n(t) < Z) \tag{1}$$

Where *Z* represents the distance a MN must travel before reaching the boundary of a cell with effective radius R_{eq}. According to the assumption of independent link failures, the path availability is given by:

$$\Pi^{k}_{m,n}(t) = \prod_{(i,j)\in k} A_{i,j}(t) \tag{2}$$

where (*i, j*) are links in the path *k* between nodes *m* and *n*.

Discrete event simulation was used to validate the analytical model. The simulation results showed excellent agreement with the analytical model and converge when large values of the product of the reciprocal of the mean epoch length and the mean epoch time (λt) were used.

Some weaknesses of this kind of approach are, in general:

- The random walk mobility models are not well suited to describe realistic situations. In reality, mobile users often exhibit correlated mobility patterns in their movements.
- The accuracy of the expressions derived for link availability depends on the satisfaction of the central limit theorem (CLT) in the derivation of the Rayleigh distribution which is assumed for the relative movement between a pair of mobile nodes. Since the criterion for satisfying the CLT in this case requires the number of independent mobility epochs to be large during the observation period, this assumption does not work well when taken over short time intervals, particularly for the path availability calculation.
- The simulation results converged after relatively long time. A more precise estimation of the link availability for shorter time periods is required since this is more in-

teresting to routing where the typical flow duration is often less than that.

- They predict the availability of the link at a certain time in the future even if the link experienced failures prior to this point. This means that no forecast can be made about the duration in which the link uninterruptedly exists. This is not practical to be used for path selection in routing protocols wherein if any link of a path breaks, a re-routing is required immediately. It is unlikely to let the related nodes wait for this broken link to become available again.

In (McDonald & Znati, 2005), an analytical framework is presented for extending the random-independent link availability model from (McDonald & Znati, 1999) to capture the effects of correlated movement. The objective was to define criteria for deciding that a pair of nodes is moving in a pattern that establishes a stable link between them, and to adaptively bias the random independent metric to reflect this assumed association. This model uses knowledge of the mobility profiles of adjacent nodes which is based on assumptions (i)-(iii) in (McDonald & Znati, 1999).

The mobility profiles can be learned by periodically acquiring global positioning systems (GPS) information, and computing the probability the nodes remain within a threshold distance, or proximity of each other at any time t in the future. The relative movement of any two nodes may be independent or correlated. The total link availability, $A^T_{m,n}(t)$ between two nodes m and n in this case is expressed as:

$$A^T_{m,n}(t) = A^i_{m,n}(t)P_i + A^c_{m,n}(t)(1-P_i) \qquad (3)$$

Where $A^i_{m,n}(t)$ is the link availability when mobility is independent as in (McDonald & Znati, 1999) and $A^c_{m,n}(t)$ is the link availability when mobility is correlated. The metric reflects independent or correlated behavior as given by the value of P_i. A value of $P_i = 1$ reflects independent movement with respect to the total link availability. The term P_i depends upon the nature of the network application, the group properties of the nodes, the size of the population and the characteristics of the environment. In very large ad-hoc network of nomadic users with few movement constraints, it is reasonable to expect large values for P_i; however, if nodes travel frequently in groups, or are constrained to move in certain directions and with certain speeds, smaller values for P_i would be expected. The value of $A^c_{m,n}(t)$ depends on the degree of correlation in movement. The idea is to initially assume that the endpoints of a link are moving independently, and to evaluate the link availability based on the independent model, $A^i_{m,n}(t)$ Any link which survives longer than its expected time-to-failure begins a gradual transition into associated mode (node movement is assumed to be correlated) based upon an exponential smoothing function. The rate of this function depends on the elapsed time since the mean time-to-failure (MTTF) and the magnitude of the independent mode availability (the default case where link availability is evaluated according to the independent mode). Assuming an approximate model for $A^c_{m,n}(t)$ such that $\acute{A}_{m,n}(t) \approx A^c_{m,n}(t)$ Equation (3) shows how to evaluate the enhanced link availability metric given that each node specifies P_i, and $A^c_{m,n}(t)$ The enhanced link availability metric is evaluated as follows:

$$\acute{A}_{m,n}(t) = \left(1-e^{-\lambda\Delta}\right) + A^i_{m,n}(t)e^{-\lambda\Delta} \qquad (4)$$

where,

$$\Delta = \max\left(0, t-\left(t_0 + MTTF\right)\right) \text{ and } \lambda = 1 - A^i_{m,n}(t) \qquad (5)$$

For all time $t \leq (t_0 + MTTF)$, link availability is evaluated directly according to the independent model. This value monotonically decreases as time progresses. However, starting at time $t = (to + MTTF)$, this decrease begins to be offset by an amount which varies from $(1 - A^{i}_{(m,n)}(t))$ to 1. The amount of this offset increases with time and the magnitude of $(1 - A^{i}_{(m,n)}(t))$. It is assumed that a pair of adjacent nodes with highly correlated movement has a correlated link availability, $A^{c}_{(m,n)}(t) = 1$, for all t. Consequently, $\acute{A}_{m,n}\left(t\right)$ is designed to approach this value as $t \rightarrow \infty$ Expressions for MTTF for the given links are derived by equating the expected value of $\vec{R}_{m,n}\left(t\right)$ to the expected value of Z (McDonald & Znati, 1999) and solving for t.

In contrast to the random-independent model in which link availability decreases with time, using the proximity model link, availability increases after the MTTF. The term $A^{T}_{(m,n)}(t)$ effectively represents a conditional probability that reflects increasing confidence in the stability of a link. With proper specification of P_i, the proximity model can achieve benefits through the detection of associated movement patterns; however, it is not possible to detect negative correlation without apriori information.

Some weaknesses of this approach are:

- The authors in (McDonald & Znati, 2005) modified the random mobility model introduced (McDonald & Znati, 1999) to reflect the future stability of a given link. However, they did not address the main disadvantage of their previous model in (McDonald & Znati, 1999) which is the non-continuous assumption of the link availability.
- The authors proposed the use of GPS to acquire information about the node mobility profile. However, to presume special location equipment for any node in an open ad-hoc network will increase the complexity by requiring additional hardware.

Link Persistence

Since link availability (non-continuous case) is not practical for ad hoc routing as mentioned earlier, an analysis of continuous link availability is given in (Jiang et al. 2005) where the authors first estimated a time period, *Tp*, that the current link will last if both nodes of the link keep their current velocity (unaffected *Tp*). Then, a probability model was applied for calculating the probability, *L*(*Tp*), that the link can last to *to* + *Tp* by considering possible changes in velocities occurring between *to* and *to* + *Tp* (affected *Tp*). In (Jiang et al. 2005), a measurement based scheme was first introduced to predict *Tp*. The accuracy of this prediction depends on the accuracy of the measurements obtained from the received signal strength or by using GPS. Assuming *Tp* is known, the authors then focused on estimating *L*(*Tp*) (*L*(*Tp*) does not depend on the scheme used to predict the unaffected *Tp*). The authors adopted the same assumptions as in (McDonald & Znati, 1999) that the node mobility is uncorrelated and epochs are exponentially distributed $(\text{mean } »)$. The term *L*(*Tp*) is calculated as follows

$$L\left(T_p\right) = L_1\left(T_p\right) + L_2\left(T_p\right) \tag{6}$$

where $L_1\left(T_p\right)$ is the probability that the epochs from *to* onwards for the two nodes are longer than *Tp* assuming constant velocity between the two nodes (both node has the same mean epoch length λ^{-1}). Therefore,

$$L_1\left(T_p\right) = e^{-2T_p} \tag{7}$$

Due to the difficulty in learning changes in link status caused by changes in velocity, the authors in (Jiang et al. 2005) estimated the value of $L_2\left(T_p\right)$. The estimation of $L_2\left(T_p\right)$ does not take into account all possible changes in velocities that may happen during period *Tp*. It only considers the first change in velocities occurring during period *Tp*, and link availability for other such changes is estimated by ∈That is,

$$L\left(T_p\right) \approx \frac{1}{2T_p} + \epsilon + e^{-2T_p}\left(pT_p - \frac{1}{2T_p} - \in\right) \quad (8)$$

where *p* is the probability for the two nodes to move close to each other after a change in velocity. For each prediction *Tp*, the node can measure the duration that the link will really last corresponding to this *Tp*, *Tr*, and can have a *measured* link availability $L_m\left(T_p\right) = \frac{T_r}{T_p}$ Then equation (8) can be used in reverse to determine parameters at the right hand by substituting

L(*Tp*) by *Lm*(*Tp*). Only is determined by measurement, and *p* is set to 0.5 intuitively for environments in which a node can freely move in any direction. Since it is very difficult to use one mobility model to describe all mobile behaviors when mobile nodes roam across different environments, using measured ∈ can make the proposed estimation adaptive to other mobility models.

A simulation study had been conducted in (Jiang et al. 2005) to validate the proposed estimation of *L*(*Tp*), which showed that the estimation can reflect the general tendency of link availability statistically. That is, *L*(*Tp*) can approximate $\bar{T}_r / T_p$, where $\bar{T}_r$ is the meantime that a link will be continuously available during period *Tp*.

The authors then assumed that for a path consisting of multiple links, path reliability mainly depends on the weakest link in terms of reliability which can be measured by T_r. Therefore, a link with a smallest T_r, T_{min}, is considered as the weakest link. An ideal path in terms of path reliability should be the one with the maximum T_{min} and the minimum number of links. Simulation results showed that when $L(T_p) \times T_p$ is adopted in path selection as reliability metric, performance can be improved significantly in mobility-sensitive environments.

Some weaknesses of this approach are:

- The intuitive *p* setting may degrade the estimation accuracy, especially in indoor environments, where a node may be forced to change its direction in the presence of obstacles.
- Distance estimation based on signal strength measurements has proven to ignore some crucial signal propagation effects especially in indoor environments which may render the stability assessment useless, and relying on GPS measurement has its shortcomings as mentioned earlier.
- In a highly mobile environment, the mobility epochs can be very small which will require a large number of estimations increasing the control overhead.

Link Residual Time

A prediction mechanism for link expiration time (LET) between any two nodes has been studied in (Su et al. 2001) to enhance various unicast and multicast ad hoc routing protocols. The authors proposed that by exploiting the non-random movement patterns of a user, we can predict the future state of a network topology and thus provide a transparent network access during the period of topology changes. In this approach, a node relies on the availability of a GPS to provide information about the current positions and velocities of the node and its neighbor in order to predict the link expiration time. Assumptions are:

- A free space propagation model is assumed, where the received signal strength solely depends on its distance from the transmitter.
- All nodes in the network have their clock synchronized (e.g., using the GPS clock).

By making use of the coordinates of the nodes $\left(\left(x_i, y_i\right);\left(x_j, y_j\right)\right)$, their speeds $\left(v_i, v_j\right)$ and direction of motion $\left(\theta_i, \theta_j\right)$, the authors predict the duration of time, D_t, the two nodes (i, j) will remain connected,

$$D_t = \frac{-\left(ab + cd\right) + \sqrt{\left(a^2 + b^2\right) - \left(ab - bc\right)^2}}{a^2 + c^2} \tag{9}$$

where

$a = v_i \cos \grave{e}_i - v_i \cos \grave{e}_j, b = x_i - x_j, \mathrm{c} = v_i \sin \grave{e}_i - v_j \sin \grave{e}_j and\, c = y_i - y_j.$

By predicting the LETs of all links of a route, the route expiration time is given as the least of all the LETs values. The route selection algorithm in (Su et al. 2001) prefers routes with longer expiration time. A modified random waypoint model was used, which requires nodes to randomly select direction until they reach a certain simulation terrain boundary, then they bounce back and continue to move. By limiting the distance of mobile nodes, the prediction value can assure its accuracy at least for a certain period of time. The proposed method offers accurate prediction support to networks with simple mobility patterns with no sudden change in direction and constant speed. Simulation results indicated that with mobility prediction enhancements, more data packets were delivered to destinations while the control packets were utilized more efficiently which results in lower overhead.

Some weaknesses of this approach are:

- This approach strongly depends on the assumption of a free space propagation model and on having GPS equipment available for distance measurements and time synchronization. However, these requirements can hardly be presumed in a realistic environment.
- The authors assumed that when two nodes become neighbors, they will not change their current velocity before the link breaks up. However, this is unrealistic since mobile nodes do change their velocity frequently in many situations. Consequently, the model in (Jiang et al. 2005) in comparison to the LET scheme better defines the mobility in an ad hoc environment by incorporating random independent mobility in the link stability metrics.

The authors in (Samar & Wicker, 2004) derived analytical expressions for the expected link lifetimes, rate of new link arrivals and probability distributions for the above quantities with the goal of optimizing the existing network protocols and designing efficient algorithms for routing. They used a constant velocity model in which nodes move with random direction and at random velocity, but both direction and velocity are kept constant over time so that the link can be analyzed using geometric observations. The authors used some of the assumptions as (McDonald & Znati, 2005).

Assumptions:

- A node has a bidirectional communication link with any other node within a distance of *R* (circular transmission range)
- The velocity of each node (constant) is uniformly distributed between $[a, b]$
- The direction of each node is uniformly distributed between $[0, 2\pi)$

- Speed, direction and location are mutually independent.
- The initial locations of nodes in the network are modeled by a two-dimensional Poisson Process with intensity such that the probability that a network region *D* with an area *A* contains *k* nodes is given by:

$$Prob\left(k\,nodes\,in\,D\right) = \frac{(\sigma A)^k e^{-\sigma A}}{k!} \qquad (10)$$

From the geometric observation of Figure 1 (adopted from (Samar & Wicker, 2004)), and some probabilistic evaluations, the expected link lifetime between node 1 and 2 as a function of node's 1 velocity, v_1, is given by:

$$\bar{T}_{link} = \frac{R}{2(b-a)} \left(\int_0^{\pi} \log \left| \frac{b + \sqrt{b^2 - v_1^2 \sin^2 \phi}}{v_1 + v_1 \cos \phi} \right| - \int_{\varphi_0}^{\pi} \log \left| \frac{a + \sqrt{a^2 - v_1^2 \sin^2 \phi}}{a - \sqrt{a^2 - v_1^2 \sin^2 \phi}} \right| d\phi \right) \qquad (11)$$

Figure 1. Transmission zone of node 1 at O with node 2 entering the zone at A and exiting at B.

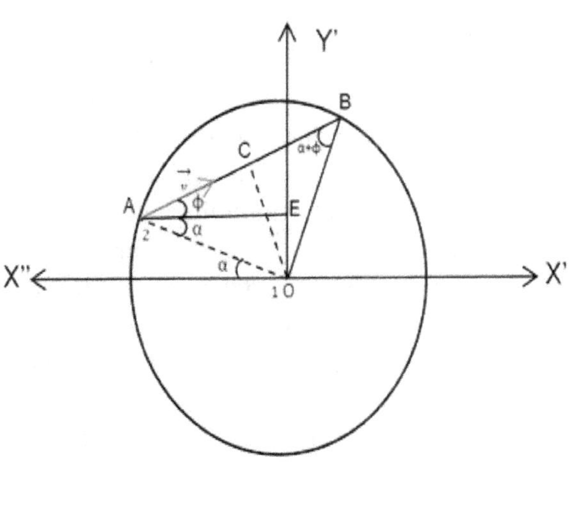

Where $\phi_0 = \pi - \sin^{-1}\left(\frac{a}{v_1}\right)$. By appropriately choosing the two parameters, *a* and *b*, the intensity of mobility being modeled can be changed and (11) can be numerically integrated to give the expected link lifetime. It is clear from (11) that the derived expression does not depend on the density or location distribution of nodes in the network. The link lifetime distributions (cumulative density function (CDF) and probability density function (PDF)) were then derived for a particular node moving with velocity v_1 (relative velocity is constant).

The authors then presented a thorough analytical study of some significant parameters of the link properties such as the expected link change rate (the sum of the expected new link arrival rate and the expected link breakage rate) and its probability distribution, as well as the probability distribution of the new link arrival, link breakage and link change inter-arrival times.

The link lifetime distribution can be used to examine the stability of the links in the network. Once communication starts over a link, its residual lifetime distribution can be calculated as a function of the link lifetime distribution. Mathematically, the probability density $r_T^{v_1}\left(t\right)$ of residual link lifetime given that the link has been in existence for *T* seconds already can be expressed as:

$$r_T^{v_1}(t) = \frac{f_{link}^{v_1}\left(T+t\right)}{1 - F_{link}^{v_1}\left(T\right)} \qquad (12)$$

where $f_l ink^{(v_1)}() and F_l ink^{(v_1)}()$ are the link life time PDF and CDF, respectively. The residual link lifetime density can be used to evaluate the lifetime of a route in the network assuming that the residual link lifetimes are independent.

A simulation study was conducted to validate the analytically derived expressions for the link properties by comparing them to corresponding statistics collected from simulations. Results showed that the theoretical curves are in fairly close agreement with the simulation results especially when the network size was increased while keeping the node density constant.

Some weaknesses of this approach are:

- Constant velocity mobility model was assumed in which the relative velocity between two nodes does not vary during the entire link connection, which is unrealistic since nodes may change their speed and direction as frequently as possible during their movement.
- The authors assumed as in (McDonald & Znati, 1999) a random mobility model where the node velocity has no correlation between two consecutive time epochs. This assumption frequently generate abrupt moving behaviors such as sudden stop and sharp turn which do not reflect the smooth motions in the real world.

Link Duration

The authors in (Sadagopan et al. 2003) studied the link and path duration for four different mobility models including the random waypoint (RW), the reference point group mobility (RPGM), the freeway (FW) and the Manhattan (MH) mobility models. They developed an approach that combines statistical analysis of simulation data and analytical modeling to get a deeper understanding of the protocol performance in the presence of mobility. They defined the link duration metrics as follows: for two nodes *i* and *j*, at time t_1, duration of the link (*i, j*) is the length of the longest time interval $[t_1, t_2]$ during which the two nodes are within the transmission range of each others (those nodes are not within transmission range at time t_1- □ and t_2+ □ for □ >0). An indicator random variable $X(i,j,t)$ is used to reflect the status of the link between node *i* and *j* at time *t*. Therefore,

$$LD(i,j,t_1) = t_2 - t_1 \tag{13}$$

iff $\forall t = t_1 < t < t_2, \epsilon > 0 : X(i,j,t_1) = 1$ and $X(i,j,t-\epsilon) = 0$ and $X(i,j,t+\epsilon) = 0$. Otherwise, $LD(i,j,t_1) = 0$

Path duration is the time interval during which all links on a path (from a source to a destination) exist. Path duration is actually the minimum link duration along a path. For each mobility model, they collected the detailed statistics of the link and path duration at both the connectivity graph level (topology) and the routing protocol level. At the connectivity graph level, the link duration was calculated as the interval between the time when the link was created and time when it broke, whereas, the path duration is the duration of shortest path between source-destination pairs based on their location information (using the mobility trace analyzer program (Bai, Sadagopan & Helmy, 2003). At routing level, the authors used the information collected under NS-2 simulation environment to calculate the path duration (the behavior of paths in NS-2 may differ from shortest path depending on the routing protocol used). After collecting a large set of samples for link and path durations, the authors in (Sadagopan et al. 2003) used the relative frequency approach (from standard probability theory) to estimate the PDFs of the link and path duration across the different mobility models as their distributions are expected to depend on the mobility models used in the simulations as well as on the range of node speeds. Their study showed that both the link and path duration have a multi-model distribution when the node speed is slow (FW and RPGM (4 groups) models). This shows the impact of traffic pattern on the path duration PDF. In high speed situations ($v_{max} \geq 10m / sec$), the path duration can be approximated by an exponential distribution (for paths of 2 or more hops) regardless of the mobility model used.

This exponential distribution (with mean $\frac{1}{\lambda_{path}}$) is parameterized by the relative speed of the mobility model, *V*, the transmission range of the node, *R*, and the number of hops in the path, *h*, according to the following equation:

$$\lambda_{path} = \lambda_0 \frac{hV}{R} \tag{14}$$

whereis a constant of proportionality. This model was verified through simulations.

Some weaknesses of this approach are:

- The authors study showed that the distribution of path duration can be approximated by an exponential distribution, however, no explanation was offered for the emergence of an exponential distribution (it was later validated by using palm calculus in (Han, La, Makowski & Lee, 2006).
- The authors analyzed the average link or path duration but no prediction about the future link availability by taking the present node location and velocity into consideration was presented.
- The authors did not mention if the minimum speed of nodes is larger than zero which is critical for RW mobility model since in this model the average speed of nodes decreases to zero over time if the minimum speed of nodes is set to zero (Yoon, Liu & Noble, 2003). Such speed decay can have a dramatic influence on time-averaged metrics where an average link duration obtained over an earlier time period may be quite different from that obtained over a later time period.

3. MOBILITY-AWARE ROUTING PROTOCOLS

The effectiveness of any routing algorithm depends to a great extent on the accuracy and the timeliness of the available network topology information. In MANETs, due to the dynamic nature of the network, mobility prediction can be used to design efficient routing protocols with minimal overhead. This can be achieved either by selecting more reliable paths or performing route maintenance before a link is disrupted. Some work focus on studying the effect of mobility on the performance of existing routing protocols (Sadagopan et al. 2003). Others focus on using mobility prediction to enhance the performance of the existing routing protocols (Sithitavorn & Qiu, 2005) or to design new routing protocols (Su et al. 2001). In this section we will discuss briefly some of the results obtained by using those different approaches.

One of the objectives of the study in (Sadagopan et al. 2003) is to find out how the routing protocol performance is related to the path duration. The authors used the case study of dynamic source routing (DSR) to derive an analytical model to study the relationship between path duration and protocol performance. They found that there exists a linear relationship between the reciprocal of the average path duration and the routing protocol performance in terms of throughput and routing overhead which is given by:

$$Throughput = r.\left(1 - \frac{t_{repair}}{PD}\right) \tag{15}$$

$$Overhead = \frac{T}{PD}\left(\left(p\right)1 + \left(1 - p\right)N\right) \tag{16}$$

where *N* is the total number of nodes, *T* is the total simulation time, t_{repair} is the time spent to repair a broken path each time, PD is the average path duration and r is the data rate which is assumed constant. The non propagating cache hit ratio, p, is the fraction of the routing requests issued by DSR in time T that are replied to by the first hop neighbors and thus needs only one route request transmission. For the remaining fraction ($1-p$), flooding of the route request is done leading to N transmissions of the request.

It is clear from (15) and (16) that the correlation is positive between the reciprocal of the average path duration and overhead while the correlation is negative between the reciprocal of the average path duration and throughput. This means that higher path duration results in a higher throughput and lower overhead. It is clear from equation (14) that this model also relates the protocol throughput and overhead to several factors including the transmission range, the average number of hops in the path, the relative speed of the mobility model used as they are related to the path duration.

To validate their model, the authors measured the Pearson coefficient of the correlation between the reciprocal of the average path duration and the DSR throughput and overhead for RW, FW and MH mobility models. The results indicated a strong correlation between the reciprocal of path duration and DSR performance. For the RPGM model, such a strong correlation does not seem to exist. The authors explained this by the fact that path changes is relatively small in RPGM model and thus the accuracy of estimation is affected. Although the simple analytical models were derived based on DSR, the authors argued that they can be extended to other reactive routing protocols such as adhoc on-demand distance vector (AODV) routing protocol with appropriate modifications.

The authors in (Sithitavorn & Qiu, 2005) suggested some modifications to DSR to explore the applicability of mobility prediction based routing towards reliable route selection. They adopted the mobility prediction method from (Su et al. 2001) to calculate lifetime estimation (LET or LLT) of wireless links according to (9). The source routing nature of DSR provides a complete picture of each hop along the path from source to destination, which helps to simplify lifetime estimation of the whole routing path.

To incorporate the mobility prediction model into DSR, a source node requires all mobile nodes along the path to the destination to calculate LLT during the routing discovery phase. These intermediate nodes who receive a route request message will calculate LLT between itself and the sender node based on mobility parameters provided by GPS. The LLT value will be collectively attached to the packet header, which will be re-broadcasted to neighbor nodes until the packet reaches the intended destination. The destination node will keep the minimum LLT value to represent the lifetime of the path. The destination node will attach these LLT values to the route-reply packet, which will be piggybacked to the source node together with its source route. In DSR, when there is more than one shortest path to the same destination, the first shortest-hop path in a route cache will be selected. This random choice of route in DSR may not pick the best path in terms of link quality resulting in frequent link breakages and unnecessary rerouting. To improve this problem, the authors used hop-count as a main routing metric, and the LLT (additional metric) was used in routing decision only when there was more than one path with the same number of hop-count to the same destination to allow a more sensible route selection.

To maintain prediction accuracy, the authors proposed a light-weight updating technique called direction tracking, which can be used to maintain accuracy of the prediction value based on changes in topology. By using direction tracking scheme, the authors were able to use the original RW model instead of the modified RW as in (Su et al. 2001). Also, by limiting the direction tracking function to active links no extra routing-update messages were needed. Direction tracking is implemented as follows: During data transmission, each node along the source route will keep track of changes in directions between itself and the next node based on the next hop information. At a sender node, any changes in directions will trigger the node to update the LLT value, of the particular sender-receiver pair, carried by the source route if the new LLT is smaller than the current LLT value (link will expire earlier than what the source node is expecting). When the LLT value is updated, the node activates a LLT updating *flag*, which is attached to the data packet header to notify all downstream nodes that the LLT value of at least one sender-receiver pair has been changed. Such updating process continues until the data packet reaches its destination, where LLT of every sender-receiver pair along the path is checked. To update the LLT value for the source node, the updated LLT value, together with the node's address and the packet's intended destination, is attached into the acknowledgement (ACK) frame of the existing 802.11 MAC layer before sending it off. Each ACK carries a flag called LLT *update flag*, which is used to differentiate itself from a default ACK message. An ACK with its LLT *flag* activated carries a new LLT value as well as the address of the node in which LLT is updated. A fresh LLT value provides a valid suggestion to the anticipated future of the network topology. By realizing the address of the soon to be broken link (where the LLT was updated), a source node can remove dead links out of the route cache and initiate early route discovery if needed.

To evaluate the performance of the proposed scheme, the authors simulated various topologies of mobile ad hoc networks using NS-2. As shown in Figure 2, DSR with direction tracking shows a reduction of average packet end-to-end delay as a result of less data packets left waiting for

unreachable hosts to be reconnected (number of nodes are 10 so network partitioning occurs).

Figure 3 shows that with direction tracking the ad hoc networks experience fewer route-error messages generated by link failure detections. The tradeoff of this scheme is a slight reduction of packet delivery ratio as illustrated in Figure 4. This is due to the introduction of the LLT restriction into path selection process such that a soon to be broken link will not be used to relay packets. Such restriction could trigger a route discovery session, resulting in more routing messages being

Figure 2. Average end-to-end delay (10 nodes).

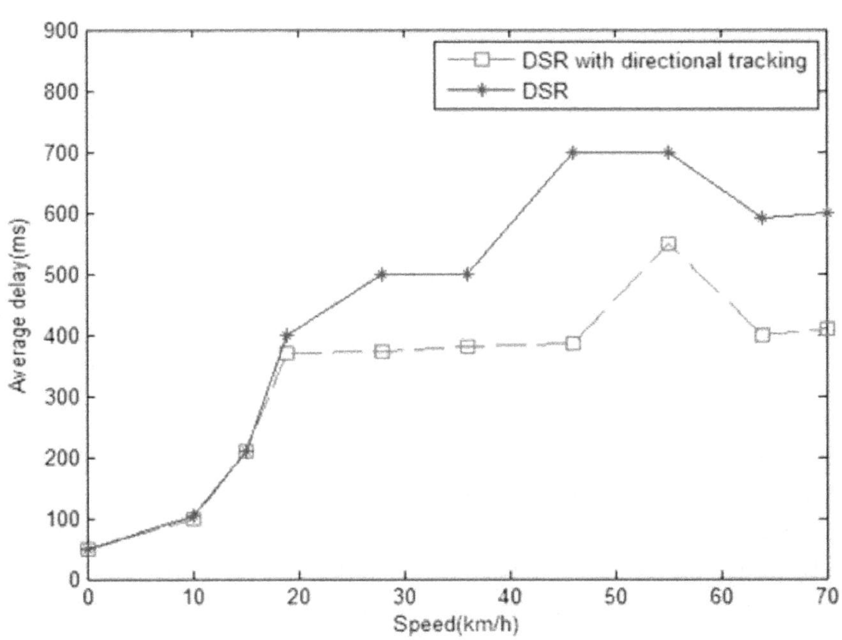

Figure 3. Number of route error messages (10 nodes)

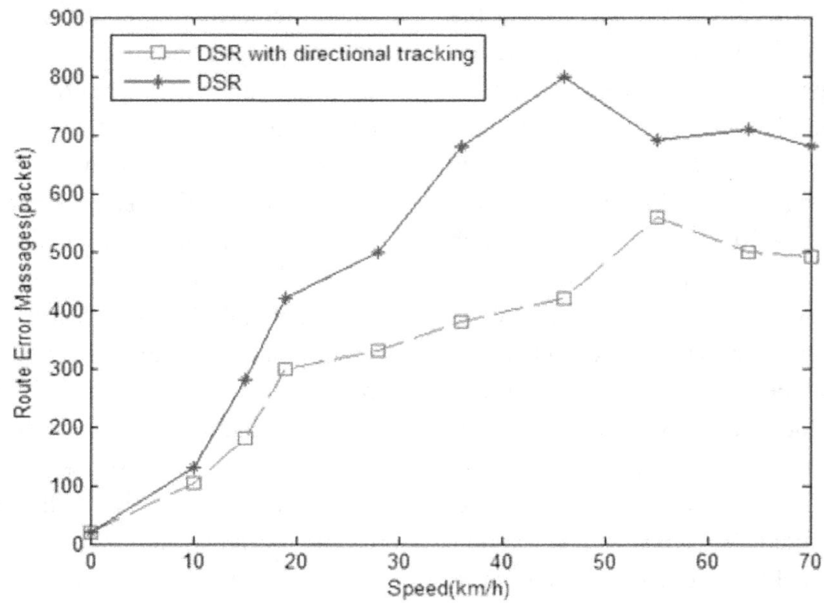

Figure 4. Packet delivery ratio (10 nodes)

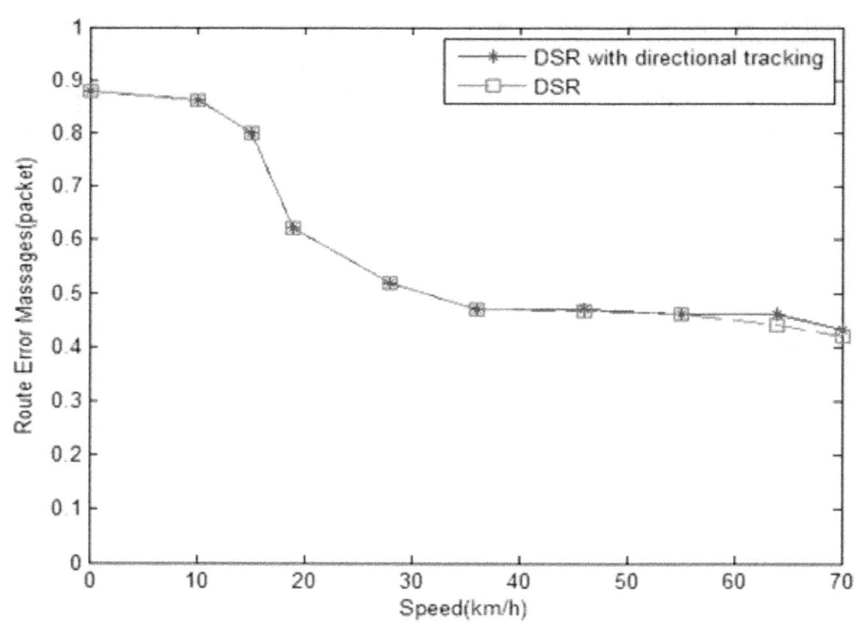

used. On the other hand, in some cases of DSR, a currently broken link might fortunately be reconnected again before the packets starting to drop resulting in slightly better performance in terms of packet delivery ratio compared to the scheme with direction tracking. The figures presented are adopted from (Sithitavorn & Qiu, 2005).

Authors in (Su et al. 2001) applied mobility prediction scheme to two unicast routing protocols namely the flow oriented routing protocol (FORP) and distance vector with mobility prediction (DV-MP). In FORP, when the sender has a flow to send, it constructs a route to the destination on-demand and injects the flow. The destination predicts the change in topology ahead of time and determines when the flow needs to be rerouted based on the mobility information contained in the data packets. A given mobile will be able to predict the link disconnection time of any of its one-hop neighbors based on the method described earlier. By predicting the LET along each hop on the route, the route expiration time (RET) can be predicted. When a source needs to send a flow to the destination, it will first check its routing table to see if it has an unexpired route to the destination. If it does, it will send data packets using this route immediately. Otherwise, it broadcasts a FLOW-REQ message, similar to route request in DSR, to find a route to the destination. The message contains a sequence number, source ID, destination ID, and the list of node IDs appended by the intermediate nodes while the message travels through the network. A neighbor, upon receiving this message, will forward it to all of its neighbors if:

A. The message contains a higher sequence number for any of the previously received source and destination ID pair.
B. The message contains the same sequence number but the message arrives from a better route (e.g., the route has a larger RET with the same or fewer number of hops).

When a node is forwarding the FLOW-REQ message, it also appends its own ID and the LET for the last link of the message. When a FLOW-REQ message arrives at the destination, it contains the list of nodes along the route it has traveled and the LETs for each hop along the route. The destination can then determine the RET for the route by using the minimum of the set of LETs along the route. If the received route is more stable than the one currently in use, the destination sends a FLOW-SETUP message back to the source along the chosen route. The intermediate nodes set up the flow state when they receive the FLOW-SETUP message. During the connection, intermediate nodes append LETs to each data packet, so the destination continues to receive RET predictions from data packets. When the destination has determined that that the route is about to expire (critical time), a Flow-HANDOFF message is generated by the destination and propagated via flooding in the same way as the FLOW-REQ message. After the source receives a Flow-HANDOFF message, it determines the best route on which to handoff the flow based on the information contained in the Flow-HANDOFF message. The source then sends a FLOW-SETUP message (same as the one used in the FLOW-SETUP phase but from source to destination) along the new route. In FORP, the critical time, *Tc*, is given by:

$$T_c = RET - T_d \tag{17}$$

where T_d is the delay experienced by the last packet which has arrived along the route.

In DV-MP, triggered updates are eliminated because routes are established based on stability.

Hence, routing update interval is relaxed and frequent updates are not required. In addition, using stable routes minimizes the disruption caused by mobility since a different route with a greater expiration time is used prior to a given route getting disconnected. Unlike the on demand approach, no route acquisition time is needed since routes are maintained proactively and the routing overhead remains constant regardless of the number of senders in the network. To utilize the prediction information (LET and RET), mobility vector field must be appended to the route update packet. In addition, the RET metric is inserted into routing table entry. Each node periodically broadcasts a route table. A sequence number is issued when generating updates, and it is incremented after each route table broadcast. The sequence number is associated with routing table entries for a particular origin of the route update.

In the proposed scheme, if node *A* receives a route table from its neighbor node *B*, the mobility vector contained in the routing table is used to calculate the LET between node *A* and *B*. The LET and the routing table received from node *B* is then used to update node A's own routing table with the following rules:

- If an entry for destination *D* with a better RET is received and the received sequence number is greater than or equal to the old entry's sequence number, node *A*'s entry for destination *D* is updated.
- If *A*'s next hop for destination *D* is node B and the incoming routing table entry for node *D* contains a higher sequence number than node *A* 's entry for *D*, node *A*'s entry for destination *D* is updated.

To evaluate the performance of the proposed protocols, the authors simulated different scenarios under various conditions using GloMoSim. They compared the two proposed protocols with location aided routing (LAR) protocol which is an on demand routing protocol that uses GPS; and the wireless routing protocol (WRP) which is a distance vector routing protocol for ad hoc networks. The packet delivery ratio as a function of mobility speed is shown in Figure 5. It is clear that as speed increases, the routing effectiveness of WRP degrades rapidly compared to the other schemes because triggered updates, and ACKs sent by nodes for every triggered update create a large amount of packets, resulting in more packet drops. FORP and DV-MP are the schemes that are the least affected by mobility, since they use mobility prediction to perform rerouting prior to route disconnection and to send data over more stable routes which minimize packet losses. Figure 6 shows the control overhead as a function of mobility speed for each protocol. The figures presented are adopted from (Su et al. 2001).

The control overhead for DV-MP does not increase as the mobility speed increases because the transmission of control packets in DV-MP is periodically triggered without adapting to mobility speed. On the other hand, the overhead of FORP becomes larger as mobility speed increases as more FLOW-REQ and FLOW-SETUP messages are sent when mobility is high. WRP has the highest ratio in mobile situations because of the small number of delivered data packets and

Figure 5. Packet delivery ratio as a function of speed

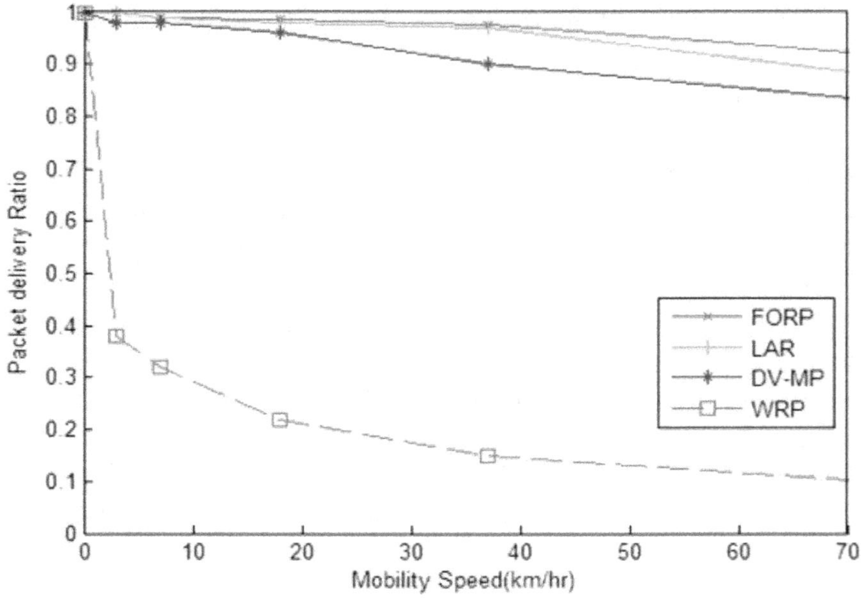

Figure 6. Number of control bytes transmitted per data bytes delivered as a function of speed

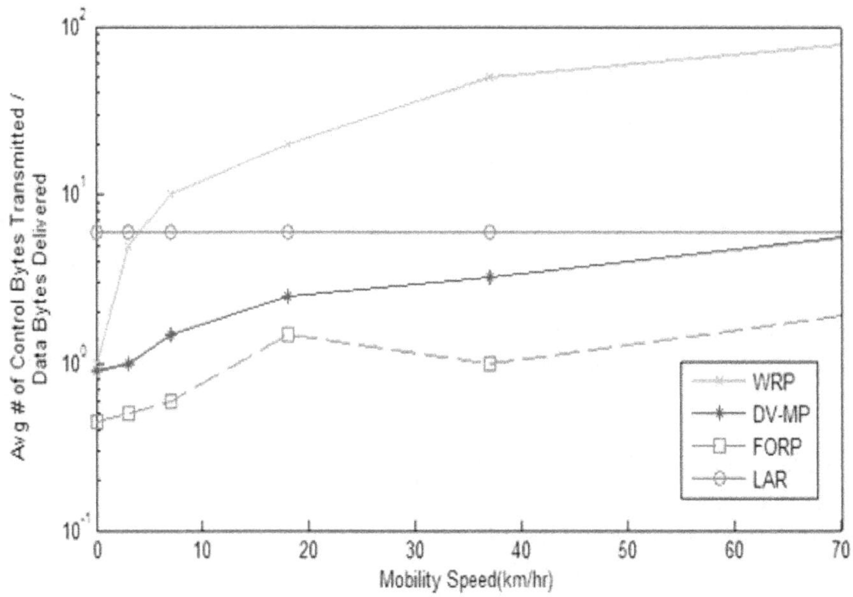

the large number of triggered updates. LAR also shows more overhead as mobility speed increases because more route breaks occur and they invoke route reconstruction procedures.

4. SUMMARY AND CONCLUSION

Among the different technology issues in MANETs, this chapter focused on the mobility issue. Since the nodes can move freely, the node mobility pattern is an important characteristic of MANETs. Frequent changes in network topology caused by mobility impose great challenges for developing efficient routing algorithms. Using mobility models to mimic the movement of mobile nodes in a certain environment makes it possible to make a relatively accurate prediction about the future state of the link. Several metrics were proposed in the literature to predict the future state of the link as a measure of network topology changes. Using those metrics, one can apply them to ad hoc routing protocols to choose more stable route and perform rerouting prior to route breaks. Mobility prediction can also be used to design new mobility aware routing protocols that exhibit better performance and lower overhead than existing routing protocols for MANETs.

Possible future work include some important performance evaluation metrics that the described mobility prediction techniques lack such as how far into the future a prediction can be made, how large the margin of error is and how often the prediction is within this margin of error for different types of applications such as in VANETs, sensor and mesh. Also, a comparative study of the different mobility prediction schemes under different conditions is needed to evaluate how well a certain scheme performs with respect to other schemes. Furthermore, most of the mobility prediction methods introduced use a random mobility model; however, a more realistic mobility model or real-world traces should be used to test the performance. In addition, the overhead caused by mobility in routing protocols should be better evaluated to decide whether it is better to incorporate the mobility prediction in existing routing protocols or design a new mobility aware routing protocols. Most of the work on mobility prediction use the average (or maximum) speed as the mobility measurement. Thus, network performance evaluations using some better metrics that capture topological changes are also needed.

REFERENCES

Bai, F., Sadagopan, N., & Helmy, A. (2003). IMPORTANT: A framework to systematically analyze the impact of mobility on performance of routing protocols for Ad hoc networks. *IEEE International Conference on Computer Communications*, (pp. 825-835).

Camp, T., Boleng, J., & Davies, V. (2002). A survey of mobility models for ad hoc network research. *International Journal of Network Management, 2*, 483–502.

Farkas, K., Hossmann, T., Ruf, L., & Plattner, B. (2006). Pattern matching based link quality prediction in wireless mobile ad hoc networks. *International Symposium on Modeling, Analysis and Simulation of Wireless and Mobile Systems,* (pp. 239-246).

Gerharz, M., de Waal, C., Martini, P., & James, P. (2003). Strategies for finding stable paths in mobile wireless ad hoc networks. *IEEE International Conference on Local Computer Networks*, (pp. 130-139).

Han, Y., La, R. J., Makowski, A. M., & Lee, S. (2006). Distribution of Path durations in mobile ad hoc networks palms theorem to the rescue. *International Journal of Computer and Telecommunications Networking, 50*, 1887–1900.

Hong, D., & Rappaport, S. S. (1986). Traffic Model and performance analysis for cellular mobile radio telephone systems with prioritized and non-prioritized handoff procedures. *IEEE Transactions on Vehicular Technology*, 77–92. doi:10.1109/T-VT.1986.24076

Jiang, S., He, D., & Rao, J. (2005). A prediction-based link availability estimation for routing metrics in MANETs. *IEEE/ACM Transactions on Networking, 13*, 1302–1312. doi:10.1109/TNET.2005.860094

McDonald, A. B., & Znati, T. (1999). A path availability model for wireless ad-hoc networks. *IEEE Wireless Communications and Networking Conference*, (pp. 35-40).

McDonald, A. B., & Znati, T. (2005). *Predicting node proximity in ad-hoc networks: A least overhead adaptive model for selecting stable routes* (pp. 39–43). First Annual Work Shop on Mobile Ad Hoc Networking and Computing.

Sadagopan, N., Bai, F., Krishnamachari, B., & Helmy, A. (2003). Paths: Analysis of PATH duration statistics and their impact on reactive MANET routing protocols. *International Symposium on Mobile Ad Hoc Networking and Computing*, (pp. 245-256).

Samar, P., & Wicker, S. B. (2004). On the behavior of communication links of a node in a multi-hop mobile environment. *ACM International Symposium on Mobile Ad Hoc Networking and Computing*, pp. 145-156.

Sithitavorn, K., & Qiu, B. (2005). *Mobility prediction with direction tracking on dynamic source routing* (pp. 1–6). IEEE TENCON.

Su, W., Lee, S., & Gerla, M. (2001). Mobility prediction and routing in ad hoc wireless networks. *Wireless Communications and Mobile Computing, 11*, 3–30.

Xu, S., Blackmore, K., & Jones, H. (2007). An analysis framework for mobility metrics in mobile ad hoc networks. *EURASIP Journal on Wireless Communications and Networking, 1*, 26–42.

Yoon, J., Liu, M., & Noble, B. (2003). Random waypoint considered harmful. *IEEE International Conference on Computer Communications*, (pp. 1312-1321).

Chapter 2
Multicast Routing Protocols in MANET

Javad Akbari Torkestani
Islamic Azad University, Iran

Mohammad Reza Meybodi
Amirkabir University of Technology, Iran

ABSTRACT

Multicast routing is an effective way to establish the group communication when the same message or the same stream of data needs to be sent to multiple receivers. Multicast routing has attracted a lot of attention in group oriented computing due to supporting data transmission from a single source node to multiple destinations concurrently. The advantage of multicast routing lies in its capability of reducing the communication cost and saving the network resources by sending only one copy of the message over the shared link leading to different destinations. Generally speaking, this survey classifies the multicast routing protocols into four categories based on the underlying routing structure: tree-based, mesh-based, hybrid, and stateless multicast routing protocols. This survey summarizes the well-known most recent protocols of each category and compares their objectives, performances, advantages, and disadvantages. This survey study helps us to choose the best multicasting protocol for each network application with respect to the requirements of application.

1. INTRODUCTION

Multicasting is a technique for data routing in networks that allows the same message is forwarded to a group of destinations simultaneously. In mobile ad hoc networks (MANETs), the most challenging issue in multicast routing is to effectively handle the frequent and unpredictable topology changes caused by host mobility, link breakage and host failure. Multicasting is intended for group-oriented computing like audio/video conferencing, collaborative works, and etc. Multicasting is an essential technology to efficiently support one-to-many or many-to-many applications. Multicast routing has attracted a lot of attention in the past decade, due to it allows a source to send information to multiple destinations concurrently. Multicasting is the transmission of packets to a group of zero or more

DOI: 10.4018/978-1-4666-0321-9.ch002

hosts called multicast group which is identified by a single destination address. A multicast group is a set of network clients and servers interested in sharing a specific set of data. A typical example of multicast groups is a commander and his soldiers in a battlefield. There are other examples in which multicast groups need to be established. Typically, the membership of a host group is dynamic: that is, the hosts may join and leave groups at any time. There is no restriction on the location or number of members in a host group. A host may be a member of more than one group at a time. A host does not have to be a member of a group to send packets to it. A multicast protocol has the objective of connecting members of the multicast group in an optimal way, by reducing the amount of bandwidth necessary but also considering other issues such as communication delays and reliability (Nadeem & Parthasarathy, 2006).

Multicast routing plays a critical role in most of the new applications such as web-base learning, video conference, and interactive multimedia games. Multicast routing in mobile ad hoc networks poses several challenges due to inherent characteristics of the network such as node mobility, reliability, and scarce resources. The main difficulty in designing a routing protocol for mobile ad hoc networks is the dynamically changing topology, due to the random movement of mobile nodes. The multicast routing protocols designed for wireless mobile ad hoc networks are fundamentally different from those for conventional infrastructure based networks in that these are self-configuring and formed directly by a set of mobile nodes without relying on any established infrastructure. In such networks, the heterogeneity of the hosts makes it difficult to achieve bandwidth efficiency and service flexibility.

In ad hoc networks, there exist several methods upon which classification of the multicast routing protocols is based. Classification based on the underlying structure of the multicast routes, based on the route acquisition time, based on multicast route initiation (based on the responsibility for route construction), and based on the forwarding state maintenance schemes are the well-known representative methods. Underlying multicast route structure is the most popular approach based on which the multicast routing protocols are classified. This classification considers the connectivity of the forwarding paths through which the multicast receivers are connected. This survey presents a new comprehensive classification of the multicast routing protocols based on the underlying route structure. In this classification, the multicasting protocols are categorized as tree-based multicast routing protocols, mesh-based multicasting protocols, hybrid protocols, and stateless protocols. This survey studies the most recent effective protocols of each category with the emphasis on the objectives, performances, costs, advantages, and drawbacks.

The rest of this paper is organized as follows. The multicast routing problem is defined in Section 2. In Section 3, the multicast routing protocols proposed for ad hoc networks are classified from several points of view. Section 4 presents a new classification of the multicast routing protocols based on the underlying routing structure and compares the existing multicasting schemes from this point of view. Section 5 concludes the paper.

2. MULTICAST ROUTING PROBLEM

Multicasting is a technique for data routing in networks that allows the same message is forwarded to a group of destinations simultaneously. Multicast routing problem (MRP) is one of the most difficult problems in communication networks in which the source node (or a set of source nodes) is connected to the set of destinations, while some cost function is simultaneously minimized. In a wireless network, due to the broadcast nature of the omnidirectional antennae, a single transmission can be received by all neighbors of the transmitting node (Nadeem & Parthasarathy, 2006). Therefore, the multicast routing protocols designed for the traditional wired networks are not applicable to the wireless networks. In a wired network, the multicast packets are forwarded along the tree edges, and so the multicast routing problem can be defined as a Steiner tree problem where the multicast group members are the terminals (leaf nodes) in the Steiner tree (Wu, Xu, Chen, & Wang, 2004). The problem of finding a Steiner tree is known to be NP-complete (Karp, 1972), even if links have unit cost (Aggarwal, Dubey, & Mehta, 2006). It should be noted that in some multicast routing protocols (Chiang, Liu, & Huang, 2007; Rodolakis et al., 2008) the minimum spanning tree problem, which is a well known approach for broadcasting, is also used to model the multicast routing problem. However, in wireless ad hoc networks, owing to the broadcast nature of the wireless channels, the topology graph of such networks can be modeled as an undirected unit disk graph G <V,E>, where the nodes represent the individual hosts and an edge connects two nodes if the corresponding hosts are within transmission range of each other. In these networks, the Steiner connected dominating set problem (Rajaraman, 2002; Bui & Zrncic, 2006) is a promising approach for solving the multicast routing problem, where the only multicast group members need to be dominated. The following represents the graph-based mathematical formulations of the minimum multicast routing problem in wireless networks.

Definition 1: Given a unit disk graph G <V,E>, a source node $s \in V$ called multicast source, and a set of destinations $D \subset V$ called multicast members, the multicast routing problem can be defined as finding the smallest set of relay (intermediate) nodes $F \subset V$ such that set $\{s\} \bigcup F \bigcup D$ is connected.

In most of the multicast routing protocols reported in the literature, it is assumed that all the hosts have the same costs, and so the proposed protocols as described in Definition 1 try to minimize the number of relay nodes for optimizing the multicast routes. In these methods, the multicast routing problem is defined as finding the minimum size Steiner connected dominating set in which a subset of relay nodes is chosen as dominators to construct a route from the multicast source to each of the multicast receivers. The host degree, average distance (the distance between the host and its neighbors), mobility characteristics (average speed or relative speed), residual energy level, and transmission power are some parameters can be applied as costs to a given host in a mobile wireless ad hoc network. In many applications of the wireless ad hoc networks, the above assumption (i.e., all hosts have the same costs) can not hold true, and reducing the number of relay nodes is not sufficient. In such networks, due to the hosts heterogeneity, host mobility, and strict resource limitations, each wireless host may have a different cost in the multicast tree. Therefore, in the following we represent a vertex-weighted definition of the minimum multicast routing problem where each host has an individual cost.

Definition 2: Given a node-weighted graph G <V,E,W> corresponding to the network topology graph, where V denotes the set of nodes, E denotes the edge set, and W denotes the set of weights which is associated with V by function $W: V \rightarrow \Re^{+}$, a source node $s \in V$, and a set of destinations $D \subset V$. Let $M = \{m_1, m_2, \ldots\}$ denotes the set of all possible multicast routes by which the multicast source is connected to all the multicast members, and

$$w_{m_i} = \sum_{\forall v_j \in m_i} w_{v_j}$$

denotes the weight of multicast route m_i. Multicast route $m^* \in M$ is the optimal solution to the MRP, if and only if we have

$$w_{m^*} = \min_{\forall m_i \in M} w_{m_i}$$

The above definition formulates the multicast routing problem as the minimum node-weighted Steiner connected dominating set of the network topology graph. Since the characteristics of the mobile ad hoc networks are stochastic, unpredictable and time-varying (Chlamtac, Conti, & Liu, 2003; Basagni et al., 2004; Mohapatra & Krishnamurthy, 2005), stochastic graphs are more appropriate data structures for modeling the ad hoc network topology graphs. Therefore, in what follows, we represent the definition of the stochastic graph to model the ad hoc networks, then, we define the MRP in such networks. A node-weighted graph is a graph in which a value (or weight) is associated with each node. A stochastic node-weighted graph is a node-weighted graph in which the weight associated with each node is a random variable.

Definition 3: Given a node-weighted graph G <V,E,W> corresponding to the network topology graph, where V denotes the set of nodes, E denotes the edge set, and W denotes the set of weights which is associated with V by probability distribution function $W : V \rightarrow \Re^{+}$, a source node $s \in V$, and a set of destinations $D \subset V$. Let $M = \{m_1, m_2, \ldots\}$ denotes the set of all possible multicast routes by which the multicast source is connected to all the multicast members, and

$$\overline{w}_{m_i} = \sum_{\forall v_j \in m_i} \overline{w}_{v_j}$$

denotes the expected weight of multicast route m_i. Multicast route $m^* \in M$ is the optimal solution to the MRP, if and only if we have

$$\overline{w}_{m^*} = \min_{\forall m_i \in M} \overline{w}_{m_i}$$

That is, the optimal (minimum cost) multicast route in a stochastic network is defined as the multicast route with the minimum expected cost. By Definition 3 the multicast routing problem in a stochastic network is formulated as the stochastic Steiner connected dominating set of the network topology graph.

3. MULTICAST ROUTING CLASSIFICATIONS

Multicast routing protocols come into play when a host needs to send the same message, or the same stream of data, to multiple destinations. Due to the unique characteristics of the mobile ad hoc networks such as host mobility, limited resources and very unreliable channel, traditional multicast protocols do not perform well in MANET scenarios. MANET multicast protocols should efficiently cope with dynamic topology changes such as fragile multicast tree structure. To compare and analyze multicast routing protocols, appropriate classification methods are important. Classification methods help researchers and designers to understand the distinct characteristics of different multicast routing protocols and find out the internal relationship among them. Therefore, we present protocol characteristics which are used to group and compare different approaches. These characteristics are mainly related to the information which is exploited for MANETs and the roles which nodes may take in the multicast routing process. There are several methods to classify multicast routing protocols for MANET. As reported in the literature, multicast routing protocols are broadly classified as tree-based and mesh-based protocols, based on the underlying structure of the multicast routes. The multicast routing protocols can be also classified as proactive and reactive protocols based on the route acquisition time, as sender-initiated and receiver-initiated protocols based on the multicast route initiation, and as source-based and group-shared protocols based on the forwarding state maintenance schemes. Some multicast routing protocols fall into one or more above mentioned categories. In the remaining of this section we briefly describe different classifications of multicast routing protocols.

3.1. Based on Underlying Routing Structure

One of the most popular methods to classify multicast routing protocols for MANETs is based on how distribution paths among group members are constructed (the underlying routing structure). According to this method, existing multicast routing approaches for MANETs can be divided into tree-based multicast protocols, mesh-based multicast protocols and hybrid multicast protocols. In tree-based multicast protocols, the members are connected through a tree structure. Tree-based protocols send the message along a tree-like data forwarding path which is rooted at the source of the multicast session, and supports a single path to each receiver. Depending on the number of trees per multicast group, tree-based multicast can be further classified as source-based multicast tree and group-shared multicast tree. Some of the tree-based multicast routing protocols are, bandwidth efficient multicast routing protocol (BEMRP) (Ozaki, Kim, & Suda, 2001), multicast zone routing protocol (MZRP) (Zhang & Jacob, 2004), multicast core extraction distributed ad hoc routing protocol (MCEDAR) (Sinha, Sivakumar, & Bharghavan, 1999), differential destination-based multicast protocol (DDM) (Ji & Corson, 2001), ad hoc multicast routing protocol utilizing increasing id-numbers (AMRIS) (Wu, Tay, & Toh, 1998), and ad hoc multicast routing protocol (AMRoute) (Xie et al., 2002).

Tree-based protocols may not perform well in the presence of highly mobile nodes because the multicast tree structure is fragile and needs to be readjusted frequently as the connectivity changes. Mesh-based multicast protocols have been proposed to address the problem by constructing a mesh structure with redundant links between mobile nodes. Indeed, mesh based protocols aims at enriching the connectivity among the multicast group members, and so they are more resilient against the network topology changes compared to tree-based protocols. Mesh based protocols generally suffer from the notorious broadcast storm problem, in which the flooding mechanism may result in excessive redundancy, contention, and collision. Furthermore, mesh-based protocols are more likely to form the routing loops. On-demand multicast routing protocol (ODMRP) (Lee, Gerla, & Chiang, 1999), Clustered Group Multicast (CGM)(Lin & Chao, 1999), forwarding group multicast routing protocol (FGMP) (Chiang, Gerla, & Zhang, 1998), and core-assisted mesh protocol (CAMP) (Madruga & Garcia, 2001) are some of the existing mesh-based multicast protocols. Hybrid approaches like AMRout (Xie et al., 2002) are combinations of tree-based and mesh-based protocols to bring together the robustness of mesh-based multicast routing protocols and the low overhead of tree-based protocols in a single scheme.

3.2. Based on Route Acquisition Time

Based on when the multicast routes are determined, the multicast routing protocols fall into two different categories, namely proactive and reactive (on-demand). Proactive multicast routing continuously makes routing decisions so that multicast routes are immediately available when packets need to be transmitted. Proactive multicast routing protocols attempt to maintain consistent, up-to-date multicast routing information between every pair of multicast members in the network by propagating, proactively, route updates at fixed time intervals. As the resulting routing information is usually maintained in tables, the protocols are sometimes referred to as table-driven protocols. Ad hoc Multicast Routing protocol (AMRoute) (Xie et al., 2002) is a proactive multicast routing protocol. Reactive multicast routing protocols, on the other hand, establish a route to the multicast members only when there is a demand for it, usually initiated by the source node through route discovery process within the network. Once a route has been established, it is maintained by the node until either the destinations become inaccessible along every path from the source or until the route is no longer used or has expired. On-demand Location Aware Multicast (OLAM) (Basagni et al., 2000), On-demand multicast routing protocol (ODMRP) (Lee, Gerla, & Chiang, 1999), and On-demand Global Hosts for mobile Ad hoc Multicast services (OGHAM) (Hu, Wu, & Chen, 2006) are some of the well known reactive multicast routing protocols. Proactive multicast routing protocols consume a great deal of radio resources to exchange routing information. Also, pre-determined multicast routes may rapidly lose their validity in an ad hoc network because its topology frequently changes. Proactive multicast routing protocols generally provide better quality of service than reactive protocols. As in proactive protocols, routing information is constantly updated, routes to every destination are always available and up-to-date, and, hence, end-to-end delay can be minimized. For reactive protocols, the source node has to wait for the route to be discovered before communication can happen. This latency in route discovery might be intolerable for real-time communications.

3.3. Based on Multicast Route Initiation

The basic schemes for reliable multicast routing can be classified as sender-initiated and receiver- initiated approach based on who (sender or receiver) is responsible for route construction. A multicast routing protocol is termed as a sender-initiated protocol when the multicast tree construction is initiated by the multicast source. That is, Sender-initiated routing protocols establish a route only when a source node requests for a route. When the multicast members take the responsibility to connect to the source, then the protocol is termed as receiver-initiated multicast protocol.

In the sender-initiated approach, the sender is responsible for the error detection. A missing piece of data at a receiver is detected if the sender does not receive an ACK from the receiver. In this case, the need to retransmit a missing packet is handled by retransmitting the missing data from the source through a unicast. When several receivers have missing packets, the sender may decide to re-multicast the missing packets to all receivers in the multicast group. When the sender-initiated approach is applied, only the sender (which keeps the history of multicast packets) is responsible for retransmitting the missing packet, and the corresponding retransmitting method is called sender-oriented. Note that when the sender receives ACK signals from all the receivers, the corresponding packet can be removed from the history. Multicast Zone Routing (MZRP) (Zhang & Jacob, 2004) protocol, and Multicast Ad hoc on-demand Distance Vector Routing (MAODV) (Royer & Perkins, 1999) protocol are two sender-initiated multicast routing protocols.

In the receiver-initiated approach, each receiver is responsible for error detection. Instead of acknowledging each multicast packet, each receiver sends a NACK once it detects a missing packet. If multicast packets are time stamped using a sequence number, a missing packet can be detected by a gap between sequence numbers of the receiving packets. WARM (Kondylis et al., 2000), short for Wireless Ad hoc Real- Time Multicast, is a receiver initiated multicast routing protocol.

4. MULTICAST ROUTING PROTOCOLS

In literature, the multicast routing protocols are mostly classified based on the underlying route structure. In such a classification, the multicast routing schemes are categorized as tree-based, mesh-based, hybrid and stateless schemes. Tree-based schemes construct a multicast tree from each of the sources to all the receivers. Examples of protocols following this approach are BEMRP (Ozaki, Kim, & Suda, 2001), MZRP (Zhang & Jacob, 2004), MCEDAR (Sinha, Sivakumar, &

Bharghavan, 1999), DDM (Ji & Corson, 2001), AMRIS (Wu, Tay, & Toh, 1998). The main advantage of using a tree as the underlying forwarding structure is that the number of forwarding nodes tends to be reduced (although not optimized). However, a tree is very fragile when there is a high mobility in the network. Mesh-based approaches like ODMRP (Lee, Gerla, & Chiang, 1999), CGM (Lin & Chao, 1999), FGMP (Chiang, Gerla, & Zhang, 1998), CAMP (Madruga & Garcia, 2001), by using additional links in their underlying forwarding structure, manage to deal with mobility very efficiently. The main drawback associated to the use of a mesh, is that the additional paths which are created can make an excessive consumption of network resources when sending data packets. Hybrid approaches like AMRout (Xie et al., 2002) try to combine the robustness of mesh-based ad hoc multicast routing and the low overhead of tree-based schemes. Figure 1 shows the new classification of the multicast routing protocols presented in this paper. In the rest of this section, the existing multicast routing protocols are classified based on the underlying routing structure and described in more detail.

4.1. Tree-based Multicast Routing Protocols

In a tree-based multicast routing algorithm, a tree-like data forwarding path is constructed which is rooted at the source of the multicast session. The multicast tree is composed of a unique path from the multicast source to each of the multicast receivers. Tree-based protocols are highly efficient (efficiency is defined as ratio of the total number of data packets received by the nodes to the total number of data packet transmissions in the network) due to the absence of multiple redundant paths to the multicast source node. The main advantage of a tree as the underlying forwarding structure is that the number of forwarding nodes tends to be reduced (although not optimized). However, multicast trees form a virtual backbone which is fragile in ad hoc networks where the mobile hosts move freely anywhere. This is due to the fact that in a multicast tree there is no alternative path between the source and destination to tolerate the frequent network topology changes. Tree-based protocols primarily focus on how to construct the tree with the minimum control overhead, and involving the minimum cost. The cost metric is generally assumed the average distance of the

Figure 1. A new classification of multicast routing protocols for ad hoc networks

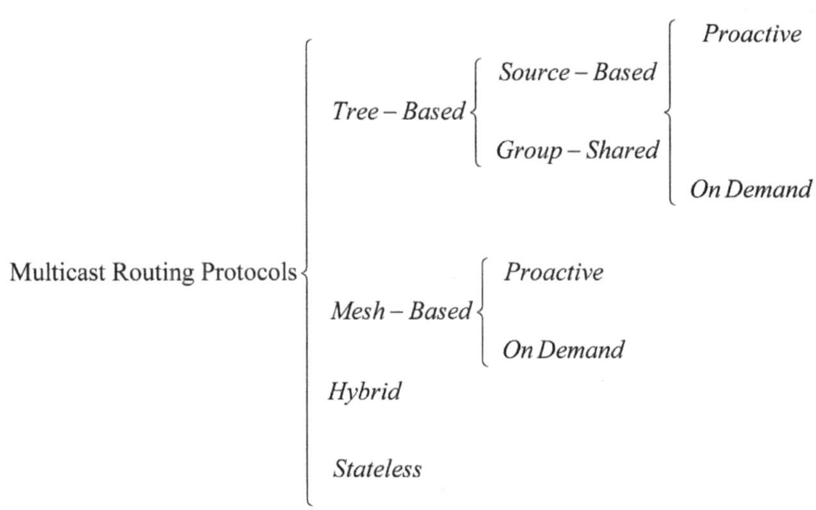

multicast members from the multicast source. The control overhead involved in tree-based protocols is low, and the performance in terms of packet delivery ratio of such protocols, decreases as the mobility increases. The major difference between the tree-based and mesh-based multicast routing protocols lies in the manner in which a multicast message is relayed. In the tree-based multicast routing protocols, each intermediate node on the tree has a well-defined list of the next-hop nodes for a specific multicast session. It will send a copy of the received multicast message to only the neighboring nodes on its next-hop list. Figure 2 shows a sample tree-like multicast route to connect the multicast source to the multicast receivers.

Tree-based multicast routing protocols can be further subdivided into group-shared tree and source-based tree. In both cases, multicast trees are constructed to interconnect all the members of the multicast group. Data is delivered along the tree paths to reach all the group members. The source-based approach maintains, for each multicast source, an individual (minimal) tree towards all its multicast receivers. Source-based tree is separately established and maintained for each multicast source node of a multicast group. The advantage of source-based tree is that each multicast packet is forwarded along the most efficient (shortest) path from the source node to each multicast group member. However, this method incurs a lot of control overhead and cannot quickly adapt to the movements of the nodes in a mobile ad hoc network. Since the construction of a minimum cost tree (for each source) spanning all the members of the multicast group is expensive, some of the tree-based multicast routing schemes use a (core-based) group-shared tree to distribute packets from all the sources. In the group-shared tree, a single tree is constructed for the whole group (e.g., regardless the sources location). Multicast packets are distributed along this shared tree to all members of the multicast group. Since the group-shared multicast tree only permits the multicast traffic to be sent out from the root to the multicast receivers, each source must forward its multicast traffic to the root. Multicast traffic of each source is then forwarded along the shared tree. The group-shared multicast tree is a well-known tree-based approach adopted by core based trees (CBT) (Ballardie, Francis, & Crowcroft, 1993). A group-shared tree is a shortest-path tree rooted at some core node. The core node is also referred to as a center node or a rendezvous point. Core nodes may be chosen from some pre-selected set of nodes or some heuristics may be employed to select core nodes. Group-shared multicast tree

Figure 2. A tree-based multicast route

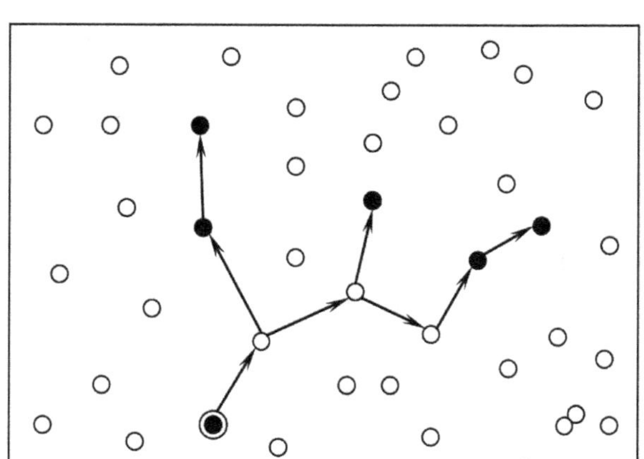

is a more scalable approach than the source-based approach in which instead of building multiple trees for each multicast group, a single shared tree is used for all multicast source nodes. Representative group-shared multicast trees are built in MAODV (Royer & Perkins, 1999) and AMRIS (Wu, Tay, & Toh, 1998).

Source-based or group-shared, the optimal multicast tree is defined as a Steiner tree, although in some multicast routing protocols (Chiang, Liu, & Huang, 2007; Rodolakis et al., 2008), the minimum spanning tree problem (MST) is used to model the multicast routing problem. Constrained Steiner tree problem is shown to be NP-complete, and so several heuristics have been proposed to solve the multicast tree problem. The rest of this section briefly reviews the existing tree-based multicast routing protocols.

4.1.1. Source-Based Multicast Tree

Distance Vector Multicast Routing Protocol (DVMRP) (Waitzman, Partridge, & Deering, 1988) is a multicast routing protocol initially designed for wired networks. Modified version of DVMRP can be also applied for multicasting in wireless ad hoc networks. DVMRP builds and maintains source-based trees for sending multicast messages. The source-based tree is created by first flooding the whole network with the multicast traffic. After that the normal prune operations are conducted. In DVMRP, when a host receives a multicast packet, it sends the packet to all attached hosts and waits for a response. Hosts with no group members return a "prune" message, which eventually prevents further multicast messages for that group from reaching the host. The prune state is soft, that is, it will time-out within a set time interval. If after sending a prune and before the state can time-out, the host wants to join the group, it has to send a "graft" message upstream. DVMRP is inefficient when the number of receivers in the group is sparsely distributed. DVMRP builds its own routing table instead of reusing the existing unicast routing table for RPF, (short for reverse path forwarding) checking of incoming packets. A packet is assumed to have arrived on the RPF interface, if a host receives it on an interface that it uses to send unicast packets to the source. If the packet arrives on the RPF interface, then host forwards it out the interfaces that are present in the outgoing interface list of a multicast routing table entry. If it does not arrive on RPF interface, it is silently discarded to avoid loop-backs. The advantage of RPF is that it does not require the host to know about spanning trees. This way, multicast adapts automatically and only is sent where it is wanted (Waitzman, Partridge, & Deering, 1988).

Akbari Torkestani and Meybodi (Akbari Torkestani & Meybodi, 2011) proposed a link stability-based multicast routing protocol for wireless MANETs in which the multicast packets are forwarded along the Steiner tree links. The weight associated with a communication link is defined as its expected duration time which is assumed to be a random variable with unknown distribution. Expected link duration time is defined as the period of time during which the link is expected to be connected, and expected duration time of a multicast route is defined as the expected duration time of the weakest link. This protocol called LLMR aims at finding the most stable multicast route against the host mobility. LLMR is composed of a number of iterations and at each iteration a multicast route is constructed by finding a Steiner tree of the network topology graph. At each iteration, the selected multicast route is rewarded, if its expected duration time is longer than those seen so far and it is penalized otherwise. As the proposed algorithm proceeds, the choice probability of the most stable multicast route converges to one. Thyagarajan and Deering (Thyagarajan & Deering, 1995) proposed a hierarchical distance-vector multicast routing protocol. This approach involves partitioning the multicast backbone (MBone) into non-overlapping regions, while using DVMRP as the inter-region routing protocol. MBone is a virtual network developed to run on top of the physical Internet. Intra-region routing protocol may be accomplished by any of the multicast routing protocols. Jetcheva and Johnson (Jetcheva & Johnson, 2001) proposed an adaptive demand-driven multicast routing algorithm called ADMR that is able to adapt its behavior based on application sending pattern, to effectively detect the link breakage, and to terminate the routes that are no longer needed. ADMR creates and maintains source-based forwarding trees connecting each source with the receivers of the multicast group. The multicast forwarding state for a given multicast group and a source is conceptually represented as a loosely structured multicast-forwarding tree routed at the source. The forwarding mechanism is based on the shortest-delay path through the tree to the receiver members of the multicast group. In this method, packet forwarding is based on two types of flooding: tree flood and network flood. Tree flood occurs among nodes of the multicast tree, while network flood is flooding among all the nodes in the network. ADMR can detect the network mobility characteristics without requiring any global positioning system or additional control traffic. Depending on the results of such a mobility detection process, ADMR multicast source switch to the flooding for a period of time when the mobility is high, and attempt to operate efficiently with multicast again in the case of low mobility.

Associativity-based ad hoc multicast (ABAM) routing algorithm proposed by Toh et al. (Toh, Guichal, & Bunchua, 2000) establishes the multicast sessions using the association stability concept introduced in ABR unicast protocol (Royer & Toh, 1999) and requiring no underlying unicast routing protocol. ABAM constructs the source-based multicast trees using this concept on-demand, which reduces the communication overhead and improves end-to-end delay.

Zhu et al. (Zhu, Parsa, & Garcia-Luna-Aceves, 1995) presented a heuristic algorithm called Bounded Shortest Multicast Algorithm (BSMA) for constructing the minimum-cost source-based multicast trees with delay constraints. BSMA starts with a Shortest Path Tree (SPT) to all destinations. It then iteratively replaces super-edges in the tree with lower cost paths until the total cost of the tree cannot be reduced any further. Although BSMA has a near optimal tree cost, it is not suitable for real networks due to its high computation cost. The proposed algorithm can set variable delay bounds on destinations and handles two variants of the network cost optimization goal: one minimizing the total cost of the tree, and another minimizing the maximal link cost. Instead of the single-pass tree construction approach used in most previous heuristics, the new algorithm is based on a feasible search optimization method which starts with the minimum-delay tree and monotonically decreases the cost by iterative improvement of the delay-bounded tree.

Rouskas et al. (Rouskas & Baldine, 1997) addressed the problem of constructing a source-based multicast tree with delay and delay jitter constraints for the one-to-many multicast paradigm and proposed a heuristic algorithm. Their algorithm can be also extended to accommodate the many-to-many. In the proposed algorithm, for each multicast source, a source-based multicast tree is initially constructed. Then, a minimum set of multicast trees is found such that for each multicast source there exists at least one feasible multicast tree.

In (Wang, Sandeep, & Gupta, 2004), Wang et al. presented a distributed algorithm called L-REMiT for enhancing the lifetime of a source-based multicast tree in wireless ad hoc networks. In this paper, the lifetime of a multicast tree is defined as the duration from the formation of the tree to the time when the first node fails due to battery energy exhaustion. The proposed multicasting algorithm assumes that the energy consumed to forward a packet is proportional to the forwarding distance and channel characteristics, and that wireless nodes can dynamically adjust their transmission power.

Protocol independent Multicast (PIM) operates in two modes: dense mode (PIM-DM) and sparse mode (PIM-SM). PIM-DM (Deering et al., 1999) proposed by Deering et al. operates similar to DVMRP. PIM-dense mode is designed for multicast communication in small networks and high group density. No rendezvous point (RP) is used. The protocol builds a source-based multicast tree for each sender as soon as it starts sending data. Multicast routing entries are kept in related routers under the mechanism of timeout. At startup, it assumes that all subnets wish to receive data. Therefore, flooding and pruning are used. Graft data units are sent to support immediate integration of new group members into the multicast tree. Due to flooding, data can end up being sent unnecessarily to network areas in which no group members are located. However, this is considered acceptable because it is assumed that the density of a group will be very high, the distance between members are short, and consequently the additional overhead will be low .

Moy (Moy, 1994) proposed an extension of OSPF unicast routing protocol called Multicast Open Shortest Path First (MOSPF) for multicast routing. MOSPF knows all the multicast groups currently residing in the network and builds an individual distribution tree for each source/group pair, i.e., it constructs source-based multicast trees. Distribution trees must be re-computed either periodically or when there is a link state change. MOSPF is a link state routing protocol that builds the map of the network topology, including location of domains and tunnels. It selects the best path to the required receivers using Djikstra's shortest path algorithm. When there are multiple sources or many groups, it is CPU intensive. It is best used when relatively few sources or groups are active at any given time. It does not work well in presence of unstable links, as it leads to frequent state update and the associated computations. MOSPF does not support tunneling. The path is calculated only on-demand and cached for later use.

4.1.2. Group-Shared Multicast Tree

Multicast Ad hoc on-demand Distance Vector Routing protocol so-called MAODV proposed by Royer and Perkins (Royer & Perkins, 1999) maintains a shared tree for each multicast group, consisting of only receivers and relays. MAODV results from the well-known unicast routing algorithm called AODV proposed by the same authors (Perkins & Royer, 1998). MAODV is an on-demand routing protocol that discovers the route only when a node has data to send. It is a hard state protocol, i.e., if a member node of a multicast group desires to terminate its group membership, it must request for termination. When the core revokes its own group membership, it selects one of its tree neighbors to become the new core. If the new core is not a group member, one of its downstream tree neighbors is selected to become the new core. This core selection process is continued until a group member is found to become the new core. When a mobile node wants to join a multicast group or send a message but does not have a route to the group, a Route Request (RREQ) is originated. All the nodes that are members of a multicast group together with the nodes that are not members of the group but their position are very critical for forwarding the multicast information, compose the tree structure. Every multicast group is identified by a unique address and group sequence numbers for tracing the freshness of the group situation. In MAODV, a node maintains multicast routing table for the group tree structure. This table contains the multicast group address, the multicast group leader address, the multicast group sequence number, hop count to the multicast group leader, next hop information and the lifetime. Nodes in a tree structure are described as downstream and upstream nodes. A downstream node is a neighborhood node, which is further from the group leader (more hop counts from the group leader). An upstream node is a neighborhood node which is nearer to the group leader (less hop counts from the group leader). It is obvious that a group leader has only downstream nodes. Whenever a node leaves the multicast group, the tree structure is pruned. If an edge in the multicast tree is broken, the downstream node of the edge is responsible for finding a path to the core and reforming the multicast tree. If this node cannot find a path to the core within a certain time period, the network is assumed to be disconnected and the subtree rooted at this node becomes another multicast tree. If it is a group member, this node then becomes the core of the newly formed multicast tree; otherwise, the core selection process is used to find the core of the newly formed multicast tree. If the network becomes connected later, the multicast trees of the same group are merged into one whose core is the same as that of the multicast tree having the largest address.

In (Gupta & Srimani, 2003), Gupta and Srimani proposed a method similar to MAODV, where the core migrates to the group member median located at the tree centroid. They presented distributed core selection and migration algorithms for mobile ad hoc networks with dynamically changing network topology. The proposed core location method is based on the notion of median node of the current multicast tree instead of the median node of the entire network. The rationale is that the mobile ad hoc network graphs are in general sparse, and so the multicast tree is a good approximation of the entire network for the current purpose. The adaptive distributed core selection and migration method uses the fact that the median of a tree is equivalent to the centroid of that tree. This leads to a significant efficiency improvement since the median (or the centroid) computation in a tree does not need distance information.

Liu et al. (Liu, Huang, & Tsai, 2008) proposed a heuristic method to reduce the cost of the core-based group-shared multicast tree by minimizing the number of non-leaves in the multicast tree, where the cost is evaluated by the total bandwidth consumption of multicasting packets among all group members. They introduced a distributed heuristic in which the multicast tree is dynamically constructed and adjusted rather than finding a new multicast tree. To do so, several non-leave nodes are transferred into leaves in an existing multicast tree.

In (Wu, Tay, & Toh, 1998), Wu and Tay presented a multicast protocol called AMRIS, short for Ad hoc Multicast Routing protocol utilizing Increasing ID numbers. AMRIS is an on-demand (reactive) multicast routing protocol which constructs a shared delivery tree to support multiple senders and receivers within a multicast session. The conceptual idea behind AMRIS is to assign every node in a multicast session an ID number. A delivery tree rooted at a particular node called Sid joins up the nodes participating in the multicast session. The relationship between the ID numbers (and the node that owns them) and Sid is that the ID numbers increase in numerical value as they radiate from the root of the delivery tree. The significance of the Sid is that it has the smallest ID number within that multicast session. Utilizing the ID numbers, nodes are able to adapt rapidly to changes in link connectivity. Recovery messages due to link breakages are confined to the region where the breakage occurred.

The Zone Routing Protocol (ZRP) (Haas, Pearlman, & Samar, 2002) is a hybrid unicast protocol that proactively maintains routing information for a local neighborhood (routing zone), while reactively acquiring routes to destinations beyond the routing zone. It is composed of IntrAzone Routing Protocol (IARP) (Haas, Pearlman, & Samar, 2002b), IntErzone Routing Protocol (IERP) (Haas, Pearlman, & Samar, 2002c) and Bordercast Resolution Protocol (BRP) (Haas, Pearlman, & Samar, 2002d) with a query-control mechanism. Zhang and Jacob (Zhang & Jacob, 2004) extended ZRP for application to multicast routing and call it the Multicast Zone Routing Protocol (MZRP). MZRP is a shared tree multicast routing protocol that proactively maintains the multicast tree membership for nodes' local routing zones at each node while establishing multicast trees on-demand. It is scalable to a large number of multicast senders and groups. In MZRP, IP tunnel mechanism is employed to improve the data packet delivery ratio during the transmission.

Canourgues et al. (Canourgues et al., 2006) proposed a shared tree multicast routing scheme called STAMP for large wireless mobile ad hoc networks. The proposed scheme clusters the network to relieve the scalability constraints. STAMP is a core-based shared tree multicast routing protocol independent from the underlying unicast routing protocol. Compared to other shared-tree multicast routing protocols, STAMP achieves noticeably better performance even under mobility conditions since it takes advantage of the broadcast capacity of the medium to deliver data on the tree.

A multicast routing protocol for mobile Ad hoc network should find a compromise between the routing overhead and data transmission efficiency so that it can efficiently use bandwidth and power. For this aim, Kaliaperumal et al. (Kaliaperumal & Jeyakumar, 2005) proposed a reactive (on-demand) multicast routing protocol called ACMP, short for Adaptive Core based Multicast Routing protocol, which constructs and maintains a group-shared tree using adaptively selected core only when the group traffic exists. The proposed multicast routing scheme attempts to react more quickly to the broken tree edge by detecting link failures during data forwarding process. In this method, once a link failure is detected, ACMP uses local route recovery to establish a temporary route and periodical tree refreshing to maintain an optimal multicast tree.

Estrin et al. (Estrin et al., 1998) proposed another version of PIM called PIM-SM (short for Protocol Independent Multicast-Sparse Mode) in which a node is selected as the RP and all group communication takes places by sending the packets to it. Each of the sources in a PIM-SM multicast group sends its packets to the RP. Since it builds unidirectional shared tree, only the RP can forward data to the members. Intermediate nodes should forward the data only to the RP. All PIM-SM traffic is transported by unicast instead of multicast. The PIM-SM router with highest IP address is Designated Router (DR) for the subnet and is responsible for sending Prune/Join messages to the RP. DR determines the RP for a group using a hash function. The tree obtained is not necessarily optimal. PIM-SM allows switching of the receiver connectivity to the tree from shared-tree path to the source-based tree path. When a group has numerous highly active sources, the bandwidth of the shared links may not be able to accommodate all the traffic. The QoS requirements of a user may not be met along the shared-tree, thus it has to switch to its shortest path to the RP. When a receiver switches from shared-tree mode to source-based tree mode, the number of packets dropped during the transition period is dependent on two factors: The delay difference between the shared-tree path and source-based tree path from the source to the receiver and the sending rate of the source.

Ballardie et al. (Ballardie, Francis, & Crowcroft, 1993) proposed a core based multicast routing algorithm called CBT (Core Based Tree) in which a single shared multicast tree is used to connect all the different multicast groups and their members. CBT builds a single bidirectional shared tree for data transmission. CBT involves having a single node, known as the core of the tree, from which branches emanate. These branches are made up of the other nodes, so-called non-core nodes, which form a shortest path between a member host and core. The core need not be topologically centered between the nodes on the tree, since multicasts vary in nature, and so can the form of a core-based tree. CBT involves having a single core tree per group, with additional cores to add an element of robustness to the model. Core selection is one of the major issues in CBT and can be handled by the various heuristics proposed for core selection. When an intermediate node receives a packet, it forwards it to the remaining members of the group that are downstream to the node. It does not need to forward it to the core. In (Gupta, 2001), Gupta and Srimani proposed a core-based shared tree for reliable multicasting in multi-hop mobile radio networks. The proposed protocol is reliable, i.e., it guarantees message delivery to all the multicast members even when the topology of the network changes during the multicasting. They introduced a notion of the forwarding region which is used to limit the scope of the flooding messages only to the vicinity of the nodes which witness topology change due to host mobility. This notion is used to glue together the fragments of the multicast trees. The proposed protocol is more efficient in conserving network bandwidth, especially when the multicast groups are small in size and the multicast group members are localized to a certain region of the network.

Since the constrained multicast routing problem is known to be NP-complete, many proposed approaches are based on heuristic methods, such as ant colony algorithms, genetic algorithms and fuzzy-based algorithms. The rest of this section reviews some of the existing heuristic methods.

Ant Colony Approach

Ant colony optimization (ACO) is a population-based meta-heuristic that can be used for finding approximate solutions to difficult optimization problems (Dorigo, 1992). ACO studies artificial systems that take inspiration from the behavior of real ant colonies and which are used to solve discrete optimization problems. In ACO, a set of software agents called artificial ants search for good solutions to a given optimization problem. To apply ACO, the optimization problem is transformed into the problem of finding the best path on a weighted graph. The artificial ants (hereafter ants) incrementally build solutions by moving on the graph. The solution construction process is stochastic and is biased by a pheromone model, that is, a set of parameters associated with the graph components (either nodes or edges) whose values are modified at runtime by the ants (Dorigo, 1992; Dorigo & Gambardella, 1997; Stutzle & HOos, 2000).

Liu et al. (Liu, Wu, & Qian, 2006) proposed a novel scheme based on ACO called DMACO for finding the minimum cost multicast tree with delay constraint, which is an NP-complete problem. The proposed algorithm is a fully distributed routing scheme which does not need a core node to possess the whole network information. In this method, the ants only use the local information to finish the task of finding the destinations and establishing the multicast tree. The proposed scheme is easy to realize by three types of ants with simple structures: forward ants, backward ants, and pheromone update ants. A forward ant is one that the source sends to look for the destinations. A forward ant will be killed when it reaches the destination, and a backward ant is generated and returned to the source node along the inverse path. When all ants find all the destinations and construct the corresponding multicast trees, the source node will send pheromone update ants to update the pheromone table of corresponding nodes. The proposed scheme is loop free and has small overhead compared with the flooding method. In (Wang & Xie, 2000), Wang and Xie combined an improved version of ant colony optimization with a heuristic algorithm to solve the multicast routing problem. In this paper, the multicast routing problem is considered as the Steiner tree problem, and to solve this problem the proposed algorithm attempts to find the minimum number of Steiner nodes that are not included in the multicast routing group. A modified ant colony system is presented by Zhang et al. (Zhang et al., 2007) to solve the single-origin-multi-destination routing problem with no constraints. This algorithm first transforms the connective network into the corresponding distance complete graph, and then develops a Prim's-algorithm-like route construction process with an inclination factor towards the multicast vertices. Tseng et al. (Tseng, Lin, & Huang, 2008) proposed an ACO-based algorithm for solving the degree-delay-constrained minimum spanning tree (DDCMST) problem. DDCMST consists of three mechanisms: the spanning tree construction, the local updating rule, and the global updating rule, stated as follows. Spanning tree construction is a fundamental mechanism to solving the DDCMST problem by evolutionary algorithms. It introduces APrim as a modified version of Prim's algorithm (Prim, 1957). APrim starts tree-constructing from a source node, and then iteratively selects a new edge from the set of candidate edges to expand the tree. Edges are selected based on a revised pseudo-random-proportional rule. After an ant has found a spanning tree, the pheromone levels on the edges of that spanning tree are updated using a local updating rule. The proposed algorithm repeatedly constructs spanning tree and updates pheromones locally and globally until the stop condition is met.

Genetic Algorithm Approach

Genetic algorithms are a particular class of evolutionary algorithms in which the techniques inspired by the evolutionary biology such as inheritance, mutation, selection, and crossover are used. Genetic algorithms are categorized as global search heuristics. A typical genetic algorithm requires a genetic representation of the solution domain, as well as a fitness function to evaluate the solution domain. The fitness function is defined over the genetic representation and measures the quality of the represented solution. The fitness function is always problem dependent. Once we have the genetic representation and the fitness function defined, genetic algorithm proceeds to initialize a population of solutions randomly. Then, it improves the population through repetitive application of genetic operations. In (Zhang & Leung, 1999), Zhang and Leung proposed an orthogonal genetic algorithm for multimedia multicast routing. Its salient feature is to incorporate an experimental design method called orthogonal design into the crossover operation. The salient feature of the proposed genetic algorithm is to incorporate an experimental design method called orthogonal array into the crossover operation, so that the resulting operation can sample the genes from n parents in a statistically sound manner to produce j offspring. As a result, the orthogonal genetic algorithm can search the solution space in a statistically sound manner and it is well suited for parallel implementation and execution. In general, a larger orthogonal array gives a better solution quality but has a higher computational complexity. They execute the orthogonal genetic algorithm to solve two sets of benchmark test problems. The results indicated that for practical problem sizes, the orthogonal genetic algorithm can find near-optimal solutions within moderate numbers of generations. Atzori and Raccis (Atzori & Raccis, 2004) proposed a genetic algorithm which focuses on the problem of network capacity assignment for multicast services. The problem is firstly formalized by defining a cost function to evaluate the goodness of a given capacity allocation configuration. Then, a novel approach based on the genetic algorithms is provided to find the near-optimal solution. The genetic algorithm proposed to solve the multicast capacity assignment problem relies on two phases. The first is the search of a set of tree solutions for each multicast session alone, and the second phase is the search of the best trees combination for all the multicast sessions together starting from the output of the first phase. A novel heuristic multicast routing algorithm was proposed by Hwang et al. (Hwang, Do, & Yang, 2000) based on genetic algorithms. The new multicast routing algorithm tries to minimize the multicast cost while maintaining a reasonable path delay. The number of generations required to reach a good solution is significantly reduced by preferring shorter routes in initializing the chromosome pool and reusing the past solutions as the initial chromosomes for the new search. To apply this algorithm to real networks, they also propose several methods for reducing the computational complexity. The main advantages of the proposed genetic algorithms are as follows. Since the solutions are coded as bit strings, referred to as chromosomes, large problems can be easily handled by using long strings. Genetic operations, such as crossover and mutation, are very easy to implement. With a pool of chromosomes (candidate solutions), genetic algorithms search the solution space at different corners in parallel, and so the algorithm can be easily implemented on multiprocessor machines to search in parallel. Randomized genetic operations, such as mutation, can keep the search from being trapped by local-optima.

Chiang et al. (Chiang, Liu, & Huang, 2007) proposed a new and efficient spanning tree encoding scheme based on genetic algorithms for solving the multicast routing problem in mobile ad hoc networks. They also proposed three genetic-based operation tailored for the new encoding scheme. Topology crossover, node mutation, and edge mutation are three operations proposed in this paper. The proposed method uses the encoding scheme proposed by Huang et al. (Huang, Chiang, & Hou, 2005) called sequence and topology encoding (ST encoding). Topology crossover performs conventional one-point crossover on the topology chromosomes of two spanning tree codes to form new topologies. The idea behind edge mutation is to cut down a randomly selected subtree, and then grafts it to a randomly selected node in the remaining tree to form a new tree. Since a chromosome is a permutation of all the tree nodes, node mutation randomly selects two genes from a chromosome except the root and swaps them. An energy-efficient genetic algorithm was proposed by Yen et al. (Yen, Chan, Chao, & Park, 2008) to resolve the multicast routing problem with multiple QoS constraints in mobile ad hoc networks. Furthermore, they designed a source-tree-based routing algorithm and built the shortest-path multicast tree to minimize delay time by using a small population size in the genetic algorithm. In this algorithm, only a few nodes are involved in the route computation. They also improved the genetic sequence and topology encoding and prolonged the lifetime of the mobile nodes which are responsible for calculating the residual energy of all nodes in a multicast tree. Zahrani et al. (Zahrani et al., 2008) introduced a genetic local search heuristic that utilizes logarithmic simulated annealing in a pre-processing step for an analysis of the landscape generated by a multicast routing problem and the associated objective function. The annealing procedure allows the algorithm to estimate the depth of the deepest local minima. The genetic local search employs the partially mixed crossover (PMX) operation in-between sequences of downward and upward search steps, where the elitist model is applied. The PMX operation seems to be particularly suited to problems like multicast routing, since the outcome of the operation is always defined for two given parent routing orders.

Fuzzy Logic Approach

In 1965, Lotfi Zadeh (Zadeh, 1965) proposed fuzzy set theory and later established fuzzy logic based on the fuzzy sets. Fuzzy logic is a form of multi-valued logic derived from fuzzy set theory to deal with reasoning that is approximate rather than precise. In contrast with binary sets having binary logic, also known as crisp logic, the fuzzy logic variables may have a membership value of not only 0 or 1. Just as in fuzzy set theory with fuzzy logic the set membership values can range (inclusively) between 0 and 1, in fuzzy logic the degree of truth of a statement can range between 0 and 1 and is not constrained to the two truth values {true (1), false (0)} as in classic propositional logic. And when linguistic variables are used, these degrees may be managed by specific functions. Fuzzy logic has been applied to diverse fields, from control theory to artificial intelligence. Oh et al. (Oh, Kim, & Bae, 2001) proposed a fuzzy-based multicast routing algorithm that provides a linguistic guarantee for disruption free service despite handoffs occurred during an active connection. The fuzzy-based multicast scheme estimates the velocity and the direction of a mobile host. Then, the fuzzy multicast determines the basic fuzzy set for each neighboring host using the membership function, composes the multicast group for the mobile host using the control rules, and computes a multicast time using the defuzzifier of linguistic guarantee for disruption free service. Then, according to the specified linguistic guarantee for a user, data packets are transmitted to the forecasted neighbors where the mobile host will be handoff potentially, after the staggered time. Chiang and Huang (Chiang & Huang, 2004) proposed a new structured representation of the multicast routing problem and improved it by using the fuzzy Petri nets with the concept of immediately reachable set. They also proposed a Fuzzy Reasoning Algorithm for finding the multicast tree and improving the efficiency of the routing scheme in ad hoc networks. A Fuzzy Logic-based secure multicast routing protocol called (FLSL) is proposed by Nie et al. (Nie et al., 2006) for mobile ad hoc networks. The basic idea of FLSL is to utilize the local multicast mechanism and the security-level to select the securest route. The proposed algorithm of Security-Level is an adaptive fuzzy logic-based algorithm that can adapt itself with the dynamic conditions of the mobile hosts. An interesting property is that every node in the MANET has the field of Security-Level based on the fuzzy logic in the route tables to select the highest Security-Level route. The FLSL routing protocol can improve MANET's security. It is feasible to the weak security character of mobile ad hoc networks. Su et al. (Su, Wang, & Huang, 2008) proposed the fuzzy modified AODV multicast routing protocol called FMAR to select two comparably stable routes by computing dynamic route lifetime for multicast routing or layered video streaming. The model compares and ranks different route lifetimes by the weighted Multi-criteria which are based on hop count, number of control packets and energy limitation. The reason for using the fuzzy logic is the uncertainties associated with node mobility and the estimation of link crash. In this protocol, the packets are multicast via multiple paths to multiple receivers of a multicast group instead of successively performing single cast in ad hoc networks.

4.2. Mesh-based Multicast Routing Protocols

In a mesh-based multicast routing protocol, multiple routes may exist between any pair of source and destination, which is intended to enrich the connectivity among group members for better resilience against topology changes. In a mesh-based multicast routing protocol, packets are distributed along the mesh structures that are a set of interconnected nodes. Route discovery and mesh building are accomplished in two ways: by using broadcasting to discover routes or by using core or central points for mesh building. Mesh-based protocols perform better in high mobility situation as they provide redundant paths from source to destinations while forwarding data packets. However, mesh-based approaches sacrifice multicast efficiency in comparison to tree-based approach. Mesh-based protocols have high packet delivery ratios compared to tree-based protocols, but incur more control overhead in route maintenance and redundant transmission. As noted before, the major difference between the tree-based and mesh-based multicast routing protocols lies in the manner in which a multicast message is relayed. In most of the mesh-based multicast routing protocols, however, relaying transmission takes a more redundant approach: each node on the mesh will rebroadcast the message upon its first reception of the message. As shown in Figure 3(a), the network-wide broadcasting (or global flooding) brings about the broadcast storm problem that occupies the network bandwidth and consumes the node energy which are scarce resources in MANET. Scoped flooding (or limited scope flooding) is an appealing solution to this problem. In this method, the network nodes along the path connecting the multicast source to the multicast receivers are only responsible for rebroadcasting the multicast messages. This significantly reduces the message overhead. Figures 3(a) and 3(b) show the global flooding and scoped flooding in a sample network having 11 nodes and 3 multicast receivers, respectively. From these figures, it can

Figure 3. (a)Global flooding, (b)Scoped flooding

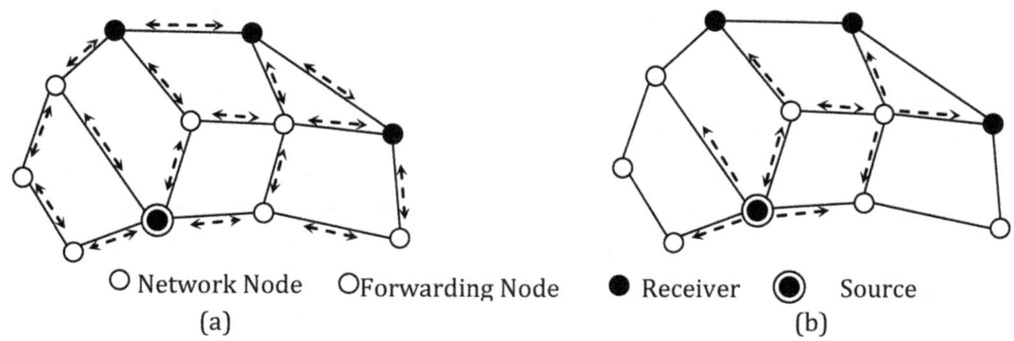

be seen that all the network nodes are involved in the global flooding mechanism, while in scoped flooding only two forwarding nodes toward the destinations rebroadcast the messages and the others only hear to the channel.

Lee et al. (Lee, Gerla, & Chiang, 1999) designed an on-demand multicast routing protocol called ODMRP for mobile ad hoc networks. ODMRP applies on-demand routing techniques to avoid channel overhead and improve scalability. It uses the concept of forwarding group (Chiang, Gerla, & Zhang, 1998), a set of nodes which is responsible for forwarding multicast data on the shortest paths between any member pairs to build a forwarding mesh for each multicast group. By maintaining and using a mesh, ODMRP avoids drawbacks of multicast trees in mobile wireless networks, such as intermittent connectivity, traffic concentration, frequent tree reconfiguration, non-shortest path in a shared tree. Indeed, ODMRP uses a forwarding mesh for data distribution. ODMRP is a reactive (on-demand) protocol that delivers packets to destination(s) on a mesh topology using scoped flooding of data within the mesh. As an on-demand routing mechanism, ODMRP reduces unnecessary channel overhead. On the other side, since routes are setup on demand, delay increases. The created mesh sub-network provides richer connectivity among multicast members compared to trees. Hence, unlike trees, frequent reconfigurations are not required. ODMRP takes a soft-state approach to maintain multicast group members. No explicit control message transmission is required to leave the group. ODMRP establishes and maintains group membership and multicast routes by the source on demand. The major strengths of ODMRP are its simplicity and scalability.

PatchODMRP (Lee & Kim, 2001) is one of the first derivatives of ODMRP. It uses the same procedure to construct the initial forwarding mesh structure, but it differs from ODMRP in that it take local repairing approach on detection of mesh destruction (i.e. link breakage) to avoid frequent mesh reconstructions. PoolODMRP (Cai & Yang, 2003) enhances route repair cost of PatchODMRP using so-called pool nodes. The pool nodes are defined as the neighbor nodes of forwarding nodes and they collect route information by overhearing data transmission. PoolODMRP reduces local route repair overhead by using the pool nodes. Cai et al. (Cai, Yang, & Wang, 2004) improved the performance of PoolODMRP by using a passive acknowledge scheme so as to reduce the overhead more. The new scheme is called PDAODMRP, short for Passive Data Acknowledge ODMRP. It removes MAC layer BEACON signal and detects a route breakage by the passive acknowledge while data transmitting. Due to the use of the passive acknowledge scheme, PDAODMRP local route recovery is initiated by a upstream node whereas a downstream node starts a local recovery in previous two protocols. An enhanced version of ODMRP called E-ODMRP was proposed by Oh et al. (Oh, Park, & Gerla, 2008). The efficiency, simplicity, and robustness of ODMRP render it one of the most widely used MANET multicast protocols. The main reason for robustness of ODMRP is the periodic route refreshing. ODMRP rebuilds the data forwarding mesh on a fixed interval and thus the route refresh interval is a key parameter that has critical impact on the network performance. If the route refresh rate is too high, the network will undergo too much routing overhead, wasting valuable resources. If it is too low, ODMRP cannot keep up with network dynamics. E-ODMRP with the refresh rate dynamically adapted to the environment adjusts the refresh interval based on the route lifetime such that the forwarding mesh is refreshed before it breaks. The Performance Enhanced On-Demand Multicast Routing Protocol (PEODMRP) (Ho, Jeon, & Lee, 2004), and the On-Demand Multicast Routing Protocol with Multipoint Relay (ODMRP-MPR) (Zhao, Xu, & Shi, 2003) are some other ODMRP-based multicast protocols. PEODMRP reduces control packet overhead via limiting the transmission area of Join Query flooding. If forwarders are set within a time, it responds with Join Reply without Join Query relay. PEODMRP has advantage when multiple sources exist in the same multicast group. ODMRPMPR minimizes the broadcasting overhead by reducing duplicated packet-forwarding. Only designated one hop neighbor nodes relay packets while packet transmitting. It significant reduces duplicated packet transmission but packet delivery ratio decreases in high mobility network.

Core-Assisted Mesh Protocol (CAMP) (Garcia-Luna-Aceves & Madruga, 1999) is a proactive mesh-based multicast routing protocol, which establishes a multicast mesh for each multicast group. CAMP avoids the need for network-wide floods (or global flooding) from each source to maintain multicast meshes by using one or more cores per multicast group. A receiver-initiated approach is used for receivers to join a multicast group by sending unicast join requests towards a core of the desired group. To join a multicast group, a node first checks if any of its neighbors already is a member of the multicast mesh. If that is the case, the node announces a membership request to its neighboring nodes. If there are no nodes to belong the mesh, node sends a join message to one of the core node. The core nodes are not necessarily needed. Because if there is no available core nodes, an expanded ring search is used to find at least one node to belong the mesh. The drawbacks of CAMP is that it needs the pre-assignment of cores to groups and a unicast routing protocol to maintain routing information about the cores, and this may incur considerable overhead in a large ad hoc network. Unlike CBT (Ballardie, Francis, & Crowcroft, 1993), in which all traffic flows through the core nodes, the core nodes in CAMP are used only to limit the control traffic when the receivers are joining multicast groups.

Lin and Chao (Lin & Chao, 1999) put forward a new protocol for multicasting in multihop wireless networks called clustered group multicast (CGM). In CGM, there is a set of forwarding nodes (called multicast virtual backbone) which are responsible for forwarding the multicast packets to the receivers. Unlike the multicasting protocols in wired networks (e.g. Internet) which construct and maintain a shortest path tree for every multicast (source, group) pair, the connectivity among the nodes in the backbone is of no longer importance. Thus, there is no tree maintenance overhead. In addition, a key feature of CGM is the use of the advertising agents to reduce advertising traffic to the system. An advertising agent acts as both a server and a client for the purpose of advertising join requests on behalf of its local clients. Because in CGM multicast traffic is only allowed to be delivered over the backbone, CGM restricts the amount of hosts participating in the backbone to decrease the impact of multicast traffic to the system.

In (Chiang, Gerla, & Zhang, 1998), Chiang et al., proposed another multicast routing protocol called Forwarding Group Multicast Protocol (FGMP) tailored for highly mobile ad hoc envi-

Figure 4. A mesh-based multicast route

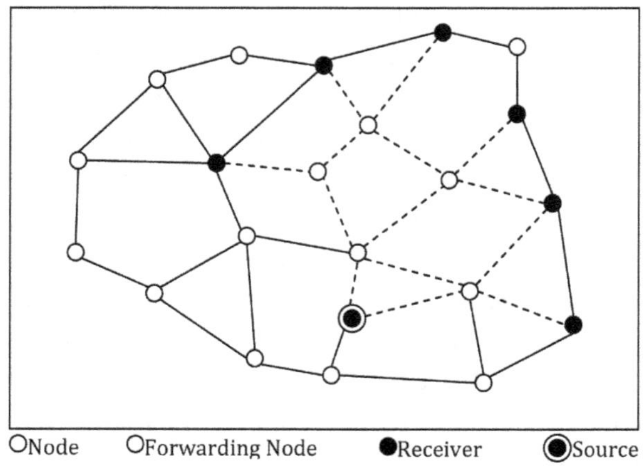

ronments. FGMP is another mesh-based routing protocol, where the mesh of nodes is formed and maintained by either receivers (FGMP-RA) or senders (FGMP-SA). To manage the maintenance overhead, FGMP-RA is used to form the mesh of nodes if the number of receivers is smaller than the number of sources. FGMP-SA is used otherwise. Like ODMRP, FGMP uses the network-wide flooding to build the mesh forwarding structure. FGMP associates a forwarding group (FG) with each multicast group. Any node in forwarding group is in charge of forwarding (broadcasting) the multicast packets. In other words, when a forwarding node receives a multicast packet, it will broadcast the packet if it is not a duplicate. All neighbors can hear it, but only the neighbors that are in FG will broadcast it, if they have not broadcasted it so far. The major problem of FGMP is how to elect and maintain the set FG of forwarding nodes. The size of FG should be as small as possible to reduce the wireless channel overhead, and the forwarding path from senders to receivers should be as short as possible to get the high throughput. This scheme can be viewed as a limited scope flooding. That is, flooding is contained within a properly selected forwarding set (mesh). It is interesting to note that with the proper selection of the forwarding group, FGMP scheme can emulate the behavior of any of the existing schemes. For example, to produce the global flooding, the FG must include all nodes in the network. For CBT, the FG is restricted to the nodes on the shared tree except the leaf nodes, and in DVMRP, FG includes all the non-leaf nodes on the source trees. Figure 4 shows a mesh-based multicast route connecting the multicast source to a set of multicast receivers having six members through a group consisting of five forwarding nodes. From Figure 4, it can be seen that since only the forwarding nodes are responsible for relaying the multicast packets, FG must cover all multicast members.

The simplest method for mesh-based multicast routing is to flood (network-wide broadcast) the multicast messages, which not only wastes the scarce resources of wireless hosts, but also diminishes the throughput of the network. This method suffers from the notorious broadcast storm problem which results in excessive redundancy, contention, and collision. One way to avoid the flooding is to use the concept of the dominating set (DS). The main idea behind the dominating set is very similar to that of forwarding group. Indeed, the aim of using the dominating set is to design an intelligent and scalable flooding scheme via limiting the flooding scope. A set is dominating if all the nodes in the network are either in the set or neighbors of nodes in the set. A connected dominating set (CDS) extracts a virtual backbone of the entire network through which the messages can be routed. Virtual backbone formation is a promising approach for multicast routing in which the routing overhead can be minimized, where the number of hosts responsible for the route discovery and data transmission can be reduced to the number of hosts in the backbone.

In (Sinha, Sivakumar, & Bharghavan, 1999), Sinha et al. proposed a multicast routing algorithm called Multicast Core Extraction Distributed Ad hoc Routing (MCEDAR) for wireless ad hoc networks. The proposed protocol bases its core-extraction criteria on the dominating set concept for the network topology, namely, a node is either a core or an immediate neighbor of a core. Besides the adopted virtual backbone procedures, the protocol itself has four key components: the mesh-based multicast structure, the join protocol, the core (broadcast-based) forwarding protocol, and the leave pruning and reconstruction protocols. In MCEDAR, a distributed minimum dominating set algorithm is applied for backbone formation purpose, and the resulting backbone has the property that all nodes are within one hop away from a core node. A core node and its dominated node set form a cluster. Once a virtual backbone is formed, the multicast operation is divided into two levels, the lower level multicasting which is within the cluster and upper level multicasting which is within the backbone.

Finding the minimum CDS problem in an arbitrary graph is a NP-Hard problem (Marathe et al., 1995; Clark, Colbourn, & Johnson, 1990), and so a host of approximation algorithms have been proposed to find a near optimal solution to this problem in a reasonable time. Guha and Khuller (Guha & Khuller, 1998) proposed two centralized greedy heuristic algorithms with bounded performance guarantees for connected dominating set formation. In the first algorithm, the connected dominating set is grown from one node outward, and in the second algorithm, a weakly connected dominating set is constructed, and then the intermediate nodes are selected to create a CDS. Guha and Khuller (Guha & Khuller, 1998) also proposed an approximation algorithm to solve the Steiner CDS problem, in which only a specified subset of vertices has to be dominated by a CDS. Butenko et al. (Butenko et al., 2004) also proposed a prune-based heuristic algorithm for constructing the small connected dominating sets. In this algorithm, the connected dominating set is initialized to the vertex set of the graph and each node is then examined to determine whether it should be removed or retained. If removing a given node disconnects the induced sub graph of the connected dominating set, then it is retained and otherwise removed. The algorithm proposed by Wu and Li (Wu & Li, 1999) first finds a CDS and then prunes certain redundant nodes from the CDS. Alzoubi et al. (2002, 2003) proposed two distributed heuristic algorithms for constructing CDS in a wireless ad hoc network. These algorithms first employ the distributed leader election algorithm (Alzoubi, Wang, & Frieder, 2002) to construct a rooted spanning tree from the original network topology. Then, an iterative labelling strategy is used to classify the nodes in the tree to be either dominator or dominatee, based on their ranks. The first heuristic uses the ID-based approach for ranking the nodes, and the second heuristic uses the level-based approach. Alzoubi et al. (Alzoubi et al., 2003) proposed another distributed algorithm for approximating the minimum CDS in a wireless ad hoc network. The proposed algorithm first constructs a maximum independent set (MIS), and then forms the CDS by adding more intermediate nodes into the MIS.

Li et al. (Li et al., 2005) proposed a greedy MIS-based algorithm for constructing CDS in wireless networks. The first step of the proposed algorithm focuses on the forming a MIS of the network topology graph. At the second step, a greedy approximation is employed for finding a Steiner tree with the minimum number of Steiner nodes to interconnect the nodes in the MIS. Gao et al. (Gao, Yang, & Ma, 2005) also presented a new distributed approximation algorithm that constructs a MCDS for wireless ad hoc networks based on a maximal independent set (MIS). In the proposed algorithm, each node only requires the knowledge of its one-hop neighbors and there is only one shortest path connecting two dominators that are at most three hops away. Xie et al. (Xie et al., 2007) proposed a novel distributed approximation algorithm to construct the MCDS in wireless sensor networks in which the network is modeled as a hierarchical graph. In this algorithm, at each level of the hierarchical graph, a selected set of nodes is served as the message hubs (to route the messages) for the other nodes in the next level of the hierarchical graph. The proposed algorithm uses a competition-based strategy to select the nodes at each level. Dai and Wu (Dai & Wu, 2004) proposed a prune-based heuristic algorithm for constructing the virtual backbone of the wireless ad hoc networks by finding the dominator nodes. In this method, the connected dominating set is initialized to the set of all vertices having two unconnected neighbors. Then, the connected dominating set is modified by removing the nodes whose neighbors are also the neighbors of the other nodes in the CDS. Wu and Li (Wu & Li, 2001) proposed a simple localized algorithm to define a connected dominating set. In this algorithm, a host is selected as a dominator if it has two unconnected neighbors. The size of the dominating set is reduced by two distributed pruning rules; a dominating node can be removed if its neighbor set is covered either by the neighbor set of an other node with higher ID or by the union of neighbor sets of two connected nodes both with higher IDs. Han (Han, 2009) proposed a novel zone-based distributed algorithm for CDS formation in wireless ad hoc networks. In this Zone-based algorithm, the network is partitioned into different zones, and a dominating tree is constructed for each zone. In this method, additional nodes are inserted into the resulting connected dominating set so as to adjust the zone borders.

The virtual backbone induced by the connected dominating set forwards the packets to all the hosts within the network, while the multicast messages only should be sent to a specified group of receivers. Therefore, sending the multicast messages along the virtual backbone (CDS) imposes extra overhead too. This overhead results from a large number of unwanted rebroadcasts for sending the multicast messages to the non-member hosts. Steiner connected dominating set (SCDS) is introduced as a well-known approach to relieve such an overhead. That is, finding the Steiner connected dominating set of the network graph is a promising solution to the multicast routing problem in wireless ad hoc networks. In this method, the SCDS includes the intermediate nodes by which the massage sent out by the multicast source is relayed. The minimum Steiner connected dominating set problem was first proposed by Guha and Khuller (Guha & Khuller, 1998), as the generalization of the well-known minimum connected dominating set problem. In this method, a (minimum size) subset of nodes is chosen as dominators to construct a route from the multicast source to each of the multicast receivers. The virtual backbone constructed by the Steiner connected dominating set includes the minimum number of dominator (forwarding) nodes in order to dominate the multicast members only. The following gives an overview of the methods proposed for solving the Steiner connected dominating set problem.

Wu et al. (Wu, Xu, Chen, & Wang, 2004) proposed two approximation algorithms based on maximal independent set (MIS) for solving the minimum SCDS (MSCDS) problem. Their former algorithm is a one-hop method for approximating the MSCDS of a unit disk graph with a constant approximation ratio at most 10. This algorithm exploits the properties of the MIS and minimum Steiner tree to form the MSCDS. The proposed algorithm first finds the MIS of the graph induced by only the vertices in multicast group. Then, the Steiner tree algorithm proposed in (Singh & Vellanki, 1998) is applied to connect the vertices of the constructed MIS. The size of the SCDS constructed by the proposed one-hop algorithm is at most 10xOPT+1. The time complexity of this algorithm is $O(D \cdot |V|)$, where D denotes the graph diameter, and V denotes the vertex set. They also proposed a d-hop algorithm in which a d-hop graph is initially constructed, where the terminals form the vertex-set of the graph. In this method, every two vertices of the graph are connected by an edge, if they are d-hop neighbors. Now, like the one-hop algorithm, an MIS of the d-hop graph is computed, and than a Steiner tree algorithm (Singh & Vellanki, 1998) is applied to connect the vertices of MIS. The d-hop algorithm computes a SCDS of subset R, whose size is at most

$$\left(8d^2 + 8d + 2 + \frac{4}{d}\right) \cdot \mathrm{OPT} - 1$$

, if d is even, and

$$\frac{\left(16d^3 + 16d^2 + 4d + 6\right) \cdot \mathrm{OPT} + 4d^3 + 4d^2 - 2d + 3}{2d - 1}$$

if d is odd. The d-hop algorithm can be implemented in a fully distributed manner, and the message and time complexity of the distributed d-hop algorithm are $O(|V|^2)$ and $O(D \cdot |V|)$, respectively. They showed the distributed d-hop algorithm can be effectively used for multicast routing in ad hoc networks by constructing a VMB with a small number of forwarding nodes. Muhammad (2006, 2007) also proposed a distributed MIS-based Steiner connected dominating set algorithm for multicast routing in wireless ad hoc networks. The first step of this algorithm constructs a maximal independent set in $O(|V|)$ time, whose size is at most $|\mathrm{MIS}| = 4 \times \mathrm{OPT} + 1$. The second step uses a distributed Steiner tree with message complexity $O(2|V|)$ and time complexity $O((|V| - 1) \cdot |E| \cdot \log |V|)$ for connecting the MIS members. The total size of the SCDS constructed by Muhammad's algorithm is smaller than $2\left(|\mathrm{MIS}| + \mathrm{OPT} - 1\right) + 1$, and the message and time complexity of this algorithm are $O(2|V|)$ and $O((|V| - 1) \cdot |E| \cdot \log |V|)$, respectively. Aggarwal et al. (Aggarwal, Dubey, & Mehta, 2006) proposed an algorithm for approximating the MSCDS in a dominating pair graph. The time complexity of the proposed algorithm is $O(|V|^8 \cdot |R|)$, where $|V|$ and $|R|$ denotes the cardinality of the vertex set and terminal-set, respectively. The authors show that the proposed algorithm computes the Steiner connected dominating set in $O(|V|^4 \cdot |R|)$ time, if the distance between the dominating pair vertices is greater than 8. The cardinality of the Steiner connected dominating set constructed by Aggarwal et al.'s algorithm is smaller than OPT+2, where dt(G) is the dominating target number of graph G. Akbari Torkestani and Meybodi (Akbari Torkestani & Meybodi, 2010b) presented three centralized approximation algorithms (called Algorithm I-III) for solving the minimum weight Steiner connected dominating set problem. In these algorithms, it is assumed that a random weight is associated with each node. At each stage of algorithms, a Steiner connected dominating set is randomly constructed by a network of cooperating learning automata. The weight of the selected Steiner connected dominating set is evaluated through the random environment, and depending on the response received from the environment, the selected Steiner connected dominating set is rewarded or penalized. The authors also designed a distributed version of the last proposed centralized algorithm (Algorithm III) for multicast routing in wireless mobile ad-hoc networks. The same authors (Akbari

Torkestani & Meybodi, 2010) designed a distributed learning automata-based algorithm called MMR-LA to solve the multicast routing problem in wireless mobile Ad-hoc networks where the network topology graph is a stochastic graph. For the first time, they introduced the concept of stochastic Steiner connected dominating set problem and used it to form the virtual multicast backbone (VMB) of the random networks where the network parameters vary over time. The proposed algorithm uses the relative mobility of each host (with respect to all its neighbors) to predict its realistic motion behavior by sampling its movement parameters during different epochs.

4.3. Hybrid Multicast Routing Protocols

Tree-based protocols provide high data forwarding efficiency at the expense of low robustness. Their advantage is their simplicity. Their disadvantage is that until the tree is reconstructed after movement of a node, packets possibly have to be dropped. Mesh-based protocols perform better in high mobility situation as they provide redundant paths from source to destinations while forwarding data packets. However, mesh-based approaches sacrifice multicast efficiency in comparison to tree-based approach. They increase the network load. Hybrid multicast routing protocols combine the both advantages of tree and mesh-based approaches, i.e., the robustness of mesh-based multicast routing protocols and low overhead of tree-based protocols. Hence, hybrid protocols address both efficiency and robustness.

In (Biswas, Barai, & Nandy, 2004), Biswas et al. proposed an efficient hybrid mesh-based multicast routing protocol for ad hoc networks. The key concept is to separate data forwarding path from join-query forwarding path by incorporating low overhead local clustering technique and forwarding data packets using DDM. DDM is a stateless multicast approach where multicast tree information is appended with each data packet header. This protocol solves the scalability issues of ODMRP by reducing the control overhead and increasing multicast efficiency for different network scenarios. They incorporated a low overhead local clustering technique to classify all nodes into core and normal categories. When multicast routes to destination nodes are unavailable, join-query messages are sent to all the nodes in the network and data packets are forwarded by the core nodes to the destination nodes using differential destination multicast. Through simulations they showed that this protocol reduces the control overhead and increases packet delivery ratio by 20-50% for different network scenarios.

An and Papavassiliou (An & Papavissiliou, 2003) proposed a Mobility-based Hybrid Multicast Routing (MHMR) protocol suitable for mobile ad hoc networks. The main features of the proposed protocol are the following: (i) mobility-based clustering and group-based hierarchical structure, in order to effectively support stability and scalability; (ii) group-based (limited) mesh structure and forwarding tree concepts, in order to support the robustness of the mesh topologies, which provides limited redundancy and the efficiency of tree forwarding simultaneously; and (iii) combination of proactive and reactive concepts that provide low route acquisition delay and low overhead. Although the MHMR creates a mesh infrastructure for multicast routing, the multicast packet forwarding is done only on a source-based tree. The trees are constructed over the given mesh structure and can be created dynamically depending on the mesh stability.

Xie et al. (Xie et al., 2002) proposed an approach for robust IP multicast in mobile ad hoc networks by exploiting user-multicast tree and dynamic logical cores, called Ad hoc Multicast Routing Protocol (AMRoute). AMRoute creates a source-based multicast distribution tree using unicast tunnels connecting the group members. As a hybrid multicast routing protocol which combines advantages of tree and mesh based approaches, AMRoute is composed of a bidirectional shared multicast tree per multicast group. Certain tree nodes are designated by AMRoute as logical cores which are responsible for initiating and maintaining the signaling component of AMRoute, such as detection of group members, tree setup, and data distribution. Each group in the network has at least one logical core. In this method, unicast tunnels are used as tree links to connect neighbors on the user-multicast tree. Thus, group state cost is incurred only by group senders and receivers, and tree structure does not need to change even in case of a dynamic network topology.

4.4. Stateless Multicast Routing Protocols

In the stateless multicast protocols, the forwarding states are included in packet header, and no protocol state is maintained at any nodes except for the multicast source node. From the information included in the packet headers, any intermediate node knows how to forward or duplicate the packet. Although packing routing information together with data traffic will enlarge data packet size, it reduces the total number of control packets generated by the protocol. Besides, when the group is idle, there is no control overhead. A recent shift toward the stateless multicasting is represented by DDM (Ji & Corson, 2001), LGT (Chen & Nahrstedt, 2002) and RDG (Luo, Eugster, & Hubaux, 2003). All these protocols do not require maintenance of any routing structure at the forwarding nodes. These protocols use different techniques to achieve stateless multicasting.

In (Ji & Corson, 2001), a stateless multicast routing protocol termed Differential Destination Multicast (DDM) is proposed. It differs from the other approaches in two ways. First, instead of distributing membership control throughout the network, DDM concentrates this authority at the data sources (i.e., senders), thereby giving sources the knowledge of group membership. Second, differentially encoded, variable-length destination headers are inserted in data packets, which are used in combination with unicast routing tables to forward multicast packets towards multicast receivers. In DDM, a source encapsulates a list of destination addresses in the header of each data packet it sends out. When an intermediate node receives the packet, its DDM agent queries the unicast routing protocol about which next-hop node to forward the packet toward each destination in the packet header. DDM is intended for small groups, therefore, it intrinsically excels only in horizontal scalability. When group size is large, placing the addresses of all members into the packet headers will not be efficient. The protocol has a caching mode, so that only the difference from the previous states is actually placed in the headers. However, as the forwarding set at the on-route nodes inevitably grow large, each intermediate node needs to keep routes for a large set of destinations. This poses a heavy burden on the supporting unicast protocol even under moderate mobility.

Chen and Nahrstedt (Chen & Nahrstedt, 2002) designed a location guided tree construction Algorithm called LGT for small group multicast in mobile ad hoc networks. LGT is also a stateless multicast protocol, where multicast data is encapsulated in a unicast packet and transmitted among group members. LGT builds an overlay packet delivery tree on top of the underlying unicast routing protocol, and multicast packets are encapsulated in a unicast envelop and unicasted between the group members. In LGT, using the location information of the group member nodes, the multicast tree is constructed without the knowledge of the network topology. The authors propose two types of heuristics, namely the location-guided k-array tree (LGK), and the location-guided Steiner tree (LGS) to construct the multicast tree with location information.

Luo and Eugster (Luo, Eugster, & Hubaux, 2003) proposed another stateless multicast routing algorithm for wireless ad hoc networks called RDG, short for Route Driven Gossip. RDG uses a probabilistically controlled flooding technique, termed as gossiping, to deliver packets to all the group members. RDG relies on a unicast protocol such as DSR (Broch, Johnson, & Maltz, 1996) to provide routing information, which is used for guiding the gossip process. Each node maintains the following data structures for each multicast group: a data buffer which stores data packets received, and a view which is a list of all other group member nodes known to this node. The view at each node is divided into two parts: active view, which contains the IDs of known members to which at least one routing path is known, and passive view which contains the IDs of known member to which no routing path is currently available. A node intending to join a group floods the network with a Group-Request message. All members receiving the message update their active view. They also return a Group-Reply to the request initiator with a certain probability. The initiator also updates its active view after receiving the Group-Reply. Each member node periodically generates a gossip message and gossips it to a set of other nodes randomly chosen from its active view. The message includes a selected subset of the data buffer, and the sequence number of the most recent missing data packets. A group member receiving a gossip message will update its view of other group members and update its data buffer with newly received data. In responding the gossip initiator's request of recovering the missing data, the receiving node will unicast the missing data back to the initiator if the data is in its data buffer.

5. CONCLUSION

Multicasting is used when the same message or the same stream of data must be forwarded to multiple destinations. Multicasting is an efficient data transmission method to support group-oriented communications in one-to-many or many-to-many applications such as audio/video conferencing, collaborative works, and so on. In MANETs, the most challenging issue in multicast routing is to effectively handle the frequent and unpredictable topology changes caused by host mobility, link breakage and host failure. This paper provided a survey of most recent multicast routing protocols for MANETs. This study showed that each multicast routing protocol may improve network performance in terms of delay, throughput, reliability or lifetime. Due to severe constraints of mobile wireless ad hoc networks such as host mobility, limited resources and very unreliable communication channel, single protocol or a set of protocols that can improve all these performance parameters is extremely hard to find. Selection of a multicast routing protocol is as much dependent on the nature of application, and different applications have diverse requirements. Stability against the host mobility, energy efficiency, low overhead, reliability, and scalability are several requirements for which the multicast routing protocols are designed. This survey enables the researchers to identify the strengths and weaknesses of different multicast routing protocols and helps them to choose the best one for a particular application.

REFERENCES

Aggarwal, D., Dubey, C. K., & Mehta, S. K. (2006). Algorithms on graphs with small dominating targets. *ISAAC 2006 . LNCS*, *4288*, 141–152.

Akbari Torkestani, J., & Meybodi, M. R. (2010). Mobility-based multicast routing algorithm in wireless mobile ad hoc networks: A learning automata approach. *Journal of Computer Communications*, *33*(6), 721–735. doi:10.1016/j.comcom.2009.11.019

Akbari Torkestani, J., & Meybodi, M. R. (2010b). Weighted Steiner connected dominating set and its application to multicast routing in wireless MANETs. *Wireless Personal Communications*, *60*(2).

Akbari Torkestani, J., & Meybodi, M. R. (2011). A link stability-based multicast routing protocol for wireless mobile ad hoc networks. *Journal of Network and Computer Applications*, *34*(4), 1429–1440. doi:10.1016/j.jnca.2011.03.026

Alzoubi, K. M., Li, X. Y., Wang, Y., Wan, P. J., & Frieder, O. (2003). Geometric spanners for wireless ad hoc network. *IEEE Transactions on Parallel and Distributed Systems*, *14*(4), 408–421. doi:10.1109/TPDS.2003.1195412

Alzoubi, K. M., Wan, P. J., & Frieder, O. (2002). Distributed heuristics for connected dominating sets in wireless ad hoc networks. *Journal of Communications and Networks*, *4*(1), 22–29.

Alzoubi, K. M., Wan, P. J., & Frieder, O. (2003). Maximal independent set, weakly connected dominating set, and induced spanners for mobile ad hoc networks. *International Journal of Foundations of Computer Science*, *14*(2), 287–303. doi:10.1142/S012905410300173X

An, B., & Papavassiliou, S. (2003). MHMR: Mobility-based hybrid multicast routing protocol in mobile ad hoc wireless networks. *Wireless Communication and Mobile Computing*, *3*, 255–270. doi:10.1002/wcm.115

Atzori, L., & Raccis, A. (2004). Network Capacity assignment for multicast services using genetic algorithms. *IEEE Communications Letters*, *8*(6), 403–405. doi:10.1109/LCOMM.2004.831328

Ballardie, T., Francis, P., & Crowcroft, J. (1993). Core based trees (CBT): An architecture for scalable inter-domain multicast routing. In *Proceedings of ACM SIGCOMM '93*, (pp. 85-95).

Basagni, S., Chlamtac, I., Syrotiuk, V. R., & Talebi, R. (2000). On-demand location aware multicast (OLAM) for ad hoc networks. In *Proceedings of Wireless Communications and Networking Conference*, Vol. 3, (pp. 1323-1328).

Basagni, S., Conti, M., Giordano, S., & Stojmenovic, I. (2004). *Mobile ad hoc networking*. IEEE Press. doi:10.1002/0471656895

Biswas, J., Barai, M., & Nandy, S. K. (2004). Efficient hybrid multicast routing protocol for ad hoc wireless networks. In *Proceedings of 29th Annual IEEE International Conference on Local Computer Networks*, (pp. 180-187).

Broch, J., Johnson, D. B., & Maltz, D. A. (1996). *The dynamic source routing in ad hoc wireless networks* (pp. 153–181). Dordrecht, The Netherlands: Kluwer Academic Publishers.

Bui, T. N., & Zrncic, C. M. (2006). An ant-based algorithm for finding degree-constrained minimum spanning tree. In *Proceedings of the 8th Annual Conference on Genetic and Evolutionary Computation*, (pp. 11-18).

Butenko, S., Cheng, X., Oliveira, C., & Pardalos, P. M. (2004). A new heuristic for the minimum connected dominating set problem on ad hoc wireless networks . In *Recent developments in cooperative control and optimization* (pp. 61–73). Kluwer Academic Publishers. doi:10.1007/978-1-4613-0219-3_4

Cai, S., & Yang, X. (2003). The performance of PoolODMRP. In *Proceedings of the 6th IFIP/IEEE International Conference on Management of Multimedia Networks and Service (MMNS2003)*, (pp. 90-101).

Cai, S., Yang, X., & Wang, L. (2004). PDAODMRP: An extended PoolODMRP based on passive data acknowledgement. *Journal of Communications and Networks*, *6*(4), 362–375.

Canourgues, L., Lephay, J., Soyer, L., & Beylot, A. L. (2006). STAMP: Shared-tree ad hoc multicast protocol. *IEEE Military Communications Conference, MILCOM 2006*, (pp. 1-7).

Chen, K., & Nahrstedt, K. (2002). Effective location-guided tree construction algorithms for small group multicast in MANET. In *Proceedings of INFOCOM*, (pp. 1180-1189).

Chiang, C. C., Gerla, M., & Zhang, L. (1999). Forwarding group multicast protocol (FGMP) for multihop, mobile wireless networks. *Journal of Cluster Computing*, *1*(2), 187–196. doi:10.1023/A:1019037500012

Chiang, T. C., & Huang, Y. M. (2004). Multicast routing representation in ad hoc networks using fuzzy Petri nets. *in Proceedings of the 18th International Conference on Advanced Information Networking and Application*, (pp. 1-4).

Chiang, T. C., Liu, C. H., & Huang, Y. M. (2007). A near-optimal multicast scheme for mobile ad hoc networks using a hybrid genetic algorithm. *Expert Systems with Applications*, *33*, 734–742. doi:10.1016/j.eswa.2006.06.020

Chlamtac, I., Conti, M., & Liu, J. (2003). Mobile ad hoc networking: Imperatives and challenges. *Journal of Ad hoc . Networks*, *1*, 13–64.

Clark, B. N., Colbourn, C. J., & Johnson, D. S. (1990). Unit disk graphs. *Discrete Mathematics*, *86*, 165–177. doi:10.1016/0012-365X(90)90358-O

Dai, F., & Wu, J. (2004). An extended localized algorithm for connected dominating set formation in ad hoc wireless networks. *IEEE Transactions on Parallel and Distributed Systems*, *15*(10). doi:10.1109/TPDS.2004.48

Deering, S., Estrin, D., Farinacci, D., Jacobson, V., Helmy, A., & Meyer, D. (1999). *Protocol independent multicast version 2 dense mode specification*. Draft-IETF-PIM-V2-DM-03.txt, 1999.

Dorigo, M. (1992). *Optimization, learning and natural algorithms*. Ph.D. thesis, Dipartimento di Elettronica, Politecnico di Milano, Milan, Italy, 1992.

Dorigo, M., & Gambardella, L. M. (1997). Ant colony system: A cooperative learning approach to the traveling salesman problem. *IEEE Transactions on Evolutionary Computation*, *1*(1), 53–66. doi:10.1109/4235.585892

Estrin, D., Farinacci, D., Helmy, A., Thaler, D., Deering, S., & Handley, M. (1998). *Protocol independent multicast-sparse mode (PIM-SM)*. Protocol specification . *RFC*, *2362*, 1998.

Gao, B., Yang, Y., & Ma, H. (2005). A new distributed approximation algorithm for constructing minimum connected dominating set in wireless ad hoc networks. *International Journal of Communication Systems*, *18*, 743–762. doi:10.1002/dac.726

Garcia-Luna-Aceves, J. J., & Madruga, E. L. (1999). Core-assisted mesh protocol. *IEEE Journal on Selected Areas in Communications*, *17*(8), 784–792. doi:10.1109/49.779921

Guha, S., & Khuller, S. (1998). Approximation algorithms for connected dominating sets. *Algorithmica*, *20*(4), 374–387. doi:10.1007/PL00009201

Gupta, S. K. S. (2001). Cored-based tree with forwarding regions (CBT-FR): A protocol for reliable multicasting in mobile ad hoc networks. *Journal of Parallel and Distributed Computing*, *61*, 1249–1277. doi:10.1006/jpdc.2001.1749

Gupta, S. K. S., & Srimani, P. K. (2003). Adaptive core selection and migration method for multicast routing in mobile ad hoc networks. *IEEE Transactions on Parallel and Distributed Systems*, *14*(1), 27–38. doi:10.1109/TPDS.2003.1167368

Haas, Z. J., Pearlman, M. R., & Samar, P. (2002). The zone routing protocol (ZRP) for ad hoc networks. Draft-ietf-manet-zone-zrp-04.txt, Internet-Draft, IETF, 2002.

Haas, Z. J., Pearlman, M. R., & Samar, P. (2002b). *The intrazone routing protocol (IARP) for ad hoc networks*. Internet Draft: draft-ietfmanet-zone-iarp-02.txt, 2002.

Haas, Z. J., Pearlman, M. R., & Samar, P. (2002c). The bordercast resolution protocol (BRP) for ad hoc networks. Draft-ietf-manet-zone-brp-02.txt, Internet-Draft, IETF, 2002.

Haas, Z. J., Pearlman, M. R., & Samar, P. (2002d). *The interzone routing protocol (IERP) for ad hoc networks*. Draft-ietf-manet-zone-ierp-02.txt, Internet-Draft, IETF, 2002.

Han, B. (2009). Zone-based virtual backbone formation in wireless ad hoc networks. *Ad Hoc Networks*, *7*(1), 183–200. doi:10.1016/j.adhoc.2008.01.003

Hu, C. C., Wu, E. H. K., & Chen, G. H. (2006). OGHAM: On-demand global hosts for mobile ad hoc multicast services. *Journal of Ad hoc . Networks*, *4*, 709–723.

Huang, Y. M., Chiang, T., & Hou, T. W. (2005). An efficient encoding tree for location-aware multicast routing protocol in wireless ad hoc networks. *Journal of Internet Technology*, *6*(1).

Hwang, R. H., Do, W. Y., & Yang, S. C. (2000). Multicast routing based on genetic algorithms. *Journal of Information Science and Engineering*, *16*, 885–901.

Jetcheva, J. G., & Johnson, D. B. (2001). Adaptive demand-driven multicast routing in multi-hop wireless ad hoc networks. In *Proceedings of the ACM International Symposium on Mobile Ad hoc Networking and Computing (MobiHoc)*, 2001.

Ji, L. S., & Corson, M. S. (2001). Differential destination multicast-A MANET multicast routing for multihop ad hoc network. *Proceedings - IEEE INFOCOM*, *2*, 1192–1201.

Kaliaperumal, B., & Ebenezer, Jeyakumar, A. (2005). Adaptive core based scalable multicasting networks. *INDICON, 2005 Annual IEEE*, (pp. 198-202).

Karp, R. M. (1972). Reducibility among combinatorial problems . In Miller, R. E., & Thatcher, J. W. (Eds.), *Complexity of computer computations* (pp. 85–103). Plenum Press.

Kondylis, G. D., Krishnamurthy, S. V., Dao, S. K., & Pottie, G. J. (2000). Multicasting sustained CBR and VBR traffic in wireless ad hoc networks. In *Proceedings of IEEE International Conference on Communications,* Vol. 1, (pp. 543-549).

Lee, M., & Kim, T. (2001). PatchODMRP: An ad hoc multicast routing protocol. In *Proceedings of the 15th International Conference on Information Networking (ICOIN '01)*, (pp. 537-543).

Lee, S. J., Gerla, M., & Chiang, C. C. (1999). *On-demand multicast routing protocol (ODMRP) for ad hoc networks*. Internet draft, draft-retf-manet-odmrp-01.txt, work in progress, 1999.

Li, Y., Thai, M. T., Wang, F., Yi, C. W., Wang, P. J., & Du, D. Z. (2005). On greedy construction of connected dominating sets in wireless networks. *Special Issue of Wireless Communications and Mobile Computing (WCMC), 5*(8).

Lin, C., & Chao, S. (1999). A multicast routing protocol for multihop wireless networks. *IEEE Global Telecommunication Conference (GlobeCom 1999)*, (pp. 235-239).

Liu, B. H., Huang, P. C., & Tsai, M. J. (2008). Distributed reformation of core-based group-shared multicast trees in mobile ad hoc networks. *Journal of Parallel and Distributed Computing*, *68*, 582–595. doi:10.1016/j.jpdc.2007.10.002

Liu, Y. P., Wu, M., & Qian, J. (2006). The distributed multicast routing scheme with delay constraint using ant colony optimization. In *Proceedings of the 6th World Congress on Intelligent Control and Automation*, (pp. 3062-3066).

Luo, J., Eugster, P. T., & Hubaux, J. P. (2003). Route driven gossip: Probabilistic reliable multicast in ad hoc networks. *Proceedings - IEEE INFOCOM, 03*, 2229–2239.

Madruga, E. L., & Garcia, J. J. (2001). Scalable multicasting: The core-assisted mesh protocol. *Kluwer Mobile Network, 6*(2), 151–165. doi:10.1023/A:1011400407314

Marathe, M. V., Breu, H., Hunt, H. B. III, Ravi, S. S., & Rosenkrantz, D. J. (1995). Simple heuristics for unit disk graphs. *Networks, 25*, 59–68. doi:10.1002/net.3230250205

Mohapatra, P., & Krishnamurthy, S. (2005). *Ad hoc networks: Technologies and protocols.* Springer Science, 2005.

Moy, J. (1994). Multicast routing extensions for OSPF. *Communications of the ACM, 37*(8), 61–66. doi:10.1145/179606.179654

Muhammad, R. B. (2006). Distributed Steiner tree algorithm and its application in ad hoc wireless networks. In *Proceedings of The 2006 International Conference on Wireless Networks (ICWN'06)*, USA, (pp. 173-178).

Muhammad, R. B. (2007). A distributed graph algorithm for geometric routing in ad hoc wireless networks. *Journal of Networks, 2*(6), 50–57. doi:10.4304/jnw.2.6.50-57

Nadeem, T., & Parthasarathy, S. (2006). Mobility control for throughput maximization in ad hoc networks. *Wireless Communication and Mobile Computing, 6*, 951–967. doi:10.1002/wcm.431

Nie, J., Wen, J., Luo, J., He, X., & Zhou, Z. (2006). An adaptive fuzzy logic based secure routing protocol in mobile ad hoc networks. *Journal of Fuzzy Sets and Systems, 157*, 1704–1712. doi:10.1016/j.fss.2005.12.007

Oh, S. J., Kim, Y. J., & Bae, I. H. (2001). A fuzzy multicast method providing linguistic guarantee of disruption free service in mobile wireless networks. *Eighth International IEEE Conference on Parallel and Distributed Systems*, (pp. 135-140).

Oh, S. Y., Park, J. S., & Gerla, M. (2008). E-ODMRP: Enhanced ODMRP with motion adaptive refresh. *Journal of Parallel and Distributed Computing, 68*, 1044–1053. doi:10.1016/j.jpdc.2008.04.004

Ozaki, T., Kim, J. B., & Suda, T. (2001). Bandwidth-efficient multicast routing for multi-hop ad hoc networks. *Proceedings - IEEE INFOCOM, 2*, 1182–1191.

Perkins, C., & Royer, E. M. (1998). *Ad hoc on demand distance vector (AODV) routing.* Internet draft, 1998.

Rajaraman, R. (2002). Topology control and routing in ad hoc networks: A survey. *SIGACT News, 33*(2), 60–73. doi:10.1145/564585.564602

Rodolakis, G., Laouiti, A., Jacquet, P., & Naimi, A. M. (2008). Multicast overlay spanning trees in ad hoc networks: Capacity bounds protocol design and performance evaluation. *Computer Communications, 31*, 1400–1412. doi:10.1016/j.comcom.2008.01.019

Rouskas, G. N., & Baldine, I. (1997). Multicast routing with end-to-end delay and delay variation constraints. *IEEE Journal on Selected Areas in Communications*, 1–9.

Royer, E. M., & Perkins, C. E. (1999). Multicast operation of the ad hoc on-demand distance vector routing protocol. In *Proceedings of MobiCom*, Seattle, (pp. 207-218).

Royer, E. M., & Toh, C. (1999). A review of current routing protocols for ad hoc mobile wireless networks. *IEEE Personal Communications*, *6*(2), 46–55. doi:10.1109/98.760423

Singh, G., & Vellanki, K. (1998). A distributed protocol for constructing multicast trees. In *Proceedings of the International Conference on Principles of Distributed Systems*, France, 1998.

Sinha, P., Sivakumar, R., & Bharghavan, V. (1999). MCEDAR: Multicast core extraction distributed ad hoc routing. *IEEE Wireless Communications and Networking Conference (WCNC)*, (pp. 1313-1317).

So, B., Jeon, H., & Lee, J. (2004). PEODMRP: Performance enhanced on-demand multicast routing protocol. In *Proceedings of the 8th Joint Conference on Communications & Informations (JCCI'04)*, Korea, 2004.

Stützle, T., & Hoos, H. H. (2000). MAX-MIN ant system. *Future Generation Computer Systems*, *16*(8), 889–914.

Su, B. L., Wang, M. S., & Huang, Y. M. (2008). Fuzzy logic weighted multi-criteria of dynamic route lifetime for reliable multicast routing in ad hoc networks. *Expert Systems with Applications*, *35*, 476–484. doi:10.1016/j.eswa.2007.07.010

Thyagarajan, A. S., & Deering, S. E. (1995). Hierarchical distance vector multicast routing for the MBone. In *ACM SIGCOMM Computer Communication Review*, (pp. 60-66).

Toh, C. K., Guichal, G., & Bunchua, S. (2000). ABAM: On-demand associativity-based multicast routing for ad hoc networks. *IEEE Vehicular Technology Conference*, (pp. 987-993).

Tseng, S. Y., Lin, C. C., & Huang, Y. M. (2008). Ant colony-based algorithm for constructing broadcasting tree with degree and delay constraints. *Expert Systems with Applications*, *35*, 1473–1481. doi:10.1016/j.eswa.2007.08.018

Waitzman, D., Partridge, C., & Deering, S. E. (1988). *Distance vector multicast routing protocol*. Internet Request for Comment 1075, November 1988.

Wang, B., Sandeep, K., & Gupta, S. (2004). Extending lifetime of multicast trees in wireless ad hoc networks. *Journal of Information Science and Engineering . Special Issue on Mobile Computing*, *20*, 219–243.

Wang, Y., & Xie, J. (2000). Ant colony optimization for multicast routing. *IEEE APCCAS 2000, The 2000 IEEE Asia-Pacific Conference on Circuits and Systems,* (pp. 54-57).

Wu, C. W., Tay, Y. C., & Toh, C. K. (1998). *Ad hoc multicast routing protocol utilizing increasing id-numbers (AMRIS) functional specification*. Internet-Draft, draft-ietf-manet-amris-spec-00.txt, 1998.

Wu, J., & Li, H. (1999). On calculating connected dominating set for efficient routing in ad hoc wireless networks. In *Proceedings of the Third ACM International Workshop on Discrete Algorithms and Methods for Mobile Computing and Communications (ACM DIALM'1999)*, (pp. 7-14).

Wu, J., & Li, H. (2001). A dominating-set-based routing scheme in ad hoc wireless networks. *Telecommunication Systems Journal*, *3*, 63–84.

Wu, Y. F., Xu, Y. L., Chen, C. L., & Wang, K. (2004). On the construction of virtual multicast backbone for wireless ad hoc networks. *IEEE International Conference on Mobile Ad hoc and Sensor Systems*, (pp. 294-303).

Xie, J., Talpade, R., McAuley, T., & Liu, M. (2002). AMRoute: Ad hoc multicast routing protocol. *ACM Mobile Networks and Applications (MONET) . Journal*, *7*(6), 429–439.

Xie, R., Qi, D., Li, Y., & Wang, J. Z. (2007). A novel distributed MCDS approximation algorithm for wireless sensor networks. *Journal of Wireless Communications and Mobile Computing*, *9*(3).

Yen, Y. S., Chan, Y. K., Chao, H. C., & Park, J. H. (2008). A genetic algorithm for energy-efficient based multicast routing on MANETs. *Computer Communications*, *31*, 858–869. doi:10.1016/j.comcom.2007.10.033

Zadeh, L. A. (1965). Fuzzy sets. *Information and Control*, *8*(3), 338–353. doi:10.1016/S0019-9958(65)90241-X

Zahrani, M. S., Loomes, M. J., Malcolm, J. A., & Dayem Ullah, A. Z. M., Steinh.fel, K., & Albrecht, A. A. (2008). Genetic local search for multicast routing with pre-processing by logarithmic simulated annealing. *Computers & Operations Research*, *35*, 2049–2070. doi:10.1016/j.cor.2006.10.001

Zhang, Q., & Leung, Y. W. (1999). An orthogonal genetic algorithm for multimedia multicast routing. *IEEE Transactions on Evolutionary Computation*, *3*(1), 53–62. doi:10.1109/4235.752920

Zhang, X., & Jacob, L. (2004). MZRP: An extension of the zone routing protocol for multicasting in MANETs. *Journal of Information Science and Engineering*, *20*, 535–551.

Zhang, Y., Cai, H., Lin, Y., & Xiao, J. (2007). *An ant colony system algorithm for the multicast routing problem.* Third International Conference on Natural Computation, ICNC 2007, 2007.

Zhao, Y., Xu, L., & Shi, M. (2003). On-demand multicast routing protocol with multipoint relay (ODMRP-MPR) in mobile ad hoc network. In *Proceedings of International Conference on Communication Technology (ICCT 2003)*, (pp. 1295-1300).

Zhu, Q., Parsa, M., & Garcia-Luna-Aceves, J. (1995). *A source-based algorithm for delay-constrained minimum-cost multicasting* (pp. 377–385). IEEE INFOCOM.

Chapter 3
Mobility Models for Ad-Hoc Networks:
A Performance Analysis Perspective

Shrirang Ambaji Kulkarni
National Institute of Ecology, India

G. Raghavendra Rao
National Institute of Ecology, India

ABSTRACT

Multihop wireless networks hold a lot of promise in terms of military based ad-hoc network applications, wireless sensor network based applications, along with vehicular and inter terrestrial applications. The simulation of these types of networks is required to analyze the working of these types of complex networks. An essential component of these simulation based studies is the modeling of mobility of nodes. Thus many mobility models have been proposed to model the movement of wireless nodes with various constraints and real world scenarios. In this chapter an insight is provided into the working of various mobility models along with its classification. The authors also analyze the performance of these mobility models by measuring their performances with appropriate mobility and connectivity metrics. The mobility models proposed to meet the challenges of vehicular ad-hoc networks and delay tolerant networks are also discussed.

1.1 MOBILITY MODELS

Mobile ad-hoc networks are characterized by wireless mobile nodes, whose movements are unpredictable. The nature of their movement's i.e. the speed, direction and rate of change can have profound effect on their performance (Yoon J, Liu M and Noble B 2003). To mimic the movement of nodes in the real world, the mobile computing community has turned towards simulating them. Thus one has to derive a model of such movement and characterize it in terms of a mobility model.

DOI: 10.4018/978-1-4666-0321-9.ch003

Figure 1. Mobility models for ad-hoc networks

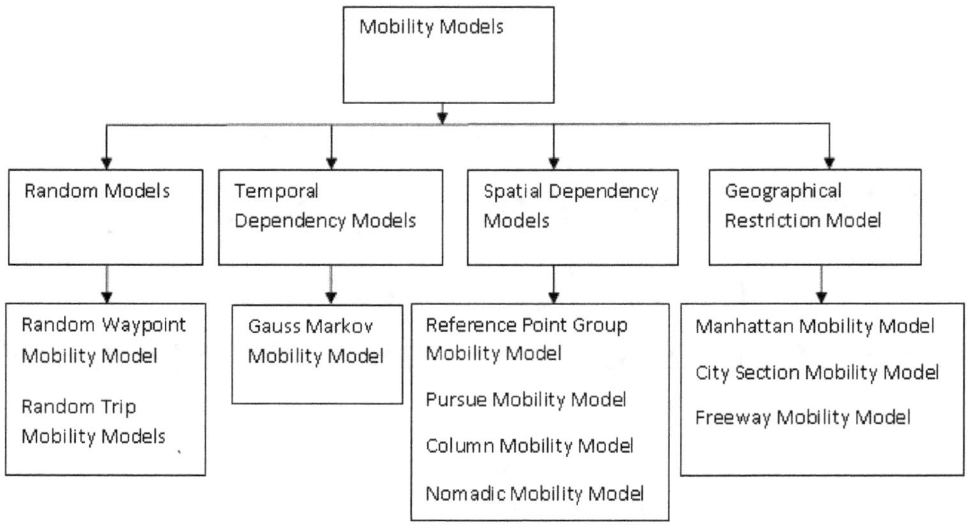

1.2 IMPORTANCE OF MOBILITY MODELS

The importance of mobility models can be characterized as follows.

- To assist in meaningful comparisons of ad-hoc routing protocols, mobility models provide movement patterns that have similarity with real world movement of mobile wireless nodes.
- Mobility models spell the main characteristics of ad-hoc networks (Sarela M 2004). These characteristics have implications on the degree of difficultness experienced by ad-hoc routing protocols. E.g. diameter of the network in hops; density of mobile nodes and mobility of nodes. Thus mobility models which exhibit less amount of network partitions, have stable links for longer durations and aid the performance of routing protocols for ad-hoc networks.
- The behavior of mobility models can be measured in terms of spatial dependency, temporal dependency and average link duration, which helps to quantify their performance and assist in predicting performance of routing protocols for ad-hoc networks.

1.3 CLASSIFICATION OF MOBILITY MODELS

The specific mobility characteristics of the mobility models lead to the following classifications (Bai F, Sadagopan N and A Helmy A 2003) as illustrated in Figure 1.

Mobility Models can also be classified as given in [4]

- Entity Mobility Model
- Group Mobility Model

1.3.1 Entity Mobility Model

- The entity mobility models represent a class of mobility models whose movements are independent of each other. Examples of entity mobility models are (Camp T, Boleng J and Davies V 2002).
- Random Waypoint Mobility Model: A model that specifies pause times whenever the changes in speed and destination occur.

- Random Walk Mobility Model: A mobility model based on random directions and speeds.
- Gauss Markov Mobility Model: A mobility model designed to adapt to different levels of randomness by tuning a parameter.
- City Section Mobility Model: Mobility model that constraints movement of nodes along streets of a map.
- Manhattan Mobility Model: A mobility model which models movements of nodes in an urban area
- Freeway Mobility Model: A mobility model which emulates the motion behavior of mobile nodes on a freeway.

1.3.2 Group Mobility Model

In case of group mobility models the movements of nodes is relative to the movement of other nodes in the group.

- Reference Point Group Mobility Model: A model which represents the random motion of a group of mobile nodes as well as the random motion of each individual mobile node within the group.
- Pursue Mobility Model: A mobility model whose mobile nodes move together from one location to another.
- Column Mobility Model: A group mobility model which encourages nodes to form a line before making forward movement.
- Nomadic Mobility Model: A group mobility model whereby mobile nodes migrate from one location to another.

The next section describes the mobility models based on the classification illustrated in Figure 1.

1.4 RANDOM MOBILITY MODELS

The random mobility models are characterized by nodes which move randomly and freely without any restrictions. These models were one of the early mobility models to appear for simulation based studies of ad-hoc networks and are popular with most of the researchers. In a study conducted in (Kurkowski S, Camp T and Navidi W), it was found that greater than 64% of researchers employed random-based mobility models for their work.

1.4.1 Random Waypoint Mobility Model

The Random Waypoint Mobility Model was proposed by (Johnson D B and Maltz D.A 1996) and soon it became a popular mobility model for evaluating mobile ad-hoc network routing protocols. The reason for its popularity was its simplicity and availability. In this mobility model a mobile node stays in one location for a certain period of time, popularly known as pause time. Once the pause time expires it chooses a new destination randomly and starts moving in that direction with a speed chosen between [min, max] speed range. The mobile node travels towards the destination with a selected speed. When it reaches the destination it pauses for a selected time and the entire process begins again.

1.4.1.1 Discussion about the Random Waypoint Mobility Model

The refined version of Random Waypoint Mobility Model considered in (Broch J, Maltz D.A, Johnson D.B, Hu Y-C and Jetcheva J 1998) has become the de-facto standard in mobile computing research. The Random Waypoint Mobility Model is assumed to have an average speed of Vmax/2 where Vmax is the maximum speed of the simulation. In a study conducted in [1], the instantaneous average speed was defined as

$$\overline{v}(t) = \frac{\sum_{i=1}^{N} v_i(t)}{N} \quad (1)$$

Where N is the total (fixed) number of nodes in the scenario, vi(t) is the speed of the ith node at time . This definition of instantaneous average speed was analyzed for 30 different scenarios and with vmax=20 and vmin=0. It was observed that the instantaneous average speed was consistently decreasing and would eventually approach zero. Their experimental analysis resulted in values of 4.07 m/s at =1000s ; 3.16 m/s at t=5000 s and 2.85 m/s at t=10000s. Thus it was observed that the decay in average speed occurs slowly and consistently. This has a direct bearing on the results which may result in misleading results. One of the solutions for the average speed decay problem was to set a non-zero minimum speed. This resulted in convergence of simulation results to constant and stable levels.

1.4.2 Random Trip Mobility Models

The Random Waypoint Mobility Model though being simple is not realistic. It suffers from non-steady state distribution at start of the simulation (Chaudhari S P, Boudec J L and Vojnovic M 2005) . The probability distribution of movement of nodes varies with time and ultimately converges to "stationary distribution". The work in (Chaudhari S P, Boudec J L and Vojnovic M 2005) emphasizes on perfect sampling of initial node mobility state so that the simulation of the node mobility is perfect, i.e. it is in steady state throughout the simulation. The next section discusses the application of perfect sampling to Random Trip Models like Random Waypoint, Random Walk with Wrapping, Random Walk with Reflection and Restricted Random Waypoint.

1.4.2.1 Steady State Random Waypoint

At a trip transition instant (Chaudhari S P, Boudec J L and Vojnovic M 2005), a node picks a trip destination uniformly at random on a rectangular area and samples numeric speed from a uniform distribution. A straight line represents the trip path that connects node positions from this to the next trip transition instant. At the arrival to the trip destination the node pauses for a random time drawn from a uniform distribution. The trip selection rules follows for other iterations. The implementation uses a perfect sampling algorithm (Boudec J-Y L and Vojnovi´c M 2005) which avoids knowing geometric constants.

1.4.2.2 Random Walk with Wrapping

This model (Chaudhari S P, Boudec J L and Vojnovic M 2005) exhibits similar behavior as Random Waypoint, but at a trip transition instant, a node picks direction, trip duration and numeric speed. These parameters then decide the nodes movements in the given direction with a given numeric speed and for the given trip duration. If a node on its trip encounters the boundary of the domain, it is wrapped around into the domain. The steady state influence on this model is to see that the node position is uniformly distributed on the domain and the node speed has the same direction as at a trip transition instant.

1.4.2.3 Random Walk with Reflection

The Random Walk with Reflection Mobility Model (Chaudhari S P, Boudec J L and Vojnovic M 2005) is similar to Random Walk with Wrapping except that whenever a node encounters a boundary of the domain, it is reflected into the domain and is not wrapped around.

1.4.2.4 Restricted Random Waypoint Model

The domain in this mobility model (Chaudhari S P, Boudec J L and Vojnovic M 2005) contains a collection of sub-domains. At some trip transition instant, a node performs Markov walk on the sub-domains, for other trip transition instants the node undergoes trips within a sub domain. The number of trips within a sub domain is drawn at random at a stage of trip transition instant when the node enters a sub domain.

1.5 TEMPORAL DEPENDENCY MODELS

In the temporal dependency models the mobility of a node may be constrained and limited by the physical laws of acceleration, velocity and rate of change of direction (Sarkar S K, Basavaraju T G and Puttamadappa C 2008). Hence the current velocity of a mobile node may depend on its previous velocity. Thus nodes a different time slots have their velocities "correlated". This characteristic of mobility models is known as the temporal dependency of velocity. The random waypoint model is inadequate to capture this phenomenon. In the next section we discuss the Gauss Markov Model as an example of temporal dependency model.

1.5.1 Gauss Markov Model

The Gauss-Markov Mobility Model was introduced (Liang, Haas B and Zygmunt J 1999) and since has been widely utilized. The mobility model makes the assumption that the velocity of a mobile node is correlated over time and modeled as a Gauss-Markov stochastic process. We consider the model implemented in (Tolety V 1999). The mobile node is represented by two values one for its speed and other for its direction. To begin with each node is assigned a current speed and direction. The movement of node at nth instance occurs by calculating speed and direction at (n-1) th instance and a random variable given by the following equations

$$s_n = \pm s_{n-1} + (1 - \pm)\bar{s} + \sqrt{(1 - \pm^2)}s_{x_{n-1}} \tag{2}$$

$$d_n = \pm d_{n-1} + (1 - \pm)\bar{d} + \sqrt{(1 - \pm^2)}d_{x_{n-1}} \tag{3}$$

where Sn and dn are the new speed and direction of node at time interval n; ± is the tuning factor and is used to vary randomness $0 \leq \alpha \leq 1$.

$\bar{s}$ and $\bar{d}$ are constants representing the mean value of speed and direction as $n \to \infty$ and $s_{x_{n-1}}$ and $d_{x_{n-1}}$ are random variables from a Gaussian distribution. Total random values are obtained by $\alpha = 0$ and $\alpha = 1$.Thus in Gauss-Markov Mobility model, the temporal dependency plays a major role in determining mobility behavior

1.6 SPATIAL DEPENDENCY MODELS

In most of the entity based mobility models, a mobile node moves independently of other nodes in terms of speed, location and direction. Thus sometimes these mobility models exhibit characteristics which are far away from realism. In most of the applications visualized for mobile ad-hoc networks (MANET), like battlefields, police officers pursuing criminals and disaster relief, some form of collaboration among the uses is required. Thus the movements of mobile nodes are correlated in "space" and this characteristic forms the basis for spatial dependency models. In the next section we discuss the popular mobility models in this category namely Reference Point Group Mobility Model (RPGM) and Pursue Mobility Model (PMM), Column Mobility Model (CMM) and Nomadic Mobility Model (NMM).

1.6.1 Reference Point Group Mobility Model

In most MANET applications like battlefields, disaster relief, etc, collaboration among users exist and the users are likely to follow the team leader. Thus the mobility of nodes is affected by their neighboring nodes. This kind of collaborative interaction among mobile nodes in group is modeled by the Reference Point Group Mobility Model (RPGM).

In RPGM (Hong X, Gerla M, Pei G and Chiang C-C 1999), each group has a logical center or a group leader. Each group is composed of one leader and a number of members. The movement of the group leader determines the mobility behavior of the group. The movement of the group leader at time t can be represented by motion vector V^{t}_{group}. Each member of the group deviates from the motion vector V^{t}_{group} by some degree. Each node mobility is assigned with a reference point that follows the group movement. The motion vector of group member i at time t, V^{t}_{group} can be described as given in Equation 4.

$$V_i^t = V_{group}^t + RM_i^t \quad (4)$$

Here the motion vector RM^{t}_{i} is an independent identically distributed (I I D) random process whose length is uniformly distributed in [0,rmax] where rmax is the maximum allowed deviation (Sarkar S K, Basavaraju T G and Puttamadappa C 2008). Since RPGM exhibits the characteristics of spatial dependency, it behaves differently from Random Waypoint Mobility Model. It also incurs less line breakage and achieves better performance for various routing protocols.

1.6.2 Pursue Mobility Model

The Pursue Mobility Model models (Sanchez M 2002), (Sanchez M and Manzoni P 1999) a mobility scenario, whereby mobile node try to tack a particular target. This model visualizes a scenario whereby police officers are trying to catch an escaped criminal. The pursue mobility model emphasizes on a single update equation for each of the mobile node.

$$new_{pos} = old_{pos} + accelration\left(target - old_{pos}\right) + random_{vector} \quad (5)$$

where acceleration(target-oldpos) represents the information of the mobile node being pursued. The random offset for each mobile node is represented by randomvector. The amount of randomness of each mobile node is restricted so as to track the mobile node being pursued. The current position (Camp T, Boleng J and Davies V 2002) of a mobile node, the random vector, and the acceleration function are taken into account in order to calculate the next position of the mobile node.

1.6.3 Column Mobility Model

Column Mobility Model represents a set of mobile nodes which form a column before moving forwards. The Column Mobility Model can be visualized as an application for search operations (Sanchez M and Manzoni P 2001). In this mobility model an initial reference grid for a column of mobile node is defined. Each mobile node is then placed in relative to its reference point in the reference grid and is n allowed to move randomly around its reference point via an entity mobility model. The new reference point is computed from the old reference point and an advance vector which serves as a predefined offset. The predefined offset that moves the reference grid is computed via a random distance and a random angle that lies between 0 to π.

1.6.4 Nomadic Community Mobility Model

The Nomadic Community Mobility Model represents a group mobility model where a crowd of mobile nodes move from one point to another Sanchez M (2002), (Sanchez M and Manzoni P 2001). Within each group of mobile nodes, individuals maintain their own personal places where they move in random ways. Example, consider a group of students touring an art museum. In the Nomadic Community Mobility Model, each mobile node employs an entity mobility model like the Random Walk Mobility Model to move around a given reference point. When the reference point changes, all mobile nodes in the group travel to the new area defined by the reference point and then begin roaming around the new reference point

1.6 GEOGRAPHICAL RESTRICTION MODEL

In Random Based Mobility Models the mobile nodes can move without any restriction and freely in the simulation area. However in many real world applications where the movements of nodes are constrained by streets, freeways and obstacles the random based mobility models cannot be applied. For such scenarios many stochastic mobility models like Manhattan Grid Mobility Model, City Section Mobility Model and Freeway Mobility Model have been proposed and are discussed in the following section.

1.7.1 Manhattan Grid Mobility Model

The node movement is affected by its environment in most real world applications. The motion of vehicles is bound by freeways or streets. The integration of paths and obstacles into mobility models has resulted into mobility models with geographic restrictions. One of the popular examples of geographic restriction mobility model is the Manhattan Mobility Model (Bai F, Sadagopan N and A Helmy A 2003), (ETSI). For achieving geographic constraints into node mobility it is necessary to restrict their movement to pathways in the map. The map is predefined and uses random graph to model the map of the city. The vertices of the graph represent buildings and edges represent the streets. Initially nodes are placed randomly on the edges of the graph. Then for each node a destination is chosen randomly and the node moves towards this destination. Upon arriving at the destination, the node pauses for some time and again chooses a new destination. The Manhattan Model can be useful in modeling movement in an urban area where pervasive devices communicate with each other (Bai F, Sadagopan N and A Helmy A 2003).

1.7.2 City Section Mobility Model

The City Section Mobility Model (Chaudhari S P, Boudec J L and Vojnovic M 2005),(PalChaudhuri) represents a street network of a section of city. The streets and speed limits on the streets mimic the type of city being simulated. City Section Model provides safe driving characteristics such as speed limit and maximum distance allowed between any two vehicles. The City Section Model provides more realistic movements drawn from a section of city and poses several constraints on traveling behavior of vehicles (Camp T, Boleng J and Davies V 2002). For the experimental analysis a steady-state version of City Section Model (Chaudhari S P, Boudec J L and Vojnovic M 2005) was used. The input to the model is a space graph defined in ASCII format that contains road_id, average road speed, coordinates of the road end points. One of the short-comings of the City Section Mobility Model is that the streets in the city section are created by users and hence is only as realistic as user's imaginations.

1.7.3 Freeway Mobility Model

The Freeway Mobility Model proposed (Bai F, Sadagopan N and A Helmy A 2003) was employed to model the movement of mobile nodes on a freeway. The model (Bai F and A Helmy A 2004) relies on several freeways of the underlying map. Each freeway is represented by lanes in both directions. As an example of geographically restricted model the movement of nodes is restricted to lanes on the freeway. The velocity of nodes in this model is temporally dependent on its previous velocity. Further, if two mobile nodes on the freeway lane are within a safety distance, the velocity of the following node cannot exceed the velocity of the preceding node. The movement of each vehicle is restricted by the following speed management rules.

Speed Update. The speed is varied by a random acceleration of maximum magnitude a. If we define · a random variable uniformly distributed in [-1, 1] then this rule is as specified in Equation 6.

$$v_i(t+\Delta t) = v_i(t) + \cdot a\Delta t \tag{6}$$

Speed bounding. The vehicle speed at a given time is bounded by vmin and vmax as specified in Equation 7.

$$v_i(t+\Delta t) = \min\left\{\max\left[v_i(t+\Delta t), v_{min}\right], v_{max}\right\} \tag{7}$$

Speed reduction. To avoid collision each vehicle maintains a safe distance as specified in Equation 8.

$$v_i(t+\Delta t) = \begin{cases} v_{i+1(t)-a/2} & \text{if } \Delta x_i(t) \le \Delta x_{safe} \\ v_i(t+\Delta t) & \text{otherwise} \end{cases} \tag{8}$$

Each vehicle starts its movement at one end of a lane with a speed selected from [vmin, vmax] uniformly till it reaches the other end of the lane. The new movement then begins on a randomly selected lane.

1.8 MOBILITY MODEL ANALYSIS METRICS

The following metrics from (Bai F, Sadagopan N and A Helmy A 2003) were employed to analyze mobility models quantitatively and qualitatively.

1.8.1 Mobility Metrics

1.8.1.1 Average Link Duration

This metric specifies the longest interval of time [t1, t2] for nodes i and j forming the link (i, j). This is then averaged for all node pairs for all existing links specifying the Equation 9.

$$LD' = \frac{\sum_{t1=0}^{T}\sum_{i=1}^{N}\sum_{j=i+1}^{N} LD(i,j,t1)}{P} \tag{9}$$

where P is no of tuples (i, j, t1) and LD (i, j, t1) $\neq$ 0

1.8.1.2 Average Relative Speed

Relative speed is given by Equation 10.

$$RST(i,j,t) = \left|V_i(t) - V_j(t)\right| \qquad 10$$

where Vi(t) and Vj(t) is the velocity vector of node i and jat time t. The average value of RST(i,j,t) is given by Equation 11.

$$RST' = \frac{\sum_{i=1}^{N}\sum_{J=1}^{N}\sum_{t=1}^{T} RST(i,j,t)}{P} \tag{11}$$

where P is no of tuples (i, j, t1) and RST (i, j, t1) $\neq$ 0

1.8.2 Connectivity Metrics

1.8.2.1 Average Degree of Spatial Dependence

It is a measure of the extent of similarity of velocities of given two nodes not so far apart, given by Ds(i, j, t) and averaged over pair of nodes and time instants and formalized by the Equation 12.

$$Ds' = \frac{\sum_{t=1}^{T}\sum_{i=1}^{N}\sum_{j=i+1}^{N} Ds\left(i, j, t1\right)}{P} \qquad (12)$$

where P is no of tuples (i, j, t1) and Ds (i, j, t1) ≠ 0

1.9 PERFORMANCE ANALYSIS OF MOBILITY MODELS

The performance analysis of the following models in terms of mobility metrics and connectivity metrics were carried out.

- Original Random Waypoint Mobility Model (ORWMM)
- Modified Random Waypoint Mobility Model (MRWMM)
- Steady State Random Waypoint Mobility Model (SSRWMM)
- Random Waypoint with Wrapping Mobility Model(RWWMM)
- Random Waypoint with Reflection Mobility Model (RWRMM)
- Random Waypoint Generalized Domain Mobility Model(RWGDMM)
- Reference Point Group Mobility Model (RPGM)
- Pursue Mobility Model (PMM)
- Column Mobility Model (CMM)
- Nomadic Mobility Model (NMM)
- Gauss Markov Mobility Model (GMMM)
- Freeway Mobility Model (FWMM)
- Manhattan Mobility Model (MHMM)
- City Section Mobility Model (CSMM)

The tools provided in (PalChaudhuri), (Bonn Motion) and (The Network Simulator – NS-2) were used to generate the mobility models. All the scenarios were generated for 5,10,20,30 and 40 maximum speed. The number of nodes was 20 and the pause time was 10 sec. The duration of each scenario file was 900 sec.

Average link duration indicated the period for which links are stable. This has a direct bearing on the performance of routing protocols for ad-hoc networks, as stable links result in less route adaptation and better performance. In Figure 2,

Figure 2. Average link duration of mobility models

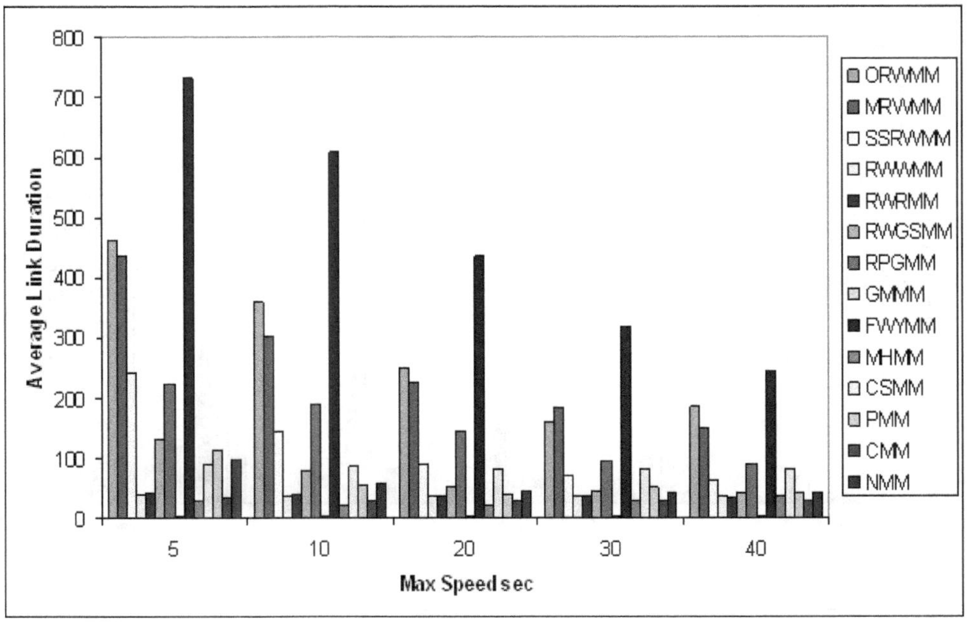

we observe that the Original Random Waypoint Mobility Model along with its modified version have a relatively superior performance as compared to other mobility models. Also the values of Freeway are similar to Random Waypoint mobility model. Reference Point Group Mobility Model has a reasonably good performance. This is because the group of mobile nodes moves at velocities that are deviated from the group leader by small fraction and the existing link between the given two nodes is expected for higher duration.

In Figure 3 we observe that the average relative speed is the least for group mobility model like Reference Point Group Mobility Model (RPGM) and other Random Based Mobility Models. However it is observed that it is relatively higher for Column Mobility Model (CMM) and Nomadic Mobility Model (NMM).It is highest for Gauss Markov, Freeway and Manhattan Mobility Model, whereby it is observed that it is more than twice as compared to Random based and Group based models. This high value for Freeway and Manhattan Mobility Model is because of motion of the mobile nodes in the opposite direction for these models. Also it is observed that Gauss Markov along with poor link duration exhibits high relative speed contributing to its poor performance.

Spatial dependency takes into account the movement pattern of a node in correlation with nodes in its neighborhood. In Random based Mobility Models the node move independent of each other. In case of Group Based Models, the movement of individual nodes is affected by the movement of the leader nodes. In Figure 4 we observe that the performance of Random Based Mobility Models is poor except for the performance of Random Waypoint in Generalized Domain (RWGDMM) and Reference Point Group Mobility Model (RPGM), where the performance is relatively better. The Pursue Mobility Model (PMM) a group based mobility model has the highest amount of average degree of spatial dependency as seen in Figure 4.

From Figure 2, Figure 3, and Figure 4, we can reasonably predict the performances of the mobility models and observe that Group Mobility Models have a reasonably good performance in terms of all performance mobility and connectivity metrics.

Figure 3. Average relative speed of mobility models

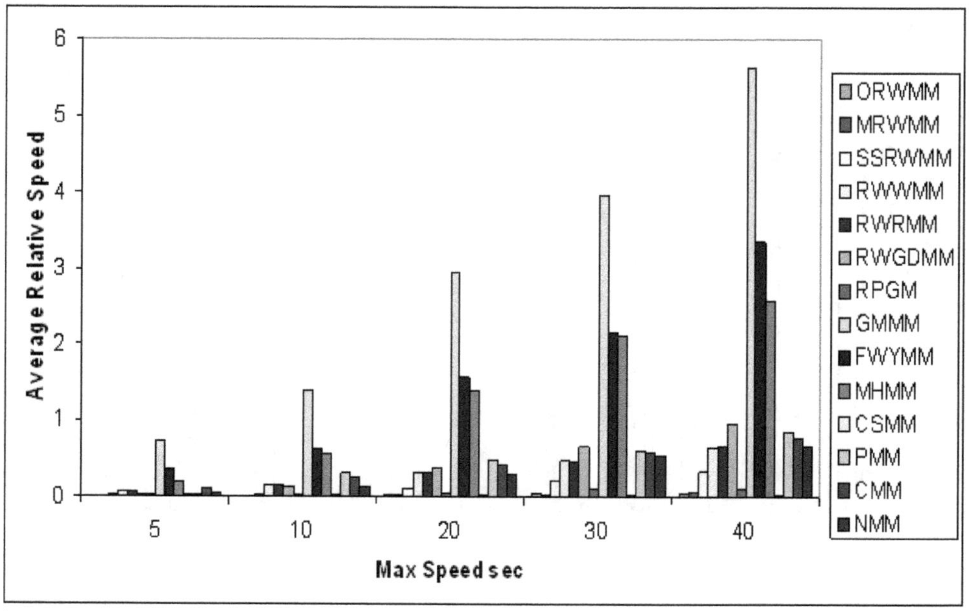

Figure 4. Average degree of spatial dependency

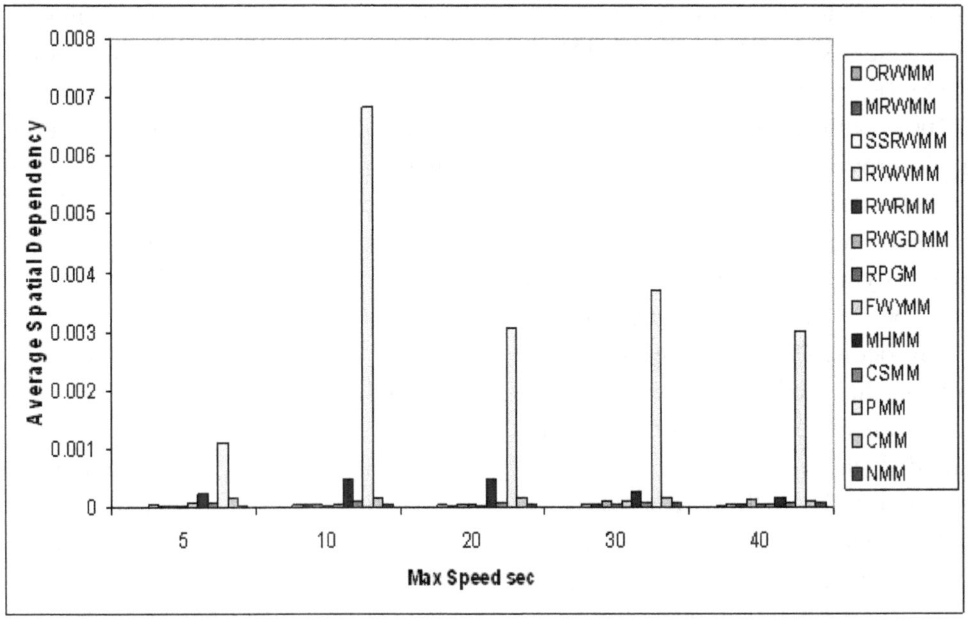

1.10 CURRENT TRENDS IN MOBILITY MODELS

The goal of most mobility models is to be as realistic as possible in modeling the real world scenarios. In future movements will be strongly affected by the needs of humans to socialize in one form or another (Musolesi M, Hailes S, Mascolo C 2004) . Fortunately, humans are known to associate in particular ways that can be mathematically modeled, and that are likely to bias their movement patterns. Thus mobility models founded on social network theory, can be worked upon as it is empirically been shown to be useful as a means of describing human relationships. In a similar work (Kim J. Kim and S. Bohacek 2005, July) experimented on a mobility model which modeled people in urban area. The model was based on extensive data collected. It composed of three sub-models, the activity model, the task model, and the agent model. The activity model was based on a recent US Labor Department study of time-use in America that collected detailed information about how people spend their time. The task model focused on the mobility of people inside office buildings and was based on work in the area of meetings analysis (N. C. Romano and J. F. Numamaker 2001). The agent model determined how mobile nodes interact with each other as they move toward their destinations and was modeled based on the work of (. Pushkarev B and. Zupan J. M 1975) . (Yang S, Zhang X Y C and Spyrou E 2010) in their work on social network theory for modeling human mobility observed that people's movement behavior are strongly affected by their social interactions with each other. They modeled a novel human mobility model based on heterogeneous centrality and overlapping community structure in social networks known as Heterogeneous Human Walk (HHW) mobility model. They managed to construct the k-clique overlapping community structure which satisfied the common statistical features observed from distinct real social networks. Simulation results demonstrated that HHW manages to exhibit statistical characteristics observed from both real social networks and mobile user traces.

Vehicular ad-hoc networks are considered as specialized form of Mobile ad-hoc networks. The research challenge posed by the study of VANETs as observed by (Fiore, M., Harri, J., Filali, F., & Bonnet, C 2007) is the definition of a vehicular mobility model providing an accurate and realistic vehicular mobility description at both macroscopic and microscopic levels. Also another challenge is to be able to dynamically alter this vehicular mobility as a consequence of the vehicular communication protocols. Thus it is a great challenge to develop a mobility model that meets the above mentioned concerns. INVENT a microscopic car-following model was proposed (INVENT) based on the Gipps model (Gipps, P. G 1981,1986). The ability of this model was that it allowed vehicles to ply at maximum speed, but by avoiding collision. The traffic generator produced scenario that included traffic lights at road intersections, bidirectional lanes and multi line traffic. The city map was derived from TIGER/ line database. Also to meet the challenge of VANET mobility a Intelligent Driver Model (IDM) was proposed (Treiber, M., & Kesting 2009), (Treiber, M., Hennecke, A., & Helbing 2000).The IDM belonged to the class of deterministic follow the leader model. It had the following advantages like; it behaved accident free because of the dependence on relative velocity. Due to it metastabilty it showed self organization characteristics. Its modeled parameters having good interpretation and it allowed a fast numerical simulation.

Another futuristic challenges of mobility models is in terms of Delay-Tolerant Networks (DTNs) which were initially developed for inter-planetary communications, and where signal propagation delay was extremely long (Burleigh S, Hooke A, Torgerson L, Fall K 2003). This concept has been visualized for terrestrial wireless mobile ad hoc networks due to the constraints of intermittent connectivity among mobile nodes as a consequence of low nodal density, node mobility and short transmission range. In their work, (Fischer D 2011) proposed a social network model for delay tolerant networks. The model named GeSoMo generalizes a number of existing social mobility models by considering the ingenious and simple concept of separating the simulation model from the structural description of the simulated social network. Evaluation results showed that GeSoMo produces simulations that are logical with a broad range of empirical data describing real-world human social behavior and mobility.

1.11 CONCLUSION

Mobility model represents an important component of simulation based studies for multi-hop wireless networks. In this chapter to provide a deeper insight we have provided taxonomy of the various mobility models. Most of the initial mobility models proposed in the ad-hoc literature were random mobility based. Mobility models which were relative to time were categorized as temporal dependent models that came at the later stages. The most famous among them was the Gauss Markov. As Mobile Ad-Hoc Networks exhibit group based collaborative applications; group based mobility models like Reference Point Group Mobility Model and Pursue Mobility Models were proposed to name a few. Mobility models may at times constraint node mobility under geographic constrained environments. To discuss these scenarios mobility models like Manhattan, Freeway and City Section Mobility Models were discussed. For an analysis of these mobility models a quantitative analysis was carried out with appropriate mobility metrics. It was observed that the Original Random Waypoint Mobility Model (ORWMM), Modified Random Waypoint Mobility Model (MRWMM) and Freeway Mobility Model (FWMM) exhibited higher average link duration. In terms of average relative speed it was observed that the Original Random Waypoint Mobility Model (ORWMM), Freeway Mobility Model (FWYMM) and Manhattan Mobility Model (MHMM) were the poor performers and exhibited higher average relative speed. In terms of spatial dependency Manhattan Mobility Model, Random Waypoint Generalized Domain and City Section Mobility Models performed well. This chapter also provided an insight into future applications of mobility models in terms of Social Network Models, Vehicular Ad-Hoc Networks and Delay Tolerant Networks along with their challenges and applications

REFERENCES

Bai, F., & Helmy, A. (2004). A survey of mobility modelling and analysis in wireless ad hoc networks. In *Wireless ad hoc and sensor network*. Kluwer Academic Publishers.

Bai, F., Sadagopan, N., & Helmy, A. (2003). The IMPORTANT framework for analyzing the impact of mobility on performance of routing protocols for adhoc networks. *Elsevier Ad Hoc Networks*, *1*, 383–403. doi:10.1016/S1570-8705(03)00040-4

Bonn Motion. (2005). *A mobility scenario generation and analysis tool*. Retrieved from http://web.informatik.unibonn.de/iv/ mitarbeiter/decvaal/bonnmotion

Boudec, J.-Y. L., & Vojnovi'c, M. (2005). Perfect simulation and stationarity of a class of mobility models. *Proceedings - IEEE INFOCOM*, 2005.

Broch, J., Maltz, D. A., Johnson, D. B., Hu, Y.-C., & Jetcheva, J. (1998). *A performance comparison of multi-hop wireless ad hoc network routing protocols. Mobile Computing and Networking* (pp. 85–97). MobiCom.

Burleigh, S., Hooke, A., Torgerson, L., Fall, K., Cerf, V., & Durst, B. (2003). Delay-tolerant networking – An approach to interplanetary Internet. *IEEE Communications Magazine*, *41*(6), 128–136. doi:10.1109/MCOM.2003.1204759

Camp, T., Boleng, J., & Davies, V. (2002). A survey of mobility models for ad hoc network research. *Wireless Communication and Mobile Computing (WCMC), special Issue on Mobile Ad Hoc Networking . Research Trends and Applications*, *2*(5), 483–502.

Chaudhari, S. P., Boudec, J. L., & Vojnovic, M. (2005). Perfect simulations for random trip mobility models. In *Proceedings of the 38th Annual Symposium on Simulations*, (pp. 72-79).

Davies, V. (2000). *Evaluating mobility models within an ad hoc network*. Master Thesis, Colorado School of Mines.

ETSI, Universal Mobile Telecommunications System (UMTS). (1998). *Selection procedures for choice of radio transmission technologies of the UMTS*, UMTS 30.03 Version 3.2.0. Retrieved from http://www.3gpp.org/ftp/specs/html-info/3003u.htm

Fiore, M., Harri, J., Filali, F., & Bonnet, C. (2007). Vehicular mobility simulation for VANETs simulation. *ANSS '07, 40th Annual Simulation Symposium*, (pp. 301-309).

Fischer, D. A. (2011). Mobility model for the realistic simulation of social context. *KiVS*, *2011*, 215–220.

Gipps, P. G. (1981). A behavioral car following model for computer simulation. *Transportation Research Part B: Methodological*, *15*, 105–111. doi:10.1016/0191-2615(81)90037-0

Gipps, P. G. (1986). A model for the structure of lane-changing decisions. *Transportation Research Board*, *20*(5), 403–414.

Hong, X., Gerla, M., Pei, G., & Chiang, C.-C. (1999). *A group mobility model for ad hoc wireless networks*. ACM/IEEE International Symposium on Modeling Analysis and Simulation of Wireless and Mobile Systems (MSWiM), Seattle, WA, USA.

INVENT. (n.d.). Retrieved from http://www.njit.edu/~borcea/invent/ invent-vehiculartraffic- generator.tar.gz

Johnson, D. B., & Maltz, D. A. (1996). *Dynamic source routing in ad hoc wireless networks in mobile computing.*

Kim, J., & Bohacek, S. (2005, July). *A survey-based mobility model of people for simulation of urban mesh networks*. In MeshNets'05, Budapest, Hungary.

Kurkowski, S., Camp, T., & Navidi, W. (2002). *Minimal standards for rigorous MANET routing protocol evaluation. Technical Report MCS 06-02*. Colorado School of Mines.

Liang, H. B., & Zygmunt, J. (1999). Predictive distance based mobility management for a PCS networks. *Proceedings of the IEEE Infocom.*

Musolesi, M., Hailes, S., & Mascolo, C. (2004). An ad hoc mobility model founded on social network theory. In *Proceeding MSWiM '04 Proceedings of the 7th ACM International Symposium on Modeling, Analysis and Simulation of Wireless and Mobile Systems.* New York, NY: ACM.

Pal Chaudhuri, S. (2011). *Ns-2 code for random trip mobility model.* Retrieved from http://monarch.s.rice.edu/ ~santa/research /mobility/

Pushkarev, B., & Zupan, J. M. (1975). *Urban space for pedestrians*. MIT Press.

Romano, N. C., & Numamaker, J. F. (2001). Meeting analysis: Findings from research and practice. in *Proceedings of the 34th Hawaii International Conference on Systems Science*.

Sanchez, M. (2002). *Mobility models*. Retrieved from http://www.disca.upv.es/ misan/mobmodel.htm

Sanchez, M., & Manzoni, P. (1999). A Java-based ad hoc networks simulator. In *Proceedings of the SCS Western Multiconference Web-based Simulation Track.*

Sanchez, M., & Manzoni, P. (2001). Anejos: A Java based simulator for ad-hoc networks. *Future Generation Computer Systems, 17*(5), 573–583. doi:10.1016/S0167-739X(00)00040-6

Sarela, M. (2004). *Measuring the effects of mobility on reactive ad hoc routing protocols*, Research Report 91, Helsinki University of Technology Laboratory for Theoretical Computer Science, 2004.

Sarkar, S. K., Basavaraju, T. G., & Puttamadappa, C. (2008). *Ad hoc mobile wireless networks* (pp. 261–270). Boca Raton, FL: Auerbach Publications.

The Network Simulator – NS-2. (n.d.). Retrieved from http://www.isi.edu/nsam/ns/

Tolety, V. (1999). *Load reduction in ad hoc networks using mobile servers*. Master's Thesis, Colorado School of Mines.

Treiber, M., Hennecke, A., & Helbing, D. (2000). Congested traffic states in empirical observations and microscopic simulations. *Physical Review E: Statistical Physics, Plasmas, Fluids, and Related Interdisciplinary Topics*, *62*, 1805–1824. doi:10.1103/PhysRevE.62.1805

Treiber, M., & Kesting, A. (2009). Models for traffic flow and vehicle motion . In Olariu, S., & Weigle, M. C. (Eds.), *Vehicular networks from theory to practice* (pp. 26–28). Boca Raton, FL: CRC Press. doi:10.1201/9781420085891-c2

Yang, S., Zhang, X. Y. C., & Spyrou, E. (2010). Using social network theory for modeling human mobility. *IEEE Network*, 6–13. doi:10.1109/MNET.2010.5578912

Yoon, J., Liu, M., & Noble, B. (2003, April). Random waypoint considered harmful. In *Proceedings IEEE INFOCOM 2003*, San Francisco, CA.

Chapter 4
Broadcasting in Wireless Ad hoc Networks:
Approaches and Challenges

Niranjan Kumar Ray
NIT Rourkela, India

Ashok Kumar Turuk
NIT Rourkela, India

ABSTRACT

Broadcasting is a process of sending message from one node to all other nodes in a network. It is used commonly in ad hoc networks, for route discovery, content distribution, data disseminations, et cetera. Most of the routing protocols in ad hoc networks use broadcasting as the common mechanism to find path between source and destination. Therefore broadcast protocols are the fundamental building block to understand basic functionality of routing in mobile ad hoc network (MANET). Moreover due to the widespread application of MANET in various scientific and military applications, broadcasting in MANET has pulled attention of many researchers in recent time. Considering its importance and growing popularity, the authors discuss different broadcasting protocols commonly used in ad hoc networks. The chapter also analyzes various issues associated with these protocols.

1. INTRODUCTION

In multi-hop ad hoc network broadcast data has to be relayed through intermediate nodes to reach every other node in the network. However, the mobility of nodes and their limited transmission range make it cumbersome to reach all the nodes. To meet this requirement each node works as a router by transmitting the message to its neighbors. This simplest approach is called *broadcasting*. Unlike point-to-point communication that follows one-to-one paradigm, broadcasting follows one-to-all transmission mechanism where all the nodes receive the message with same reliability. Here reliability refers to the ability of message reaching to all neighbors of the source node in a bounded time. Broadcasting performs route discovery, updating network knowledge, trans-

DOI: 10.4018/978-1-4666-0321-9.ch004

mitting critical signals, etc. Most of the routing protocols in wireless ad hoc networks like AODV (Perkins, & Royer, 1999), DSR (Johnson & Maltz., 1996), ZRP (Haas, Pearlman & Samar, 2002), and LAR (Ko & Vaidya, 1998) use broadcasting for their route discovery process. Although the operating procedure of each protocol is different, the most common among them is the use of broadcast techniques to find a path to destination. When a route to destination is needed, the node broadcasts a route request (RREQ) message. The *RREQ* message contains the source and destination address. Each node other than destination on receiving RREQ broadcast it. The broadcasting in MANET is a multi-objective problem that tries to maximize the number of nodes reached satisfying some constraints like energy consumption in the network, message delay, communication overhead etc. The most common problem associated with broadcasting is the *broadcast storm problem*. This is the main cause of miss utilization of valuable resources and network congestion in MANET. This problem is discussed more precisely with other issues in section 2.

Several broadcasting protocols have been proposed in literatures (Williams & Camp, 2002, Lou & Wu, 2007, Khan *et al.*, 2008, Zhang & Jiang, 2005). They are broadly categorized into two types; *centralized* and *localized*. The centralized protocols are operated on global topology information; in contrast localized protocols require information which can be acquired at node level. Hence localized protocols are suitable for multi-hop ad hoc networks and final topology can be formed by sharing the local information among the nodes. This chapter is dedicated to various broadcasting issues and challenges in wireless ad hoc networks, focusing on localized protocols. The localized protocols are categorically discussed in this chapter based upon their mode of operation. The major limitation to most of the broadcast protocols is their higher energy consumptions; as a significant amount of energy is consumed for performing the broadcast task. Considering this issue we discussed some energy efficient broadcast techniques which minimize the energy consumptions by implementing different techniques. The remaining part of the chapter is organized as follows; the basic concepts, major challenges of broadcasting mechanism are analyzed in section-2. Section-3 emphasizes on various broadcasting protocols. Conclusions are made in section-4.

2. BACKGROUND AND PRELIMINARIES

This section briefly discusses the basic concepts, design guidelines, issues and applications of broadcast approaches in wireless ad hoc network.

2.1. Basic Concepts

Simple Flooding

The most common approach of broadcast is called *blind flooding*. It does not make any attempt to reduce the number of redundant transmission and does not require any knowledge of state information. In this approach a node on receiving the message for first time forward the message. All the nodes in the network perform the same task. It is assumed that the message will reach to all nodes in the network. Unfortunately this leads to network clogging due to the participation of all nodes in this process. This not only causes loss of precious battery power also increases interference in the network. Blind flooding is the main cause of *broadcast storm problem*. However this simple flooding approach is still used in many routing protocols. This approach is not suitable for networks with higher node density because it would then generate too much network traffics and sometimes could even lead to network congestion. Some variants of simple flooding are proposed for specific network. Two such flooding methods are *hybrid flooding* and *adaptive flooding*.

RAD

It is a common approach used by most of the broadcast protocols to determine which packets to send and which to drop. In this approach a node on receiving a broadcast message doesn't transmit immediately. It waits for some time which is randomly selected by the node between values [0, T_{max}], where T_{max} is the highest possible delay intervals in seconds. The time introduced is called *random access delay* (RAD). This delay helps to reduce the collision of data packets. RAD implementations can be done at MAC layer or at network layer.

MST

A minimal spanning tree (MST) is a tree that covers all nodes in the network and minimizes the total tree cost, which is considered as power. Finding an MST in a multi hop wireless network is a known NP - complete problem. In a network where all nodes operate at the same uniform transmission power, finding an MST is equivalent to finding the minimum *connected dominating set* of the network. In contrast, in a network where nodes can adaptively adjust their transmission power, the problem is to find the min - power spanning tree. Different variants of MST based broadcast protocols for wireless ad hoc network are discussed in this chapter.

Localize and Globalize Broadcasting Methods

In globalized method nodes need entire network topology information. In acquiring these information nodes experience higher delay and higher power consumption in a heavily dense network. However in localize methods each node uses the information of its one-hop or within its two-hop neighbor. The information is shared among nodes in a distributed manner to form the final topology.

2.2. Design Guidelines

Few design guidelines for broadcast mechanisms in MANETs are discussed below.

Scalability

A broadcast mechanism should have low protocol overhead. When the network size increases it should be able to accommodate dynamic network environments.

Energy Efficiency

An efficient mechanism should be able to minimize the total power consumed for accomplishing the task of broadcasting and accordingly prolong the network lifetime. The mechanism should adaptively adjusts the transmission power and have features to reduce the numbers of retransmission.

Localized Operation

To forward a packet, it is preferable to perform localized operations such that each node can take forwarding decision based on the available local information or the information available from its neighborhood. Localized information requires only neighborhood information which can be exchanged among the neighbors periodically.

Reliable Transmission

Broadcast is mostly unreliable due the broadcast storm problem. However in many cases, reliable transmissions are necessary, which need to be considered the broadcast and lossy features of the wireless transmission medium. Simple flooding rarely guarantees perfect message reliability. However reliable transmission requires more control overheads. So an efficient broadcast protocol should consider these facts.

2.3. Applications of Broadcasting Protocols

Broadcasting has numerous applications in ad hoc network. Two most common broadcasting applications are discussed here.

- **Content Distribution:** News spreading is a typical application for broadcasting in MANETs. News spreading is commonly used in some applications. For example: broadcasting of *aid information* in disaster management system, *accident information* through VANET in a city scenarios, transmission of *critical information* in environment monitoring systems etc. Content distribution is also used in sensor networks where real-time sensor data are disseminated to other nodes in order to realize a fully-replicated database, such that every database node has a consistent view of real time events.
- **Resource Discovery and Advertisement:** Resource discovery by broadcasting is commonly used in wireless ad hoc and sensor networks. Nodes in MANET have no or little knowledge about other nodes. Nodes periodically advertise by sending a beacon message to prove its existence. Beacon message contains the different useful information about the sender node such as: *location information, neighbor list, residual battery power* etc. This information may be utilized by other nodes to determine the two-hop neighbor list, multi-point relay nodes, finding intermediate node, selecting transmission power, etc.

2.4. Broadcast Problems

Broadcast Storm

Already we have discussed that simple flooding is considered as the most common approach to perform broadcasting in ad hoc wireless networks. In this approach a sender node broadcast the message which will be reachable to all the neighbors within its transmission range. Nodes upon receiving the message transmit if it has not been received before. This way broadcast packet is transmitted in the whole network causing serious network problems in terms of *redundancy transmission*, *packet collision* and *medium contention*. Redundant broadcasting not only causes increased energy consumptions in the network but also responsible for decreased packet delivery ratio as well as decreased network throughput. This is referred as *broadcast storm problem* (Ni *et al.*, 1999). Following are the effects caused due to simple flooding in a CSMA/CD network.

I. Redundant Rebroadcast

It causes when a node decides to rebroadcast a message to its neighbor. However if all neighbors have already received the message then the transmission is redundant. This redundancy transmission is the major cause of energy consumption.

II. Packet Collision

It occurred due to deficiency of backoff mechanism, lack of RTS/CTS handshaking and absence of collision detection mechanism. This is also the main cause of throughput degradation.

III. Medium Contention

The contention occurs in the network when neighboring nodes want to retransmit at same time. The nodes within the overlapping radio range compete with each other to access the wireless medium after receiving the broadcast message. The node which gets access to the medium will transmit the message and other will be in wait state. The contention becomes a major issue for network with higher node density.

Broadcast storm problem can be well understood from the scenario shown in Figure 1. When broadcast message is generated and forwarded by

Figure 1. Broadcasting storm problem: (a) Broadcasting by node A; (b) Redundant broadcasting

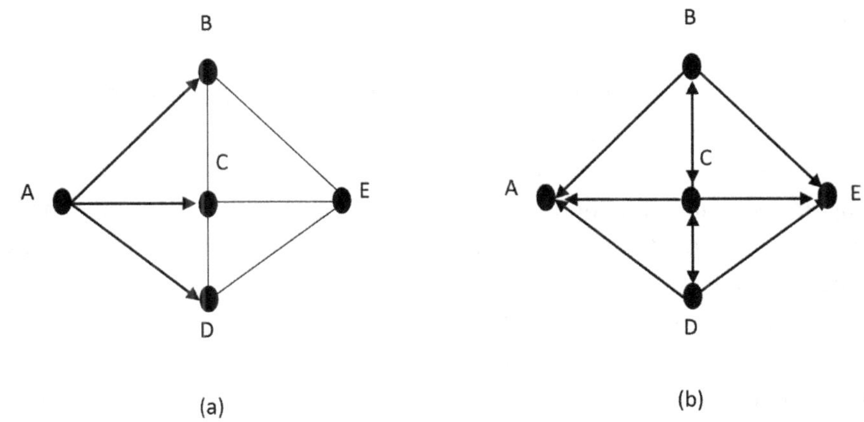

node A, the neighbor nodes B, C and D receives the message as shown in Figure 1(a). On receiving the message each neighbor will try to forward the message. When they decide to forward the message following problems may be occurred.

A. Node A will receive three redundant messages from node B, C and D as shown in Figure 1(b)
B. The node B, C, and D may contend with each other for wireless medium; if they are within the overlapping transmission range to each other. If the overlapping area contains more number of nodes; there may be an increased contention for the broadcast medium.
C. There may be a greater chance of packet collisions at E if node B, C and D broadcast at the same time. Even if node B, C, and D broadcast the message at different time; node E will receive three redundant broadcast messages.

The above discussion gives some basic idea about broadcast storm problem in wireless ad hoc network caused by simple flooding mechanism. The broadcast storm problem can be avoided by merely reducing the number of redundancy transmissions of broadcast messages. Several such schemes are discussed in section 3.

Unreliability

Message drop mechanism used for reduction of redundancy transmission is the main cause of unreliable message transmission in wireless multi-hop network. The message drop may effectively resolve the broadcast storm problem caused by simple flooding. However it affects end-to-end reliability and increases latency in the network. Simple flooding rarely guarantees perfect message reliability. Broadcast can be the called reliable when it ensures that all the nodes in the network receive the messages. Reliability can be achieved by message transmission with acknowledgment. However acknowledgement may cause serious medium contention surrounding the sender node for which no acknowledgement mechanisms are used for broadcasting messages. Also due to MANET characteristics some nodes in the network may not decode the message properly. The nodes placed in *carrier sensing zone* can't read the broadcast message properly.

Mobility and Partitioning

Mobility of nodes affects routing which in term affects the performance of broadcast protocols. Spanning tree based techniques for MANETs may provide effective solutions to broadcast storm problem at low mobility networks. However, for networks where topology changes very frequently they fail to provide effective solutions. Some algorithms for dynamic network have been proposed in the literature but they have their own pitfalls. So it is observed that mobility associated with broadcasting needs greater attention.

Network partitioning constitutes another greatest issue when it comes to broadcasting in MANETs. Network partitioning occurs when some nodes are not participating in the network traffic. Range adjustment is necessary to consider network portioning and energy saving in a network. Topology control protocols try to satisfy these constraints.

3. BROADCASTING SCHEMES IN MANET

Broadcast protocols can be classified based on different parameters. Some protocols are based upon the amount of state information required for performing broadcast task while other are based on network knowledge to perform the same task. Most protocols are deterministic in nature while some are probabilistic. An algorithm is said to be deterministic if it behaves predictably. Based on the state information a protocol can be called global or local. Global protocols are also called centralized protocols and their main drawback is that they are not scalable. They are hence not suitable for MANETs. This section discussed some common broadcasting approaches.

3.1. Heuristic Based Broadcasting

Heuristic based schemes require careful selection of different network parameters and thresholds which are closely related to ad hoc network environments. The performance of these approaches is highly dependent on selected parameters and thresholds. Several heuristics are proposed in (Ni, *et al.*, 1999) Here some heuristic based broadcasting approaches are discussed briefly.

Probability Based Scheme

A broadcast method can be called probabilistic when a node retransmits the message depending on a random probability values. Probabilistic based approach can be efficiently used to reduce the redundancy message transmissions. A node on receiving the broadcast message for first time; broadcast the message with a random probability p. The schemes became simple flooding when $p=1$. The value of p may be adapted according to the topological properties of the network.

Sasson *et al.* (2003) use a probabilistic algorithm for relaying broadcast messages. A node has p probability for rebroadcast and $1-p$ probability of taking no action in the rebroadcasting of the message. The approach explores the possibility of applying phase transitions for selecting the probability of rebroadcasting a message. The authors proclaimed that their probabilistic approach is more suitable to highly dynamic ad hoc networks as compared to other deterministic based approach. The paper showed there is a difference between the ideal behavior and the results of the actual simulations. The method found that probabilistic flooding does not exhibit a bimodal behavior as percolation and graph theory would suggest. Probabilistic flooding did increase the success rate of transmissions when network is densely populated.

Some possible future work is to determine an algorithm to dynamically adjust the value of p. Other area of future research in probabilistic scheme is to understand the effect that modifying the transmission range with regard to p.

Distance and Location Based Scheme

In *distance-based* scheme a node on receiving the broadcast message calculates the distance between sender and itself. Distance to sender can be calculated by receiver from the location information appended in the message or from the received signal strength. When the distance between sender and receiver is greater than the threshold distance set by receiver then the node (receiver) will forward that message. In all other case it will not broadcast the message. The functioning of distance based scheme is given below.

- Source sends the message
- The receiver determines the threshold distance *D*. Redundant message will be stored until the RAD expires.
- When the RAD expires the distance to source node *d*, will compare with the threshold distance *D*

If the $d < D$, message will be dropped otherwise it will be forwarded.

Location-based approach estimates the additional coverage area more precisely. Ni, *et al.* (1999) proposed a method, which directly calculates the expected area covered, from the positions of senders. This calculation of the additional coverage becomes complicated when several copies of the same message are received. The location-based approach also requires the assistance of position based technology like GPS for every node.

The shortcomings to distance and location based scheme is the additional cost associated for calculating the distance and location.

Counter Based Schemes

The scheme is based on restricting some nodes to broadcast the messages when the expected additional coverage is too low. For this purpose a counter (C) is used to keep the track of the broadcast message received by a node. A threshold (T) is chosen by the node which plays a deciding factor for broadcasting the message. When the value of C is greater than T, broadcast is restrained. The scheme is sequentially given below.

- Initialize counter C = 1 on hearing broadcast message for the first time. If message heard again, interrupt step (ii) and perform (iv).
- Wait for a RAD. Then submit message for transmission and wait until the transmission actually starts.
- Exits the procedure if message is transmitted.
- Increase C by one. If C < T, resume the interrupted waiting in (ii). Otherwise C= T, proceed to (v)
- Cancel the transmission of message if it was submitted in (ii). The node is prohibited from rebroadcasting.

3.2. Neighbor Coverage Based Schemes

To prove the existence of a node in the network, node advertises itself at a regular interval. Node broadcast the beacon message for that purpose. The beacon messages contain address of broadcasting node and one-hop neighbor list. Two-hop neighbor information can be obtained by exchange of beacon message. This information determines the link state topology of a network. This approach also helps to retrieve information about nodes in indoor model where GPS doesn't work properly. Additional information such as residual battery power, location information, signal to noise ratio (SNR), etc. may be included in the broadcast message. These information can improve the network performance and energy saving. However use of beacon in neighbor discovery may suffer from various problems such as:

- *Message complexity*:- It may be in term of *O* (n^2) to determining the two-hop neighbor information. Where each node is surrounding by *n* number of neighbors.
- *Time constraint*:- Multiple exchange of beacon message over an extended period of time may be required to discover the

two-hop topology as nodes do not transmit simultaneously.

- *Loss of information*: - This may be a case when two or more nodes send the beacon to specific nodes at the same time. As a result information may be lost due to collision.
- *Mobility*: - The nodes mobility may result in link state errors.
- *Overhead*:- Frequent movement of nodes may require more exchange of information in a quicker period of time. Thus introduces additional overhead and increase chance of packet collision.

From the above information we concluded that neighbor coverage based broadcasting is not being preferable over two or more hops.

Self-Pruning

In this approach (Lim & Kim, 2000), two neighbor sets are maintained at each node. Suppose node *u* transmits a broadcast message. Neighbor node *v* on receiving the message determines its covering set as:

$$Cover(v) = N_v - N_u - \{u\}$$

where N_u and N_v are the neighbor set of *u* and *v* respectively. Here *Cover* (*v*) denotes that some neighbors of *v* are not covered by node *u* yet. When *Cover* (*v*) is empty, it implies that all the neighbors of node *v* are covered by node *u;* so *v* will not broadcast the message. In this way redundancy transmission can be reduced.

Scalable Broadcast Algorithm

It is another technique based on neighbor information (Peng & Lu, 2000). It uses two-hop neighbor information and a broadcast delay timer to determine about the broadcast. A node upon receiving the broadcast message determines its cover as stated in the self pruning scheme. If some nodes are not covered then node broadcast the message otherwise node will schedule the broadcast procedure by putting some delay between the message transmission. The delay is calculated as the ratio between the maximum node degrees of neighbors to its own node degree. The delay calculated by the node *v* is given as:

$$T_{dely}(v) = \frac{Max\left[Degree\left(N_v^1\right)\right]}{Degree(v)}$$

where N_v^1 is the one-hop neighbor of node *v*.

Multipoint Relay

This is a flooding-reduction mechanism in which only subset nodes are selected as multipoint relay (MPR) (Qayyum *et al.*, 2001) which participates in the broadcasting process. This process reduces the number of retransmissions for broadcasting operation while retaining resilience against transmission errors (inherent when using wireless links) and against obsolete two-hop neighbor information. Each node periodically broadcast *hello* message. Hello message permits each node to learn topology up to two-hops. The objective of MPR selection is to select a subset of its neighbor such that a packet, transmitted by these selected neighbors, will be received by all two-hop neighbors. This property is "*coverage criterion*". The MPR set of a node is computed such that, it satisfies this criterion. Information required to perform this calculation (i.e., link sensing and neighborhood information) is acquired through periodic exchange of hello messages. Smaller is the set, lower is the overhead. However, it is not essential that the MPR set is minimal; the "coverage criterion" must be satisfied by the selected MPR set.

Figure 2. Multipoint relay selection

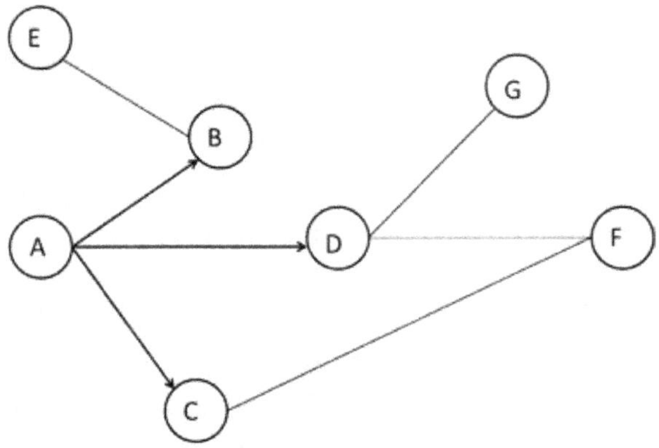

In Figure 2 it shows that node B, C and D are the one-hop neighbors and E, F and G are the two-hop neighbors of A. Node A determines MPR as follows:

- Determines one-hop and two-hop neighbors
- Determines the resultant cover set. The set of two-hop neighbors that will receive the message from the MPR set.
- From one-hop neighbor set those are not chosen as member of the MPR set; determine the ones that cover the most two-hop neighbors not in the cover set.
- Repeat the step (ii) till all two-hop neighbors are explored.

From the above procedure node A finds one hop neighbors (B, C, D) and two-hop neighbors (G, E, F) in step (i). Node B is selected as the MPR for node E, as two-hop neighbors of E is only reachable through B. The other two-hop nodes G and F are similarly covered by D. Now D joins the MPR with B. Node C is not added to the MPR set as its neighbor node F is already reachable through node D. So the MPR set of node A contains {B, D}. MPR is a good approach to restrict rebroadcasting.

3.3. Double Covered Broadcast

The objective of double covered based broadcast (DCB) is to reduce the redundancy in broadcasting. Redundancy broadcasting can be useful in some environment. Advantage of redundancy broadcasting is explained by Lou and Wu (2004) to improve the packet delivery ratio in a network that has high transmission error rate. In an error prone environment message acknowledgement is an important consideration as compared to other requirements. The broadcast initiator node selects few nodes from its one-hop neighbor list to forward the broadcast message. These nodes are called *forwarding nodes*. A node is selected as a forwarding node from the one-hop neighbors based upon two criteria.

- All two-hop neighbors of the sender node are covered by them
- All one hop nodes other than those are selected as forward node must be covered by at least two forwarding nodes.

A sender node on transmitting the broadcast message waits for the acknowledgment from its forwarding nodes. When the forwarding nodes retransmit the message one copy of message also reach at sender. Sender treated that message as an acknowledgment of message transmission by the forwarding nodes. If sender does not receives all the acknowledgments from its each forwarding nodes it broadcast till it reaches the maximum re-transmission limit or get reply message from that node. Authors claim that DCB is a reliable broadcasting mechanism for wireless ad hoc networks. The simulation results show that the double covered broadcast algorithm has higher delivery ratio, lower forwarding ratio, low overhead and low end-to-end delay for a broadcast operation under high transmission error ratio environment. However it is very sensitive to node's mobility. The delivery ratio may be dropped when node's mobility increases. This may be due to the frequent change of neighbor information at high mobility.

DCB is most suitable for a network where medium is error prone; but it consumes more energy due to its redundancy transmission. A modification is required to achieve energy efficiency in DCB. This can be possible by transmitting beacon message in an independent low bandwidth channel.

3.4. Dominating Set Based Broadcasting

Dominating sets based approach has been used to solve varieties of problem in MANETs. The problems caused by simple flooding can be minimized by transmitting the message through connected dominating set (CDS) backbone. Most of the routing protocols are based on blind flooding. The broadcast storm problem caused by blind flooding not only depletes the precious battery power but also the major cause of bad carrier sensing. These issues can be easily manageable by CDS. Dominating sets based applications includes routing, media access coordination, energy efficiency, etc. Some basic terms frequently used in dominating set is stated below. In a graph G= (V, E) where *V* is the set of vertex and *E* is the set of edges.

- *Dominating Set* (D): is defined as a subset of *V* such that each node either in D or at least one of its neighbor node is in D.
- *Connected dominating Set* (CDS): A set is said to be CDS if it is a dominating set of G and it includes the connected sub graph of G.
- *Minimum Connected Dominating Set* (MCDS): It is a CDS with minimum cardinality. Finding minimum sized dominating set is NP-Hard, it should not be surprising that finding the MCDS is also NP-Hard.

Protocols in dominating set based approach either seek a MCDS or a CDS that provides good resource conservation property. A good CDS algorithm, must consider following parameters: *performance bounds*, *degree of localization*, *message complexities*, and *stability with respect to node movement*. Few protocols considering these parameters are discussed here.

Das *et. al* (1997) proposed dominating set based routing. In that approach message routed from the source to other neighbor in a dominating set, along the CDS to the dominating set member closest to the destination node, and then to the destination node. Other than routing, the virtual backbone formed by dominating set also used to propagate "*link quality*" information for route selection for other network traffic. CDS is also beneficial for location based routing, in which messages are forwarded based on node's location information, rather than topological connectivity. Intermediate nodes are selected based on their proximity to the destination.

Wu and li (1999) proposed a simple broadcasting protocol based on CDS. The protocol introduced two rules to minimize the number of intermediate nodes. A node *u* can be an "*intermediate node*" if there exists two neighbors *v* and *w* of *u* such that they are not direct neighbors to each other. That means communication between two nodes is only possible through intermediate node. The protocol decides the criteria of selecting intermediate node as follows:

- For two intermediate node *u* and *v* if $N(u) \subseteq N(v)$ and $id(v) < id(u)$ then node *v* cannot be the part of CDS backbone.
- For nodes *u*, *v*, and *w*. If $N(v) \subseteq N(u) \cup N(w)$ and $id(v) = \min\{id(v), id(u), id(w)\}$ then *v* may be removed from the CDS sets.

Here N(*u*), N(*v*) and N(*w*) is the neighbor set of intermediate node *u, v* and *w* respectively and *id*(*u*), *id*(*v*) and *id*(*w*) are their respective *node-id*. Authors also suggest that if nodes are GPS enabled then they can determine the intermediate node easily.

Stojmenovic *et al.* (2001) proposed a method to replace the node-id with a record *key (degree, x, y)* criteria from the above rule (i) Where (*degree, x, y*) represents the node degree and coordinate position of a node. The selection criteria for intermediate node are based on node degree. Node with higher degree will be selected as the intermediate node and in case of ties the node with higher coordinate position will be selected as intermediate node. One problem observed in selecting the intermediate node is that only few nodes are always selected as intermediate node due to their geographic position and they are frequently used by all other nodes to forward the broadcasting message. This may lead to quick network partition as the battery powers of these nodes are limited. One possible solution to this may be addressed by considering residual battery power of a node along with node degree while selecting intermediate nodes. The node with maximum remaining batter power will be more likely in CDS.

3.5. Cluster Based Broadcasting

In this scheme broadcast message is transmitted by all cluster-heads (and gateway nodes). The nodes in the network are grouped together to form a *cluster*. Few nodes from each cluster are selected as cluster-head. In a single cluster there may be one or many cluster-heads which depends upon the selection criteria of cluster-head. Cluster-heads can support *inter cluster* and *intra cluster* communications. Some nodes are common to many clusters and are called as gateway nodes. Only cluster nodes and gateway nodes can retransmits the broadcast message. Figure 3 depicts the simplest network formed by two clusters. In Figure 3, node A and E are two cluster head and D is the gateway node.

A cluster can be an *active* or *passive* which is based on their way of cluster formation. In active cluster formation all nodes elect a cluster heads which is done by exchanging periodic hello messages among them. Lin and Gerla (1997) proposed an active clustering algorithm to select cluster-head based on node-ID. Node with lower or higher ID can be selected as cluster head. Some other algorithm (Lou & Wu, 2002) construct cluster based on static and dynamic backbone. The static backbone is formed by independent connected dominating set where dynamic backbone is formed by dependent connected dominating set.

Passive clustering is based on background data traffic. This data traffic is used to propagate cluster information and also collects neighbor

Figure 3. Network with two clusters, node C initiates the broadcasting process; the message is propagated in the entire network by cluster node A, and E through gateway node D.

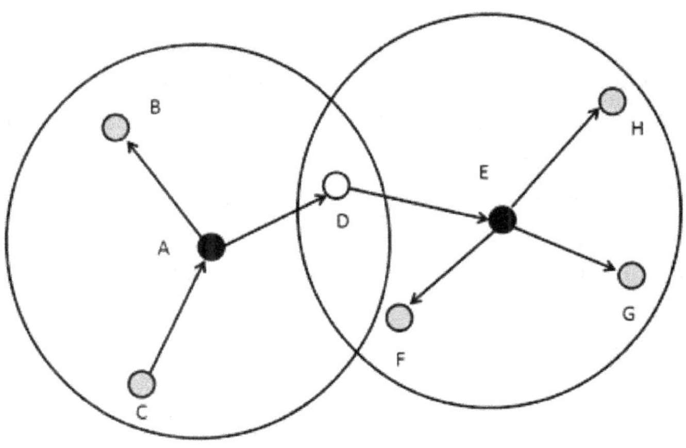

information through packet reception. Packet reception can be achieved by allowing MAC layer to send all received packet to TCP/IP. Passive clustering has major advantage in term of lower massage overheads as compare to active clustering. However without existing data traffic it can't form the cluster.

3.6. Energy Efficient Scheme

Different broadcasting protocols aiming to reduce energy consumption in the network are categorically discussed here based upon their energy saving approach. We focused on *transmission based* approaches and *direction based* approaches. In the first approach nodes select transmission radius based on some parameter. In second approach directional antennas are used for energy saving purpose.

I. Transmission Based Broadcasting

These protocols adjust transmission range efficiently to minimize the energy consumptions. Power control based approaches are commonly applied in ad hoc wireless network for energy saving purpose. The messages transmitted with maximum power not only consume more power but also increases interference in the network. For which most of the power control protocols adaptively adjust its transmission power to a certain level depending on the requirements. When a node transmit message with minimum power its coverage area shrink as result its neighbor list also changes along with its node degree Nodes with lower node degree are preferred for energy saving point of view but that sometimes that adversely affects the other network parameters such as that causes: increase in hop-counts, delay and jitter, decreased in throughput etc. Thus adaptive adjustment of transmission power is necessary in this situation. Some protocols based on incremental power are discussed below.

The most common protocol of this family is *broadcast incremental power* (BIP) protocol (Wieselthier, *et. al.* 2000). It constructs a broadcast tree from the source node to every other node in an ad hoc network using multicast advantage. Multicast advantage is indicates as the broadcast characteristics of a radio transmission; simply it means when a node transmits the broadcast message with certain power all nodes within that range will receive the message. BIP is a centralize protocol which requires global knowledge of network topology. The global topology information introduces more delay in acquiring the topology information and significantly increases the network overhead. To resolve these problems some distributed version of BIP has been proposed in literatures. Ingelrest & Simplot-Ryl (2005) proposed a broadcasting protocol based on BIP. The protocol locally applies the BIP heuristics and calculates its own BIP within its k-hop neighborhood. The localized tree is formed at node which is then broadcast for other node to form the final broadcast tree.

Kang and Poovendran (2003) presented an algorithm to constructs a static routing tree. The algorithm maximizes network lifetime by considering residual battery power at the nodes. The algorithm however, does not really maximize the possible network lifetime, if nodes are mobile.

Ingelrest *et al.* (2006) proposes two localized protocol based on target radius. The author demonstrates an optimal transmission radius (target radius) which minimizes the total power consumption for broadcasting task. Based on that observation they proposed two localized protocols TR-LBOP and TRDS. The first protocol computes the minimal radius needed for connectivity and second protocol considers only neighbors whose distance is greater than the target radius.

I. Direction Based Broadcasting

Several protocols use directional antenna for saving energy in wireless ad hoc networks. Wieselthier *et al.* (2002) proposed two broadcast algorithms based on directional antenna. First algorithm is reduced beam BIP (RB-BIP) and the second protocol is directional BIP (D-BIP).

RB-BIP uses one-to-one communication model to join the neighbors in the BIP tree. However in D-BIP each node sends only one message. D-BIP adjusts the angle of the beam or radio range to join more neighbors. At each step it takes decision about whether to extend the beam or not. Due to this the tree formed by D-BIP and RB-BIP are different. When cost associated to the angle beam and its related overhead are very high D-BIP transmits the message with large radii and beam angle to avoid retransmissions by every node.

Relative neighborhood graph (RNG) is the most popular geometric structure used in wireless ad hoc networks. Several RNG based broadcasting protocols are presented in literatures due to their good geometric property. Borbash and Jennings (2002) proposed a distributed RNG (DRNG) protocol for wireless ad hoc network to use in connected network topology to minimize node degrees, hop-diameter, maximum transmission radius and the number of bi-connected components. However, DRNG do not describe the use of RNG in solving any specific problems. It is found that average node degree of RNG based protocols are 2.5. RNG based protocols maintain connectivity in worst-case.

Li, Hou and Sha (2003) proposed LMST protocol based on minimum spanning tree algorithm. In this protocol each nodes sends beacon message to their one-hop neighbors. Upon receiving all beacons from its one-hop neighbors node constructs the local MST. The MST formed in this way may not be unique. For that purpose LMST uses a *link weight function* to obtain a unique MST. Message exchange takes place among the neighbors to discover edges that belong to the computed MST. Final topology is obtained by the intermediate neighbors of the local MST. Transmit power required to send a message to any neighbor is generated from the final topology. The node also measures the broadcast power to reach the farthest node. Experimental results show that the average node degree of LMST is 2.04 which are closer to RNG and BIP.

Cartigny *et al.* (2003) proposed two broadcast protocols based on RNG and LMST. The RBOP is the variant of RNG while LBOP is based on LMST. Both protocols uses Omni-directional antenna for reducing energy consumptions. Both protocols preserve network connectivity at worst case. Iguchi-Cartigny *et al.* (2009) proposed directional antenna based protocols D-LBOP and D-RBOP which are the extension of LBOP and RBOP respectively. Both protocols uses one to one communication pattern. Nodes send message to their neighbors with minimal beam or range. Neighbor elimination scheme is not used in D-LBOP as they suffered with minimal beam communication due to LMST. This benefit of LMST (average node degree of LMST lower than RNG) reduces the number of broadcasting message used by direction based LMST protocols. Through simulation authors found that energy saving by D-LBOP is superior as to D-RBOP.

Both D-RBOP and D-LBOP are suffered from high message overheads at higher node density. To overcome this problem; same author proposed two protocols based on one-to-many communication patterns. The OM-DRBOP is the one to many communication pattern of D-RBOP while OM-DLBOP is of D-LBOP. It is observed that message transmitted with angle $4\Pi/3$ minimizes the overlap transmission range.

4. CONCLUSION AND FUTURE DIRECTIONS

This chapter discussed broadcast mechanism for wireless ad hoc networks emphasizing on different issues and requirements. We present different broadcasting protocols to provide a complete overview of the one-to-many communication patterns in ad hoc wireless networks. Protocols are differing from each other in term of their mode of operation to support broadcasting; however the basic objective of each protocol is same. i.e. to reduce redundancy transmission of broadcast message, contention of broadcast medium and collision of packets. It is observed that more emphasis is given to local broadcast scheme. In localize scheme node requires only neighbor information within its one or two-hop. This chapter also identified how broadcast is used to optimize routing protocols and introduce new research on energy efficiency. We summarize here highlighting important points.

- Broadcast storm is an important issue in MANETs. Several schemes namely probabilistic, counter based, location based, distanced based, cluster based, etc have been proposed to alleviate the problem associated with broadcast.
- Broadcast protocols use RAD to differ their transmissions. RAD may reduces number of collisions of broadcast messages but it induces more delay. RAD approaches are suitable for networks with lower node density but for network with higher node density they may not ideal.
- Distance and location based approaches work perfectly for deriving the additional coverage of node. If a node observed that its broadcasting will not improve coverage substantially it drop the message. When nodes are GPS enabled determining coverage becomes easy, but GPS based approach has certain limitation in indoor model. Counter based schemes eliminates much redundancy transmission when node density is high.
- Up to two-hop neighborhood information is sufficient for neighbor based approach. Message complexity is a major issue to maintain neighbor information above two-hop. This approach affected when network topology changes frequently.
- DCB is the only approach which utilizes rebroadcasting efficiently in error prone environment.
- Cluster based and CDS based approach are very much useful for energy saving as well as for reducing collisions in the networks.
- Some energy conserving broadcasting protocols are presented, as most of the nodes in MANETs are battery operated. Most of the energy efficient schemes uses directional antenna and properly adjust transmission radius of a node to save energy.

We concluded that significant research on broadcast protocol has been carried out in the last few years focusing on basic problem (broadcast storm) but none of them considers on the quality of service, security and reliability properly. In this modern era of communication quality of service is necessary for many applications, and reliable broadcasting has become indispensable for successful deployment of MANETs and security is another greater issue which is still lacking in ad hoc broadcast protocols. Thus broadcasting in ad hoc network is a significant problem that merits further explorations.

REFERENCES

Borbash, S., & Jennings, E. (2002). Distributed topology control algorithm for multihop wireless networks. In *Proceedings of IEEE International Joint Conference on Neural* Networks, Honolulu, HI, (pp. 355-360).

Cartigny, J., Simplot, D., & Stojmenovic, I. (2003). Localized minimum energy broadcasting in ad-hoc networks. In *Proceedings of IEEE INFOCOM.*

Das, B., Sivakumar, R., & Bharghavan, V. (1997). Routing in ad hoc networks using a virtual backbone. In *Proceedings of IC3N'97*, (pp.1-20).

Haas, Z. J., Pearlman, M. R., & Samar, P. (2002). *The zone routing protocol (ZRP) for ad hoc networks.* Internet-Draft (work in progress) draft-ietf-manet-zone-zrp-04.txt, July 2002.

Iguchi-Cartigny, J., Ruiz, P. M., Simplot-Ryl, D., Stojmenovic, I., & Yago, C. M. (2009). Localized minimum-energy broadcasting for wireless multihop networks with directional antennas. *IEEE Transactions on Computers, 58*(1), 120–131. doi:10.1109/TC.2008.125

Ingelrest, F., & Simplot-Ryl, D. (2005). Localized broadcast incremental power protocol for wireless ad hoc networks. In *Proceedings of 10th IEEE Symposium on Computer and Communication*, ISCC 200, (pp. 28-33).

Ingelrest, F., Simplot-Ryl, D., & Stojmenovic, I. (2006). Optimal transmission radius for energy efficient broadcasting protocol in ad hoc and sensor networks. *IEEE Transactions on Parallel and Distributed Systems, 17*(6), 536–547. doi:10.1109/TPDS.2006.74

Johnson, D. B., & Maltz, D. A. (1996). Dynamic source routing in ad hoc wireless networks . In Imielinski, T., & Korth, H. F. (Eds.), *Mobile computing* (*Vol. 353*, pp. 152–181). Kluwer Academic Publishers. doi:10.1007/978-0-585-29603-6_5

Kang, I., & Poovendran, R. (2003). A comparison of power efficient broadcast routing algorithms. In *Proceedings of IEEE Global Telecommunications Conference (GLOBECOM),* December.

Khan, A. A., Stojmenovic, I., & Zaguia, N. (2008). Parameterless broadcasting in static to highly mobile wireless ad hoc, sensor and actuator networks. In *Proceedings of* 22nd IEEE *International Conference on Advanced Information Networking and Applications (AINA2008)*, Ginowan, Okinawa, Japan, March 25-28.

Ko, Y.-B., & Vaidya, N. H. (1998). Location-aided routing (LAR) in mobile ad hoc networks. *Proceedings of 4th ACM International Conference on Mobile Computing and Networking (ACM MobiCom 1998)*, Dallas, Texas, United States, 1998, (pp. 66-75).

Li, N., Hou, J., & Sha, L. (2003). Design and analysis of an MST-based topology control algorithm. In *Proceedings of IEEE INFOCOM'03*, San Francisco, CA, (pp. 1702-1712).

Lim, H., & Kim, C. (2000). Multicast tree construction and flooding in wireless ad hoc networks. In *Proceedings of the 3rd ACM International Workshop on Modeling, Analysis and Simulation of Wireless and Mobile Systems*, (pp. 61-68).

Lin, C., & Gerla, M. (1997). Adaptive clustering for mobile wireless networks. *IEEE Journal on Selected Areas in Communications, 15*(7), 1265–1275. doi:10.1109/49.622910

Lou, W., & Wu, J. (2002). On reducing broadcast redundancy in ad hoc wireless networks. *IEEE Transactions on Mobile Computing, 1*(2), 111–122. doi:10.1109/TMC.2002.1038347

Lou, W., & Wu, J. (2004). Double-covered broadcast (DCB): A simple reliable broadcast algorithm in MANETs. In *Proceedings of the 23st Annual Joint Conference of the IEEE Computer and Communications Societies (INFOCOM '04),* Hong Kong.

Lou, W., & Wu, J. (2007). Toward broadcast reliability in mobile ad hoc networks with double coverage. *IEEE Transactions on Mobile Computing, 6*(2), 148–163. doi:10.1109/TMC.2007.31

Ni, S., Tseng, Y. C., & Sheu, J. (1999) "The broadcast storm problem in a mobile ad hoc network. In *Proceedings of the ACM/IEEE International Conference on Mobile Computing and Networking (MOBICOM)*, (pp. 151–162).

Peng, W., & Lu, X. C. (2000). On the reduction of broadcast redundancy in mobile ad hoc networks. In *Proceedings First Annual Workshop on Mobile and Ad Hoc Networking and Computing*, (pp. 129-130).

Perkins, C. E., & Royer, E. M. (1999). Ad-hoc on-demand distance vector routing. *Proceedings WMCSA '99- Second IEEE Workshop on Mobile Computing Systems and Applications,* (pp. 90-100).

Qayyum, A., Viennot, L., & Laouiti, A. (2001). Multipoint relaying: An efficient technique for flooding in mobile wireless networks. In *Proceedings of the 35th Annual Hawaii International Conference on System Sciences*.

Sasson, Y., Cavin, D., & Schiper, A. (2003). Probabilistic broadcast for flooding in wireless mobile ad hoc networks. In *Proceedings IEEE Wireless Communications & Networking Conference (WCNC 2003)*, (pp. 1124-1130). March 2003.

Stojmenovic, I., & Seddigh, M. (2001). Dominating sets and neighbor elimination based broadcasting algorithms in wireless network. *IEEE Transactions on Parallel and Distributed Systems*, *2*(12), 1–12.

Wieselthier, J. E., Nguyen, G. D., & Ephremides, A. (2000). On the construction of energy-efficient broadcast and multicast trees in wireless networks. *Proceedings - IEEE INFOCOM, 00*, 585–594.

Wieselthier, J. E., Nguyen, G. D., & Ephremides, A. (2002). Energy limited wireless networking with directional antennas: The case of session-based multicasting. In *Proceedings of IEEE INFOCOM.*

Williams, B., & Camp, T. (2002). Comparison of broadcasting techniques for mobile ad hoc networks. *In Proceedings of MOBIHOC'02*, June 2002

Wu, J., & Li, H. (1999). A dominating set based routing scheme in ad hoc wireless network. In *Proceedings of Third International Workshop Discrete Algorithm and Methods for Mobile Computing and Communication*, (pp. 7-14).

Zhang, H., & Jiang, Z. P. (2005). Theoretic analysis of broadcasting schemes for two-dimensional MANETs. In *Proceedings of the Workshop on Performance of Wireless Networks and Communication Systems (WiNCS)*, July.

Chapter 5
Connectivity as a Fundamental Characteristic of Mobile Ad Hoc Networks

Jiwa Abdullah
Universiti Tun Hussein Onn Malaysia, Malaysia

ABSTRACT

In this chapter, the idea of formulating the node connectivity into a form suitable for the purpose of monitoring the dynamic of any given node pair was proposed. It had been able to show that the node connectivity time constitutes the length of time where node-pairs are connected and communication sessions can be established. By analytical calculation and simulation, the node connectivity time depends on the relative velocity of the node-pair. It shows that there is an upper limit and a lower limit for the connection time. Between these limits, communication sessions can be carried out. In order to monitor the quality of connection between a node-pair and for the whole route which is made up of a collection of node-pairs, npem, npcm, and nci are introduced. The nci is shown to be a viable way to estimate the quality of node-pair connection in terms of its timeliness. The index, nci, could be used as one of the metrics to identify and select the best path to the destination.

INTRODUCTION

Mobile ad hoc network (MANET) is formed by a group of nodes that communicate with each other over a single or multiple wireless channels. They operated in a decentralized and self–organizing manner without the necessity for a fixed wireless networking infrastructure. An outstanding feature of MANET is the multihop mode of communications, beyond the direct radio connection amongst the nodes. Each node could be a transmitter, a receiver or a router forwarding data traffic toward its destination. The fundamental idea of such networks goes back to the 1980s (I. Chlamtac, M. Conti, and J. J.-N. Liu (2003)), but have shown an increase in research interest recently, due to emerging technology in wireless communications and the microelectronic industries. A node is typically equipped with a radio transmitter and receiver, which allows communication sessions

DOI: 10.4018/978-1-4666-0321-9.ch005

with other nodes. Salient features of MANET are; (1) its implementation could increase the mobility and flexibility, as it can be activated in a very short time; (2) it can be more economical due to limited infrastructure costs; (3) more robust than conventional wireless networks because of their non-hierarchical distributed control mechanisms; (4) due to its multi-hop character, communication could be established beyond the Line of Sight (LOS); (5) low power consumption of wireless devices, attributed to less power is required for sending a signal over any distance in multiple shorter hops. Several application scenarios, such as mobile computers for conferencing and home networking, wireless sensor networks (I.F. Akyildiz, W. Su, Y. Sankara subramaniam and E. Cayirci (2002)), multihop extensions of cellular telecommunication systems (X. Jiang et al (2004)), and vehicular networking (K. J. Wong *et al.*(2003), C. Lochert *et al.*(2003)) are the favourites amongst the researchers today. This technology has been proposed as a complementary to the fourth-generation wireless networks. Nodes typically rely on each other to establish communication sessions. Hence, network connectivity could be regarded as a fundamental characteristic for the proper functionality of the network. The node density, number of nodes, transmission range and the spatial nodes distribution generally affect node connectivity.

Nodes in MANET are typically mobile with some mobility characteristics. The dynamicity of MANET topology is due to nodes mobility. Mobile nodes also perform as router to the packet transmission, hence each single node contributes to the node and network connectivity. The connectivity failed, if a node inadvertently fails and secondly when the spatial density of the nodes is too low. Connectivity is one of the most fundamental aspects of MANET. Network connectivity of MANET is a complex problem due to the uncertainty of the network topology. A number of studies concerning network connectivity modeling and analysis of a particular MANET have been reported in the recent technical literature (P. Santi and D. M. Blough (2002), P. Santi (2005), C. Bettstetter (2002), O. Dousse, P. Thiran, and M. Hasler (2002), M. Desai and D. Manjunath (2002), C. H. Foh and B. S. Lee (2004)). All these research works aim at providing a mean to calculate the Quality of Services (QoS) measured in terms of the probability of network connectivity.

Route reliability issues have been addressed in several works. In the Associativity-based Routing (C. K. Toh (1997), the association stability of the links is accounted when choosing routing paths. It attempts a measure of goodness using associativity ticks which indicates how stable a route is. The Signal Strength Adaptive protocol (R. Dube, C. D. Rais, K. Y. Wang and S. K. Tripathi (1997)) further considers the signal strength in choosing the good paths. With the Route-Lifetime Assessment Based Routing protocol (S. Agarwal, A. Ahuja, J. Singh and R. Shorey(2000)), a parameter called affinity is defined to characterise the signal strengths and stability of a route. An AODV(C.E. Perkins and E. M. Royer (1999)) tries to preempt a link failure by using periodic hello packets and propagating failure messages to source using the route. Pre-emptive routing (T. Goff, N.B. Abu-Ghazaleh, D.S. Phatak and R. Kahvecioglu(2001)) is an enhancement to DSR that enables nodes to estimate the time to link failure and propagate route failure information in advance. All these mechanisms suffer from the fact that they are local to the node and not necessarily good strategy for the whole route. An attempt to view the whole route was carried out (W. Su, S. J. Lee and M. Gerla(2000)) using some kind of mobility prediction. This requires a global clock synchronisation and a positioning system for the network. These mechanisms do not capture the aspect of the relative velocity for the route.

The chapter addresses the issue of connectivity in MANET and considers it as fundamental characteristics. Ideally it is expected to consolidate the so many research specifically on MANET and connectivity. It would be very useful since it gathers and streamlined the workable ideas into a collection of theories, analytical and experimental experience with respect to node connectivity. The chapter finally described and proposed the Node Connectivity Index, a dynamic and measureable quantity. It is based on the node movement that is characterised by the expansion and contraction model. The performance evaluation shows evidence of feasible and workable model.

NETWORK MODEL

The dynamicity of MANET requires that the understanding of the node connectivity be based on the following three primary models: (a) model for the spatial node distribution, (b) model for wireless channel between the nodes, and (c) model for nodes mobility. For a set of n nodes, which are independently and randomly placed in a simulation area A, either one or two dimensional, with uniform random distribution, the node density is defined as, $\rho = n/A$. It represents the expected capacity of nodes per unit area. Wireless transmission between nodes is model as a radio link where each node has a certain transmission range of r utilizing omnidirectional antennas.

Figure 1 illustrated two nodes are able to communicate directly wirelessly if they are within transmission range of each other. This connectivity model corresponds to wireless propagation with a particular signal path loss. Let P_0 denote the transmitted signal power at the sending node and P_r the received power at a distance r from the transmitter. The received power dropped off as $P_r \, \alpha \, r^{-\gamma} P_0$, where γ is the path loss exponent which depends on the environment, normally taken as, $2 \leq \gamma \leq 5$. The wireless transmission range r can then be mapped to the equivalent transmission power P_0 using a threshold for receiver sensitivity P_s. A node can receive properly if $P_0 \geq P_s$. All nodes are considered stationary or mobile within the specified area. For mobile nodes, their movement typically related to a certain mobility model (e.g., random waypoint, random direction, highway one-dimensional model, etc.). Generally, such a network model is less complex, suitable for researching into the fundamental properties of node connectivity. For the node and network connectivity, the folowing propagation models are considered; free-space propagation model, two-ray ground model and log-distance path loss model. The free-space propagation model assumed the ideal propagation condition with only one clear line-of-sight path between transmitter and receiver. The model could be used to predict radio-signal propagation when there exist a clear and unobstruct path between transmitter and receiver. Formally, it is stated as,

$$P_r(d) = \frac{P_t G_t G_r \lambda^2}{(4\pi)^2 d^2 L}, \tag{1}$$

where Pr(d) is the power of the radio signal received by the receiver located at the distance d from the transmitter, P_t is the power generated by the transmitter, G_t is the gain of transmitter antenna, G_r is the gain of the receiver antenna, $L \geq 1$ is the system loss factor and λ is the wavelength in meters. We can simplify the equation for the propagation model as follows,

Figure 1. A simple MANET model. All nodes are connected to at least one other node, if the node pairs are within transmission range

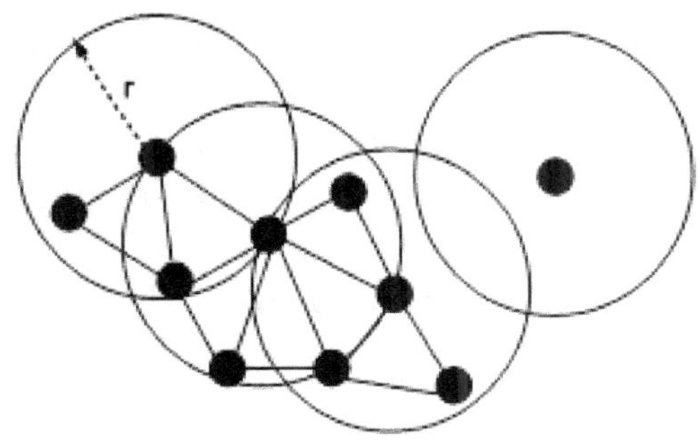

$$P_r = \frac{C_f P_t}{d^2}, \quad (2)$$

where C_f is a constant depending on the nature of transmitter and receiver. When the sensitivity of the receiver node is given as P_s and the background noises and interferences are denoted as N_0, then the transmitted message can be correctly received if the distance d satisfies;

$$d \leq \sqrt{\frac{C_f P_t}{P_s N_0}}. \quad (3)$$

The model represents the communication range as a disk or a sphere around the transmitter with radius of $\sqrt{\left(C_f P_t / P_s N_0\right)}$. If a receiver is within the disk, it receives all packets, otherwise it losses all packets. Thus, this model provides a single direct path between the transmitter and the receiver as the only physical means of radio signal propagation. The two-ray ground model considers both the direct path and the ground reflection path. It is shown by Rappaport (Rappaport, T.(1996)), that this model gives a more accurate prediction at long distance than the free space model. The received power at distance *d* is predicted by;

$$P_r = \frac{P_t G_t G_r h_t^2 h_r^r}{d^4 L}, \quad (4)$$

where h_t and h_r are the heights of the transmitting and receiving antennas respectively. When the distance *d* between the transmitter and the receiver is relatively large, the formulation of the received power can be simplifed as,

$$P_r = \frac{C_t P_t}{d^4}, \quad (5)$$

where C_t is a constant depending on the transmitter and receiver. When the sensitivity of the receiver node is given as P_s and the background noise and interference are denoted as N_0, it can be stated that the transmitted message can be correctly received if and only if the distance d satisfies;

$$d \leq \sqrt[4]{\frac{C_t P_t}{P_s N_0}}. \quad (6)$$

This means that the model represents the communication range as a disk or a sphere around the transmitter with a radius of $\sqrt[4]{\frac{C_t P_t}{P_s N_0}}$. If a receiver is within the disk, it receives all packets, otherwise it losses all packets. When the sensitivity of the receiver node is given as P_s and the background noise and interference are denoted as N_0, then the transmitted message can be correctly received if the distance d satisfies;

$$d \leq \sqrt{\frac{C_f P_t}{P_s N_0}}. \quad (7)$$

This means that the model represents the communication range as a disk or a sphere around the transmitter with a radius of $\sqrt{\left(C_f P_t / P_s N_0\right)}$. If a receiver is within the disk, it receives all packets, otherwise it losses all packets. The log-distance model is derived from the combination of analytical and emperical methods. The log-distance path-loss model is an indoor radio-propagation model that predict the path loss for signals encounters inside a building over a distance, expressed as;

$$L = L_0 + N.\log\frac{d}{d_0} + X_g, \quad (8)$$

where L is the total path loss inside a building, L_0 is the path loss at the reference distance, N is the path loss distance exponent times 10 and X_g is a Gaussian random variable with zero mean and σ standard deviation. The log-distance model can be seen as a generalization of both the free space-space and the two-ray-ground propagation model, known also as path-loss exponent, given by,

$$P_r \ \alpha \ \frac{P_t}{d^\alpha}. \quad (9)$$

The model represents the communication range as a disk or a sphere around the transmitter with a radius of $\sqrt[\alpha]{\frac{C_f P_t}{P_s N_0}}$. If a receiver is within the disk, it receives all packets, otherwise it losses all packets. All these propagation models predict the average received power at a certain distance from the transmitter. Typically, the intensity of the received signal is denoted by a random variable and the actual value varies from the predicted average value. Thus, probablistic models have been used to account for the time- and location-dependent wireless channel.

NODE CONNECTIVITY INDEX

Quantifiable Connectivity Metric for QoS Route Determination

The vision of MANET is characterised by mobile, lightweight and personalised computing devices, being collaborative in nature towards a self-establishing network. When network topology changes due to the mobility of nodes, this will result in three node pair scenarios, least-connected, highly-connected or breaking-up. The changes in the set of links of a node affect not only the node's ongoing communication, but may impede the communication of other nodes as well. As the capacity and communication ability of MANETs are dependent on node connectivity, it is important to understand how node connectivity behaves in MANETs. Routing protocols may cause the route to change in response to the variation of quality of node connectivity. The resultant route changes may alter the traffic distribution in the network, causing congestion at a number of nodes with less traffic at other nodes. The network dynamics require a mobility metric which can quantify the node movements and eventual relationship with the neighbours. The final outcome would be how reliable inter-nodal communication is. The stable and durable routes are selected based on the connectivity criteria from a node to its neighbours and successively in pairs towards the target nodes. Hence, QoS routing in this case is the mechanism to find a feasible route which has overall good connectivity criteria from amongst the routes discovered.

For effective performance of QoS routing protocols, it requires a connectivity metric that can accurately capture the networking dynamics of the node. Effectively it indicates the resulting performance of the QoS routing protocol. Such a metric could provide feedback information which shows the quality of node connectivity. The connectivity metric must be applicable to a real mobile network with real nodes. The first step is to define a set of requirements that must be met to enable effective function of QoS routing protocols. The metrics must be (1) able to compute in a distributed environment, (2) able to adapt a locally measured performance, (3) a numerical quantity which indicates the quality of node-pair connectivity, and (4) capable of processing in a real time implementation. Ultimately, the metric should describe quantitatively the dynamic of the nodes such that it can be of use in the route discovery and selection process to choose the most reliable, longer duration and stable routes. It can also be used to select the cost-effectiveness of the longer duration and stable routes. Lastly, it is expected to generate the available QOS routes.

Characteristics of Node Movements

The main challenge of QoS routing in mobile ad hoc networks is to handle the topology changes appropriately. The performance of a protocol is greatly determined by its ability to adapt to these changes. Hence, it would be useful to have metrics that characterise the effect of mobility on the connectivity of node pair. Mobility characteristics that affect the connectivity metrics need to be defined. These are the link lifetime, route lifetime, the number of link changes, node degree and route availability. Link lifetime is the duration of connectivity for a single hop node pair. The route lifetime is the duration of full connectivity from source to destination for a certain number of hops. Node degree is the number of neighbouring nodes within the transmission range of source node. Link availability is the fraction of time where there exists a connection between two neighbouring nodes. It would be necessary to consider a metric in terms of a single positive value as node-pair connectivity index, to be incorporated into the route selection algorithm. The metrics are all culminated from the effect of mobility with certain velocities. The velocities of nodes within MANET have the characteristics of spatial and temporal dependence. Spatial dependence is the extent of similarity of the velocities of two nodes that are within the transmission range. Temporal dependence is the extent of similarity of the velocities of a node at two time instances that are not too far apart. Relative velocity is the velocity difference between two nodes considered. Node pairs moving closer, and node pairs moving apart, experience higher relative velocity. Node pairs moving along side each other or moving in the same direction experience low relative velocity. The relative velocity is chosen to impact on our QoS routing algorithm as it is governed by the strength of connectivity between any set of node pairs within the route. The strength of connectivity between the node pair can be derived from the fact that node-pairs may be moving towards each other in a contraction model or they may be moving away from each other in an expansion model.

Contraction and Expansion Model of Node Movement

Ideally the contraction model emulates the movement of mobile nodes towards a logical centre from all directions. In practical scenarios, at $t = 0$, a node may enters another node transmission range at an angle and move in a direction not towards the other node. Contraction and expansion rates increase as velocity increases. This results in a node reaching the boundary of transmission range after a short time duration. Consequently, this affects the node connectivity time, route lifetime and the quality of node connectivity. Figure 2 shows the contraction model. If the movement pattern of the nodes is away from each other, than it is expansion model. Ideally, if at $t = 0$, a node is within a transmission range of the other node, then it moves away from the centre to the edges. In a practical scenario, the mobile nodes may move away from each other at an inclination angle. Incidentally, the starting point of the node

Figure 2. Contraction model of node connectivity

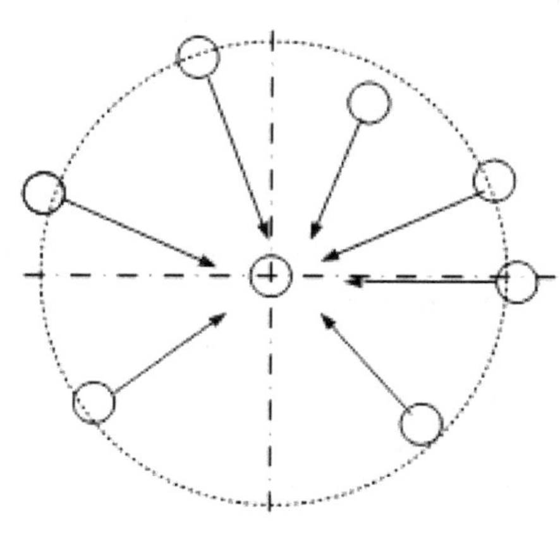

Figure 3. Expansion model of node connectivity

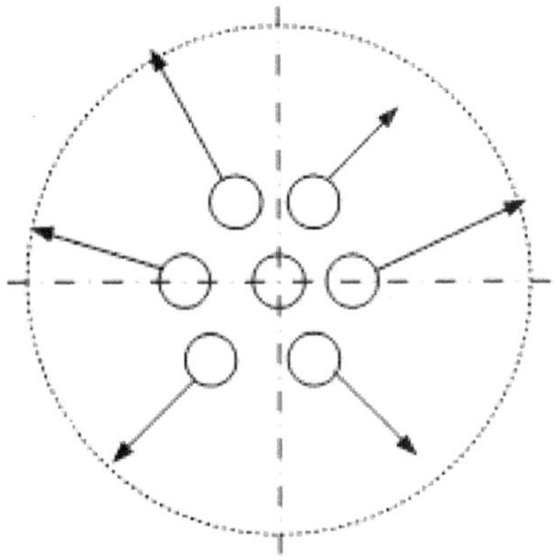

is within the transmission radius, in contrast to the contraction model. Figure 3 shows the expansion mobility model.

Hybrid Model Characteristics

In this mobility model, the movements of a node may switch from a contraction to an expansion model. These switchable characteristics are due to the relative distance between the node-pair and the time a node is within transmission range of its neighbours. Initially, the node is in the contraction model, having entered the transmission range at an angle of arrival, until it reaches a perpendicular point towards the other node where it enters the expansion model.

Finally, the node reaches the edge of the transmission range at a point of departure. Considering that the node moves in a straight line motion, at a point in time, the node will exhibit the expansion mobility characteristics. There exists a finite time between the node pair when they will always be in connectivity. The node connectivity time is an important QoS routing parameter for mobile nodes. The connectivity time depends on the angle of arrival and velocity. Figure 4 shows the hybrid model, illustrating that the node connectivity time is higher when it is at a particular position within the range for contraction, compared to if it were at a position within the range for expansion.

Analytical Consideration of Node Connectivity Time

Expected Node Connectivity Time

Consider two nodes *i* and *j*, at time t_1, when the duration of the node connectivity $L(i, j)$ is the time interval (t_1, t_2) during which the two nodes are within the transmission range of each other. How-

Figure 4. Hybrid of contraction and expansion model

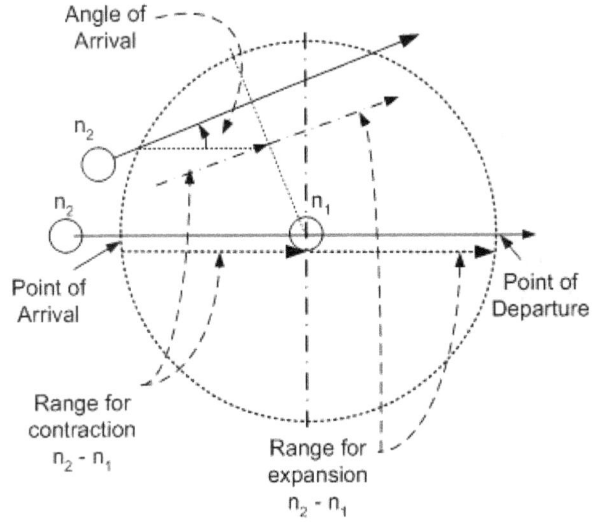

ever, these two nodes could not be within the transmission range at time $t_1 < t$ and also at time $t > t_2$. Formally, $L(i, j) = (t_1, t_2)$ if and only if, for all values of t, $t_1 \leq t \leq t_2$. An analytical model is presented to determine the mean connectivity time of a node-pair. The mean connectivity time of a node pair shows the estimated length of time a node pair is in contact with respect to the relative velocity of the nodes. Hence the lower and upper bound of node pair connectivity time which is useful in ensuring a complete communication session can be obtained.

The analytical estimate of the node pair connectivity time is obtained using the approach taken by Hong et. al., (D. Hong and S. S. Rappaport. (1986)), in the cellular network where the authors derived the expected time of a node residing within a cell. In this case, the connectivity time of a node pair, is estimated, which is the time that a node is always in contact wirelessly with the other node at a velocity and angle of arrival assumed to be uniformly distributed. The transmission range is set to 250m. Next, it is assumed that during the connectivity time, the mobile node moves along the specific direction, with uniform velocity. The velocity distribution of different nodes at different time intervals conforms to a velocity profile, with uniform distribution. Figure 5 shows a mobile node, n_2 moving in the vicinity of another node, n_1 which is assumed to be stationary. In reality, both nodes may be mobile. The mobile node arrives at the boundary of the circular range of the stationary node, at time of arrival, TOA, and it keeps on moving until the boundary is reached again, at departure time, TOD. The velocity of the mobile node is assumed to be uniformly distributed between 0 and V_{max} m/s. The direction of arrival of node n_2 is given by θ, the value of which is between $-\pi/2$ and $+\pi/2$. The mobility of the node is then characterised by the velocity PDF, $f_v(v)$ and directional PDF $f_\theta(\theta)$. In order to determine analytically the time range where n_2 resides within the range of n_1, the two PDF are defined. Velocity PDF is given as,

Figure 5. Estimating the expected node connectivity time

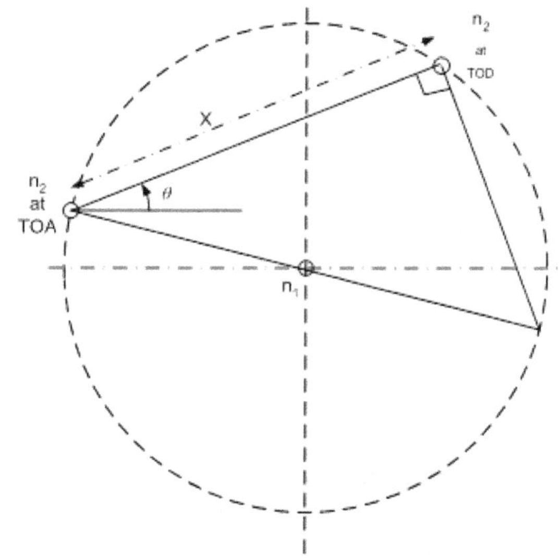

$$f_v\left(V\right) = \begin{cases} \dfrac{1}{V_{\max}} & for \quad 0 \leq V \leq V_{\max} \\ 0 & otherwise. \end{cases} \tag{10}$$

When the mobile node arrives at time TOA and may transverse a distance X in any direction with equal probability, the random variable θ, has PDF as,

$$f_\theta\left(\theta\right) = \begin{cases} \dfrac{1}{\pi} & for -\dfrac{\pi}{2} \leq \theta \leq \dfrac{\pi}{2} \\ 0 & elsewhere \end{cases} \tag{11}$$

Consider node n_2 from Figure 5, which traverses a distance X, thus,

$$X = 2R \cdot \cos\ \theta \tag{12}$$

For node n_2 traversing a distance x with time (where $0 \leq x \leq 2R$) then the density function of X,

$$F_X(x) = \Pr(X \leq x). \quad (13)$$

Resolving, for $F_x(x)$,

$$F_X(x) = \begin{cases} 0, & \text{for } x<0, \\ 1 - \frac{2}{\pi} \cdot \cos^{-1}\left(\frac{x}{2R}\right), & \text{for } 0 \leq x \leq 2R, \\ 1, & \text{for } x > 2R. \end{cases} \quad (14)$$

Then the PDF of X,

$$f_x(x) = \frac{d}{dx} \cdot F_X(x) \quad (15)$$

reduced to the form,

$$f_X(x) = \begin{cases} \frac{1}{\pi} \cdot \frac{1}{\sqrt{R^2 - \left(\frac{x}{2}\right)^2}} & \text{for } 0 \leq x \leq 2R \\ 0, & \text{elsewhere} \end{cases} \quad (16)$$

Node pair connectivity time is defined as the mean time; a node is within the transmission range of the other node and both are fully connected. The node connectivity time for the mobile node to travel from the point of arrival to the point of departure, a distance of X, with the velocity V, generally is given as,

$$T_{NCT} = \frac{X}{V} \quad (17)$$

The PDF of the node connectivity time can be found using a standard method.

$$f_{NCT}(t) = \begin{cases} \frac{4R}{\pi V_m} \cdot \frac{1}{t^2}\left[1 - \sqrt{1 - \left(\frac{V_m t}{2R}\right)^2}\right] & \text{for } 0 \leq t \leq \frac{2R}{V_m}; \\ \frac{4R}{\pi V_m} \cdot \frac{1}{t^2} & \text{for } t \geq \frac{2R}{V_m}; \end{cases} \quad (18)$$

The expected mean node connectivity time for the 2-node is given as,

$$\overline{T_{NCT}} = \int_0^\infty t.f_{NCT}(t)\, dt \quad (19)$$

By solving Eqn.(19), the mean node connectivity time, T_{NCT} is now a function of velocity. Figure 6 shows the plotting of T_{NCT} against velocity. From the plot the time, T_{NCT} for any given velocity can be explicitly estimated. The lower bound and the expected upper bound of the node connectivity can also be determined. From the graph, it can be infered that, node connectivity time depends on the velocity of the node. The estimate of T_{NCT} for any velocity, can then be obtained. There is also a limit to the velocity of each node in the system. For this QoS routing simulation, the appropriate velocity is chosen.

Node Connectivity Time Obtained Through Simulation

Figure 6 shows the plot of T_{NCT} against velocity. Each reading was taken for a specific angle of arrival. The time difference between the Time of Arrival (TOA) and Time of Departure (TOD) is taken as the node connectivity time in seconds. Different angles of arrival give a slightly different curve. Low velocity gives higher node connectivity time. The result from Opnet simulation is similar to the one obtained analytically, with a slightly higher connectivity time for a given velocity. Hence we

Figure 6. Plot of $\boldsymbol{T}_{NCT}$ against velocity from analytical model

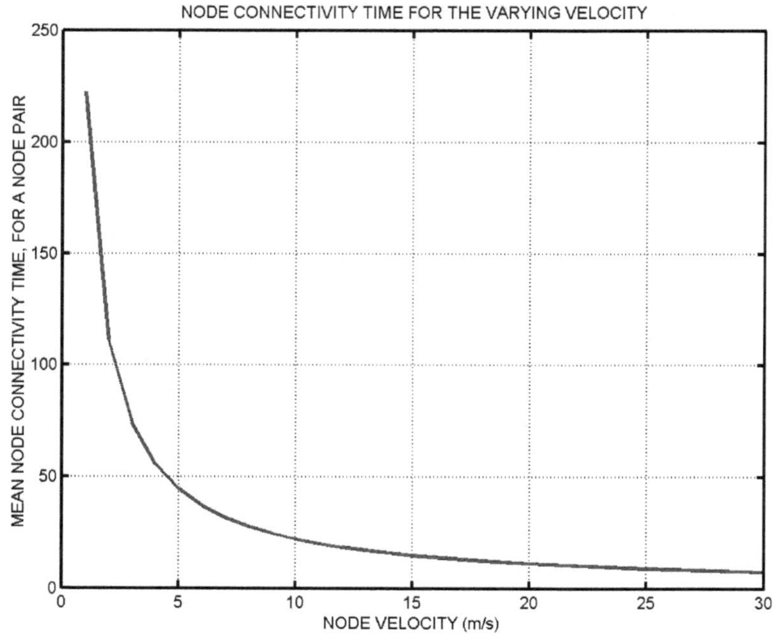

it can be concluded that Opnet simulation can be used to simulate the node mobility model and extract a suitable node connectivity time. The model can be used to estimate node connectivity time for a given velocity and the maximum time for the purpose of QoS routing implementation. Velocity of 25 m/s and 1 m/s resulted in the connectivity duration of about 10 seconds and 225 seconds respectively. Hence the velocity which gives a reliable connectivity duration would be between 0.5 m/s (3.6 km/h) and 25 m/s (90 km/h). The velocity of 0.5 m/s depicts a walking scenario whereas the velocity of 25 m/s depicts a car travelling on the highway. A metric which can describe the relative difference in connectivity time for different node pairs is needed. By comparing the values from different node pairs, those which have surplus connectivity time left can be selected. In the next section a node connectivity index is proposed.

Estimation of the Normalised Relative Velocity

It is not enough to view the network dynamics as having nodes connected or unconnected, but rather how long the connection will last. The longer the connection the better it contributes to the reliability and stability of the whole route. It is proposed to derive a metric which describes the dynamicity of the mobile nodes. In order to compute this metric, the connectivity between two adjacent nodes must be measured. Then the connectivity strength of every node pair is computed to produce a positive numerical index. The index can proportionately describe how reliable a route is. Those low speed nodes which have longer duration and are more reliable are represented by a low value index. Stationary nodes produce a zero index. The rate of separation, expanding and contracting measures, between each adjacent pairs of nodes can be estimated by computing the relative velocity. Consider node n_1 receiving packets from node n_2

Figure 7. Node connectivity time against velocity

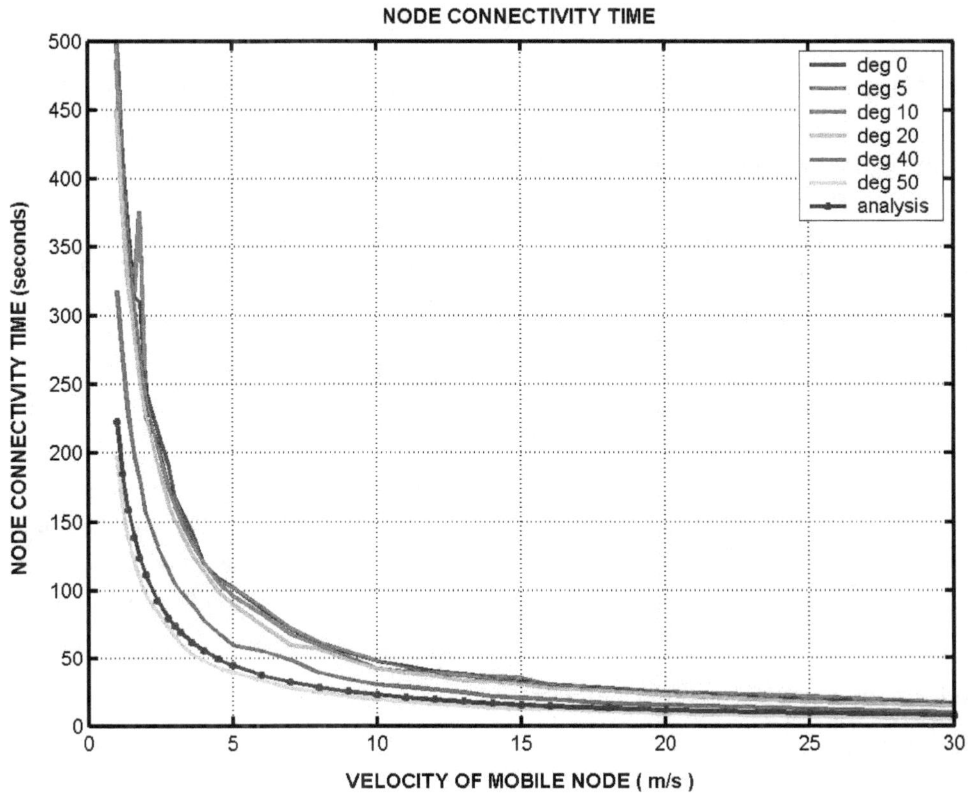

in Figure 7. Within the transmission range, node n_2 will always be on the circular disk centred at node n_1. Node n_2 is said to be within the virtual concentric region. The virtual

concentric region is a circular region within the node's (i.e n_1) transmission range, where at any instant its adjacent node (i.e n_2) will always be before reaching point of departure. Let t_1 and t_2 be the times at which the last two packets from n_2 were received. The received power from node n_2 when packets have arrived is denoted as P_1 (at time t_1) and P_2 (at time t_2). From this there exist two possibilities: that nodes are moving closer when $P_2 > P_1$ and nodes moving further apart when $P_2 < P_1$.

The free space path loss propagation model is descridbed in (D. Hong and S. S. Rappaport. (1986)), hence power received by the receiver antenna due to transmission from the transmitter is given by $P_r = P_t G_t G_r \lambda^2 / (4\pi)^2 d^2 L$ where, P_r is the received power; P_t is the transmitted power; G_t is the transmitter antenna Gain; G_r is the receiver antenna gain; d is the distance between transmitter and receiver; L is the system loss factor and λ is the wavelength in metres. It is also assummed that all the parameters above, except distance, are the same for all nodes and the received power is inversely proportional to the square of distance, d. Hence, the power is deduced as $P_r = k / d^2$, where $k = P_t G_t G_r \lambda^2 / (4\pi)^2 L$. In order to simplify further, nodes are assumed to move in a linear motion and posses unlimited battery power. Additionally, the maximum transmission range is assumed to be 250 m.

Contraction and Expansion Mobility Model

Expansion Model

Referring to the Figure 8, consider nodes moving apart at a constant velocity and in linear motion. Two extreme scenarios may occur, that is, n_2 moves from position 2 to position 4 or from position 2 to position 5. Consider, n_2 moves from 2 to 4; distance = d_3 - d_1, then,

$$\left(d_3 - d_1\right)/\sqrt{k} = (1/\sqrt{P_2}) - (1/\sqrt{P_1}) \quad (20)$$

$$v / \sqrt{k} = (1 / (t_2 - t_1))((1/\sqrt{P_2}) - (1/\sqrt{P_1})) \quad (21)$$

Next, consider, n_2 moves from position 2 to position 5; distance = $\sqrt{\left(d_3^2 - d_1^2\right)}$ then,

$$\sqrt{(d_3^2 - d_1^2)}/\sqrt{k} = \sqrt{(1/P_2) - (1/P_1)} \quad (22)$$

$$v/\sqrt{k} = (1/(t_2 - t_1))\sqrt{(1/P_2) - (1/P_1)} \quad (23)$$

Eqn. (23) gives a worst-case scenario, hence this equation is being considered in the calculation of *nci*.

Contraction Model

Figure 8 shows nodes moving closer with a constant speed in linear motion. Similarly, two scenarios can be used to find the relationship that will deduce the speed of the node. Consider, n_2 moves from position 2 to position 3, distance = d_1 - d_2, then,

Figure 8. The graphical illustration of nci calculation

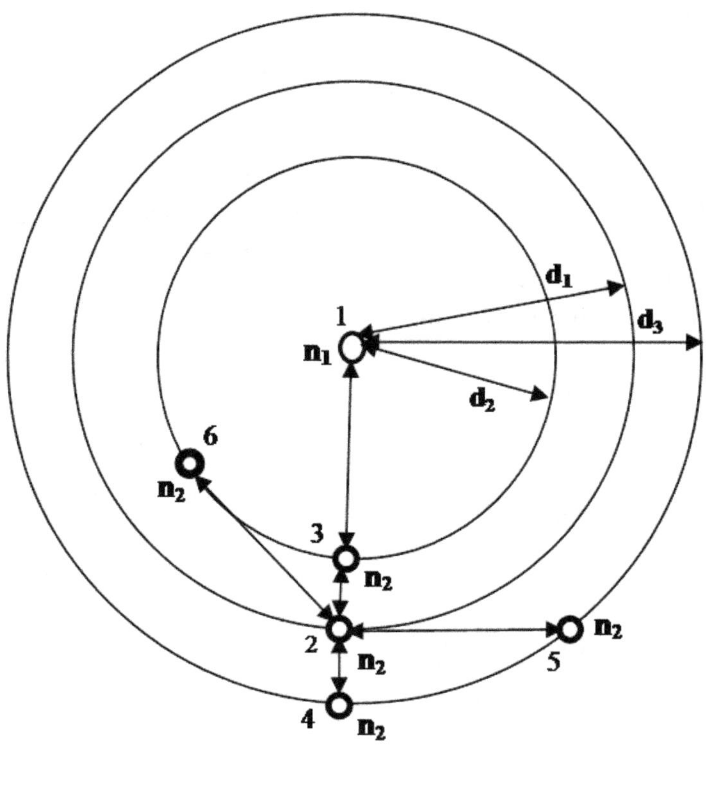

$$(d_1 - d_2) / \sqrt{k} = (1 / (\sqrt{P_1}) - 1 / (\sqrt{P_2})) \tag{24}$$

and hence,

$$v/\sqrt{k} = 1 / (t_2 - t_1)\left((1 / \sqrt{P_1}) - (1 / \sqrt{P_2})\right) \tag{25}$$

Next, consider n_2 moves from position 2 to position 6, distance $= \sqrt{d_1^2 - d_2^2}$, then

$$v/\sqrt{k} = 1 / (t_2 - t_1)((1 / \sqrt{P_1}) - (1 / \sqrt{P_2})) \tag{26}$$

The equation Eqn.(26) gives a worst-case scenario, hence it is also being considered in the calculation of *nci*. From these equations we have the estimates of the node-pair relative velocity normalised by the constant due to antenna and propagation factors. The relative velocity can be used to estimate the aggregate values for all node-pairs that are within the source-destination route.

Definition of npem, npcm, and nci

Definition 1: Consider a route R which consists of a set of nodes, which are in motion with a velocity, ***v*** and with varying angle of arrival, then R= {n_0, n_1, n_2,... n_{n-1} }. For R to be a reliable route, each node must be connected to its adjacent node for the whole period of message transmission. Let E denote the node pairs, (n_j, n_k) such that they are adjacent nodes. If the pairs are moving away from each other, the node pair expansion metric (*npem*) is given by,

$$npem = (1 / (t_2 - t_1))\sqrt{((1 / P_1) - (1 / P_2))} \tag{27}$$

Definition 2: Consider a route R which consists of a set of nodes, which are in motion with a velocity, v and with varying angle of arrival, R= {n_0, n_1, n_2,...n_{n-1}}. For R to be a reliable route, each node must be connected to its adjacent node for the whole period of message transmission. Let E denote the node pairs, (n_j, n_k) such that they are adjacent nodes. If the pairs are moving toward each other, the node pair contraction metric (*npcm*) is given by,

$$npcm = (1 / (t_2 - t_1))((1 / \sqrt{P_2}) - (1 / \sqrt{P_1})) \tag{28}$$

Observe that *npcm* and *npem* are positive quantities but in different directions, *npcm* towards high positive values and *npem* towards low positive values. These two values must be combined to form a single metric to indicate the quality of connectivity between the two adjacent mobile nodes. We note that the node with *npcm* lasts longer than that with *npem*. Hence we need a weighted form of these two values in a single index quantity. This value will indicate a strong connectivity if its value is low and weak connectivity if its value is high. A high value shows that the node is more dynamic and will soon reach the point of departure from the other node's transmission range.

Definition 3: Consider a set E which consists of the node pairs, (n_j, n_k) such that they are adjacent nodes. If the pairs are moving towards each other or away from each other, the node pair connectivity index, *nci* is a positive value which describes the quality of connectedness between any two adjacent nodes. The least *nci* value indicates a good quality connection, in which the node pair connectivity time is larger compared to high *nci* value, then *nci* is defined as,

$$nci = \begin{cases} 0.25 - \left[\dfrac{1.0 \times 10^5}{8.0 \times 10^5 - npem} \right], & for\ P_2 < P_1. \\ \dfrac{1.0 \times 10^5}{8.0 \times 10^5 + npcm}, & for\ P_2 > P_1. \\ 0, & for\ P_2 = P_1. \end{cases} \tag{29}$$

The algorithm for calculation of *npem, npcm* and *nci* is part of the whole algorithm for monitoring of MANET. The algorithm for calculation of *npem, npcm* and *nci* is label as Box 1.

Performance Evaluation of *nci*

Various simulation experiments are carried out to show the viability of *nci* as a potential comparative measure to compare the connectedness among all the potential node pairs connectivity.

First Scenario

Consider a scenario which consist of two nodes, one stationary and the other is mobile, as shown in Figure 9. Initially the mobile node is positioned at an angle of arrival to the stationary node. Each node produces a periodic node connectivity packet, (for example, CONN packet), which is a one-hop transmission packet, design to identify nodes that are within transmission range. This node and their parameters are recorded in a form of node state and saved into node state cache. Readings of *nci* were taken for various speeds and various angles of arrival. The results were tabulated for *nci* with speeds of 1 m/s, 2 m/s, 3 m/s, 4 m/s, 5 m/s, 6 m/s, 10 m/s and 15 m/s and graphically presented as shown in Figure 10 through to Figure 18 respectively. For each graph values of *nci* for each angle of arrival of 0, 30, 45, 60, 70 and 80 degrees were drawn. For each speed, the time of

Box 1. Calculating the npcm, npem, nci

$S \leftarrow \{i, H_i, B_{AVA}, D_{ij}, nci, t_{CUR}, t_{PRV}, P_{CUR}, P_{PRV}\};$

If ($P_{PRV} > P_{CUR}$) then

$$npem = \left(\frac{1}{t_{CUR} - t_{PRV}} \right) \times \left(\frac{1}{P_{PRV}} - \frac{1}{P_{CUR}} \right);$$

$$nci = 0.25 - \left(\frac{1.0 \times 10^5}{8.0 \times 10^5 - npem} \right)$$

elseif ($P_{CUR} > P_{PRV}$) then

$$npcm = \frac{1}{t_2 - t_1} \left(\frac{1}{\sqrt{P_{CUR}}} - \frac{1}{\sqrt{P_{PRV}}} \right);$$

$$nci = \frac{1.0 \times 10^5}{8.0 \times 10^5 + npcm};$$

endif;

$S \leftarrow \{i, H_i, B_{AVA}, D_{ij}, nci, t_{CUR}, t_{PRV}, P_{CUR}, P_{PRV}\}$

departure is less for small angles of arrival. The *nci* on the other hand showed that as the time of departure is decreasing, the value of *nci* is increasing. For the purpose of data transmission, it would be convenient to choose a node for which *nci* is the lowest among the entire neighbouring nodes. As can be seen, as *nci* increases, the expected connectivity time is reduced.

The results for the *nci* values for a number of velocities that were tabulated in Figure 10 through to Figure 17 are now redrawn to show the relationship between *nci* and different velocities. Figure 18, shows the redrawn graph. It illustrates the effect of velocity on *nci*. It can be concluded that the greater the velocity, the higher the *nci*. If there are two node-pairs with different relative velocities, then the one which has the least *nci* value will be chosen at that instant as its performance is the best in terms of how long the node-pair connection lasts. Hence this confirms that *nci* could be a unique comparative index, to select which node pairs that provides longer node connectivity.

Second Scenario to Demonstrate the Effect of nci

The second scenario consists of one mobile node, n1 and two stationary nodes, n2 and n3 as shown in Figure 19(a). The mobile node is set to move in a straight line in the direction where it enters the radio transmission range of n2 and n3 at time 0 second and 280 seconds respectively. At time 0 second, n1 and n2 are connected since n1 is just within the transmission range of n2. At that instant *nci* is about 0.01. From 0 second to 280 seconds n1 is always in connection with n2 such that the value of *nci* (n1<->n2) varies from ~0.01 to ~0.23. If the selection process occurs during this period, no change of path will take place. Upon reaching the 280th second, n1 enters n3 connectivity range. The value *nci* (n1<->n3) is ~0.01. At this stage n1 has two choices of connection. In the event there is a selection process, *nci* (n1<->n3) is much less than *nci* (n1<->n2), hence the node-pair connection changes to n1<->n3. The graph in Figure 19(b) shows the reading of *nci* for node n1. The *nci* values for node n1, are the n1 node states for neighbours n2 (n1<->n2) and n3 (n1<->n3). From the graph in Figure 19(b), at time ~280 seconds,

Figure 9. Scenario to determine nci

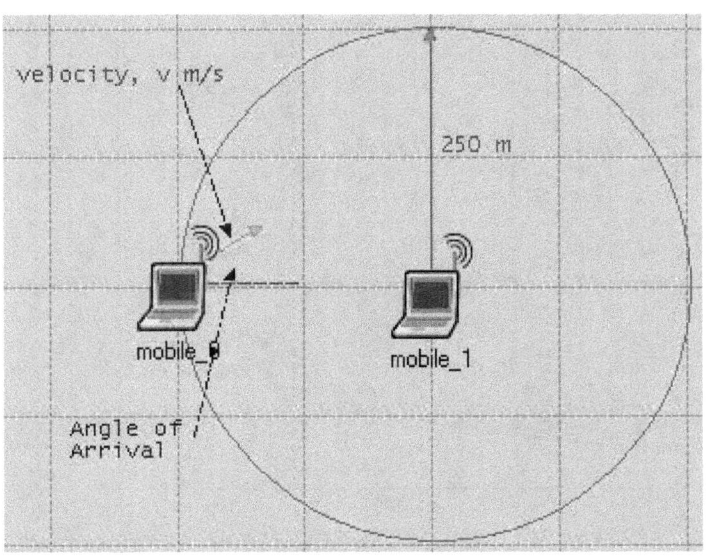

Figure 10. nci for velocity 1 m/s

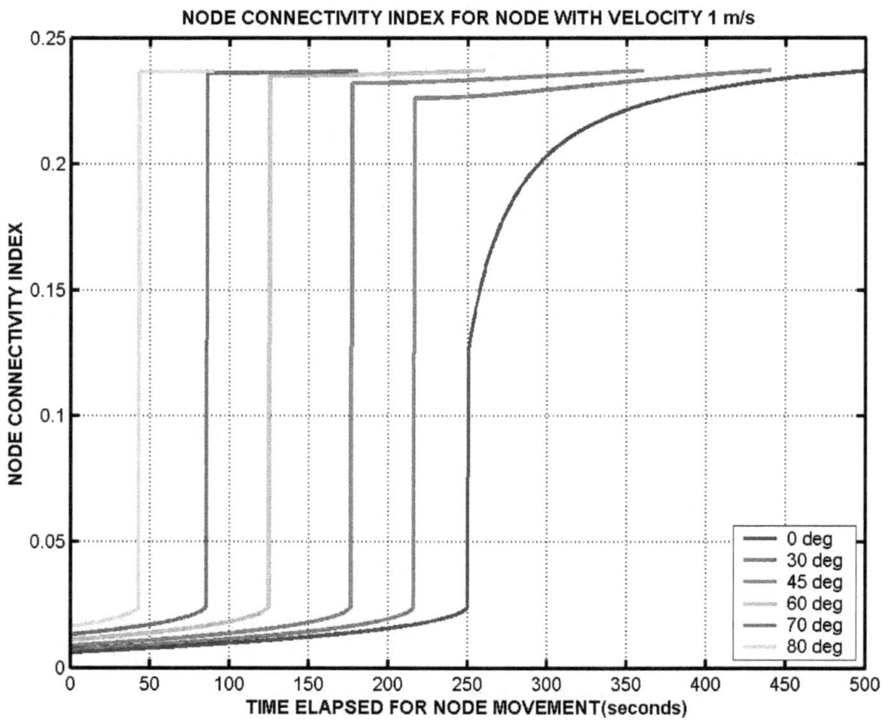

Figure 11. nci for velocity 2 m/s

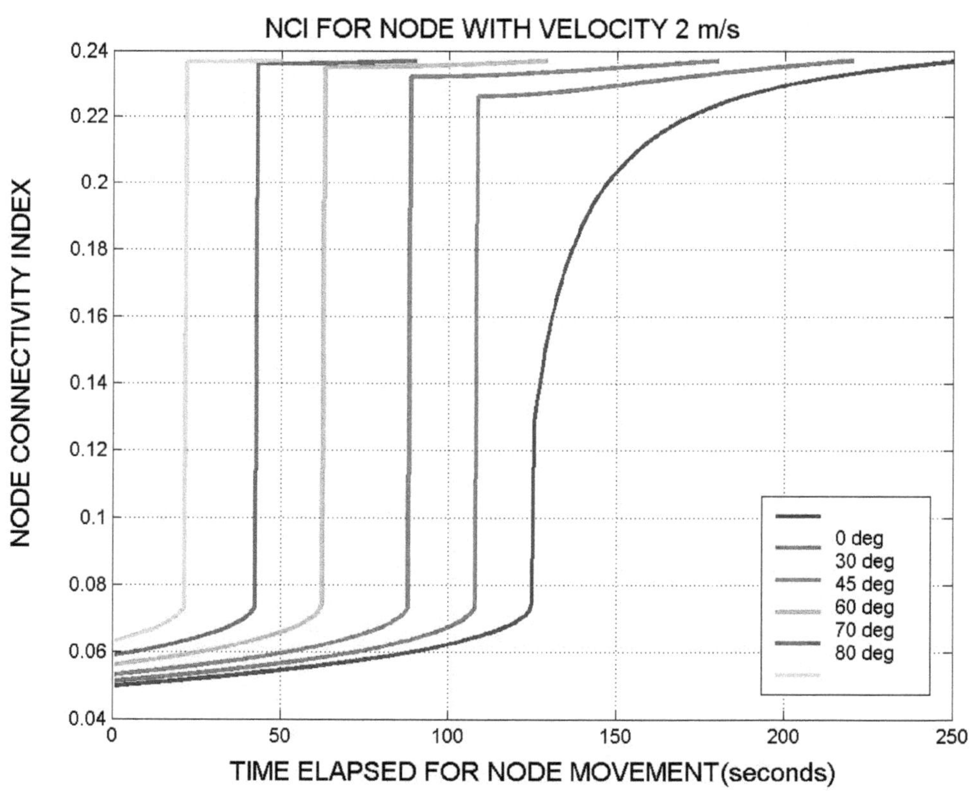

Figure 12. nci for velocity 3 m/s

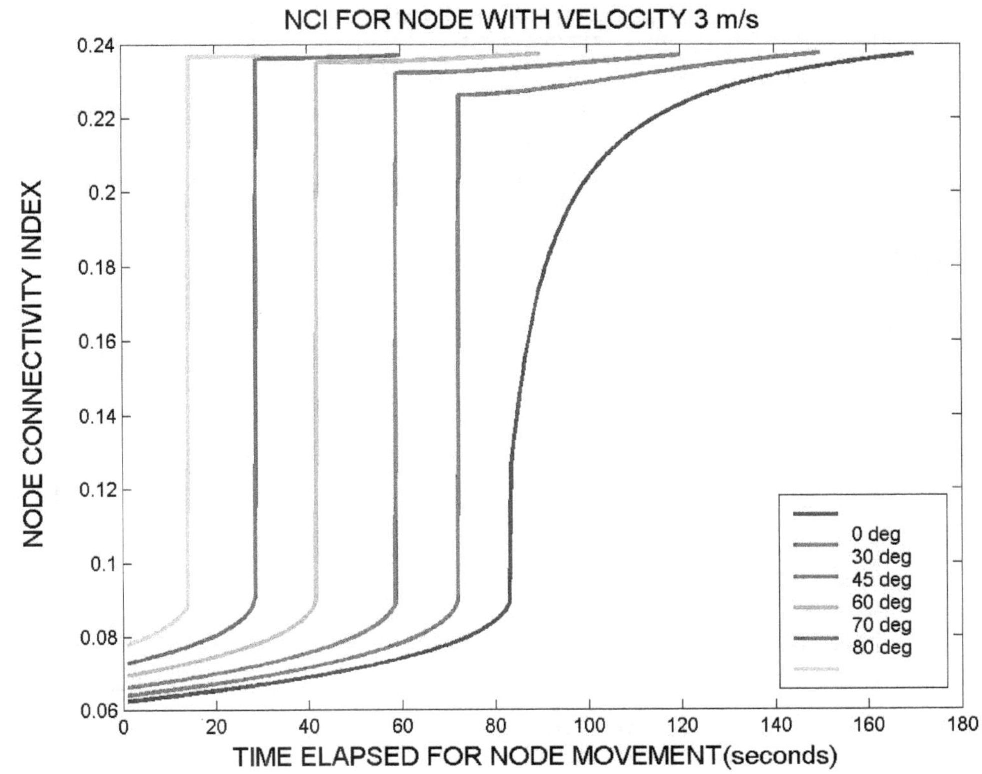

Figure 13. nci for velocity 4 m/s

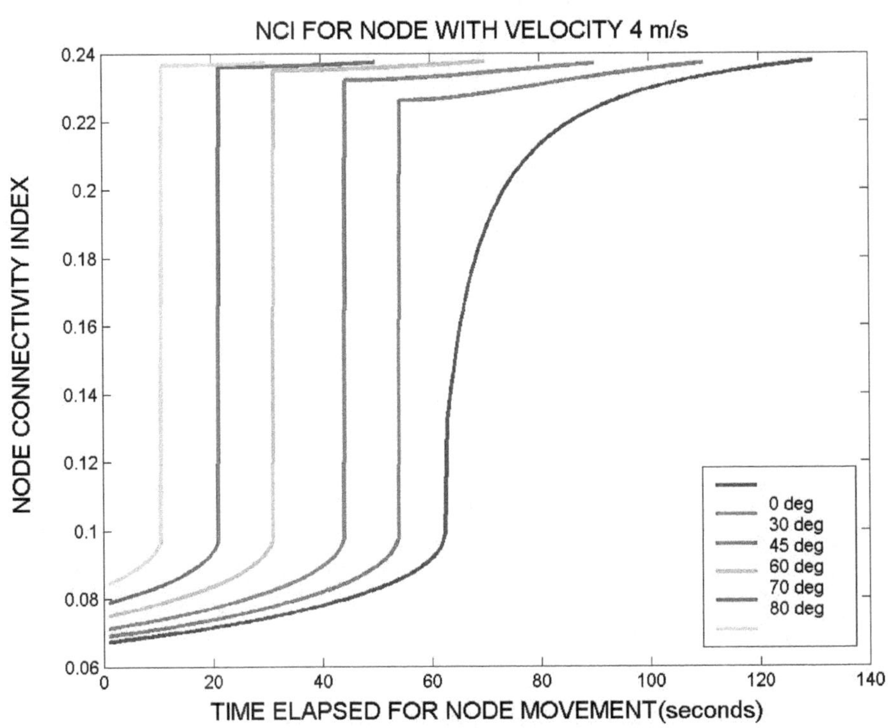

Figure 14. nci for velocity 5 m/s

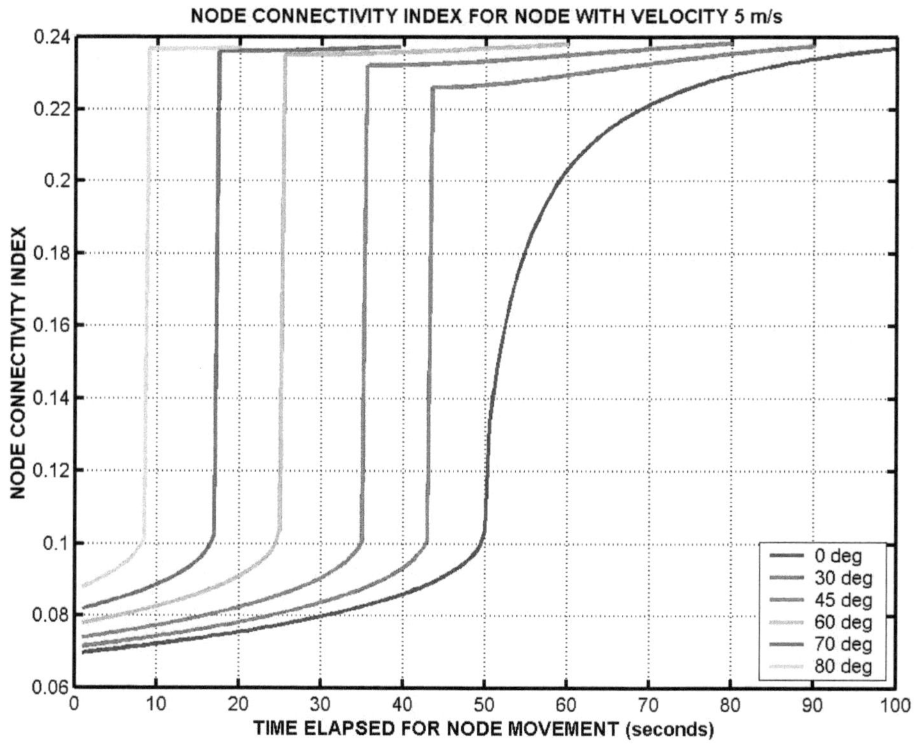

Figure 15. nci for velocity 6 m/s

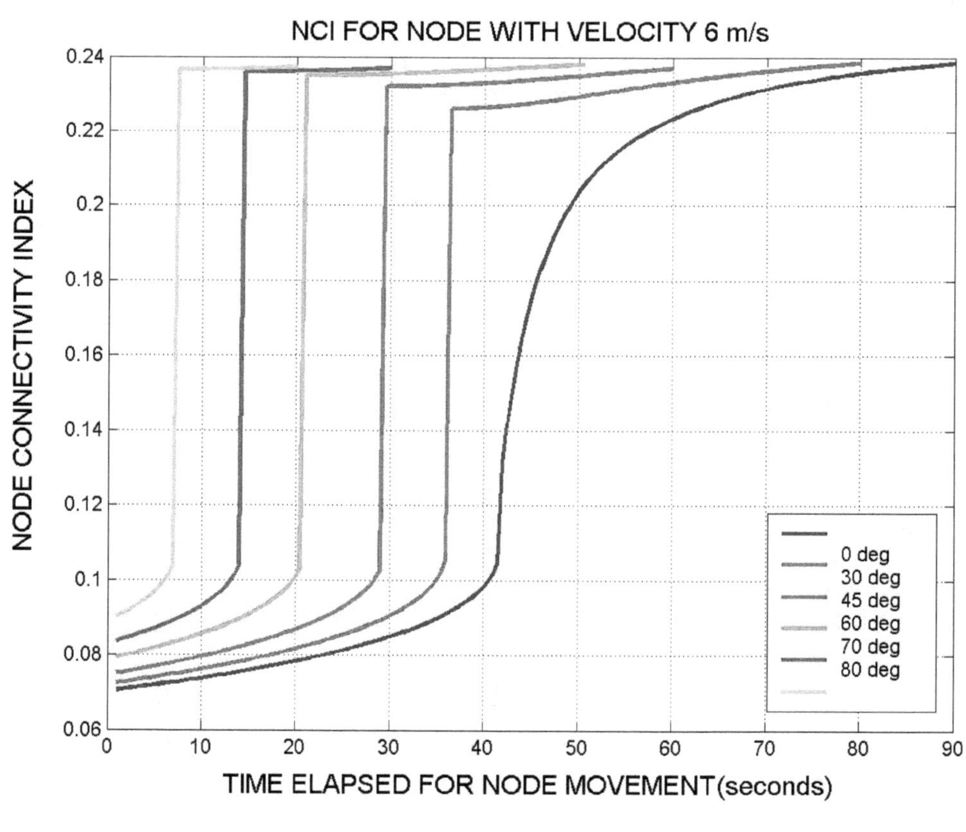

Figure 16. nci for velocity 10 m/s

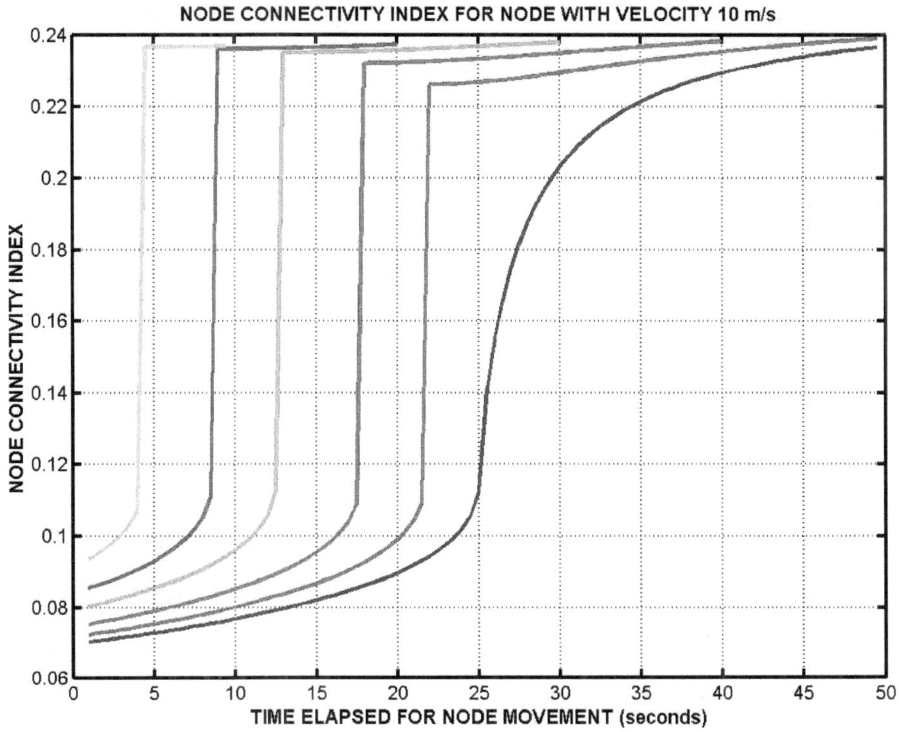

Figure 17. nci for velocity 15 m/s

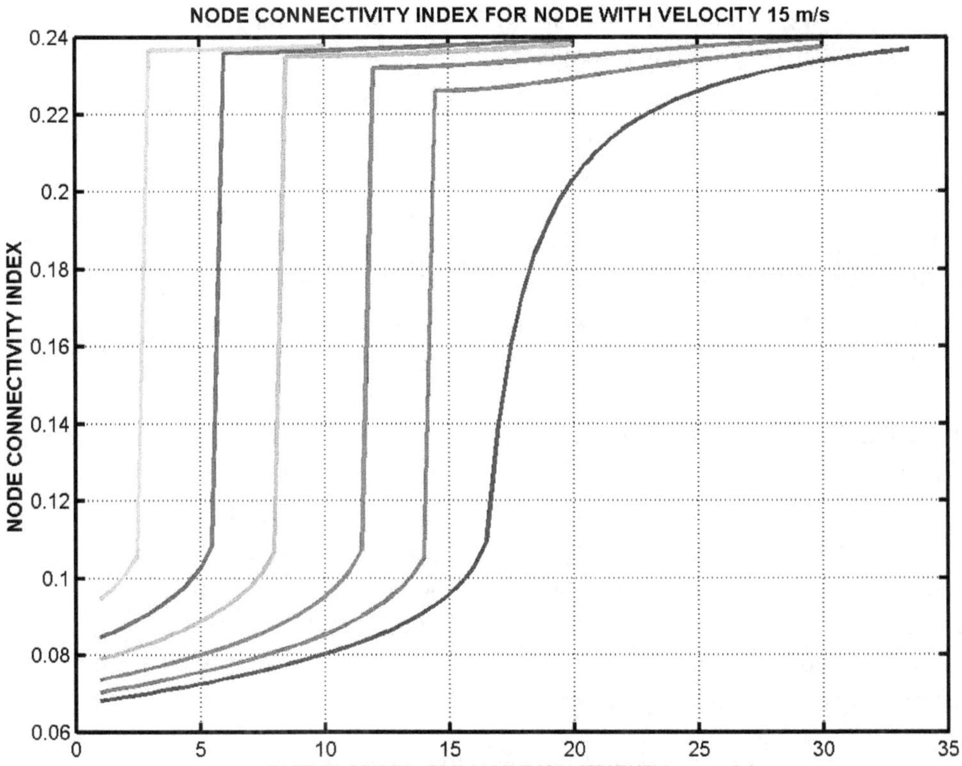

Figure 18. nci for various node velocities

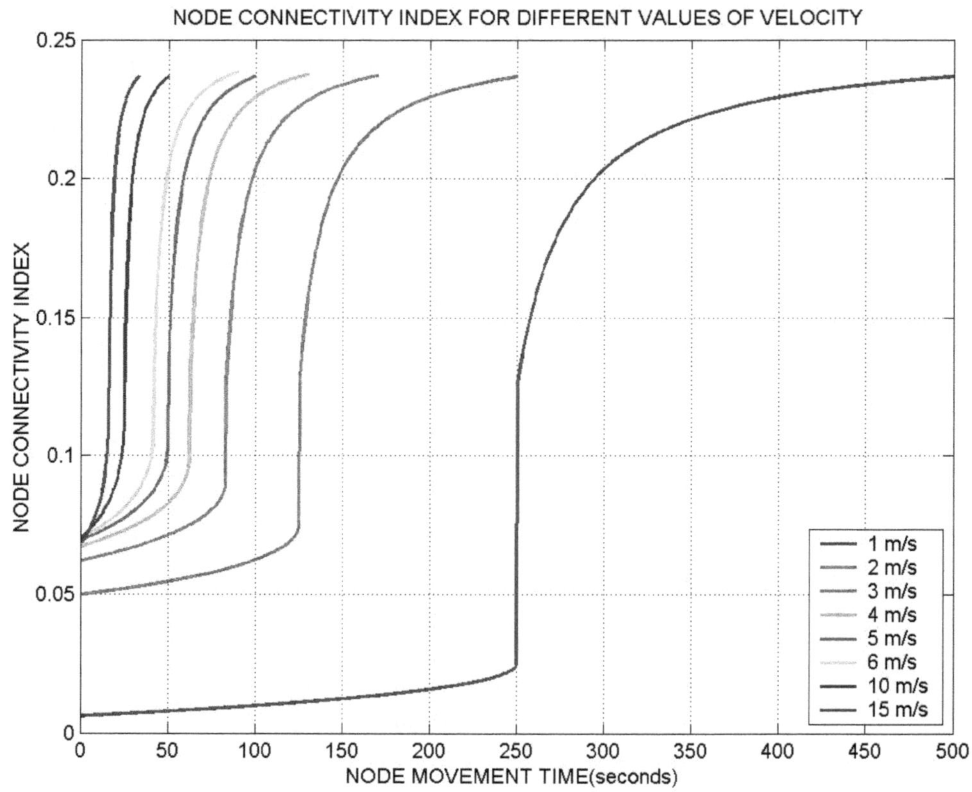

the *nci* (n1<->n2) is ~0.23 and *nci* (n1<->n3) is ~0.01. If the line of *nci* (n1<->n2) is extrapolated from that point until the end, the time left is ~180 seconds. On the other hand, if the line of *nci* (n1<->n3) is extrapolated from that point until the end, the time left is ~ 460 seconds. Hence it can be deduced that the element of node state represented by *nci,* does infer connectivity between node-pair. Next, any node-pair connectivity may or may not be part of the potential routes.

Third Scenario to Demonstrate the Effect of nci

The third scenario consists of three nodes, one mobile node and two stationary nodes, with the setup as shown in Figure 20(a). The mobile node, n1 initiates its movement in a straight line with a speed of 1 mps. It immediately enters the transmission range of n2 and then a little while later into the range of n3. The graph in Figure 20(b) shows the readings of *nci* against the time elapsed for mobile node n1. From the graph *nci* (n1<->n2) is less than the *nci* (n1<->n3) from the time it enters the range of n3 until the time of departure at 200 seconds. Node n1 connected to n3 traversed a shorter distance compared to n1 connected to n2. Connectivity time for n1 to n2 is much longer, hence when *nci* for both connections are compared, the one which has longer connectivity in the expansion region offers the least *nci*. In the event, the selection process occurs at any instance between 0 second and 200 seconds, the connection of (n1<->n2) would be chosen. It is noted that the value of *nci* (n1<->n3) just before n1 is within the transmission range of n3, is not zero but that it simply did not exist, since both are

Figure 19. Two stationary nodes and one mobile node

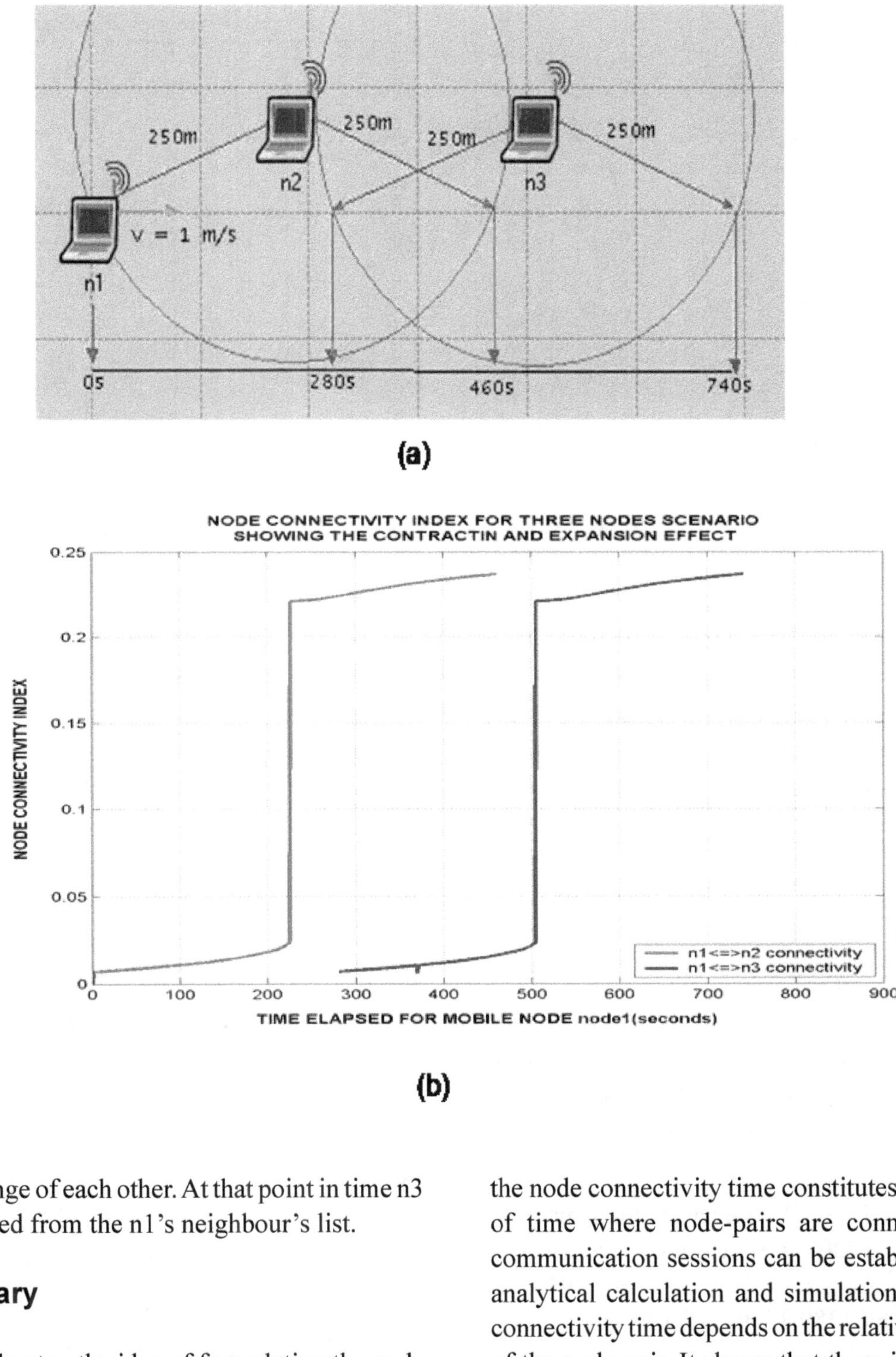

out of range of each other. At that point in time n3 is removed from the n1's neighbour's list.

Summary

In this Chapter, the idea of formulating the node connectivity into a form suitable for the purpose of monitoring the dynamic of any given node pair was proposed. It had been able to show that the node connectivity time constitutes the length of time where node-pairs are connected and communication sessions can be established. By analytical calculation and simulation, the node connectivity time depends on the relative velocity of the node-pair. It shows that there is an upper limit and a lower limit for the connection time. Between these limits communication sessions can be carried out. In order to monitor the qual-

Figure 20. The effect of two stationary nodes in vicinity of one mobile node

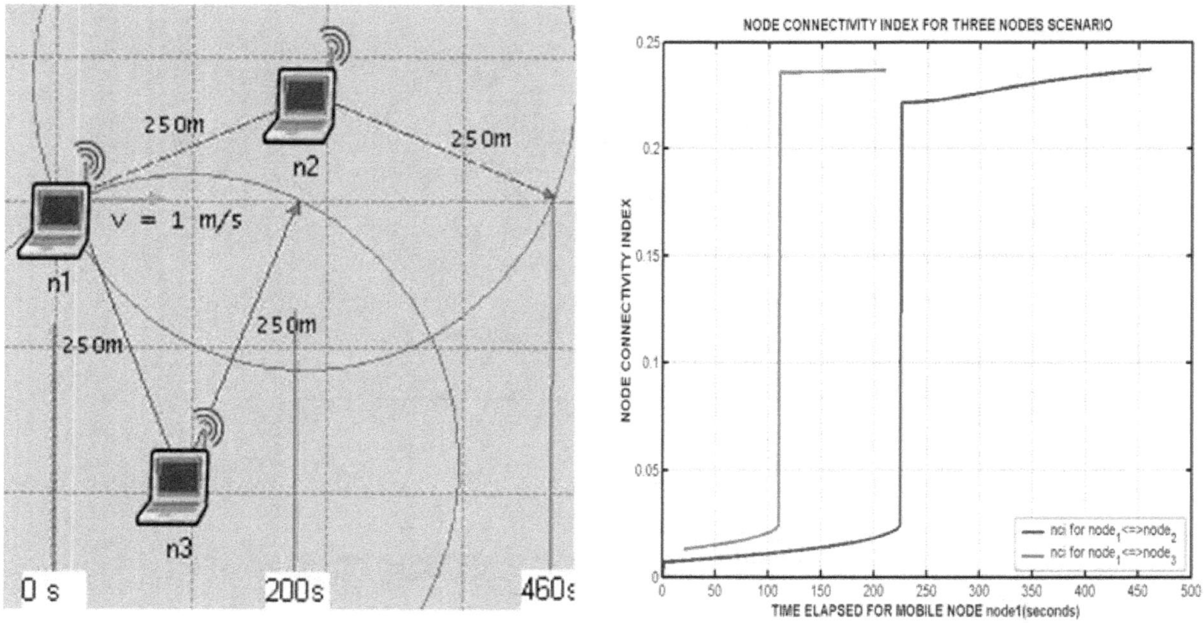

(a) **(b)**

ity of connection between a node-pair and for the whole route which is made up of a collection of node-pairs, *npem*, *npcm* and *nci* are introduced. The *nci* is shown to be a viable way to estimate the quality of node-pair connection in terms of its timeliness.

REFERENCES

Agarwal, A., Ahuja, A., Singh, J., & Shorey, R. (2000). *Route-lifetime assessment based routing (RABR) protocol for mobile ad-hoc networks* (pp. 1697–1701). IEEE ICC.

Akyildiz, A. F., Su, W., Sankarasubramaniam, Y., & Cayirci, E. (2002). Wireless sensor networks: A survey. *Computer Networks*, *38*, 393–422. doi:10.1016/S1389-1286(01)00302-4

Bettstetter, C. (2002). On the minimum node degree and connectivity of a wireless multi-hop network. *Proceedings ACM MobiHoc, 2002*, 80–91.

Chlamtac, I., Conti, M., & Liu, J. J. N. (2003). Mobile ad hoc networking: imperatives and challenges. *Ad Hoc Networks*, *1*, 13–64. doi:10.1016/S1570-8705(03)00013-1

Desai, M., & Manjunath, D. (2002). On the connectivity in finite ad hoc networks. *IEEE Communications Letters*, *6*, 437–439. doi:10.1109/LCOMM.2002.804241

Dousse, O., Thiran, P., & Hasler, M. (2002). Connectivity in ad-hoc and hybrid networks. *Proceedings - IEEE INFOCOM*, (March): 2002.

Dube, R., Rais, C. D., Wang, K. Y., & Tripathi, S. K. (1997). *Signal stability adaptive routing (SSA) for ad hoc mobile networks.* IEEE PC, 1997.

Foh, C. H., & Lee, B. S. (2004). A closed form network connectivity formula for one-dimensional MANETs. *Proceedings IEEE ICC*, June 2004.

Goff, T., Abu-Ghazaleh, N. B., Phatak, D. S., & Kahvecioglu, R. (2001). *Preemptive routing in ad hoc networks* (pp. 43–52). ACM/IEEE MobiCom.

Hong, D., & Rappaport, S. S. (1986). Traffic model and performance analysis for cellular mobile radio telephone systems with prioritized and nonprioritized handoff procedures. *IEEE Transactions in OVT*, *35*, 77–92.

Jiang, X. (2004). Siren: Context-aware computing for firefighting. [Springer Verlag.]. *IEEE Pervasive Computing / IEEE Computer Society [and] IEEE Communications Society*, *2004*, 87–105.

Lochert, C., et al. (2003). A routing strategy for vehicular ad hoc networks in city environments. *Proceedings IEEE Intelligent Vehicular Symposium*, June 2003, (pp. 156-161).

Perkins, C. E., & Royer, E. M. (1999). Ad-hoc on-demand distance vector routing. *Proceedings of IEEE Mobile Computer Systems and Applications,* (pp. 90-100).

Rappaport, T. (1996). *Wireless communications: Principles and practice*. Englewood Cliffs, NJ: Prentice Hall.

Santi, P. (2005). The critical transmitting range for connectivity in mobile ad hoc networks. *IEEE Transactions on Mobile Computing*, *4*, 310–317. doi:10.1109/TMC.2005.45

Santi, P., & Blough, D. M. (2002). An evaluation of connectivity in mobile wireless ad hoc networks. *Proceedings IEEE DSN*, June 2002.

Su, W., Lee, S. J., & Gerla, M. (2000). Mobility prediction and routing in ad hoc wireless networks. *International Journal of Network Management*, *11*(1).

Toh, C. K. (1997). Associativity based routing for ad hoc mobile networks. *International Journal on Wireless Personal Communciations*, *4*, 103–139. doi:10.1023/A:1008812928561

Wong, K. J., et al. (2003). Busnet: model and usage of regular traffic patterns in mobile ad hoc networks for inter-vehicular communications. *Proceedings of International Conference on Information and Communication Technologies (ICICT)*, 2003.

Chapter 6
Overview of Temporally Ordered Routing Algorithm and QOS Components in MANETS

Jahangir Khan
Sarhad University of Science and Information Technology Peshawar, Pakistan

Abou Bakar Nauman
Sarhad University of Science and Information Technology Peshawar, Pakistan

ABSTRACT

TORA is a highly adaptive routing protocol that provides loop free routing of data packets in ad hoc network at every instant. Each node of network has the capability to contain information about its local topology changes and regular coordination with their neighbour nodes in order to detect partitions occurs in network. In Ad-Hoc networks, multicast communication is not ideal due to the frequent nodes mobility and failures. Such problems in transmission of data become reasons for increased delays and decreased throughput, and also sudden changes in delay, traffic, and link breakage occur due to nodes mobility. The network parameters like network size, number of nodes, number of hops per route, traffic patterns (bits/packets), link capacity, mobility trajectory, and frequency of sleeping nodes do have great results on the performance of routing protocols in the mobile ad hoc networks. TORA has three basic route functions: establishment, maintenance, and erasing. In TORA routing protocol, each node has capable to contain a structure of network describing node's height and status of all connected links. The importance of these heights in TORA routing protocol are that a node in network may only forward data packets downstream instead of upstream. TORA are more appropriate for QOS in ad hoc networks with a higher mobility and data transfer rate.

DOI: 10.4018/978-1-4666-0321-9.ch006

OVERVIEW OF TORA

Temporally Ordered Routing Algorithm (TORA) is a source initiated on-demand routing protocol presented by Park and Corson in 1997 for wireless mobile ad hoc network. TORA is an efficient, highly adaptive, and scalable routing protocol based on link reversal algorithm. TORA provide multiple routes to transmit data packet between source and destination nodes of mobile ad hoc network. TORA is an adaptive routing protocol for multi-hop networks that contains different attributes such as; distributed execution, loop free and multi-path routing. This routing protocol is designed to reduce communication overhead by adapting local topological changes in ad hoc network. Another smart function of TORA routing protocol is the localization of control packets to a small region (set of nodes) near the occurrence of a topological changes due to route break. Therefore each node of network needs to contain its local routing and topology information about adjacent (one hop) nodes. [Anna, 2003]

TORA has three basic functions

1. Route Establishment
2. Route Maintenance
3. Route Erasing

1. ROUTE ESTABLISHMENT

TORA routing protocol performed route establishment function when a source node in network wants a route to transmit data packets to destination but does not contain information about directed link. Route establishment between source and destination nodes are carried out by two types of packet, i.e.

- QUERY (QRY)
- Update (UPD)

For route discovery TORA establishes a destination oriented directed acyclic graph (DAG) by assigning directions to links in ad hoc network rooted at destination. Thus TORA assigns directional heights to links in network so as to transmit data packets from a higher source node to a lower destination node.

TORA establish multiple routes quickly to destination. To establish a route, source node broadcast a QRY packet containing destination address through network. The QRY packet propagates through mobile nodes of ad hoc network until it reaches either destination or an intermediate node which has a rout to destination. Each node in network has a height with respect to the destination that is computed by TORA routing protocol. These heights are assigned to links based on direction towards destination. The destination or intermediate node that terminates QRY packet broadcasts UPD packet containing its distance/height with respect to destination (the height is zero at destination node). Each node that receives UPD packet sets its height to a value greater than height of neighbour node from which it received UPD packet. This series of directed links creates DAG in network rooted to destination.

In TORA routing protocol, each node has capable to contain a structure of network describing node's height and status of all connected links. The importance of these heights in TORA routing protocol are that a node in network may only forward data packets downstream instead of upstream. Each node height metric is represented by a quintuple (τ, oid, r, δ, i) which includes the following values.

- τ : Logical Time of a link failure
- oid : Unique ID of the node that defined the reference level
- r : Reflection indicator bit
- δ : Propagation ordering parameter
- i : Unique ID of the node

Figure 1. Directed acyclic graph rooted at destination (Larsson, 1998)

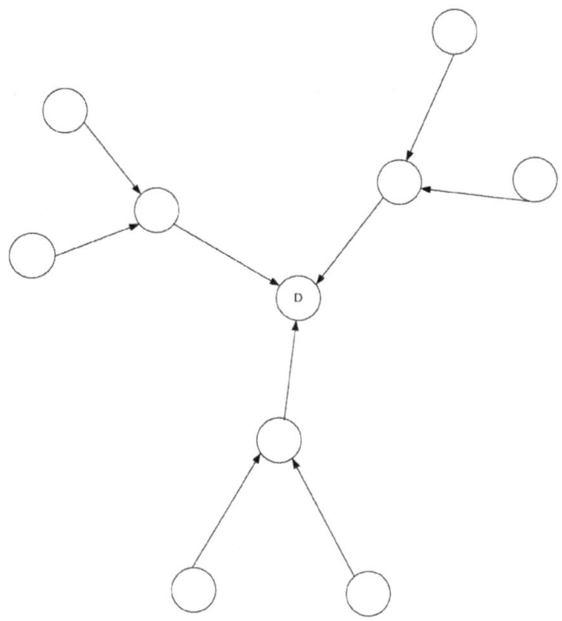

Consider following Figure 2 where source node A wants to establish a route to destination node G to send data packets. A QRY packet is propagated by source node A through network. The QRY packet is forwarded by intermediate nodes B, C, D, E, and F to reach destination node G, or any other node which has route information to destination.

As soon as destination node or intermediate node which contains route to destination sends UPD packet back to source node. In following Figure 3 node G is destination node which sends UPD message back to source node through their neighbour's node. Thus source node receives UPD packet from different links and create four different routes from source A to destination G as shown in the following figure. The source node selects shortest route from available four routes and start forwarding data packets.

2. ROUTE MAINTENANCE

When the DAG route is not available any more for communication due to link failure or node movement, TORA routing protocol needs route maintenance to re-establish a DAG route to destination. When a link is broken between nodes the height of the nodes is set to a height greater than that of any neighbouring nodes. The height of the node at broken link is set greater for purpose that data packet will now transmit backward and find alternative path to destination. [Anna, 2003]

Consider the following Figure 4 where link between node E and destination node G is broken.

The broken link node E reverses the path between itself and neighbour nodes and sends UPD packet to source node telling about broken link. If source node contains information about other alternative routes to destination, it selects one of them and forward data packet otherwise it initiates a new QRY/UPD procedure to discover a new route.

Figure 2. QRY packet route establishments in TORA

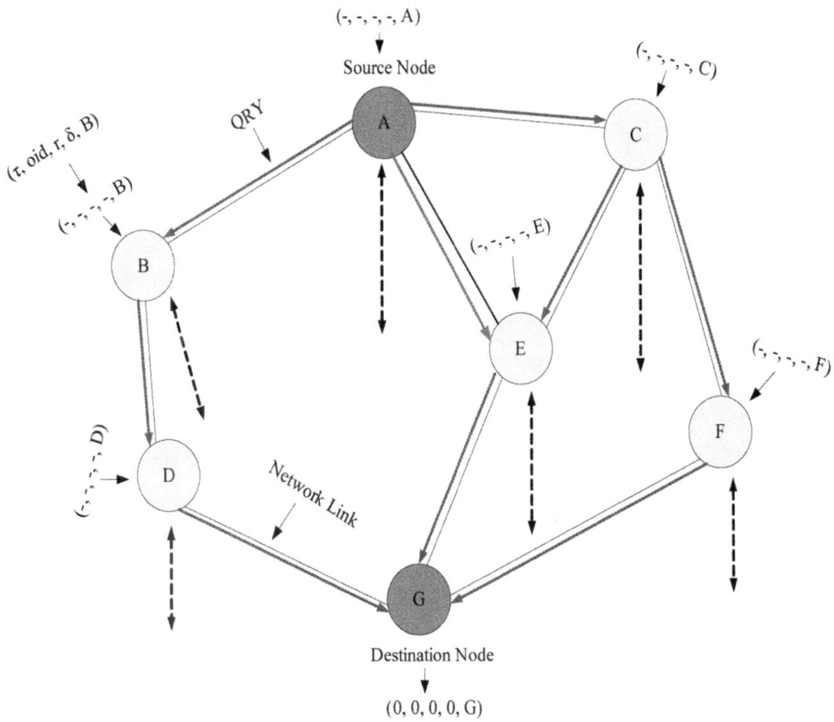

Figure 3. UPD packet route establishments in TORA

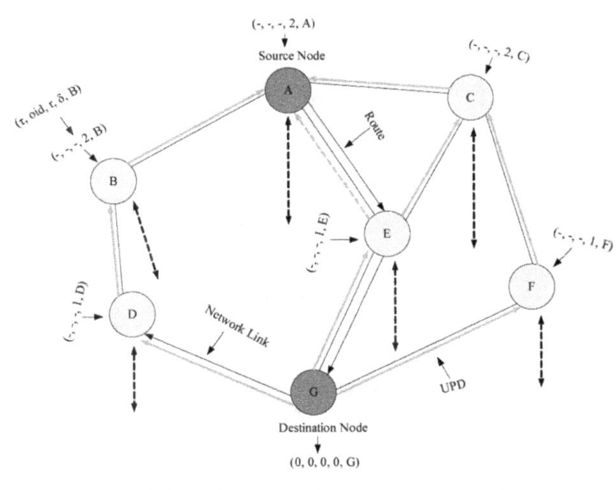

Figure 4. Route maintenance in TORA

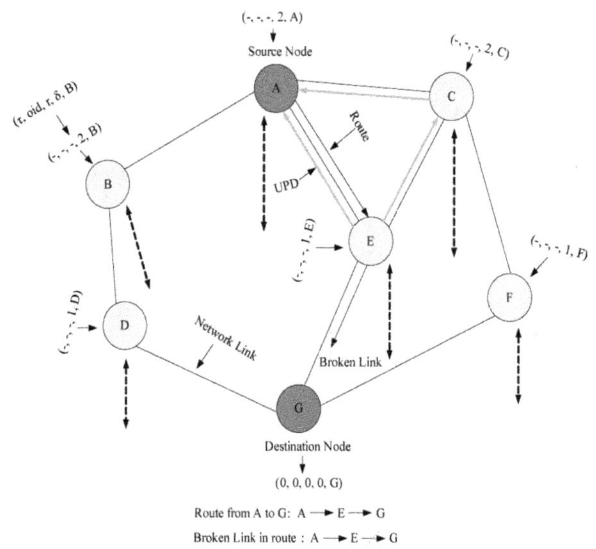

Figure 5. Maintaining routes decision tree (Park, 2001)

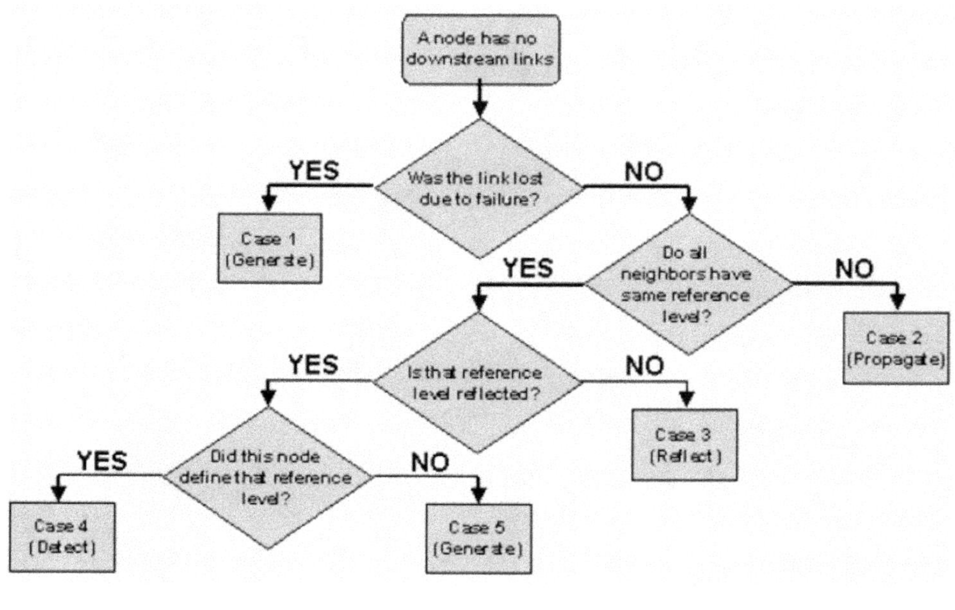

The maintenance of broken route in TORA routing protocol is carried out by the following way as shown in the Figure 5.

TORA routing protocol maintaining routes only for those nodes whose height is not null. During route discovery TORA does not use intermediate node whose height is null. In TORA a node i in network is said to have no downstream links if height of node i is less than height of non-null neighbour node k.

Height[i] < Height Neighbour[k]

One in five different cases are possible depends on node status and event occurrence as shown in the above Figure 5. [Park, 2001]

Case 1: Generate

In TORA routing case1, i.e. Generate occurs for mobile node i when there is no downstream links from node i due to link failure. i.e.

$(\tau[i], oid[i], r[i]) = (\tau, i, 0)$

$(\delta[i], i) = (0, i)$

If node i contains any upstream neighbour it defines a new reference level otherwise node i sets its height to null.

Case 2: Propagate

This case occur when node i in the network has no downstream link due to link reversal following receipt of an UPD packet and all neighbour nodes k have not the same reference level, i.e.

$(\tau[i], oid[i], r[i]) = \max\{ (\tau[k], oid[k], r[k]) \}$

$(\delta[i], i) = (\delta [m-1], i)$, m is the lowest neighbour with max reference level

Thus in propagate case node i propagates its highest neighbour reference level and select a lower height in all neighbour related to the propagated reference level.

Case 3: Reflect

The reflect case in TORA maintaining routes decision tree has occurred when node i has no downstream link in network due to link reversal following receipt of an UPD packet and all neighbours node k have same reference level but not reflected, i.e.

$(\tau[i], oid[i], r[i]) = (\tau[k], oid[k], 1)$

$(\delta[i], i) = (0, i)$

Thus the neighbour nodes propagated the same non reflected reference level to node i. When node i receive these it reflects back with a higher sub level.

Case 4: Detect

During TORA route maintenance detect case has occurred when node i has no downstream link due to link reversal following the receipt of an UPD packet and all the neighbours node k have the same reflected reference level defined by node i, i.e.

$oid[k] = i \rightarrow$ node i defined the level

$(\tau[i], oid[i], r[i]) = (-, -, -)$

$(\delta[i], i) = (-, i)$

TORA routing protocol detect partition in the network when the last reference level defined by node i has been reflected and propagated back in the neighbours node of the network.

Case 5: Generate

The last case Generate in TORA route maintenance decision tree has occurred when node i has no downstream link due to link reversal following the receipt of an UPD packet and all the neighbours node k have the same reflected reference level but not defined by node i, i.e.

oid[k] != i → node I did not define the reference level

(τ[i], oid[i], r[i]) = (τ, i, 0)

(δ[i], i) = (0, i)

Thus in such situation node i defines a new reference level for route discovery and flooded again in network.

3. ROUTE ERASING

In route erasing phase of TORA routing when a node in the ad hoc network has detected a partition, it assigns a null value to its own height and heights of its entire neighbour's node for the destination in its table. The node also floods a broadcast clear packet (CLR) through the network to erase invalid routes to destination and updates all entries in its link state table. The CLR packet consists of reflected reference level, destination ID, and reflected reference bit which is always 1 (CSC, TORA).

CLR(τ, oid, 1)

The intermediate nodes that receive the CLR packet set the heights in the same way and re-broadcast the CLR packet to next neighbours node. By this process the height of each node in the partitioned portion of the network is set to null and all invalid routes are erased from link table. (refer to Figure 6)

Figure 6. TORA route erasing

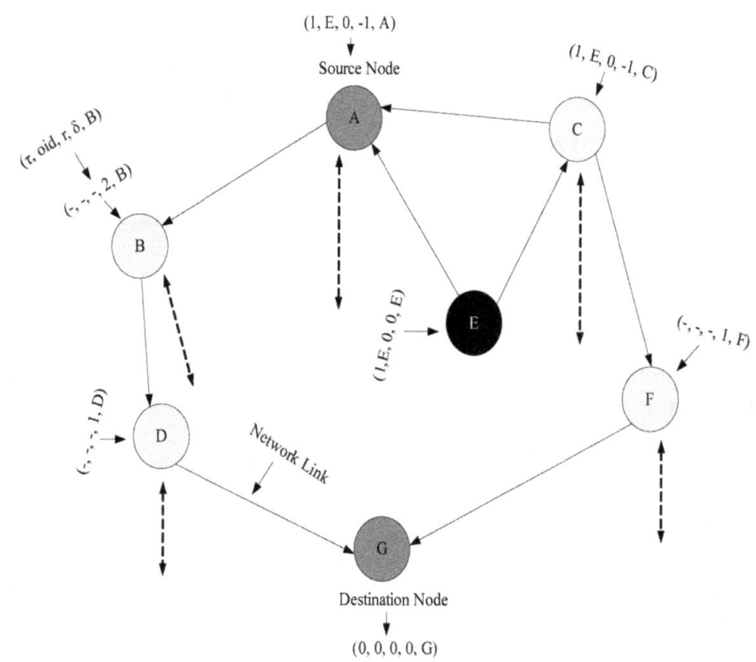

PROPERTIES OF TORA

TORA routing protocol is designed for dynamic environment of ad hoc network where topologies are changing rapidly. To discover a route to destination TORA maintains state information sufficient to form a DAG and does not contains extra information to calculate shortest path to destination. It is distributed routing protocol and establishes multiple routes quickly. The maintenance of Directed acyclic graph in TORA enables a loop free communication. TORA protocol minimizes total amount of control and data traffic by efficient use of bandwidth. These characteristics of TORA provide greater reliability in ad hoc network along with possible QoS extension support.

QUALITY OF SERVICE IN MANETS

The widely accepted definition of QoS is defined by the consultative committee for international telephony and telegraph (CCIT) recommendation E.800 as "the collective effect of service performance which determines satisfaction degree of a service user".

The maturity of wireless mobile technologies and the evolution of different applications provide a reason for the introduction of QoS in wireless ad hoc networks.

The goal of QoS routing in MANET is to select routes with sufficient resources for data packets with QoS requirements to increase possibility that network will be capable of supporting and maintaining them. The following Figure 7 shows the QoS components.

The QoS routing in mobile ad hoc network is measured with respect to the following attribute.

- Available bandwidth
- Control message overhead
- Data throughput
- Probability of packet loss
- Table storage overhead
- Delay (setup connection)
- Delay (end-to-end)
- Service coverage area
- Low power operation

Thus from above discussion it is clear that achieving QoS in mobile ad hoc network corresponds to a real need and is difficult as compared with traditional wired networks. QoS is essential element in routing which informs source node about successful availability of destination node. QoS guarantees in mobile ad hoc networks are difficult because of the following elements;

- Dynamic network topology
- Imprecise state information

Figure 7. QoS components in routing (Agha, 2003)

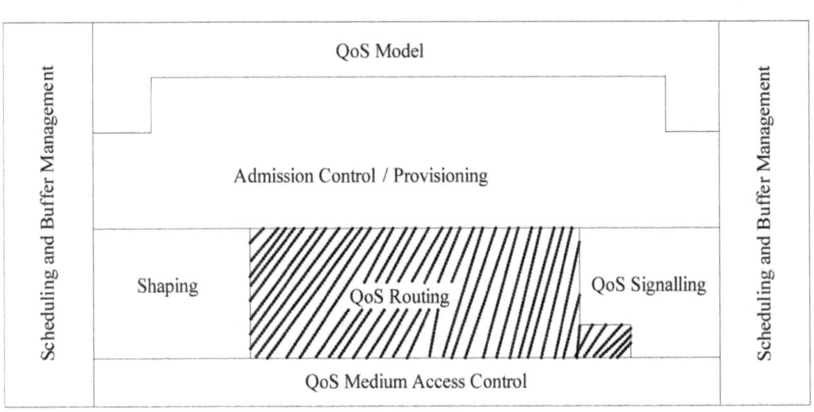

- Lack of central administration
- No pre-existing base station / infrastructure less network
- Hidden terminal problem
- Limited resources availability
- Energy constrained operation
- Multicasting
- Variable capacity link
- Weak physical security

To simulate different routing protocols in mobile ad hoc network models some nodes of the above object palette are used. The commonly used nodes in MANET network models are;

- **Wireless LAN Workstations and Servers**

To generate application traffic such FTP, E-mail, and HTTP over TCP over IP over wireless LAN in a MANET network model these node models of object palette are used. For each MANET routing protocol these nodes can be configured to run in a specific way.

- **MANET Stations**

To generate raw packet over IP over wireless LAN these node models of MANET object palette are used. For each MANET routing protocol these nodes can be configured to function as traffic sources or destination.

- **Wireless LAN routers and MANET gateway**

These node models play very important role in MANET network models by performing function as access point in a mobile ad hoc network. By enabling MANET gateway, these nodes of object palette can also connect mobile nodes of ad hoc network to IP based networks.

MANET NODE MODEL STRUCTURE OF OPNET

Routing protocols AODV, DSR, and TORA are implemented at IP layer in MANET model structure as shown in the following figure. Models of AODV, DSR, TORA, and OLSR are available in OPNET version 12. The following figure 8 explains node model architecture of TORA in MANET.

ip_dispatch is root process for IP in MANET node and creates a child process manet_mgr. The manet_mgr function as manager process and

Figure 8. MANET node model structure (OPNET model documentation)

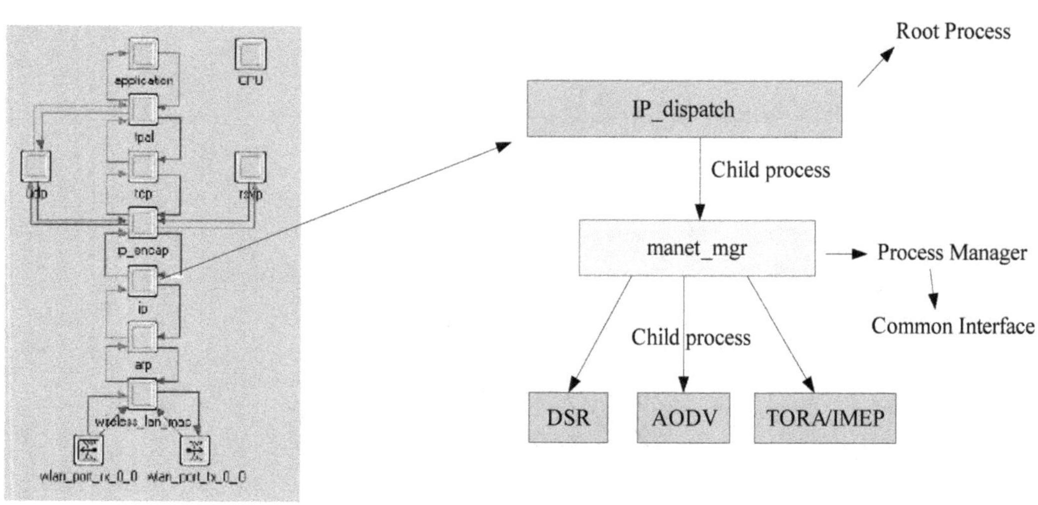

provides a common interface to multiple mobile ad hoc routing protocols in OPNET. The manet_mgr is further child process for required MANET protocol as configured in parameter.

This OPNET ad hoc routing model does not support multiple routing protocols on a single node and redistribution between MANET routing protocols. The model also limits router node interfaces to be running same MANET routing protocol.

CONCLUSION

MANET cannot be trusted for auto configuration in case of break down. Therefore support of distributed operation is required by routing protocols to solve such type of problems. Due to lack of physical security, MANET protocols are highly exposed to different types of attacks. In MANET it is very easy for attacker to disturb network traffic, corrupt packet header, change addresses of routing messages, and increase traffic to waste bandwidth. To avoid all these threats a sufficient security protection is highly desirable in all MANET routing protocols. As MANET nodes are battery powered and the energy used in both sending and receiving of packets. The routing protocol should be smart enough to adjust sleep period by itself without user command. As discussed, that nodes of MANET are generally limited in resources such as bandwidth, energy, storage capacity, security, and CPU capacity. Therefore the routing protocol should design to eliminate or minimize control traffic and other periodic messages. TORA can contribute to reduce the constrained issues.

REFERENCES

Agha, A. K., & Omidyar, G. C. (2003). *Mobile and wireless communications networks* (2nd ed.). Singapore: World Scientific.

Anna, H. (2003). *Mobile telecommunication protocol for data networks* (1st ed.). John Wiley & Sons Ltd.

Bani Yassein, M., Ould Khaoua, M., Mackenzie, L. M., Papanastasiou, S., & Jamal, A. (2006). *Improving route discovery in on-demand routing protocols using local topology information in MANETs. PM2HW2N'06, October 2, 2006, Torremolinos*. Malaga, Spain: ACM.

Feeney, M. L. (2005). *Introduction to MANET routing*. Swedish Institute of Computer Science. Retrieved from http://www.nada.kth.se/kurser/kth/2D1490/ 05/lectures/feeney_mobile_adhoc_routing.pdf

Forouzan, A. B. (2005). *TCP/IP protocol suite* (3rd ed.). McGraw Hill.

Halgamuge, K. S., & Wang, P. L. (2005). *Classification and clustering for knowledge discovery* (1st ed.). Springer.

He, C. (2003). *Throughput and delay in wireless ad hoc networks. Final report of EE359 Class project*. Stanford University.

Khan, J., & Hayder, S. I. (2010). *A comprehensive performance comparison of on-demand routing protocols in mobile ad-hoc network*. FGCN2010 Jeju Island South Korea 13-15 December 2010.

Khan, J., Shah Khattak, N., & Jan, H. (2011). A step of mobile ad-hoc on-demand routing protocols towards 4G cellular networks. *International Journal of Computers and Applications*, *6*(4). Retrieved from http://www.ijcaonline.org/archives/volume16/number6/2017-2721

Kim, T. H. (Eds.). (2010). *FGCN 2010, Part II, CCIS 120* (pp. 354–369). Heidelberg, Germany: Springer.

Lang, D. (2006). *On the evaluation and classification of routing protocols for mobile ad hoc networks*. Retrieved from http://home.leo.org/~dl/dissertation.pdf

Lee, S. (2000). *Routing and multicasting strategies in wireless mobile ad hoc networks*. Retrieved from http://www.sigmobile.org/ phd/2000/theses/ sjlee.pdf

Mukija, A. (2001). *Reactive routing protocols for mobile ad hoc network.* Retrieved from http:// www.ifi.unizh.ch/ ~mukhija/papers/rrp_thesis.pdf

OPNET. (2007). *Modeler wireless suite for defence*. Retrieved from http://www.opnet.com/ products/modeler/home.html

Park, V., & Corson, S. (2001). *IETF MANET working group.* Retrieved from http://tools.ietf.org/id/ draft-ietf-manet-tora-spec-04.txt

Ram, S. C., & Manog, S. B. (2004). *Ad hoc wireless networks: Architecture and protocol* (1st ed.). Prentice Hall.

Tavel, P. (2007). *Modeling and simulation design*. Natick, MA: AK Peter Ltd.

Yih-Chun, H., & Johnson, D. B. (2002). *Ensuring cache freshness in on-demand ad hoc network routing protocol*. POMC'02, October 30–31, 2002, Toulouse, France. ACM.

Section 2
MANET Protocols

Chapter 7
Topology-based Classification of Multicast Routing Protocols for Mobile Ad hoc Networks

Natarajan Meghanathan
Jackson State University, USA

ABSTRACT

This chapter presents an exhaustive survey on the different categories of topology-based multicast routing protocols that have been proposed for mobile ad hoc networks (MANETs). Due to the inherent broadcast nature of the wireless medium, one-to-many multicast communications are characteristic of most of the MANET applications. MANET multicast routing protocols are primarily classified as tree-based and mesh-based protocols depending on the underlying topology used for communication. Tree-based protocols are further classified to source-tree based and shared-tree based depending on whether the tree is rooted at the multicast source or a common node. Further, there are several sub-categories of source-tree based and shared-tree based multicast routing protocols. The mesh-based multicast protocols are primarily categorized to source-initiated and receiver-initiated protocols depending on which entity (the source node or the receiver nodes) initiates the mesh formation. The book chapter explains in detail the working of eleven different multicast routing protocols, covering all of the sub-categories of tree-based and mesh-based routing protocols.

DOI: 10.4018/978-1-4666-0321-9.ch007

INTRODUCTION

A mobile ad hoc network (MANET) comprises of wireless nodes that move independent of each other forming a dynamically changing distributed resource-constrained system. The wireless nodes are often limited in their transmission range and multi-hop routing is a common phenomenon that has been of significant interest. There are several application domains in which MANETs have been deployed. These include: disaster recovery, military operations in a battlefield, outdoor entertainment activities, crowd control, conferences, and etc. One-to-many multicast communication is a characteristic feature of all these applications. Several multicast routing protocols have been proposed in the literature to support these MANET applications.

Depending on the underlying topology used for communication, the multicast protocols are mainly classified as: tree-based and mesh-based protocols. In tree-based protocols, only one route exists between a source and a destination and hence these protocols are efficient in terms of the number of link transmissions. There are two major categories of tree-based protocols: source tree-based (the tree is rooted at the source) and shared tree-based (the tree is rooted at a core node and all communication from the source nodes to the receiver nodes is routed through this core node). Even though shared tree-based multicast protocols are more scalable with respect to the number of sources, these protocols suffer under a single point of failure, the core node. On the other hand, source tree-based protocols are more efficient in terms of traffic distribution. The source tree-based protocols are further classified into the following sub-categories: (i) Minimum hop-based, (ii) Minimum link-based, (iii) Stability-based, and (iv) Zone-based protocols. The shared tree-based multicast protocols are further classified into: (i) Cluster-based, (ii) Session-specific, and (iii) IP multicast session-based protocols.

In mesh-based multicast routing, multiple routes exist between the source node and each of the receivers of the multicast group. A receiver node receives several copies of the data packets, one copy through each of the multiple paths. Mesh-based multicast routing protocols provide robustness in the presence of node mobility; however, at the expense of a larger number of link transmissions leading to inefficient bandwidth usage. The mesh-based protocols are classified into source-initiated and receiver-initiated protocols depending on the entity (the source node or the receiver nodes) that initiates mesh formation.

The book chapter will review widely studied, characteristic, representative multicast routing protocols for each of the above sub-categories and explain the salient features of the particular sub-categories through these protocols. The book chapter will also discuss the pros and cons of each of these multicast routing protocols and the categories to which they belong. A classification tree of the different categories and their representative multicast routing protocols that would be discussed in the book chapter is given in Figure 1. As the routing protocols are often referred using their acronym, Table 1 provides the complete name of the multicast protocols listed in Figure 1.

The rest of the book chapter is organized as follows: The next section presents background information on MANETs, routing and multicast protocols. In the subsequent sections, the chapter discusses representative routing protocols for each of the 4 categories of source tree-based MANET multicast protocols, for each of the 3 categories of shared tree-based multicast protocols; followed by representative protocols for the source initiated and receiver-initiated categories of the mesh-based multicast protocols. The last two sections present the conclusions and future research directions. Throughout the chapter, the terms 'node' and 'vertex', 'path' and 'route', 'edge' and 'link', 'message' and 'packet' are used interchangeably. They mean the same.

Figure 1. Topology-based classification of MANET routing protocols

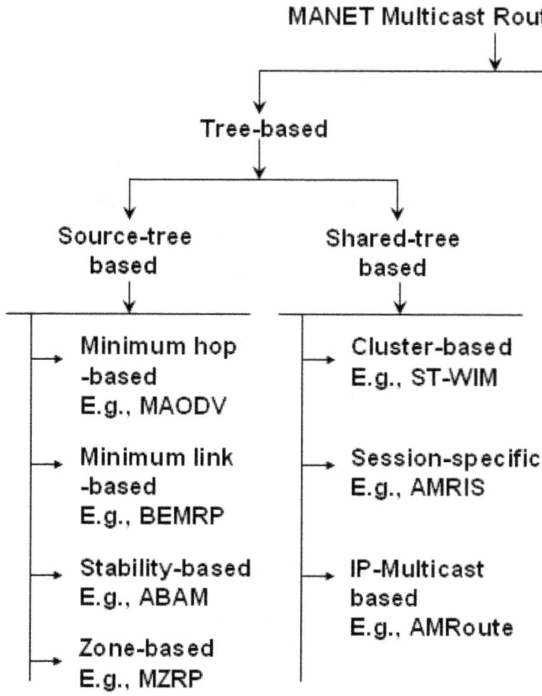

Table 1. Acronym and complete name of the MANET multicast routing protocols

Acronym	*Complete Name*
ABAM	Associativity-based Ad hoc Multicast Routing Protocol
AMRIS	Ad hoc Multicast Routing protocol utilizing increasing ID Numbers
AMRoute	Ad hoc Multicast Routing Protocol
BEMRP	Bandwidth Efficient Multicast Routing Protocol
CAMP	Core-Assisted Mesh Protocol
IOD-MRP	Intelligent On-Demand Multicast Routing Protocol
MAODV	Multicast-extension to the Ad hoc On demand Distance Vector Protocol
MZRP	Multicast Zone-based Routing Protocol
NSMP	Neighbor Support multicast Mesh Protocol
ODMRP	On Demand Mesh-based multicast Routing Protocol
ST-WIM	Shared Tree-based Wireless Multicast Protocol

BACKGROUND

Wireless networks provide location-independent access to information for users on the move and hence are known for ubiquitous communication capability. A mobile ad hoc network (MANET) is a self-organizing wireless network that can dynamically reconfigure itself without any centralized administration and lacks a fixed infrastructure. Nodes that are part of a MANET often function in a "peer-to-peer" mode – being the source, destination and intermediate forwarding node for different communication sessions.

Multicasting has become a de facto standard for wired and wireless networks deployed for applications requiring one-to-many communication. A sample list of applications for which multicasting is a desirable and essential technology includes: distance learning, collaborative and groupware applications, corporate communications, audio/video conferencing, stock quotes, news distribution and etc. Multicasting is different from multiple unicasting. With multicasting, a single stream of data is distributed to multiple recipients and data is duplicated only when required; on the other hand, with uncasting, data is transferred independently from the source to each of the receivers and this could lead to network clogging.

MANETs are resource-constrained networks: limited bandwidth, battery operated nodes, limited CPU and memory capacities and a shared medium leading to interference. MANET topology changes frequently with time due to random node movement. Because of such unique characteristics and constraints, multicast routing protocols that have been developed for traditional wired and even wireless cellular networks cannot be directly applied for MANETs. The routing algorithms need to be more robust to dynamically changing topology of MANETs.

Ad hoc routing protocols can be either proactive or reactive, depending on when the routes are determined. Proactive routing protocols predetermine routes to destinations by continuous exchange of routing information among nodes. Routes are immediately available when packets are to be transmitted. Examples of proactive ad hoc routing protocols are the Destination Sequenced Distance Vector (DSDV) routing protocol (Perkins & Bhagwat, 1994) and the Wireless Routing Protocol (Murthy & Garcia-Luna-Aceves, 1996). Reactive routing protocols determine routes only upon demand. A node queries the network for a route when it has a packet to transmit. Examples of reactive ad hoc routing protocols are TORA (Park & Corson, 1997), DSR (Johnson, 1994) and AODV (Perkins & Royer, 1999). But, neither pure proactive nor pure reactive routing protocols are suitable for MANETs. In MANETs, the changes in network topology can be more frequent than routing requests. In such situations, the routing information generated by proactive routing protocols becomes stale. Reactive routing protocols perform a global-search on-demand route discovery process that requires significant control traffic. As a result, the bandwidth-constrained MANETs can easily become saturated in a short interval of time. Hence, several ad hoc multicast routing protocols, that are not just extensions of the above unicast routing protocols, are proposed in the literature. This is contrary to what is seen in wired networks where many multicast routing protocols are just multicast extensions of their corresponding unicast routing protocols. Examples are: DVMRP (Waitzman et. al., 1988) and MOSPF (Moy, 1994). A robust and efficient ad hoc multicast routing protocol must be independent and should not rely on any underlying unicast routing protocol for route determination and updates.

SOURCE TREE-BASED MANET MULTICAST ROUTING PROTOCOLS

In this section, a representative protocol from each of the following major categories of source-tree based multicast routing protocols is described: (i) Minimum hop-based, (ii) Minimum link-based, (iii) Stability-based, and (iv) Zone-based protocols.

Minimum Hop-based Multicast Protocols

The minimum hop-based multicast routing protocols aim for a minimum hop path between the source node and every receiver node that is part of the multicast group. Each receiver node is connected to the source node on the shortest (i.e., the minimum hop) path and the path is independent of the other paths connecting the source node to the rest of the multicast group. The Multicast extension to the Ad hoc On-demand Distance Vector (MAODV) protocol (Royer & Perkins, 1999) is a classical example of minimum hop-based multicast routing protocols for MANETs. In this sub-section, the working of MAODV is described in detail.

Tree Formation (Expansion) Phase

In MAODV, a receiver node joins the multicast tree through a member node that lies on the minimum hop path to the source. A potential receiver wishing to join the multicast group broadcasts a *Route-Request* (*RREQ*) message. If a node receives the *RREQ* message and is not part of the multicast tree, the node broadcasts the message in its neighborhood and also establishes the reverse path by storing the state information consisting of the group address, requesting node id and the sender node id in a temporary cache. If a node receiving the *RREQ* message is a member of the multicast tree and has not seen the *RREQ* message earlier, the node waits to receive several *RREQ* messages and sends back a *Route-Reply* (*RREP*) message on the shortest path to the receiver. The member node also informs in the *RREP* message, the number of hops from itself to the source. The prospective receiver receives several *RREP* messages and selects the member node which lies on the shortest path to the source. The receiver node sends a *Multicast Activation* (*MACT*) message to the selected member node along the chosen route. The route from the source to receiver is set up when the member node and all the intermediate nodes in the chosen path update their multicast table with state information from the temporary cache.

Working Example

Figure 2 illustrates tree formation (expansion) under the MAODV protocol using an example. Here, the multicast tree is already established between the source node S and the two receivers R1 and R2 of the multicast group. A node is considered a multicast tree node if it is a source, receiver or an intermediate node of the multicast tree. Now, consider a new member node R3 joining as the receiver of the multicast group. To become part of the multicast tree, R3 broadcasts a Route Request (*RREQ*) control packet to its neighbors. If a neighbor node is a multicast tree node, it does not further propagate the *RREQ* packet; otherwise, it broadcasts the packet to its neighbors. The multicast tree node that receives a *RREQ* packet waits for a certain time period to receive any more *RREQ* packets and then responds back with a Route Reply (*RREP*) packet on the shortest (minimum hop) path to the initiator (R3). In the *RREP* packet, the tree node also includes the number of hops between itself and the source. In our example, intermediate tree node I1 (on the shortest path from S to R2) and the receivers R1 and R2 respond back with *RREP* packets to R3. The number of hops on the shortest path from R3 to each of R1 and R2 is 4; whereas the number of hops from I1 to R3 is 3. Also, the number of hops from S to R1 and R2 on the shortest path is respectively 2 and 3; whereas, the number of hops from S to I1 is 1. Considering all these, the shortest path from R3 to the source S would be the path that goes through the intermediate tree node I1. Hence, R3 decides to join the multicast tree through I1 and sends a Multicast Activation (*MACT*) message to I1. The intermediate nodes I5 and I8 that forwarded all of the three control packets (*RREQ*, *RREP* and *MACT*) now become part of the multicast tree.

Figure 2. MAODV - Receiver joining a multicast tree

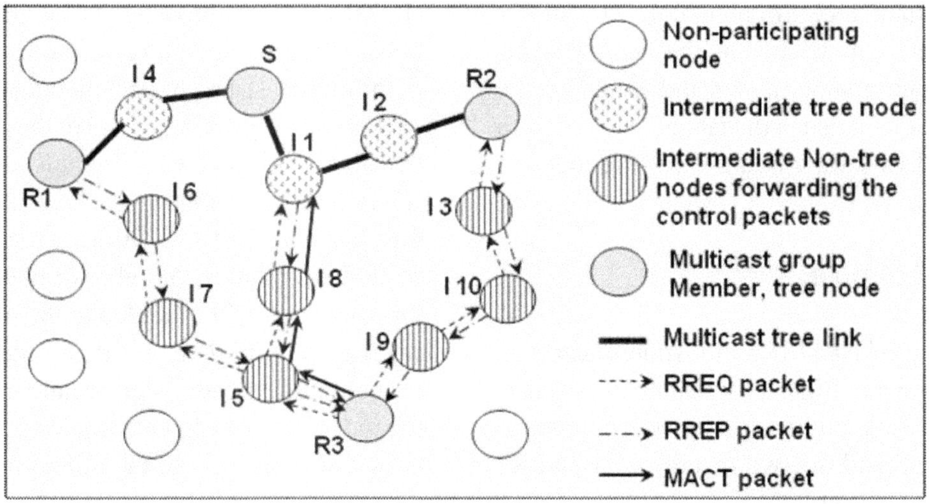

Tree Maintenance Phase

Tree maintenance in MAODV is based on the expanding ring search (ERS) approach, using the *RREQ*, *RREP* and *MACT* messages. The downstream node of a broken link is responsible for initiating ERS to issue a fresh *RREQ* for the group. This *RREQ* contains the hop count of the requesting node from the multicast source node and the last known sequence number for that group. It can be replied only by the member nodes whose recorded sequence number is greater than that indicated in the *RREQ* and whose hop distance to the source is smaller than the value indicated in the *RREQ*.

Minimum Link-based Multicast Protocols

The minimum link-based multicast protocols aim for an overall minimum number of links in the multicast tree connecting a source node to all the receiver nodes of the multicast group. Such a tree would use the bandwidth efficiently and facilitate simultaneous use of the wireless channel for several node pairs whose communication will not interfere with each other. The Bandwidth Efficient Multicast Routing Protocol (BEMRP) is a classical example of the minimum link-based source-tree multicast routing protocols as the protocol attempts to minimize the number of additional links that get incorporated when a new receiver node joins an already existing multicast tree. In this sub-section, the working of BEMRP is described in detail.

Tree Formation (Expansion) Phase

According to BEMRP (Ozaki et. al., 1999), a newly joining node to the multicast group opts for the nearest forwarding node in the existing tree, rather than choosing a minimum hop count path from the source of the multicast group. As a result, the number of newly added links in the multicast tree is minimum leading to savings in the network bandwidth. Multicast tree construction is receiver-initiated. When a node wishes to join the multicast group as a receiver, it initiates the flooding of *Join* control packets targeted towards the nodes that are currently members of the multicast tree. On receiving the first *Join* control packet, the member node waits for a certain time before sending a *Reply* packet. The member node sends a *Reply* packet on the path, traversed by the

Join control packet, with the minimum number of intermediate forwarding nodes. The newly joining receiver node collects the *Reply* packets from different member nodes and would send a *Reserve* packet on that path that has the minimum number of forwarding nodes from the member node to itself.

Working Example

The working of BEMRP is illustrated using an example shown in Figure 3. Assume there exists a minimum link multicast tree connecting the source node S and the two receivers R1 and R2 of the multicast group (the darkened links shown in Figure 3 form the multicast tree). Similar to our terminology used for the MAODV example, a tree node is referred to be either a multicast group member (source and receiver nodes) or an intermediate node in the tree. Now, when a third receiver node, R3, wants to join the multicast group, it broadcast the *Join* control packets in its neighborhood and the packets get forwarded further until they are received by a tree node. After receiving the first *Join* control packet, the tree node waits for a while and sends back a *Reply* packet on the path that has the minimum hop count to the initiator of the *Join* control packet (R3). However, unlike MAODV, the number of hops from the responding tree node to the source of the multicast group is not included in the *Reply* packet and only the hop count of the path from the tree node to the receiver is included. The receiver R3 collects all the *Reply* packets and decides to join the multicast tree through the closest tree node (I1) so that the number of new links (3, in this case) that will be added to the multicast tree will be the minimum. R3 then sends a *Reserve* packet to the chosen intermediate tree node and the end nodes of the links traversed by this *Reserve* packet are now part of the minimum link-based multicast tree.

Note that if R3 had chosen to send the *Reserve* packet directly to the source S by responding to its *Reply* packet, R3 would have been connected to the source on a minimum hop path. However, including such a path to the already existing multicast tree would create an addition of 4 new links to the tree and hence R3 prefers to go through I1 (that resulted in only 3 new links getting added to the tree). The tradeoff is that the number of hops from the source S to the receiver R3 is now 5 and this is larger than the minimum number of hops between S and R3 in the network.

Figure 3. BEMRP - Receiver joining a multicast tree

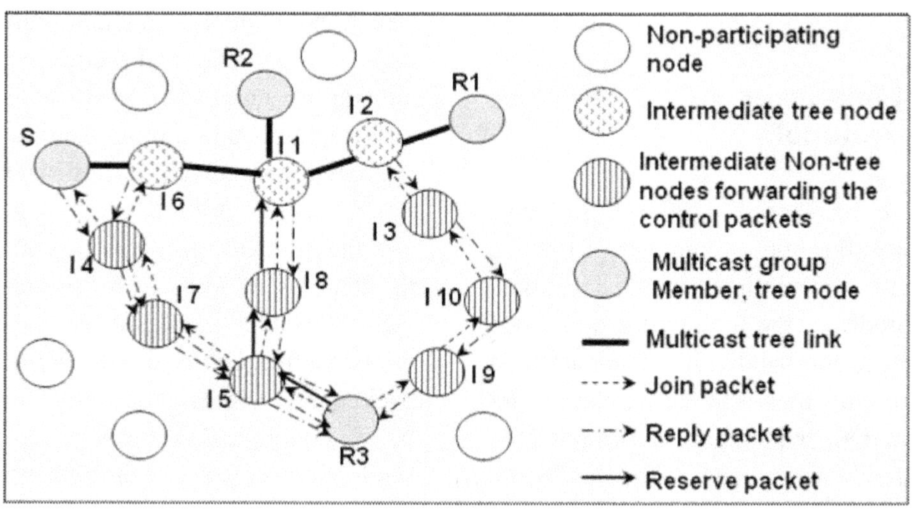

Tree Maintenance Phase

To provide more bandwidth efficiency, the tree maintenance approach in BEMRP is hard-state based, i.e. a member node transmits control packets only after a link breaks. BEMRP uses two schemes to recover from link failures: *Broadcast-multicast scheme* – the upstream node of the broken link is responsible for finding a new route to the previous downstream node; *Local-rejoin scheme* – the downstream node of the broken link tries to rejoin the multicast group using a limited flooding of the *Join* control packets.

Stability-based Multicast Protocols

The stability-based multicast protocols aim for a long-living tree connecting the source node to the receiver nodes of the multicast group. In pursuit of this, at the time of joining the tree, each receiver node selects the most stable path to the source node. This would minimize the number of tree reconfigurations. In order to determine stable paths and trees, routing protocols use metrics that are a measure of the longevity of the links in the network. Metrics such as predicted link expiration time (Su & Gerla, 1999), link affinity (Agarwal et. al., 2000), associativity ticks (Toh, 1997) have been used by the routing protocols for determining stable paths as well as stable trees. This sub-section discusses such multicast routing protocol called the Associativity-Based Ad hoc Multicast (ABAM) routing protocol (Toh et. al., 2000), which along with its unicast version, Associativity-Based Routing, ABR (Toh, 1997), have been quite popular and well-studied in the MANET literature for stability-based routing.

Tree Formation (Expansion) Phase

The stability of a link for a node with its neighbor is characterized by the associativity ticks, which is the number of beacons periodically received from that neighbor since the link was formed. Each node stores the value of the associativity ticks with its neighbors. Multicast tree construction is source-initiated and it could be repaired or expanded through receiver-initiated broadcast queries. In this section, the source-initiated tree formation is explained. The receiver-initiated tree expansion and repair is explained in the subsequent sub-section as part of tree maintenance.

The source node initiates the tree construction phase by broadcasting a Multicast Broadcast Query (*MBQ*) message in the network to inform all prospective receiver nodes. When an intermediate node receives the *MBQ* message, it appends its node ID, the associativity ticks with the upstream node and then rebroadcasts it. Upon receiving several *MBQ* messages through different paths, a receiver node of the multicast group selects the most stable path and sends a *MBQ-Reply* packet along the selected path. The most stable path is the path with the largest proportion (i.e., percentage) of stable links. A link is said to be stable if the value of the associativity ticks is greater than or equal to the associativity threshold, calculated based on the node velocity and transmission range per node. In case of a tie (i.e., if two or more paths have the same largest proportion of stable links), then the minimum hop path among the contending paths is chosen. After receiving *MBQ-Reply* packets from each receiver of the group, the source sends *MC-Setup* messages to all receivers in order to establish the multicast tree.

Working Example

The working examples presented so far in the previous sections (on MAODV and BEMRP) are receiver-initiated. Receiver-initiated tree repair and expansion of ABAM would also be similar. The only difference is that the route selection metric adopted by the tree node receiving the *Join Query* packets would not be the number of hops between itself and the initiating receiver node; rather, it would be based on the associativity ticks of the links traversed by the *Join Query* packets.

The tree node chooses the path with the largest proportion of stable links and sends a *Join Reply* packet. So, for a change, in this sub-section, a working example is presented that is source-initiated and is based on the concept of choosing paths with the largest proportion of stable links.

As shown in Figure 4, the source node S initiates a broadcast query reply cycle of the *MBQ* packets. The link weight shown in the figure represents the associativity tick value. The associativity threshold value is assumed to be 6. Accordingly, if the associativity tick value for a link is greater than or equal to the threshold, the link is said to be "stable"; otherwise, unstable. When a multicast receiver node receives *MBQ* packets across several paths, it calculates the proportion (i.e., percentage) of stable links on each of these paths and chooses the path that has the largest proportion of stable links. In the example, R3 receives *MBQ* packets across three paths from S: S → I4 → I6 → I7 → I5 → R3 (3/5); S → I1 → I8 → I5 → R3 (2/4) and S → I1 → I2 → I3 → I10 → I9 → R3 (4/6). The ratios included inside the parenthesis represent the proportion of stable links on each of the above paths. R3 chooses the path S → I1 → I2 → I3 → I10 → I9 → R3 that has the maximum proportion (4/6 = 67%) of stable links and sends a *MBQ-Reply* packet along this path back to the source. Likewise, the receiver nodes R1 and R2 choose the paths S → I4 → I11 → R1 (2/3) and S → I1 → I2 → R2 (2/3) respectively. For simplicity in the illustration, the figure does not show the source S sending the *MC-Setup* message on each of the paths through which the *MBQ-Reply* packets were received.

Tree Maintenance Phase

Tree maintenance is by using a *Local Query Reply* cycle. The upstream node of the broken link attempts to fix a route to the receiver node by broadcasting a *Local Query* message with TTL value of 1. When the receiver node receives the *Local Query* message, it responds with a *Local Reply* message. The upstream node then sends the *MC-Setup* message to the receiver. If the upstream node could not find a route to the receiver, then it transfers the responsibility of fixing the route to its immediate upstream node on the path from the receiver to the source. This upstream node then initiates a broadcast of the *Local Query* message with TTL value of 2. This procedure is continued

Figure 4. ABAM – Source initiated multicast tree construction

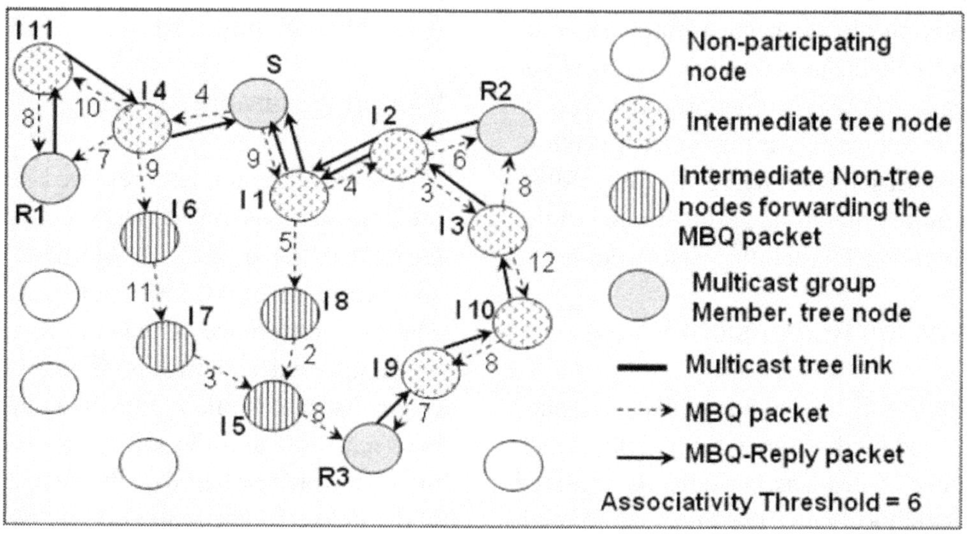

until the timer at the receiver node expires and it broadcasts a *Join Query* message to join the multicast group.

The *Join Query* message is broadcast by a newly joining receiver node or a receiver node that got cut off from the multicast tree and could not join the tree using a *Local Query Reply* cycle. The *Join Query* message propagates until it reaches a tree node. During its propagation, the forwarding nodes append their node ID and the associativity tick values with the downstream node from which the message was received. When a tree node receives a specific *Join Query* message (identified using a sequence number) for the first time, it waits for a while to receive the *Join Query* messages along different paths; chooses the path with the largest proportion of stable links and sends a *Join Reply* message on the selected stable path. The receiver node confirms its participation in the multicast session by responding back with a *Reserve* message on the path traversed by the *Join Reply* message.

Multicast Zone-based Routing Protocol (MZRP)

MZRP (Zhang & Jacob, 2003) is the multicast extension of the unicast Zone Routing Protocol, ZRP (Haas & Pearlman, 2001), which is a hybrid of both proactive and reactive routing strategies. A zone in the network comprises of nodes that are within 2 or 3 hops from each other. There exists multiple such zones in the network and often these zones overlap with each other. Nodes that are part of more than one zone are referred to as border nodes. ZRP employs proactive routing for intra-zone communication and a combination of proactive and reactive routing protocols for inter-zone communication. For example, if a source node *s* has to communicate with a destination node *d* that is in the same zone of *s*, then it utilizes the services of the proactive routing protocol for that zone; on the other hand, if the destination is in some other zone, then the source node has to utilize the proactive routing protocols implemented in their respective zones and a reactive routing protocol that is implemented for inter-zone communication. ZRP is generic in the sense that it doesn't depend on any specific proactive and reactive routing protocol for intra-zone and inter-zone communication. Likewise, MZRP does not depend on any underlying unicast routing protocol. The proactive routing mechanism within a zone is realized through periodic beacon-exchange and the reactive routing mechanism for communication across zones is realized through on-demand flooding and multicast tree construction.

Node Advertisement and Zone Initialization

Each node advertises itself to its zone members by broadcasting an *Advertisement* packet whose propagation is controlled using a time-to-live (*TTL*) value that is normally set to the zone radius. Nodes receiving the *Advertisement* packet create a route entry for the source in their *zone routing table* and update their *neighbor table* with the upstream node that sent the *Advertisement* packet. ZRP operates in soft state i.e., nodes expunge route entries for a node that fails to send periodic *Advertisement* packets to indicate its presence in the zone. Thus, topology changes within a zone are often localized and not broadcast to other zones.

Multicast Tree Creation inside a Zone

Multicast tree creation within a zone (illustrated in Figure 5) is done as follows: The source node broadcasts a *Tree-Create* packet, uniquely identified using a session id, with a *TTL* value set to the zone radius. An intermediate node upon receiving the *Tree-Create* packet creates a multicast route entry with an empty list of downstream nodes and the upstream node is set to the node that sent the *Tree-Create* packet. Any zone node that is interested to be a receiver of the multicast session responds with a *Tree-Create-ACK* packet

Figure 5. Multicast tree creation inside a zone (Zhang & Jacob, 2003)

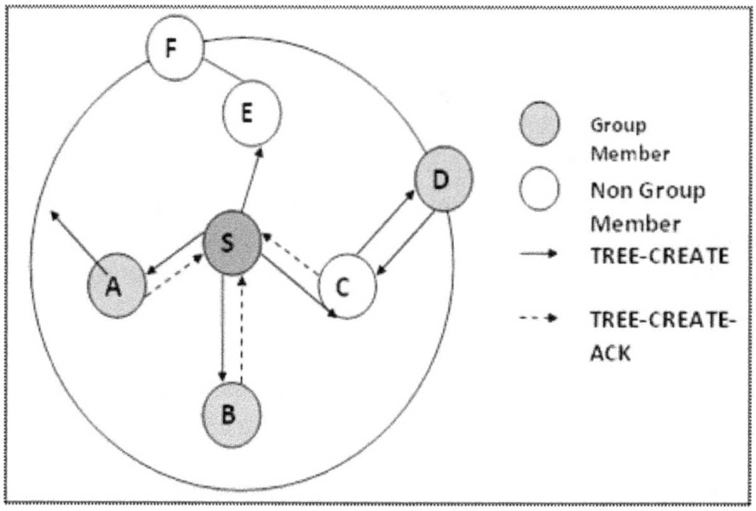

that travels back to the source on the reverse path traversed by the *Tree-Create* packet. Intermediate nodes that receive the *Tree-Create-ACK* packet, for a multicast session for which an entry has been already created in the routing table, updates the downstream node list by adding the ID of the neighbor node from which the *Tree-Create-ACK* packet was received.

Extension of the Multicast Tree to the Entire Network

To extend the multicast tree to the entire network (refer Figure 6), the source node unicasts a *Tree-Propagate* packet to the border nodes of its own zone. After receiving a *Tree-Propagate* packet, a border node creates a route entry for the session in its multicast table and initiates the zone-wide broadcast of a *Tree-Create* packet. Any interested node in the zone of the border node now responds back with a *Tree-Create-ACK* packet that is duly forwarded by the border node to the multicast source. The source specifies, in the *Tree-Propagate* packet, a network-wide *TTL* value that indicates the number of zones to which the packet has to be forwarded by a border node. A border node decrements this *TTL* value by 1 and if it is still above zero, the *Tree-Propagate packet* is sent to the other border nodes within that zone. Once the multicast tree is constructed, the source node starts sending data on the tree and is propagated across zones, as per the downstream node list maintained at the intermediate forwarding nodes.

Zone and Multicast Tree Maintenance

Multicast route entries are maintained in soft state: The source node periodically, for every *Refresh-Interval*, refreshes the entries at the intermediate nodes and receiver nodes (collectively referred to as member nodes) by sending a *Tree-Refresh* packet that propagates down the tree across zones. If a member node gets disconnected from the tree, it sends a *Join* packet, identifying the multicast session, to all of its zone nodes. Any member node that is in the multicast tree, and has not suffered any disconnection, responds back with a *Join-ACK* packet and adds the neighbor node that sent the *Join* packet to the list of downstream nodes. A similar procedure is adopted at all the nodes that receive the *Join* and/or *Join-ACK* packets. Figure 7 illustrates the rejoin process within a zone. If a disconnected member node fails to receive any *Join-ACK* packets from its

Figure 6. Extension of the multicast tree to the entire network (Zhang & Jacob, 2003)

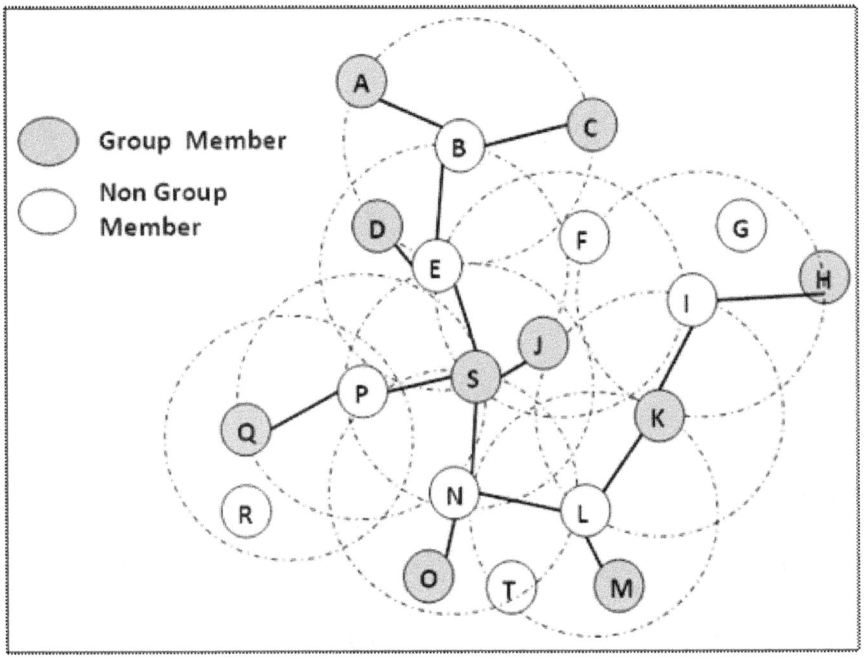

zone nodes within a certain time, then it sends a *Join-Propagate* packet to all of its border nodes, which in turn send *Join* packets to all their zone nodes. This process continues until a border node gets a *Join-ACK* packet that is duly forwarded to the disconnected member node. If a receiver node no longer wants to be part of the multicast session, it can get disconnected by sending a *Tree-Prune* message to its upstream node in the tree. The upstream node will remove the receiver node from its list of downstream nodes for the tree; if the list becomes empty after the removal and the node is not an interested receiver node (i.e., has been just an intermediate node), then the node sends a *Tree-Prune* message to its upstream node further up in the tree.

SHARED TREE-BASED MANET MULTICAST ROUTING PROTOCOLS

Shared tree-based protocols construct a single tree that is rooted at a central control point called the Rendezvous Point (*RP*). Typically, MANETs operate on a 2-level mobility model with only a fraction of the nodes moving fast and the rest of the nodes moving relatively very slowly (Gerla et. al., 1999). The *RP* is often chosen to be the slowest moving node in a MANET so that the shared tree can be maintained irrespective of the mobility of the source nodes. The *RP* is connected to the receiver nodes of the multicast session and the source nodes send the data packets to the *RP* based on the underlying unicast routing protocol.

As shared tree-based protocols are vulnerable to make data packets take a circuitous route (source to *RP*; *RP* to receiver), shortest paths between the source nodes and the RP is usually the preferred unicast routing strategy. Still, shared tree-based protocols are known to yield a lower throughput compared to per source tree-based protocols. A compromise solution called "adaptive-tree multicast" is applicable wherein if the receivers request the source, the source node may construct its own rooted tree to deliver the packets on the shortest path (Gerla, et. al., 1999). This section reviews

Figure 7. Zone and multicast tree maintenance (Zhang & Jacob, 2003)

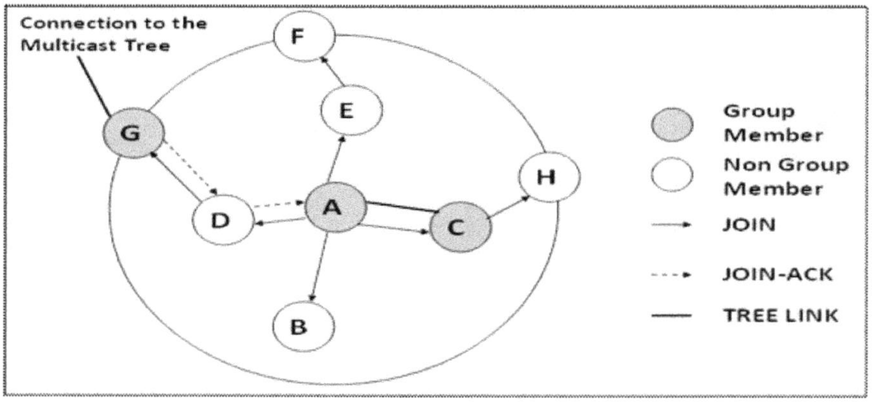

the features and performance of the Shared-Tree WIreless Multicast (ST-WIM) protocol, Ad hoc Multicast Routing protocol utilizing Increasing id numberS (AMRIS) and the Ad hoc Multicast Routing protocol (AMRoute) as representatives of the Cluster-based, Session-specific and IP-multicast session based protocols.

Cluster-based Shared-Tree Wireless Multicast Protocol (ST-WIM)

Shared-tree wireless multicast protocol, ST-WIM (Chiang et. al., 1997a), is based on sparse PIM (Deering et. al., 1996), a unicast protocol for wired networks. ST-WIM (Chiang et. al., 1997a) is portable to different wireless platforms as it is independent of the underlying wireless routing protocol. Each intermediate node in a shared-tree must keep track of the state of its downstream members. ST-WIM uses two schemes for shared-tree maintenance: hard state and soft state. ST-WIM is implemented in an architecture that uses cluster routing and the token access protocol for inter-cluster and intra-cluster communication (Chiang et. al., 1997b). So, the cluster networks that form the backbone of ST-WIM are briefly described here.

Clusters are aggregation of nodes under a cluster-head. Each node in a cluster can directly communicate with the cluster-head and possibly with each other. Gateways are nodes located in more than one cluster and provide support for adjacent clusters to communicate. Cluster-heads are commonly selected using the lowest-ID or highest-connectivity algorithms. But frequent changes to the cluster-head affect the performance of the protocols using them as well as in scheduling and resource allocation. Chiang et. al. (1997b) proposed a least-cluster-head-change algorithm that advocates only two conditions for change in cluster-head: (a) when two or more cluster-heads come within the range of each other and (b) a node gets disconnected from any cluster. Clustering provides efficient allocation of wireless channels among different clusters. It permits spatial reuse of different spreading codes across different clusters, which reduces inter-cluster interference (Chiang et. al., 1997a). Orthogonal codes are used across adjacent clusters. Clustering algorithms converge in O(1) time and hence the presence of clusters does not deter the convergence of the routing algorithm.

Hard State Approach

In the hard state approach, a node that wishes to join the multicast group sends an explicit *Join-Request* message to the *RP* and waits for an acknowledgement to become a member. An upstream node maintains the link with its downstream node until it receives an explicit *Quit-Request* from the downstream node or the downstream link is disconnected. The hard-state approach relies heavily on the underlying medium access control (*MAC*) protocol to provide information regarding the connectivity of the downstream links. The *MAC* protocol uses a periodic beaconing mechanism to detect connectivity of nodes. Downstream nodes can also disconnect from one upstream node and send an explicit *Join-Request* message to the *RP* to join another upstream node. When a parent link is disconnected, a downstream node recovers by either sending a *Join-Request* message to rejoin the tree via a new parent node or by broadcasting a *Flush-Tree* message to its children (Chiang et. al., 1997a). In the latter scenario, the *Flush-Tree* message propagates down the entire subtree and the downstream leaf nodes rejoin the tree individually. Rejoin by an internal node is bound to sometimes create loops when it forms a link with its own downstream node. Hence, in MANET, rejoin using the flush scheme is preferred.

Soft State Approach

In the soft state approach, a node wishing to remain in the multicast group must send a periodic *Join-Request* packet to the *RP*. No acknowledgment packets are sent back. An upstream node receiving a *Join-Request* packet from the downstream node creates a state for the downstream neighbor and attaches a timestamp to it. Nodes receiving the *Join-Request* packet forward them directly to the *RP* using the underlying routing scheme and there is no need to keep track of the upstream nodes. This is different from the hard state approach where a node needs to keep track of both its upstream and downstream nodes.

Hard State vs. Soft State

Simulation experiments show that the hard state acknowledgment delay is less than that of the soft state join latency which is less than the hard state data delay (Chiang et. al., 1997a). The values for the $T_{refresh}$ and $T_{timeout}$ time periods must be carefully chosen to account for mobility and channel access overhead. Even though short refresh periods are desirable in a highly mobile network, it will cause an increase in the channel overhead. Large values of $T_{timeout}$, will result in the network having stale links with wasteful duplicate transmissions. With decreasing $T_{refresh}$, the control traffic for soft state has been found to be higher than that of hard state.

Session-Specific Ad hoc Multicast Routing Protocol Utilizing Increasing ID Numbers (AMRIS)

AMRIS (Wu & Tay, 1999) provides a unique session-specific multicast session member id (*msm id*) to each participant. The *msm id* indicates the logical height of a node in the multicast delivery tree rooted at the sender that has the smallest *msm id* (*Sid*) in the tree. All other nodes in the tree have an *msm id* that is higher than that of its parent. In case of a multiple-sender environment, the *Sid* is elected among the senders. AMRIS uses the underlying *MAC* layer beaconing mechanism to detect the presence of neighbors.

Tree Initialization Phase

During the tree-initialization phase, the *Sid* broadcasts a *New-Session* message (containing the *Sid*, *msm id* and other routing metrics) in its neighborhood. A node, receiving the *New-Session* message, updates the *msm id* in the message with a newly computed larger value that is also used to identify the node in the tree. If a node receives more than one *New-Session* messages from several neighbors within a random jitter amount of time, then the message with the best routing metric is selected and updated with a newly computed *msm id* value. The updated *New-Session* message is then rebroadcast. The above strategy prevents broadcast storms. Note that the newly computed *msm ids* are not consecutive and the gaps can be used to locally repair the delivery tree.

To join the multicast group, a downstream node X sends a unicast *Join-Request* message to one of a randomly chosen neighbor node that is also a potential parent node (having a lower *msm id*), say Y. If Y is already part of the multicast tree, then Y sends a *Join-ACK* message to X. Otherwise, Y forwards the *Join-Request* message to a set of potential parents of itself. The above process continues until the *Join-Request* message reaches a parent node that is part of the multicast delivery tree. The parent node responds back with a *Join-ACK* message to X, which can now confirm its participation in the multicast session by sending back a *Join-Conf* message to the parent node.

Tree Maintenance Phase

Upon link failure, the downstream node (with a relatively higher *msm id*) is responsible to rejoin the multicast tree (Wu & Tay, 1999). Before broadcasting a *Join-Request* message, the downstream node of a broken link attempts to find potential parent nodes by locally repairing the route using an expanded-ring search process. Even the *Join-Request* message is broadcast with a *TTL* value that restricts the number of hops the message can propagate. If a node has no valid *msm id* to (re)join a multicast tree, it computes an *msm id* for itself based on the *msm ids* of its neighboring nodes and then joins the multicast tree through the branch reconstruction process explained above (Wu & Tay, 1999). The beacon update period at the *MAC* layer has to be properly chosen to avoid detection of micro-term breakages that can unnecessarily trigger the branch reconstruction process and incur a lot of control overhead.

IP Multicast Session-based Ad hoc Multicast Routing Protocol (AMRoute)

The ad hoc multicast routing protocol, AMRoute (Xie et. al., 2002), is an attempt to enable the use of IP multicast in MANETs. AMRoute make use of the underlying unicast routing protocol to detect network dynamics while it takes care of the frequent tree reconfigurations. The AMRoute distribution tree assumes unicast connectivity among member nodes and continues to function despite network changes. The two key features that make AMRoute robust and efficient in MANETs are: (1) User-multicast trees that allow group members to replicate and forward over unicast tunnels and (2) Core node migrating dynamically resulting in changes to the group membership and network connectivity. Since non-member nodes in the tree need not be even multicast enabled, user-multicast trees eliminate the need to reconfigure the tree frequently in highly dynamic MANETs. But, the tradeoff is a reduction in bandwidth efficiency as non-member nodes cannot replicate packets on user-multicast trees.

The AMRoute protocol consists of a mesh creation phase following a tree creation phase. Each group member forms a 1-node mesh identifying itself as the core. The core node in pursuit of other group members periodically broadcasts *Join-Request* messages in an expanded ring fashion. A member node, receiving *Join-Request* messages from a core node in a different mesh, responds back with a *Join-ACK* establishing a bi-directional tunnel between the two nodes. Multiple cores result due to mesh mergers. All the core nodes elect a single core for the unified mesh using a core resolution algorithm (for e.g., the node with the highest IP address could be selected as the core node). The core node need not be static. It can migrate dynamically according to group membership and network connectivity (Xie et. al., 2002).

The core node of the unified mesh is responsible for the tree creation process. The core node periodically sends *Tree-Create* messages along the links incident on it in the mesh. The mesh size and node mobility in the mesh determines the periodicity of *Tree-Create* messages. Group members that receive non-duplicate *Tree-Create* messages forward it to all the links except the incoming link.

If the link is not used as part of the tree, a *Tree-Create-NAK* is transmitted through the incoming link. Upon receiving a *Tree-Create-NAK*, the group member marks the incoming link as a mesh link. A member node forwards data received on a tree link but discards it when received on a mesh link and sends a *Tree-Create-NAK* along that link.

Simulation studies of AMRoute show that the broadcast traffic is independent of group size. This reflects on the *JOIN-REQ*s being generated at a fixed rate by the core. When the network size is constant and the group size increases, join latency decreases. The major disadvantage of the protocol is it forms temporary loops during tree creation and non-optimal trees at high mobility (Lee et. al., 2000). At high mobility and small group size, critical links (unidirectional links) are formed among the group members. Prolonged existence of critical links may sometime isolate the sources. A performance comparison of the major ad hoc multicast routing protocols shows AMRoute to be the least effective (Lee et. al., 2000).

SOURCE-INITIATED MESH BASED MULTICAST PROTOCOLS

A mesh is a set of nodes in the network such that all the nodes in the mesh forward multicast packets via scoped flooding (Lee & Cho, 2001). As stated before, mesh-based protocols are more robust to link failures than tree-based protocols. Mesh-based protocols can be either source-initiated or receiver-initiated. In most of the cases, the forwarding mesh in source-initiated protocols is a union of per-source meshes, while receiver-initiated mesh protocols form a single shared mesh for all the sources. This section discusses the source-initiated mesh based multicast routing protocols and the next section discusses the receiver-initiated mesh based protocols. In the category of source-initiated mesh based multicast routing protocols, the well-known On-Demand Multicast Routing Protocol (ODMRP) along with its extensions to handle high mobility and low node density (i.e., sparse networks), and the Neighbor Supporting multicast Mesh Protocol (NSMP) are discussed.

On-Demand Multicast Routing Protocol (ODMRP)

ODMRP (Lee et. al., 1999) is a mesh-based multicast routing protocol based on the notion of a forwarding group (shown in Figure 8) – a set of nodes that forward data on the shortest paths between any two multicast members (Burri et. al., 2006; Chiang et. al., 1998). Multicast group membership and routes are established and updated by the source in an on-demand basis. This leads to reduction in channel/ storage overhead and an increase in scalability. A soft-state approach is used for mesh maintenance and member nodes are not required to explicitly send leave messages while quitting a group. A performance comparison of the major ad hoc multicast routing protocols in shows ODMRP to be the most advantageous and preferred protocol in mobile wireless networks (Lee et. al., 2000). ODMRP can also operate independently as an efficient unicast routing protocol.

ODMRP operates through a request phase and a reply phase. Multicast sources, which are not aware of the routes or membership, broadcast a *Join-Data* packet. When a node receives the *Join-Data* packet for the first time, it updates its routing table by storing the upstream node id (backward learning process) and rebroadcasts the packet. Upon receiving a non-duplicate *Join-Data* packet, a multicast receiver creates and broadcasts a *Join-Reply* packet in its neighborhood. When a node receives the *Join-Reply* packet, it checks if it is listed as the next node ID in the packet. If so, the node is located on the path to the source and becomes part of the forwarding group. A FG (forwarding group) flag is set in its routing table and the node broadcasts its own *Join-Reply* packet. Likewise, the *Join-Reply* packet gets forwarded by the FG member nodes until it reaches the multicast source on the shortest path.

Figure 8. Concept of forwarding zone (Bae et. al., 2002)

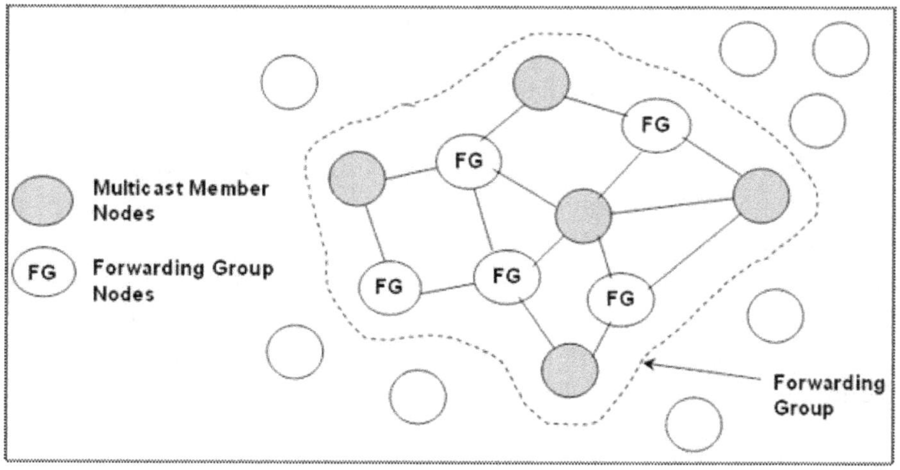

As a result of this source-receiver route construction and update process, a mesh of nodes referred to as the *forwarding group* is built.

ODMRP makes use of the IEEE 802.11 *MAC* protocol to reliably transmit the *Join-Reply* packets that are critical to establishing and refreshing the forwarding groups and the associated multicast routes. Once the forwarding group is established and multicast routes are constructed, a source node sends down data packets to the receivers (Lee et. al., 1999). To leave a multicast group, the source node and receiver nodes can respectively stop sending the *Join-Query* and *Join-Reply* packets meant for that group. If a FG member node does not receive a *Join-Reply* packet before a timeout period, the node demotes itself to a non-forwarding node of the mesh.

The performance of ODMRP heavily depends on the values selected for the route refresh interval and the FG timeout interval (Lee et. al., 1999). With small route refresh interval, there will be frequent broadcast of fresh route and membership information that may lead to a deluge of packets causing network congestion. On the other hand, if a larger route refresh interval is selected, nodes may not be aware of the latest route and multicast membership changes and this could lead to poor performance. The FG timeout value must be at least 3-5 times larger than the route refresh interval value. For networks with a heavy traffic load, a smaller value for the FG timeout interval will help to reduce redundancy of membership messages (Lee et. al., 1999); whereas, in the case of high mobility, a larger FG timeout value can make way for more alternative paths.

ODMRP Based on Mobility Prediction

Since ODMRP requires periodic flooding of *JOIN-QUERY* messages to refresh routes and group membership, finding an optimal refresh interval is critical for its performance. Bae et. al. (2002) proposed an improved version of ODMRP that adapts the refresh interval to mobility patterns and speeds. The duration of time the routes will remain valid is predicted using the location and mobility information provided by Global Positioning System, GPS (Hofmann-Wellenhof, 2004). Using the predicted time of route disconnection, *Join-Query* messages are sent only when there are imminent route breaks of ongoing data sessions (Bae et. al., 2002).

The prediction method used by Bae et. al. (2002), originally proposed by Su & Gerla (1999), assumes a free-space propagation model, in which the received signal strength solely depends on its distance to the transmitter. Also, all nodes in the network are assumed to have their clocks synchronized using the network time protocol (NTP) or the GPS clock itself. With all these assumptions, if the motion parameters of two neighbors (e.g., speed, direction and radio propagation range) are known, the duration of time the two nodes will remain connected can be determined (Su & Gerla, 1999). Let two mobile nodes *i* and *j* be within the transmission range *r* of each other. Let (x_i, y_i) and (x_j, y_j) be the co-ordinates of *i* and *j* respectively. Let v_i and v_j be the speeds and θ_i and θ_j $(0 < \theta_i, \theta_j < 2\pi)$ be the moving directions of *i* and *j* respectively. The amount of time, the two nodes will remain connected is predicted as (Su & Gerla, 1999):

$$D_t = \frac{-(ab+cd)+\sqrt{(a^2+c^2)r^2-(ad-bc)^2}}{a^2+c^2}$$

where: $a = v_i\cos\theta_i - v_j\cos\theta_j$, $b = x_i - x_j$, $c = v_i\sin\theta_i - v_j\sin\theta_j$, $d = y_i - y_j$

The structure of the *Join-Query* and the *Join-Reply* packets need to be slightly modified to accommodate the information derived from mobility prediction. The *Min-LET* (minimum link expiration time) value field in the *Join-Query* packet indicates the minimum expiration time of the links traversed by the packet, starting from the source. The *Min-LET* value set by the source in the *Join-Query* packet is the *Max-LET* value (theoretically ∞), as the source is not connected to any upstream node. An intermediate node, upon receiving a non-duplicate *Join-Query* packet, will predict the *LET* of the link with the immediate upstream sender of the packet, and set the *Min-LET* value in the packet to be the minimum of its current value and the predicted *LET*. The same procedure is followed at the member nodes too and the final value of *Min-LET* in the *Join-Query* packet is referred as to the Route Expiration Time (*RET*) of the path from the source to the receiver node. A member node waits for a certain time before responding back with a *Join-Reply* packet. If more than one *Join-Query* packet is received, the member node chooses the (stable) path with the largest *RET* and sends a *Join-Reply* packet (with the *RET* value included) on this path. If a forwarding group node receives several *Join-Reply* packets, it chooses the path with the minimum *RET* and sends its own *Join-Reply* packet with the chosen *RET* value. After receiving *Join-Reply* packets across several paths, the source sets the predicted expiration time of the mesh to be the minimum of the *RET* values of these packets. The source starts sending data packets using the constructed stable mesh. Before the predicted mesh expiration time approaches, the source node floods a *Join-Query* packet.

Some of the drawbacks of the mobility prediction technique is that if a node suddenly changes direction and/or speed, the predicted expiration time of the mesh and the *RET* values of the paths involving that node may be no longer valid. Also, since route and mesh construction is source-initiated, a new receiver node wishing to become a multicast member, has to wait for the next sequence of *Join-Query* packets coming from the source. However, the performance of ODMRP has been reported to be significantly improved (Bae et. al., 2002) with the use of mobility prediction.

PatchODMRP

In ODMRP, the forwarding group (FG) for a multicast group is an aggregate of the per-source meshes. For multicast groups with fewer sources, the forwarding mesh will be sparsely populated and may require frequent reconfiguration using flooding of *Join-Query* packets. To handle this problem, the PatchODMRP protocol has been proposed by Lee & Kim (2001), according to which whenever a FG member node sees the possibility of it getting separated from the mesh, it starts a patching procedure to reconnect itself to the mesh through a temporary path. With PatchODMRP, the *Join-Query* interval can be sufficiently long (the tradeoff is multicast group join latency for new receivers) even in the presence of high mobility. Simulation results illustrate that for larger *Join-Query* intervals, the packet delivery ratio of PatchODMRP increases while that of the original ODMRP decreases. For a given degree of node mobility, the control overhead of PatchODMRP is 6 to 7 times lower than that of original ODMRP (Lee & Kim, 2001). The routing table at a FG node contains a *HopCount* field (not in the original ODMRP protocol) specifying the number of hops from the destination to the node and the field is updated when a FG node processes a *Join-Query* packet. The forwarding group table also contains two additional fields (compared to ODMRP): the *Upstream FG nodes* (has the next hop information) and *Source list* (list of multicast source nodes); these two fields are processed when an FG node processes the *Join-Reply* packets. When a node is no longer a neighbor, an FG node checks its forwarding group table whether the lost neighbor is in its *Upstream FG nodes* list.

If an FG node loses connectivity to an upstream FG node, it initiates a local patching procedure by flooding an advertisement (*ADVT*) packet with a limited *HopCount* value (typically set to 2 or 3) for propagation. The *ADVT* packet contains the 3-field tuple {*MGID*, *SrcAddr*, *SrcHopCount*} where *MGID*, *SrcAddr*, *SrcHopCount* respectively refer to the multicast group ID, the multicast source address and the number of hops from *SrcAddr* to the initiator of the *ADVT* packet. When an FG node receives the *ADVT* packet, it checks whether it has a table entry for the *MGID* specified in the packet – if so, the FG node checks the value for the *HopCount* field in its routing table for the path from itself to the *SrcAddr* listed in the *ADVT* packet. If the *HopCount* value in the routing table is less than the *SrcHopCount* specified in the *ADVT* packet, it means that the FG node lies on a non-cyclic path from the initiator of the *ADVT* packet to the multicast source – *SrcAddr*. The FG node sends back a *Patch* packet to the initiator of the *ADVT* packet and the nodes located in the path traversed by the *Patch* packet become the new FG nodes of the multicast group. If any or all of the above conditions are not met, then the FG node decrements the *HopCount* field value in the *ADVT* packet by 1 and if the value is still above zero, the packet is broadcast to the neighbor nodes with the *PreHopAddress* set to the address of the FG node.

Neighbor Supporting Multicast Mesh Protocol (NSMP)

NSMP (Lee & Kim, 2000) is an efficient mesh-based multicast routing protocol that reduces the number of control message broadcasts as much as possible. NSMP resorts to network-wide flooding only during initial route establishment and during network partition repair. For normal periodic mesh maintenance, a local route discovery is initiated with control messages sent and forwarded only by nodes that are part of the mesh. When a node wants to join the multicast mesh as a receiver, it selects a path that contains a maximum number of the existing FG nodes. By reducing the number of forwarding nodes, NSMP enhances route efficiency (less contention, lower end-to-end delay). Simulation experiments indicate that irrespective of the number of sources, NSMP (compared to ODMRP) reduces the control overhead and the number of redundant data transmissions by more than 15% and 30% respectively (Lee & Kim, 2000).

The multicast mesh creation process of NSMP is illustrated in Figure 9. The initial network is shown in Figure 9.a and the final multicast mesh (with node 4 as the source; nodes 6 and 13 as the receivers) is shown in Figure 9.b. To start with, node 4 broadcasts a *Flood-REQ* packet to its neighbors. Upon receiving the *Flood-REQ* packet, node 5 creates a reverse path entry for node 4 in its routing table and forwards the packet to its neighbor nodes. Receiver node 6, upon receiving the *Flood-REQ* packet, responds with a *REP*

Figure 9. NSMP – Multicast mesh creation (Lee & Kim, 2000)

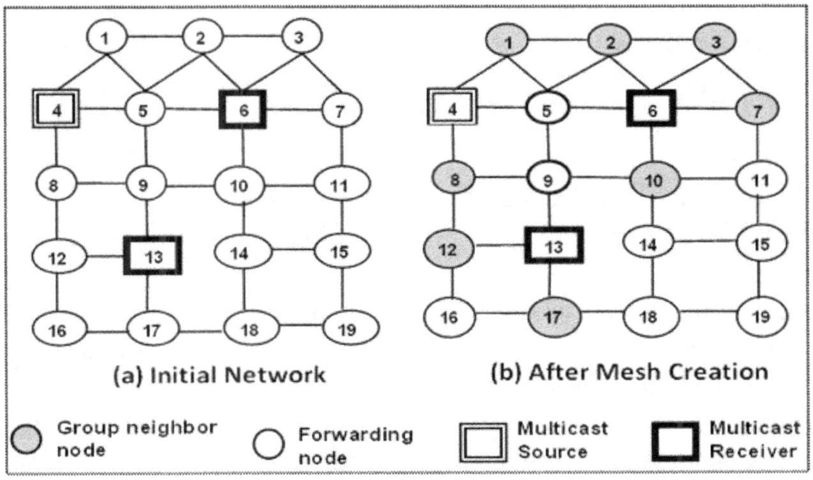

packet that reaches the source node 4 through node 5. Upon receiving the *REP* packet, node 5 updates its routing table by creating an entry for the multicast group and becomes part of the mesh. Similarly, node 13 responds to the *Flood-REQ* packet through node 9 that becomes a forwarding node for the mesh.

Group neighbor nodes are nodes that are adjacent to at least one forwarding node in the mesh for the group. Nodes 1, 2, 3, 7, 8, 10, 12 and 17 are the group neighbor nodes for the mesh shown in Figure 9.b. If a neighbor node 17 is interested in becoming a receiver for the multicast group, then it chooses a path (4, 5, 9, 13, 17) that will introduce the least number of new forwarding nodes to the mesh. Thus, NSMP offers route efficiency at the cost of route robustness (Lee & Kim, 2000). However, NSMP can be configured with a cost function that will enable a receiver node to choose a robust path over a more efficient path or vice-versa. The cost function is $(1-a)*FC+a*NC$ where $0 \leq a \leq 1$ and FC, NC are respectively the number of forwarding nodes and the number of non-forwarding nodes traversed by the *Local-REQ* packets (used for mesh maintenance, explained below) along different paths from the multicast source to the interested receiver. Larger the value of the weight factor '*a*,' more robust is the path chosen.

NSMP mesh maintenance is through a source-initiated local route discovery involving *Local-REQ* packets that are forwarded only by the mesh nodes and the group neighbor nodes. Nodes that are two-hops away from the mesh receive these *Local-REQ* packets and could join the mesh as receiver or as a forwarding/ group neighbor node by responding back with *REP* packets (similar to the approach adopted to respond for the *Flood-REQ* packets). However, a non-member node that is more than two hops away from the mesh can join the mesh only by flooding a *MEM-REQ* (membership request) packet towards the mesh. NSMP operates in soft-state: the forwarding nodes and group neighbor nodes of a mesh are to be refreshed within a pre-defined timeout period in order for them to stay committed to the mesh.

An example for NSMP mesh maintenance is illustrated in Figure 10 (Lee & Kim, 2000). Consider the link (9, 13) between a FG node 9 and a receiver node 13 to fail. The source node 4 eventually sends out a *Local-REQ* packet that is forwarded down the mesh by the group neighbor nodes 8 and 12. When node 13 receives the *Local-REQ* packet, it responds with a *REP* packet through node 12 to build a new route to the source.

The repaired mesh is shown in Figure 10.b. Note that local route discovery cannot repair all link failures – for example, the failure of link (8, 12) in Figure 10.b cannot be repaired.

RECEIVER-INITIATED MESH BASED MULTICAST PROTOCOLS

Receiver-initiated mesh protocols are more robust to node mobility as they attempt to maintain a shared mesh involving all the senders of the multicast group. The multicast source nodes forward packets on the reverse shortest path from the receiver nodes to the source. This section reviews the working of two receiver-initiated multicast mesh protocols: the core-assisted mesh protocol, CAMP (Garcia-Luna-Aceves & Madruga, 1999), and the intelligent on-demand multicast routing protocol, IOD-MRP (Wang & Chang, 2001).

Core-Assisted Mesh Protocol (CAMP)

CAMP designates nodes into three categories: duplex, simplex and non-members. Duplex nodes are the regular mesh nodes having the functionality seen in the routing protocols discussed in the previous sections. Simplex members can forward packets from the source-nodes to the rest of the mesh; but cannot respond for any membership query packets. CAMP maintains one or more core nodes per mesh. A node interested to join the multicast mesh first queries its neighbor nodes to see if any of them are part of the mesh. If none of the neighbors are part of the mesh, the prospective receiver node initiates a scoped flooding of the *Join-Request* messages targeted towards the core nodes of the mesh. Upon receiving a non-duplicate *Join-Request* message, a duplex member node or the core node responds with a *Join-ACK* message that is propagated back to the initiator of the *Join-Request* message. CAMP lets core nodes to leave the multicast mesh if these nodes are not required to provide efficient dissemination paths. A core node leaves the multicast group by broadcasting a *Quit* notification message to its neighbors.

A receiver node periodically checks whether the multicast data packets traverse the reverse shortest path back to the source. If it is not the case, the receiver node sends a *Heart-Beat* or the *Push-Join* message along the reverse shortest

Figure 10. NSMP – Multicast mesh maintenance (Lee & Kim, 2000)

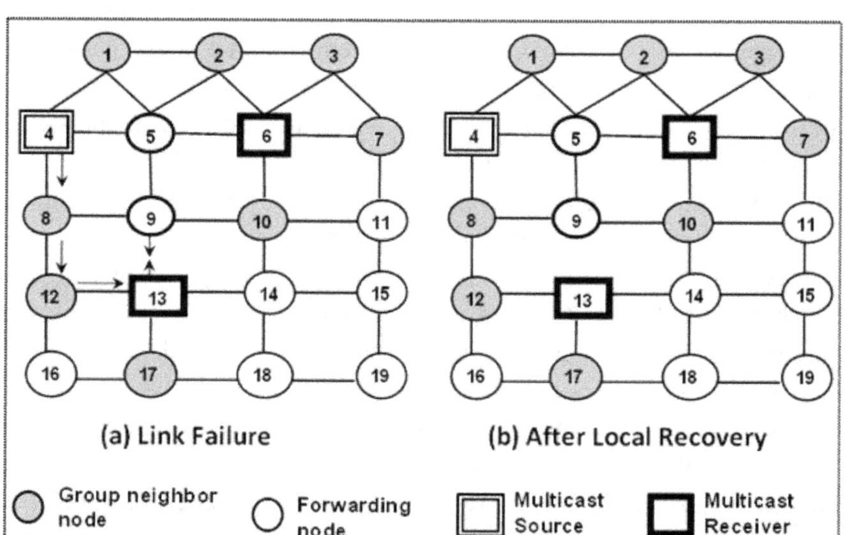

path to the source. A member node (including the receiver node) forwards the *Hear-Beat* message to its successor node on the reverse shortest path if the latter is already a member of the mesh; otherwise, the member node sends a *Push-Join* message to the neighbor successor node asking it to join the multicast mesh and waits for an *ACK* from that node. Duplex members respond with a regular ACK, while simplex members send an *ACK-Simplex* message. If no *ACK* is received within a certain time, the *Push-Join* message is propagated further until a member node that is directly connected to the source is reached. CAMP allows the member nodes of a mesh to select their "anchor" nodes – defined as neighbor nodes that are required to re-broadcast any non-redundant data packets received. The member nodes can periodically refresh their "anchor" nodes by broadcasting updates. A neighbor node can deny the "anchor" request or discontinue from that role if it is no longer interested at.

CAMP heavily needs the support of the underlying unicast routing protocol to provide route updates. In this context, CAMP prefers to co-exist with an underlying unicast routing protocol called wireless routing protocol that marks a subset of destinations as unreachable during periods of network re-convergence. CAMP piggybacks its control messages onto the back of WRP updates and the control overhead is bound to increase exponentially with traffic. Also, CAMP cannot be used directly with unicast routing protocols based on the Bellman-Ford algorithm (Cormen et. al., 2001) as well as extensions are needed to work with on-demand routing protocols.

Intelligent on-Demand Multicast Routing Protocol (IOD-MRP)

Even though, the concept of receiver-initiated joining in ad hoc multicast routing protocols seemed to have more advantages than source-initiated joining, CAMP has its own set of disadvantages that require serious modifications in order to be competent with ODMRP and its extensions. In this direction, Wang & Chang (2001) proposed an intelligent on-demand multicast routing protocol (IOD-MRP) that removes the concept of cores from CAMP and manages to reduce the control overhead, using an intelligent mobility management scheme, to 3-13% to that of CAMP. In CAMP, prospective receiver nodes unicast a *Join-Request* message to the core node to join the mesh. However, this may not be successful due to the dynamically changing network topology and routing information need not be always accurate and available. IOD-MRP proposes an intelligent mobility management procedure according to which receivers can search for the shortest path to join the multicast mesh in an on-demand fashion.

IOD-MRP designates nodes into two categories: I-nodes (nodes interested in the multicast session) and U-nodes (nodes not interested in the multicast session). A node opening the multicast session, floods a *Broadcast-Join-REQ* message if it cannot find any information about the session in its multicast routing table. Upon receiving a non-duplicate *Broadcast-Join-REQ* message, an I-node responds with a *Unicast-Join-REQ* message to the sender of the *Broadcast-Join-REQ* message and joins the multicast session.

IOD-MRP allows prospective receiver nodes to join the multicast mesh through two modes: *Unicast-Join-REQ* mode (preferred) and *Broadcast-Join-REQ*. An I-node first sends a *Unicast-Join-REQ* message to join the multicast session. When an active member node receives the *Unicast-Join-REQ* message for which it is also the targeted destination, the intermediate node responds with a positive *ACK* message sent back to the I-node. When a non-member node receives the *Unicast-Join-REQ* message, it will either send a negative *ACK* to the I-node or propagate the *Unicast-Join-REQ* message further towards the intended destination node. If an I-node does not receive a positive *ACK* within a certain time, it enters into the *Broadcast-Join-REQ* mode and initiates a scoped flooding of the *Broadcast-Join-REQ* message using expanding ring search. A non-member node receiving the *Broadcast-Join-REQ* message creates an entry for the multicast group in its routing table, decrements the *TTL* flag in the message by 1 and forwards the message further if the *TTL* is still positive. If a U-node receives either the *Join-REQ* packets or the multicast data packets, it can simply record the multicast session information in its routing table. Later, if the U-node becomes an I-node, it can simply initiate a *Unicast-Join-REQ* message to join the multicast group.

After a node joins the multicast session, the quality of the path traversed by the data packets sent from the multicast source is continuously monitored through an intelligent procedure (described below) so that the receiver node stays connected to the multicast source on the best path, called the *anchor* path. A path between the multicast source and a receiver is said to be an *anchor* path if (1) the path is traversed by most of the multicast data packets sent from the source to the receiver and (2) the path has a low packet loss rate. The receiver periodically monitors the quality of the paths traversed by the multicast data packets and notifies the multicast source of the *anchor* path it has chosen by sending a *Set-Anchor* message through that path. Intermediate nodes along the chosen *anchor* path are forced to stay in the mesh. If the receiver node determines a new path that is better than the current *anchor* path, then it sends a *Reset-Anchor* message on the recently used *anchor* path and sends a *Set-Anchor* message on the newly chosen best path. If no anchor paths can be found, the receiver resorts to unicast communication with the multicast source. When a source node leaves the multicast group, it sends a *Session-Leave* message that is forwarded down the mesh so that the member nodes can purge any table entries associated with the leaving multicast source.

CONCLUSION

This book chapter survey reviewed various multicast routing protocols for MANETs. Most of the multicast routing protocols for MANETs are either mesh-based or tree-based. Tree-based protocols can be either shared-trees or per-source based shortest trees. Mesh protocols can be either source-initiated (union of per-source meshes) or receiver-initiated (shared mesh). Tree-based protocols are efficient in data transmission while mesh-based protocols are robust to topology changes and are more stable than tree-based protocols. The routes chosen using tree-based protocols are fragile and require frequent reconfiguration. Each of the protocols discussed in this survey has its own advantages and disadvantages. An ideal protocol that satisfies all the requirements of MANETs is tough to design. For example, a protocol that aims at routing efficiency has to sacrifice for route robustness.

FUTURE RESEARCH DIRECTIONS

Future research would involve a comprehensive simulation analysis of all of these representative multicast routing protocols in a discrete-event simulator and study their performance under identical scenarios. A lot of research papers that are currently available on performance comparison studies of MANET multicast routing protocols do not conduct a comprehensive simulation analysis of the different categories of multicast routing protocols (identified in this chapter) under identical scenarios. So, it is not easy to generalize the performance tradeoffs between the different categories and the representative multicast routing protocols. This could lead to development of more hybrid tree and mesh based protocols that could neutralize the tradeoffs and yield optimal performance simultaneously for more than one performance metric.

Another potential research could be to study the behavior of these multicast routing protocols under different topology control graphs. Examples are: Relative neighborhood graph, Gabriel graph, Yao graph and Delaunay triangulation (Santi, 2005). Currently, the simulation studies available in the literature model the underlying network topology to be a unit disk graph wherein there is a link between any two nodes if the Euclidean distance between the two nodes is less than or equal to the transmission range of the nodes. While the unit disk graph model can yield significantly high network connectivity, each node will have a relatively larger number of neighbors. In a shared medium, typical of wireless networks, a larger neighborhood could severely drain the energy resources of a node as well as the network capacity due to the receipt of unintended packets and redundant transmissions. As the wireless nodes are battery charged, it is imperative to extend the operational lifetime of these devices as much as possible. Topology control is a potential solution to enhance the energy-efficiency and the capacity of the network. The focus of the topology control algorithms is to maintain the connectivity of the network with a reduced number of links as well as with a reduced transmission power per node (rather than the transmission power corresponding to the maximum transmission range) and thus optimize node lifetime.

REFERENCES

Agarwal, S., Ahuja, A., Singh, J. P., & Shorey, R. (2000). Route-lifetime assessment based routing protocol for mobile ad hoc networks. In *Proceedings of the International Conference on Communications* (pp. 1697-1701). New Orleans, LA: IEEE.

Bae, S., Lee, S., Su, W., & Gerla, M. (2002). The design, implementation, and performance evaluation of the on-demand multicast routing protocol in multi-hop wireless networks. *IEEE Network*, *4*(1), 70–77.

Burri, N., Rickenbach, P., Watternhofer, R., & Weber, Y. (2006). Topology control made practical: Increasing the performance of source routing. *Lecture Notes in Computer Science*, *4325*, 1–12. doi:10.1007/11943952_1

Chiang, C. C., Gerla, M., & Zhang, L. (1997a). Shared tree wireless network multicast. In *Proceedings of the 6th International Conference on Computer Communications and Networks* (pp. 28-33). Las Vegas, NV: IEEE.

Chiang, C. C., Gerla, M., & Zhang, M. (1998). Forwarding group multicast protocol (FGMP) for multi-hop, mobile wireless networks. *Baltzer Cluster Computing*, *1*(2), 187–196. doi:10.1023/A:1019037500012

Chiang, C. C., Wu, H. K., Liu, W., & Gerla, M. (1997b). Routing in clustered multi-hop, mobile wireless networks with fading channel. In *Proceedings of the Singapore International Conference on Networks* (pp. 197-211). Singapore: IEEE.

Cormen, T. H., Leiserson, C. E., Rivest, R. L., & Stein, C. (2001). *Introduction to algorithms*. Cambridge, MA: MIT Press.

Deering, S., Estrin, D. L., Farinacci, D., Jacobson, V., Liu, C. G., & Wei, L. (1996). The PIM architecture for wide-area multicast routing. *IEEE/ACM Transactions on Networking*, *4*(2), 153–162. doi:10.1109/90.490743

Garcia-Luna-Aceves, J. J., & Madruga, E. L. (1999). The core-assisted mesh protocol. *IEEE Journal on Selected Areas in Communications*, *17*(8), 1380–1394. doi:10.1109/49.779921

Gerla, M., Chiang, C. C., & Zhang, L. (1999). Tree multicast strategies in mobile, multi-hop wireless networks. *ACM Baltzer Mobile Networks and Applications*, *4*(3), 193–207. doi:10.1023/A:1019150932356

Haas, Z. J., & Pearlman, M. R. (2001). The performance of query control schemes for the zone routing protocol. *IEEE/ACM Transactions on Networking*, *9*(4), 427–438. doi:10.1109/90.944341

Hofmann-Wellenhof, B., Lichtenegger, H., & Collins, J. (2004). *Global positioning system: Theory and practice*. Berlin, Germany: Springer Publishers.

Johnson, D. B. (1994). Routing in ad hoc networks of mobile hosts. In *Proceedings of the Workshop on Mobile Computing Systems and Applications* (pp. 158-163). Santa Cruz, CA: IEEE.

Lee, C. Y., & Cho, H. K. (2001). Multicast routing considering reliability and network load in wireless ad hoc networks. In *Proceedings of the 53rd VTS Spring Vehicular Technology Conference* (pp. 2203-2207). Rhodes, Greece: IEEE.

Lee, M., & Kim, Y. (2001). PatchODMRP: An ad hoc multicast routing protocol. In *Proceedings of the 15th International Conference on Information Netwrking* (pp. 537-543). Beppu City, Japan: IEEE.

Lee, S., Gerla, M., & Chaing, C. C. (1999). On-demand multicast routing protocol. In *Proceedings of the Wireless Communications and Networking Conference* (pp. 1298-1302). New Orleans, LA: IEEE.

Lee, S., & Kim, C. (2000). Neighbor supporting ad hoc multicast routing protocol. In *Proceedings of the 1st Annual Workshop on Mobile and Ad hoc Networking and Computing* (pp. 37-44). Boston, MA: ACM.

Lee, S. J., Su, W., Hsu, J., Gerla, M., & Bagrodia, R. (2000). A performance comparison study of ad hoc wireless multicast protocols. In *Proceedings of the 19th Annual Joint Conference on Computer Communications* (pp. 565-574). Tel Aviv, Israel: IEEE.

Moy, J. (1994). Multicast extensions to OSPF. *Request for Comments, 1584*.

Murthy, S., & Garcia-Luna-Aceves, J. J. (1996). An efficient routing protocol for wireless networks. *Mobile Networks and Applications*, *1*(2), 183–197. doi:10.1007/BF01193336

Ozaki, T., Kim, J. B., & Suda, T. (1999). Bandwidth-efficient multicast routing protocol for ad hoc networks. In *Proceedings of the 8th International Conference on Computer Communications and Networks* (pp. 10-17). Boston, MA: IEEE.

Park, V. D., & Corson, M. S. (1997). A highly adaptive distributed routing algorithm for mobile wireless networks. In *Proceedings of the 16th Annual International Conference on Computer Communications* (pp. 1405-1413). Las Vegas, NV: IEEE.

Perkins, C. E., & Bhagwat, P. (1994). Highly dynamic destination sequenced distance vector routing for mobile computers. In *Proceedings of the International Conference on Special Interest Group on Data Communications* (pp. 234-244). London, UK: ACM.

Perkins, C. E., & Royer, E. M. (1999). Ad hoc on-demand distance vector routing. In *Proceedings of the 2nd International Workshop on Mobile Computing Systems and Applications* (pp. 90-100). New Orleans, LA, USA: IEEE.

Royer, E., & Perkins, C. E. (1999). Multicast operation of the ad hoc on-demand distance vector routing protocol. In *Proceedings of the 5th International Conference on Mobile Computing and Networking* (pp. 207-218). Seattle, WA: ACM.

Santi, P. (2005). *Topology control in wireless ad hoc and sensor networks*. Hoboken, NJ: Wiley Publishers. doi:10.1002/0470094559

Su, W., & Gerla, M. (1999). IPv6 flow handoff in ad hoc wireless networks using mobility prediction. In *Proceedings of the Global Telecommunications Conference* (pp. 271-275). New Orleans, LA: IEEE.

Toh, C. K. (1997). Associativity-based routing for ad hoc mobile networks. *IEEE Personal Communications*, *4*(2), 103–139.

Toh, C. K., Guichal, G., & Bunchua, S. (2000). ABAM: On-demand associativity-based multicast routing for ad hoc mobile networks. In *Proceedings of the 52nd Fall Vehicular Technology Conference* (pp. 987-993-97). Boston, MA: IEEE.

Waitzman, D., Partridge, C., & Deering, S. (1988). Distance vector multicast routing protocol. *Request for Comments, 1075*.

Wang, K., & Chang, C. T. (2001). An intelligent on-demand multicast routing protocol in ad hoc networks. In *Proceedings of the 15th International Conference on Information Networking* (pp. 909-914). Beppu City, Japan: IEEE.

Wu, C. W., & Tay, Y. C. (1999). AMRIS: A multicast protocol for ad hoc wireless networks. In *Proceedings of the IEEE Military Communications Conference* (pp. 25-29). Atlantic City, NJ: IEEE.

Xie, J., Talpade, R. R., Mcauley, A., & Liu, M. (2002). AMRoute: Ad hoc multicast routing protocol. *Mobile Networks and Applications*, *7*(6), 429–439. doi:10.1023/A:1020748431138

Zhang, X., & Jacob, L. (2003). Multicast zone routing protocol in mobile ad hoc wireless networks. In *Proceedings of the 28th International Conference on Local Computer Networks* (pp. 150-159). Bonn, Germany: IEEE.

ADDITIONAL READING

Bae, S., Lee, S., Su, W., & Gerla, M. (2000). The design, implementation and performance evaluation of the on-demand multicast routing protocol in multi-hop wireless networks. *IEEE Network*, *4*(1), 70–77.

Bettstetter, C., Hartenstein, H., & Perez-Costa, X. (2004). Stochastic properties of the random-way point mobility model. *Wireless Networks*, *10*(5), 555–567. doi:10.1023/B:WINE.0000036458.88990.e5

Bianchi, G. (2000). Performance analysis of the IEEE 802.11 distributed coordination function. *IEEE Journal on Selected Areas in Communications*, *18*(3), 535–547. doi:10.1109/49.840210

Bononi, L., & Felice, M. D. (2006). Performance analysis of cross-layered multi-path routing and MAC layer solutions for multi-hop ad hoc networks. In *Proceedings of the International Workshop on Mobility Management and Wireless Access* (pp. 190-197). Terromolinos, Spain: ACM.

Broch, J., Maltz, D. A., Johnson, D. B., Hu, Y. C., & Jetcheva, J. (1998). A performance comparison of multi-hop wireless ad hoc network routing protocols. In *Proceedings of the 4th International Mobile Computing and Networking Conference* (pp. 85-97). Dallas, TX: ACM.

Camp, T., Boleng, J., & Davies, V. (2002). A survey of mobility models for ad hoc network research. *Wireless Communication and Mobile Computing*, *2*(5), 483–502. doi:10.1002/wcm.72

Chen, Y. P., & Liestman, A. L. (2002). Approximating minimum size weakly-connected dominating sets for clustering mobile ad hoc networks. In *Proceedings of the International Symposium on Mobile Ad hoc Networking and Computing*. Lausanne, Switzerland: ACM.

Creixell, W., & Sezaki, K. (2007). Routing protocol for ad hoc mobile networks using mobility prediction. *International Journal of Ad Hoc and Ubiquitous Computing*, *2*(3), 149–156. doi:10.1504/IJAHUC.2007.012416

Das, S. K., Manoj, B. S., & Murthy, C. S. R. (2002). Weight-based multicast routing protocol for ad hoc wireless networks. In *Proceedings of the Global Telecommunications Conference* (pp. 117-121). Taipei, Taiwan: IEEE.

Farago, A., & Syrotiuk, V. R. (2003). MERIT: A scalable approach for protocol assessment. *Mobile Networks and Applications*, *8*(5), 567–577. doi:10.1023/A:1025193929081

Fasolo, E., Zanella, A., & Zorzi, M. (2006). An effective broadcast scheme for alert message propagation in vehicular ad hoc networks. In *Proceedings of the International Conference on Communications, Vol. 9* (pp. 3960-3965). Istanbul, Turkey: IEEE.

Feeney, L. M. (2001). An energy consumption model for performance analysis of routing protocols for mobile ad hoc networks. *Journal of Mobile Networks and Applications*, *3*(6), 239–249. doi:10.1023/A:1011474616255

Gil, H.-R., Yoo, J., & Lee, J.-W. (2003). An on-demand energy-efficient routing algorithm for wireless ad hoc networks. *Lecture Notes in Computer Science*, *2713*, 302–311. doi:10.1007/3-540-45036-X_31

Gunes, M., Sorges, U., & Bouazizi, I. (2002). ARA: The ant colony based routing algorithm for MANETs. In *Proceedings of the ICPP Workshop on Ad Hoc Networks* (pp. 79-85). Vancouver, Canada: IEEE.

Keiss, W., Fuessler, H., & Widmer, J. (2004). Hierarchical location service for mobile ad hoc networks. *ACM Mobile Computing and Communications Review*, *8*(4), 47–58. doi:10.1145/1052871.1052875

Lim, G., Shin, K., Lee, S., Yoon, H., & Ma, J. S. (2002). Link stability and route lifetime in ad hoc wireless networks. In *Proceedings of the International Conference on Parallel Processing Workshops* (pp. 116-123), Vancouver, Canada: IEEE.

Lin, X., & Stojmenovic, I. (2003). Location-based localized alternate, disjoint and multi-path routing algorithms for wireless networks. *Journal of Parallel and Distributed Computing*, *63*, 22–32. doi:10.1016/S0743-7315(02)00037-0

Meghanathan, N. (2007). Impact of broadcast route discovery strategies on the performance of mobile ad hoc network routing protocols. In *Proceedings of the International Conference on High Performance Computing, Networking and Communication Systems* (pp. 144-151). Orlando, FL, USA: Promote Research.

Meghanathan, N. (2007). Path stability based ranking of mobile ad hoc network routing protocols. *ISAST Transactions Journal on Communications and Networking*, *1*(1), 66–73.

Meghanathan, N. (2009). Survey and taxonomy of unicast routing protocols for mobile ad hoc networks. *The International Journal on Applications of Graph Theory in Wireless Ad hoc Networks and Sensor Networks, 1*(1), 1-21.

Meghanathan, N. (2009). A location prediction based reactive routing protocol to minimize the number of route discoveries and hop count per path in mobile ad hoc networks. *The Computer Journal*, *52*(4), 461–482. doi:10.1093/comjnl/bxn051

Meghanathan, N. (2010). An algorithm to determine minimum velocity-based stable connected dominating sets for ad hoc networks. *Communications in Computer and Information Science Series*, *94*, 206–217. doi:10.1007/978-3-642-14834-7_20

Meghanathan, N., & Sugumar, M. (2009). A beaconless minimum interference based routing protocol for mobile ad hoc networks. *Communications in Computer and Information Science Series, 40*, 58–69. doi:10.1007/978-3-642-03547-0_7

Mueller, S., Tsang, R. P., & Ghosal, D. (2004). Multi-path routing in mobile ad hoc networks: Issues and challenges. *Lecture Notes in Computer Science, 2965*, 209–234. doi:10.1007/978-3-540-24663-3_10

Ni, S., Tseng, Y., Chen, Y., & Sheu, J. (1999). The broadcast storm problem in a mobile ad hoc network. In *Proceedings of the International Conference on Mobile Computing and Networking* (151-162). Seattle, WA: ACM.

Sheu, P., Tsai, H., Lee, Y., & Cheng, J. (2009). On calculating stable connected dominating sets based on link stability for mobile ad hoc networks. *Tamkang Journal of Science and Engineering, 12*(4), 417–428.

Son, D., Helmy, A., & Krishnamachari, B. (2004). The effect of mobility-induced location errors on geographic routing in mobile ad hoc sensor networks: Analysis and improvement using mobility prediction. *IEEE Transactions on Mobile Computing, 3*(3), 233–245. doi:10.1109/TMC.2004.28

Viswanath, K., Obraczka, K., & Tsudik, G. (2006). Exploring mesh and tree-based multicast routing protocols for MANETs. *IEEE Transactions on Mobile Computing, 5*(1), 28–42. doi:10.1109/TMC.2006.11

Wang, H., Ma, K., & Yu, N. (2005). Performance analysis of multi-path routing in wireless ad hoc networks. In *Proceedings of the International Conference on Wireless Communications, Networking and Mobile Computing, Vol. 2* (pp. 723-726). Maui, HI: IEEE.

Ye, Z., Krishnamurthy, S. V., & Tripathi, S. K. (2003). A framework for reliable routing in mobile ad hoc networks. In *Proceedings of the International Conference on Computer Communications* (pp. 270-280). San Francisco, CA: IEEE.

KEY TERMS AND DEFINITIONS

Beacon: Each node in the network periodically broadcasts a small control message called 'beacon' to all the nodes within its transmission range to announce its presence in the neighborhood.

Broadcast: Communication initiated from one source node and targeted to all the other nodes in the network.

Multicast Communication: Communication between a source node and multiple receiver nodes.

Multicast Group: A set of receiver nodes that are interested in receiving packets sent as part of a multicast communication from a particular source node.

Multicast Mesh: A connected sub graph of a network graph connecting one or more multicast source nodes and the receiver nodes of the multicast group through more than one path, providing robustness at the cost of efficiency.

Multicast Tree: A connected sub graph of a network graph connecting the multicast source node or a core node and the receiver nodes of the multicast group through exactly one path; a tree can get disconnected if any link in the tree fails.

Mobile Ad hoc Network: A dynamically changing infrastructureless and resource-constrained network of wireless nodes moving arbitrarily, independent of each other.

Network Simulation: The process of imitating the working of a network through a software program that implements one or more interaction models between the different entities (nodes, links and etc)

Routing Protocol: A protocol used for communication between any two nodes or a set of nodes in a network through one or more hops

Unicast: Communication between a single source node and a single destination node.

Chapter 8
Performance Comparison of AODV and DSDV Routing Protocols of MANET

Bhaskar N. Patel
B.S. Patel Polytechnic, India

S.G. Prajapati
B.S.Patel Polytechnic, India

ABSTRACT

A mobile ad hoc network is a collection of mobile nodes communicating without wired network and infrastructure or centralized control. Due to limited transmission range of wireless network, more than single hops may be required to transfer data across the network. In order to reliable communication within the network, a routing protocol is used which are call MANET routing protocol. The major function of such an MANET routing protocol is to establishment short and real route between a pair of nodes so that messages may be delivered in a timely manner. This chapter examines two routing protocols for mobile ad hoc networks– the Destination Sequenced Distance Vector (DSDV), the pro-active routing protocol, and the Ad hoc On- Demand Distance Vector routing (AODV), an re-active routing protocol. The authors compare both protocols based on packet delivery fraction and average delay with changing number of source nodes and pause time.

1. INTRODUCTION

With the recent advances in mobile and wireless technology, more and more mobile devices with wireless capability are covering the global market. It has become necessary to set up networks without costly infrastructure and investment. This is the driving force behind wide-scale use of MANETs (Mobile Ad-Hoc Networks). An Ad-Hoc mobile network is a collection of nodes that are capable of changing on a continual basis. A routing protocol is used to discover routes between the nodes, so as to facilitate communication between these nodes.

DOI: 10.4018/978-1-4666-0321-9.ch008

- **Description of Chapter**

This paper examines two routing protocols for mobile ad hoc networks. the Destination Sequenced Distance Vector (DSDV), the table-driven protocol and the Ad hoc On- Demand Distance Vector routing (AODV), an On –Demand protocol. It evaluates both protocols based on packet delivery fraction, average delay and throughput while varying number of nodes and pause time using NS-2 (Network Simulator). Traffic pattern and node mobility is generated using ns-2 utility and performance metrics are calculated using a java module.

2. PROACTIVE ROUTING PROTOCOL

The nodes maintain a table of routes to every destination in the network, for this reason they periodically exchange messages. At all times the routes to all destinations are ready clear and as a consequence initial delays before sending data are small(MANET, n.d.). Keeping some routes to all destinations up-to-date, even if they are not used, is a disadvantage with regard to the usage of bandwidth and of network resources. It is also possible to delays data packets due to control packets traffic and there are more packet collisions due to more network traffic. Proactive protocols do able to change in the topology. Therefore the proactive protocol is appropriate for a low mobility network.

3. REACTIVE ROUTING PROTOCOL

These protocols were designed to overcome the wasted effort in maintaining Unused routes. Routing information is acquired only when there is a need for it. The needed routes are calculated on demand (MANET, n.d.). This saves the overhead of maintaining unused routes at each node, but on the other hand the latency for sending data packets will considerably increase. It is obvious it wil take long time before data transmission because it has to wait until a route to the destination is ready. As reactive routing protocols flood the network to discover the route, they are not optimal in terms of bandwidth utilization, but they scale well in the frequency of topology change. Thus this strategy is suitable for high mobility networks.

4. DESTINATION SEQUENCE DISTANT VECTOR PROTOCOL

The destination sequenced distance vector routing protocol is a proactive routing protocol which is a modification of conventional Bellman-Ford routing algorithm. every time adds a new attribute, sequence number, to each route table entry at each node. Each node maintained Routing table and with this table, node transmits the packets to other nodes in the network. This protocol was motivated for the use of data exchange along changing and arbitrary paths of interconnection which may not be close to any base station(Ramachandran, 2006).

4.1 Advantage

- DSDV protocol guarantees loop free paths .
- Count to infinity is less .
- Avoid extra traffic with incremental updates instead of full dump updates.
- DSDV maintains only the best path instead of maintaining multiple paths to every destination. so the space in routing table is reduced.

4.2 Disadvantage

- Wastage of bandwidth due to periodically hello message even if no change in the network topology .
- Doesn't support Multi path Routing.

- Difficult to find a time delay for the advertisement of routes.
- It is difficult to maintain the routing table's advertisement for larger network. which consumes more bandwidth.

5. ADHOC ON DEMAND DISTANCE VECTOR ROUTING PROTOCOL

AODV is a very simple, efficient, and effective routing protocol for MANET. which do not have fixed topology. This algorithm was motivated by the limited bandwidth that is available in the media that are used for wireless communications. It has all the advantage which are in DSR and DSDV. The on demand route discovery and route maintenance from DSR and hop-by-hop routing, usage of node sequence numbers from DSDV make the algorithm cope up with topology and routing information (Perkins, Belding-Royer, & Das, 2002). Obtaining the routes purely on-demand makes AODV a very useful and desired algorithm for MANET.

5.1 Advantage

- Because of its reactive nature, AODV can handle highly dynamic behavior of MANET
- Used for both unicasts and multicasts using the 'J' (Join multicast group) flag.

5.2 Disadvantage

- *Requirement on broadcast medium:* The algorithm expects/requires that the nodes in the broadcast medium can detect each others' broadcasts.
- *Overhead on the bandwidth:* Overhead on bandwidth will be occurred compared to DSR, when an RREQ travels from node to node in the process of discovering the route info on demand, it sets up the reverse path in itself with the addresses of all the nodes through which it is passing and it carries all this info all its way.
- *No reuse of routing info:* AODV lacks an efficient route maintenance technique. The routing info is always obtained on demand, including for common case traffic .

6. COMPARISION OF ROUTING PROTOCOLS

Up to here we have compare theoretically. Table 1 summarizes and compares the result from those theoretical analyses and shows what properties the protocol have and do not have (Yadav & Yadav, 2007). As it can be seen from table, none of the protocol supports power conservation. This is however working in progress and probably be added to the protocols. All protocols are distributed, thus none of the protocols is depended on a centralized node and can there fore easily reconfigure in the event of topology changes.

6.1 Differences between AODV and DSDV

- DSDV is a table driven routing protocol, which maintains routes to each and every node in the network, while AODV is a reactive routing protocol, which finds the path on demand, or whenever the route is required.

Table 1. Parameter compare

Parameter	DSDV	AODV
Loop Free	Yes	Yes
Reactive	No	Yes
Multicast	No	Yes
Power Consevation	No	No
Periodic Broadcast	Yes	Yes
Require reliable or sequences data	No	No

- Broadcasting in DSDV is done periodically to maintain routing updates and in AODV, only hello messages are propagated to its neighbors to maintain local connectivity.
- DSDV routing algorithm maintains a sequence number concept for updating the latest information for a route. Even, the same concept is adapted by AODV routing protocol.
- For sending data to a particular destination, there is no need to find a route as DSDV routing protocol maintains all the routes in the routing tables for each node. While, AODV has to find a route before sending a data.
- Overhead in DSDV is more when the network is large and it becomes hard to maintain the routing tables at every node. But, in AODV overhead is less as it maintains small tables to maintain local connectivity.
- DSDV cannot handle mobility at high speeds due to lack of alternative routes hence routes in routing table is stale. While in AODV this is the other way, as it find the routes on demand.

7. NETWORK SIMULATOR

Network Simulator (NS) is an object-oriented, discrete event simulator for networking research (NS-2, n.d.A; NS-2, n.d.B). NS provides substantial support for simulation of TCP, routing and multicast protocols over wired and wireless networks. NS is written in C++, with an OTcl (object oriented Tcl) interpreter as a command and configuration interface. The C++ part, which is fast to run but slower to change, is used for detailed protocol implementation. The OTcl part, on the other hand, which runs much slower but can be changed very quickly, is used for simulation configuration. One of the advantages of this split-language programming approach is that it allows for fast generation of large scenarios.

7.1 System Requirements:

This result developed on a machine running fedora 7 and NS 2. The NS-2 version used in this project is "ns-allinone-2.28", which is a single tar file with all the requisite packages that easily installs with one command (NS-2, n.d.A).

8. PERFORMANCE EVALUATION METRICS

Three important performance metrics are evaluated:

- *Packet delivery fraction:* The ratio of the data packets delivered to the destinations to those generated by the CBR sources.
- *Average end-to-end delay of data packets:* This includes all possible delays caused by buffering during route discovery latency, queuing at the interface queue, retransmission delays at the MAC, and propagation and transfer times.

8.1 Simulation Model

The simulation model consists of 50 mobile nodes. The topology is a square area with 500 m length and 500 m width. All simulations are run for 100 seconds of simulated time. In one scenario 15 of the 50 mobile nodes are constant bit rate traffic sources (cbr). They are distributed randomly within the mobile ad hoc network. Once the simulation starts all the sources start sending data to receivers. After these time the sources continue sending data until one second before the end of the simulation. The destination of each of the sources is chosen randomly. Two other scenarios executed with 30 and 45 number of sources respectively.

8.2 Simulation Parameters

9. RESULTS

In simulation model of this paper, each packet starts its journey from a random location to a random destination with a randomly chosen speed (uniformly distributed between 0–25 m/s). Once the destination is reached, another random destination is targeted after a pause. The pause time, which affects the relative speeds of the mobiles, is varied. Simulations are run for 100 simulated seconds. Identical mobility and traffic scenarios are used across protocols to gather fair results (refer to Table 2).

9.1 Packet Delivery Fraction

Data packets in Figure 1a, 1b, and 1c show a comparison between both the routing protocols on the basis of packet delivery fraction as a function of pause time and using different number of traffic sources. Both the protocols deliver a greater percentage of the originated when there is little or no mobility. So PDF is almost 100% when there is no mobility i.e. pause time is equal to simulation time.

Table 2. Simulation parameters

Simulator	Ns 2.28
Study Protocol	DSDV, AODV
Sumulation time	100
Simulation area	500 X 500
Movement model	Random waypoint
Speed	0-25 m/s with 5 interval
Traffic type	CBR
Pause time	0-100 with 20th interval
Medium access protocol	IEEE 802.11
Plat form used	Fedora 7.0
System Configuration	Core 2 duo processor with 1 GB Ram

- The On demand protocol, AODV performed particularly well, delivering very high percentage of the data packets regardless of the mobility rate. The packet delivery of AODV is almost independent of the number of sources that is varying number of sources does not affect the AODV that much.
- DSDV performance is worst when mobility is high. This poor performance is because of the reason that DSDV is not an on demand protocol and it keeps one route per destination, there for lack of alternate routes and presence of stale routes in the routing table when nodes moving at higher rate leads to packet drops. The packet delivery of DSDV protocol depends on the number of sources, as it is obvious from Figure 1.

9.2 Average End-End Delay

Figure 2 shows comparison between both the routing protocols on the basis of average end-to-end delay as a function of pause time, using different number of sources. DSDV performed pretty stable and the delay kept about 0.04 seconds when pause time increased from 0 seconds to 100 second. The reason is that it is a table driven protocol, so a node does not need to find a route before transmitting packets. So the delay is quite stable.

For AODV the delay is much more then the DSDV. As AODV is On-demand protocol, with an increased number of sources and high mobility there are more link failures therefore there are more route discoveries. AODV takes more time during the route discovery process as first it finds the route hop by hop and then it gets back to the source by back tracking that route. All this leads to delays in the delivery of data packets.

Figure 1. Packet delivery

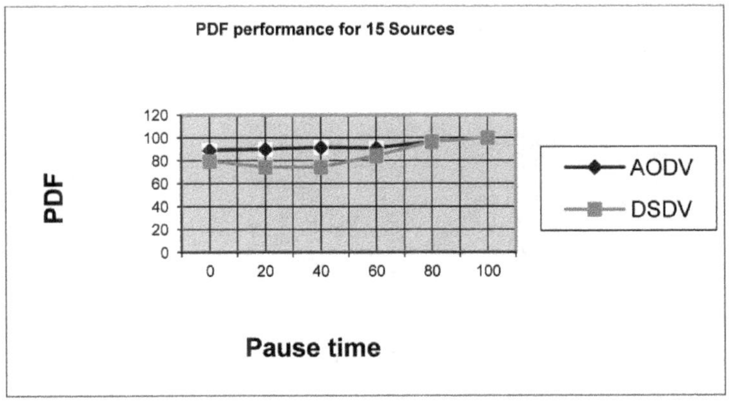

(a) for 15 sources

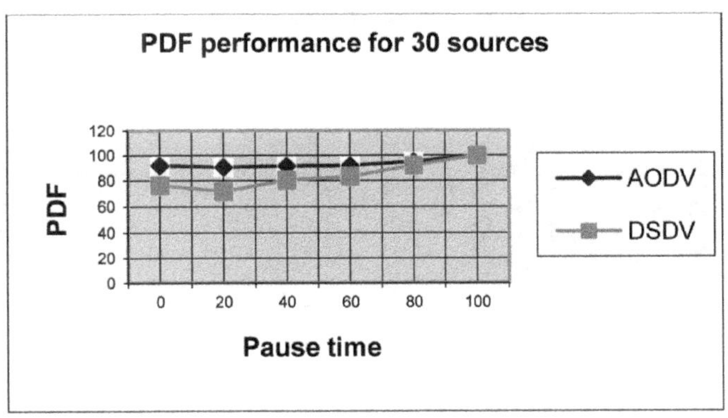

(b) for 30 sources

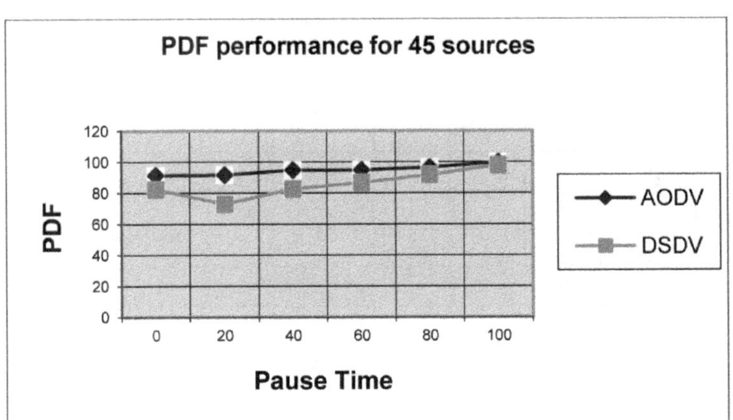

(c) for 45 sources

Figure 2. AVG end to end delay

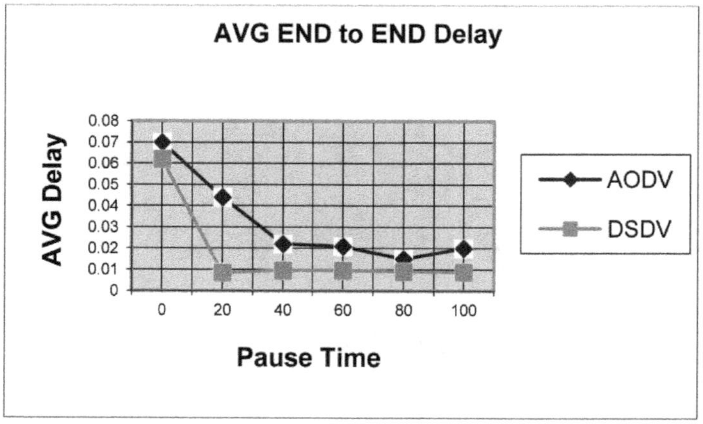

(a) for 15 sources

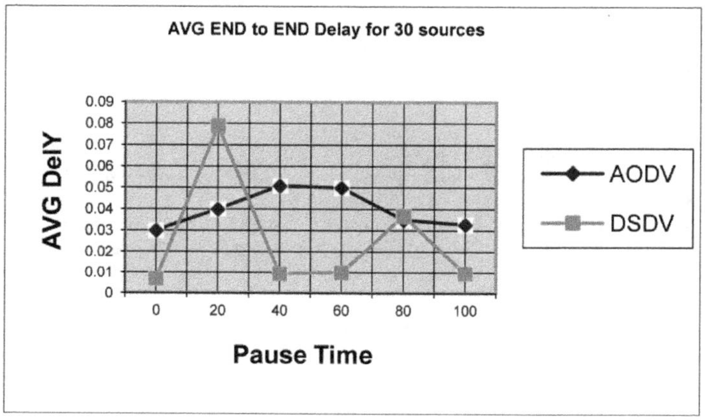

9.3 (b) for 30 sources

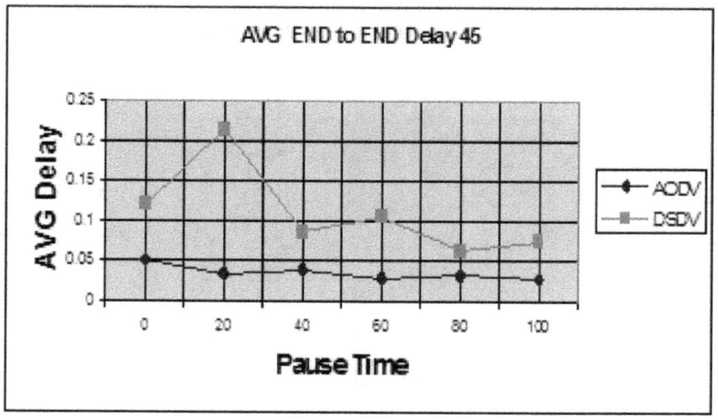

9.3 (c) for 45 sources

10. CONCLUSION

This paper compared the two ad hoc routing protocols. AODV an On – Demand routing protocol, and DSDV a table driven protocol. Simulation results show that both of the protocols deliver a greater percentage of the originated data packets when there is little or no node mobility, and gives 100% pdf when there is no node motion. The packet delivery of AODV is almost independent of the number of sources. AODV suffers from end to end delays. DSDV packet delivery fraction is very low for high mobility scenarios.

The results presented are valid for the specific scenario used in this paper. Therefore, one cannot tell this will be the final result for the different scenarios. There are many factors that can be changed and their impact should be investigated. The aim in future work will be to examine them in greater detail. For example, changing the number of mobile nodes and the size of the topology changes the mobile node density. Its impact should be investigated. Other issues that should be examined are the impact of the. Other factor like normalize routing load can be considered for simulation.

- The number of packets sent per second
- The speed of the mobile nodes.

REFERENCES

MANET. (n.d.). *The MANET web page*. Retrieved from http://www.ietf.org /html.charters/manet-charter.html

NS-2. (n.d.A). *Documentation*. Retrieved from http://www.isi.edu/nsnam/ ns/ns-documentation.html

NS-2. (n.d.B). *Tutorial*.

Perkins, C. E., Belding-Royer, E. M., & Das, S. R. (2002). *Ad hoc on-demand distance vector (AODV) routing*. Internet Engineering Task Force, Internet Draft (Work in Progress), June 2002.

Perkins, C. E., & Bhagwat, P. (1994). Highly dynamic destination –Sequence distant vector routing (DSDV) for mobile compute. *Proceedings of the Conference on Communications Architectures, Protocols and Applications*, England, August 1994, (pp. 234-244).

Ramachandran, K. (2006). *Technical report, University of California, Santa Barbara, USA*. Retrieved April 15, 2006, from http://www.cs.ucsb.edu/krishna/aodv-st

Yadav, N. S., & Yadav, R. P. (2007). Performance comparison AND Analysis of on demand & table driven routing protocols. *International Journal of Information Technology, 4*(2). ISSN 13052403

Chapter 9
Analyzing Performance of Ad hoc Routing Protocols under Various Constraints

Lalitsen Sharma
University of Jammu, India

Supriya Gupta
University of Jammu, India

ABSTRACT

The mobility of nodes in mobile ad hoc networks and absence of any centralized control cause unpredictable changes in the network topologies. This makes routing a challenging task. Several routing protocols for mobile ad-hoc networks have come into existence. The protocols are classified in mainly in three categories: proactive, reactive, and hybrid. In this chapter, a study of one of each of the proactive and reactive protocols (respectively, Destination Sequence Distance Vector routing [DSDV], and Dynamic Source Routing [DSR]) is presented. The performance of above said protocols has been measured under varying mobility environment using NS-2 simulator based upon three quality metrics: average end-to-end delay, throughput, and jitter.

INTRODUCTION

Wireless networking is an emerging technology that allows users to access information and services electronically, regardless of their geographic position. There are currently two variations of mobile wireless networks. The first is known as the *infrastructure network* (i.e., a network with fixed and wired gateways). The bridges for these networks are known as *base stations*. A mobile unit within these networks connects to, and communicates with, the nearest base station that is within its communication radius. When it goes out of range of one base station, it connects with new base station and starts communicating through it. This is called handoff. In this approach

DOI: 10.4018/978-1-4666-0321-9.ch009

the base stations are fixed. The second type of mobile wireless network is the infrastructure less mobile network, commonly known as mobile *ad hoc* network (MANET). Ad hoc networks have no fixed routers; all nodes are movable and can be connected dynamically in an arbitrary manner. As the range of each host's wireless transmission is limited, so to communicate with hosts outside its transmission range, a host enlists its nearby hosts and uses them in forwarding packets to the destination. So all nodes of these networks behave as routers and take part in discovery and maintenance of routes to other nodes in the network. The mobile ad hoc networks are becoming increasingly popular due their less cost and rapid deployability, inherent support for mobility and the potential to provide ad hoc connectivity to devices, to form temporary networks in areas where there is no fixed infrastructure. Nodes in mobile ad-hoc network are free to move and organize themselves in an arbitrary fashion. Each user is free to roam about while in communication with others and operates as a host as well as a router by forwarding packets for other mobile nodes in the network, that are not in direct wireless transmission range of each other (Scott, Joseph, & Gregory, 1999).

An ad hoc routing protocol is a standard that controls how nodes decide which way to route packets between hosts in a mobile ad hoc network. The primary goal of such an ad hoc network routing protocol is to establish correct and efficient route between a pair of nodes so that messages may be delivered in a timely manner. Route construction should be done with a minimum of overhead and bandwidth consumption. Development of efficient routing protocols for mobile ad hoc networks is a challenging task as due to the limited resources in MANETs, the node movements, the change in network topology randomly and rapidly at unpredicted times and also the bandwidth is limited and can vary due to fading, noise and interference. A routing algorithm is required that can efficiently use the limited resources while at the same time be adaptable to the changing network conditions such as network size, traffic density and network partitioning. The traditional link-state and distance-vector algorithms do not scale in large MANETs (Perkins, Belding-Royer, & Das, 2003). The periodic or frequent route updates in large networks may consume a significant part of the available bandwidth, increase channel contention, and require each node to frequently recharge its power supply. To overcome the problems associated with the link-state and distance-vector algorithms, a number of routing protocols have been proposed for MANETs. In this chapter we initially present the main classes of routing protocols viz proactive, reactive and hybrid. Then the performance of two major routing protocols for mobile ad hoc networks, 1) Dynamic Source Routing (DSR) and 2) Destination Sequence Distance Vector routing (DSDV), is measured under varying mobility environment using NS-2 simulator. The performance is analyzed on basis of three quality metrics average end-to-end delay, throughput and jitter.

CLASSIFICATION OF ROUTING PROTOCOLS FOR MOBILE AD HOC NETWORKS

The mobile ad hoc network routing protocols are generally classified on the basis of how routing information is acquired and maintained by mobile nodes. According to this criterion, mobile ad hoc network routing protocols are divided, as discussed above, into proactive or global routing, reactive or demand routing, and hybrid routing.

Proactive Routing Protocols

A proactive routing protocol is also called a "table-driven" routing protocol. Using a proactive routing protocol like DSDV (Perkins & Bagwat, 1994), CGSR (Liu et al., 1997), OLSR (Clausen & Jacquet, 2003), GSR (Chen & Gerla, 1997), nodes in a mobile ad hoc network continuously evaluate routes to all reachable nodes and attempt to maintain consistent, up-to-date routing information from each node to all other nodes by exchanging routing information periodically. When a network topology change occurs, respective updates must be propagated throughout the network to notify the change. Using proactive routing algorithms, mobile nodes proactively update the network state and maintain a route regardless of whether data traffic exists or not, and the overhead to maintain up-to-date network topology information is high. The main advantage of proactive routing is when a source needs to send packets to a destination; the route is already available with no latency. But it consumes a lot of the scarce resources like bandwidth, buffers and other overheads to maintain up-to-date information of all the routes even if some routes may never be used. The next section will introduce the proactive mobile ad hoc network routing protocol viz. DSDV, the Destination Sequenced Distance Vector routing protocol.

Reactive Routing Protocols

Reactive routing protocols for mobile ad hoc networks are also called "on-demand" routing protocols. In reactive routing protocols such as DSR (Johnson, Hu, & Maltz, 2007), AODV (Scott, Joseph, & Gregory, 1999), SSA (Dubey et al., 1996), routing paths are searched only when needed. A route discovery operation invokes a route-determination procedure. The discovery procedure terminates when either a route has been found or no route is available after examining all route permutations. In a mobile ad hoc network, active routes may be disconnected due to node mobility. Therefore, route maintenance is an important operation of reactive routing protocols. Compared to the proactive routing protocols, there is less control overhead in reactive routing protocols. However, when using reactive routing protocols, source nodes may suffer from long delays for route searching before they can forward data packets. A reactive routing protocol Dynamic Source Routing (DSR) Protocol will be discussed in next section.

Hybrid Routing Protocols

Hybrid routing protocols are proposed to combine the merits of both proactive and reactive routing protocols and overcome their shortcomings. The examples of hybrid routing protocols for mobile ad hoc networks are the Zone Routing Protocol (ZRP), Zone-Based Hierarchical Link State (ZHLS) Routing Protocol.

PROTOCOL DESCRIPTIONS

In this chapter the two protocols DSDV, a proactive routing protocol, and DSR, a reactive routing protocol, are briefly reviewed and a comparative performance analysis is presented. Our results are based on simulations of an ad hoc network of 50 wireless mobile nodes randomly moving about and communicating with each other. The Protocols are compared on the basis of the performance metrics viz throughput, average end-to-end delay and jitter. A short description and working of these two protocols is presented here under.

Destination Sequenced Distance Vector: DSDV

DSDV (Perkins & Bagwat, 1994) is a hop-by-hop distance vector proactive routing protocol in which packets are transmitted between the stations of the network by using routing tables which are stored at each station. Each routing table lists all available destinations, and the number of hops to each destination. Each route table entry is tagged with a sequence number which is originated by the destination station. To maintain the consistency of routing tables in a dynamically varying topology, each station periodically transmits updates immediately when significant new information is available. The DSDV protocol requires each mobile station to advertise, to each of its current neighbors, its own routing table by broadcasting its entries. The entries in this list may change fairly dynamically over time therefore advertisement must be made often enough to ensure that every mobile computer can almost always locate every other mobile computer. In addition, each mobile computer agrees to relay data packets to other computers upon request. This agreement places a premium on the ability to determine the shortest number of hops for a route to a destination. In this way a mobile computer may exchange data with any other mobile computer even if the target of the data is not within range for direct communication. The data broadcast by each mobile computer will contain its new sequence number and the following information for each new route:

- The destination's address;
- The number of hops required to reach the destination;
- The sequence number of the information received regarding that destination,

The transmitted routing tables will also contain the hardware address, and the network address, of the mobile computer transmitting them, within the headers of the packet. The routing table will also include a sequence number created by the transmitter. Routes with more recent sequence numbers are always preferred as the basis for making forwarding decisions, but not necessarily advertised. Of the paths with the same sequence number, those with the smallest metric will be used. Routes received in broadcasts are also advertised by the receiver when it subsequently broadcasts its routing information; the receiver adds an increment to the metric before advertising the route, since incoming packets will require one more hop to reach the destination. Mobile Hosts cause broken links as they move from place to place. The broken link may be detected by the layer-2 protocol, or it may instead be inferred if no broadcasts have been received for a while from a former neighbor. A broken link is described by a metric of ∞ (i.e., any value greater than the maximum allowed metric). When a link to a next hop is broken, any route through that next hop is immediately assigned an ∞ metric and assigned an updated sequence number. Since this qualifies as a substantial route change, such modified routes are immediately disclosed in a broadcast routing information packet.

The parameter values used for DSDV in the simulation are:

- Periodic route update interval -15s
- Periodic updates missed before link declared broken -3
- Route advertisement aggregation time -1s
- Maximum packets buffered per node per Destination -5

Dynamic Source Routing: DSR

Dynamic Source Routing (Johnson, Hu, & Maltz, 2007) is a reactive routing protocol designed specifically for use in multi-hop wireless ad hoc networks of mobile nodes. DSR allows the network to be completely self-organizing and self-configuring, without the need for any existing network infrastructure or administration. The protocol is composed of the two main mechanisms of "Route Discovery" and "Route Maintenance", which work together to allow nodes to discover and maintain routes to arbitrary destinations in the ad hoc network. In DSR, Route Discovery and Route Maintenance each operate entirely "on demand". In particular, unlike other protocols, DSR requires no periodic packets of any kind at any layer within the network. The protocol allows multiple routes to any destination and allows each sender to select and control the routes used in routing its packets, for example, for use in load balancing or for increased robustness. The DSR protocol is designed mainly for mobile ad hoc networks of up to about two hundred nodes and is designed to work well even with very high rates of mobility.

DSR Route Discovery

Route Discovery is the mechanism by which a node, say S, wishing to send a packet to a destination node D obtains a source route to D. Route Discovery is used only when S attempts to send a packet to D and does not already know a route to D. When some source node originates a new packet addressed to some destination node, the source node places in the header of the packet a "source route" giving the sequence of hops that the packet is to follow on its way to the destination. Normally, the sender will obtain a suitable source route by searching its "Route Cache" of routes previously learned; if no route is found in its cache, it will initiate the Route Discovery protocol to dynamically find a new route to this destination node. To initiate the Route Discovery, a source node A transmits a "Route Request" as a single local broadcast packet, which is received by all nodes currently within wireless transmission range of A. Each Route Request identifies the initiator and target of the Route Discovery, and also contains unique request identification, determined by the initiator of the Request. Each Route Request also contains a record listing the address of each intermediate node through which this particular copy of the Route Request has been forwarded. This route record is initialized to an empty list by the initiator of the Route Discovery. When another node receives this Route Request, if it is the target of the Route Discovery, it returns a "Route Reply" to the initiator of the Route Discovery, giving a copy of the accumulated route record from the Route Request; when the initiator receives this Route Reply, it caches this route in its Route Cache for use in sending subsequent packets to this destination. Otherwise, if this node

Figure 1. Resultant simulation parameters

Table 1. Resultant Simulation Parameters

Parameters	DSR	DSDV
Number of generated packets	15599	14308
Number of sent packets	15506	14220
Number of forwarded packets	3197	28
Number of dropped packets	189	92
Number of lost packets	159	9
Minimal delay(sec)	0.001936380	0.001937026
Maximal delay(sec)	36.86088256	72.69449107
Average delay(sec)	0.4021218357	0.3557508646

receiving the Route Request has recently seen another Route Request message from this initiator bearing this same request identification and target address, or if this node's own address is already listed in the route record in the Route Request, this node discards the Request. Otherwise, this node appends its own address to the route record in the Route Request and propagates it by transmitting it as a local broadcast packet (with the same request identification).

DSR Route Maintenance

Route maintenance is the mechanism by which node S is able to detect, while using a source route to D, if the network topology has changed such that it can no longer use its route to D because a link along the route no longer works. An acknowledgement can provide confirmation that a link is capable of carrying data; the acknowledgement request should be retransmitted up to a maximum number of times. After the acknowledgement request has been retransmitted the maximum number of times, if no acknowledgement has been received, then the sender treats the link to this next-hop destination as currently "broken". It should remove this link from its Route Cache and should return a "Route Error" to each node that has sent a packet routed over that link since an acknowledgement was last received.

The parameter values used in the DSR simulations are:

- Time between retransmitted requests- 500 ms
- Size of source route header carrying n addresses bytes *4n+4*
- Time-out for non propagating search- 30 ms
- Time to hold packets awaiting routes- 30s
- Maximum rate for sending replies for a route- 1/s

Figure 2. End2End delay frequency distribution

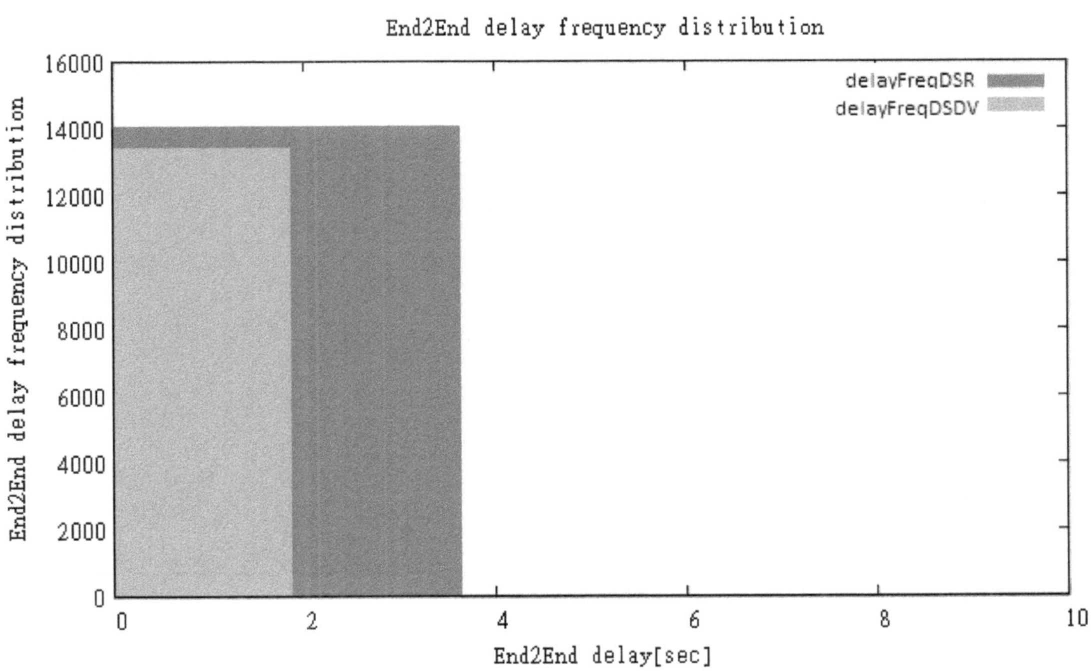

Figure 2:End2End delay frequency distribution

Figure 3. End2End delay obtained using DSDV routing protocol at pause time 20s, 50s and 100s

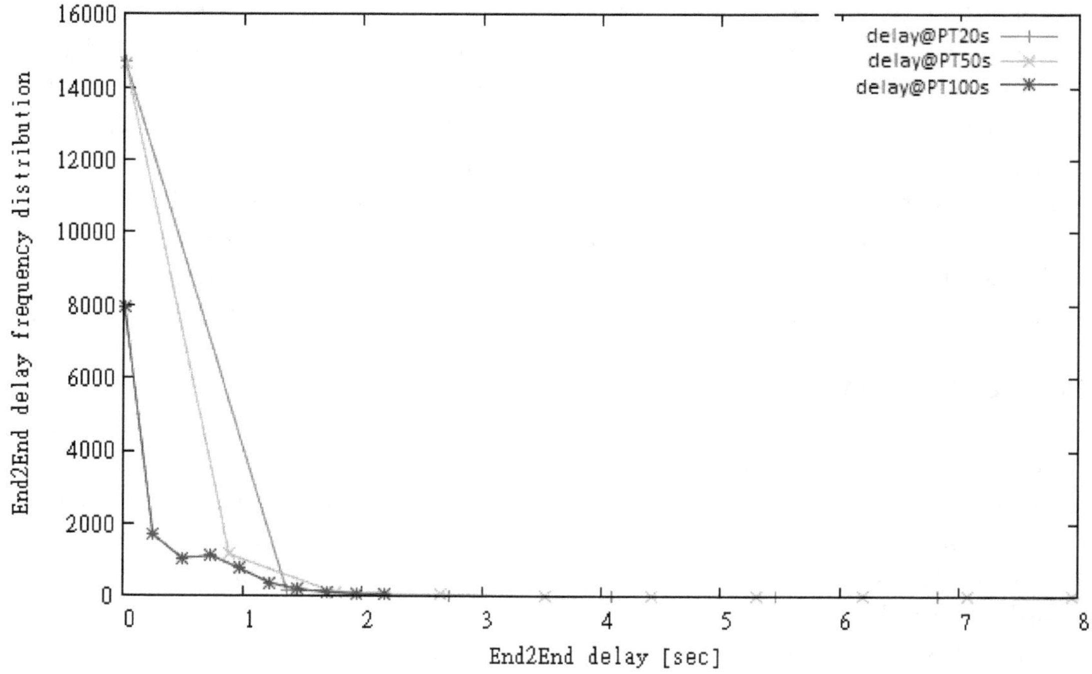

Figure 4. End2End delay obtained using DSR routing protocol at pause time 20s, 50s, 100s

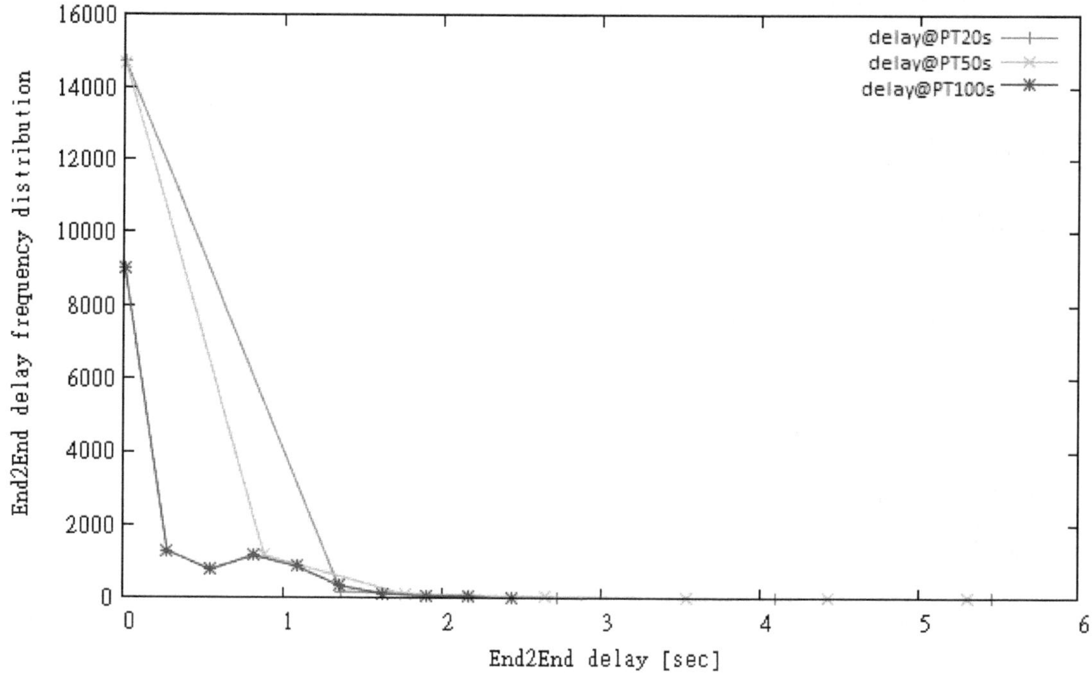

SIMULATION MODEL

Network Simulator

The entire simulations were carried out using ns2.34 network simulator which is a discrete event driven simulator developed at University of California Berkeley (NS-2, n.d.) as a part of the VINT project (The CMU Monarch Project, 1998). It provides substantial support for simulation of TCP, routing, multicast protocols over wired and wireless (local and satellite) networks, etc. The simulator is event-driven and runs in a non-real-time fashion. It consists of C++ core methods and uses Tcl and Object Tcl shell as interface allowing the input file (simulation script) to describe the model to simulate. Arbitrary network topologies composed of nodes, routers, links can be defined. A rich set of protocol objects can be attached to nodes, usually as agents. It has become the *de facto* standard in networking research (Pan, 2007).

Traffic Generation Models

Random traffic connections of TCP and CBR (constant bit rate) can be setup between mobile nodes using a traffic-scenario generator script. This traffic generator script is available under ~ns/indep-utils/cmu-scen-gen and is called cbrgen.tcl. It can be used to create CBR and TCP traffic connections between wireless mobile nodes. In order to create a traffic-connection file, we need to define the type of traffic connection (CBR or TCP), the number of nodes and maximum number of connections to be setup between them, a random seed and incase of CBR connections, a rate whose inverse value is used to compute the interval time between the CBR packets. So the command line looks like the following:

```
ns cbrgen.tcl [-type cbr|tcp] [-nn
nodes] [-seed seed] [-mc connections]
[-rate rate] > [outdir/movement-file]
```

Figure 5. Comulative sum of numbers of all dropped packets

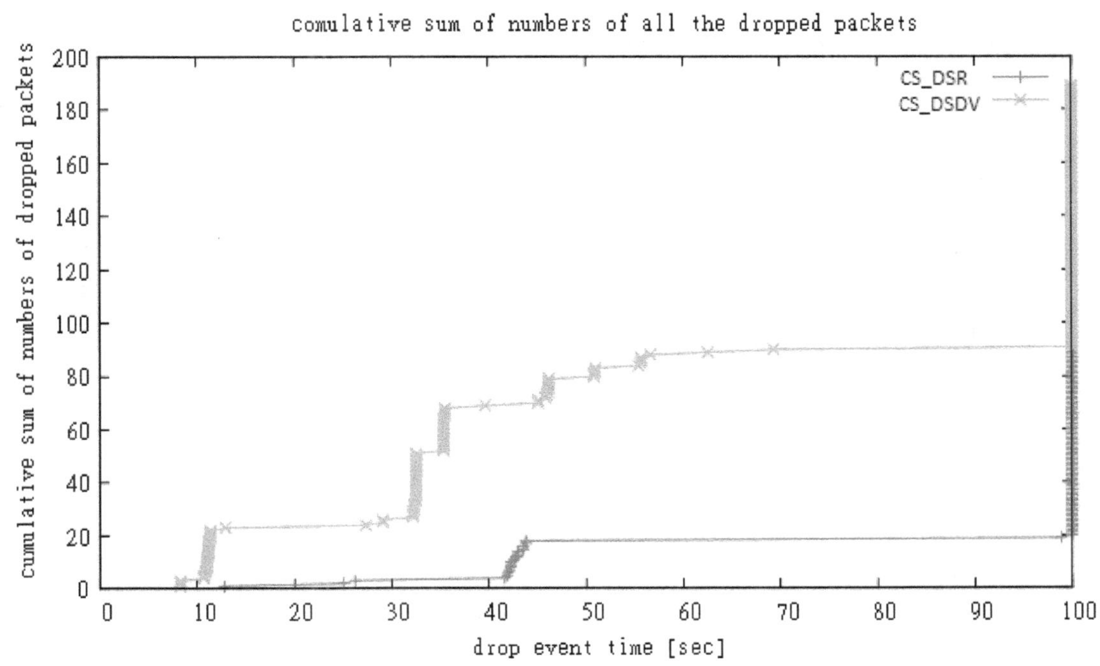

Figure 6. Cumulative sum of numbers of dropped packets using DSDV routing protocol at pause time 20s, 50s, 100s

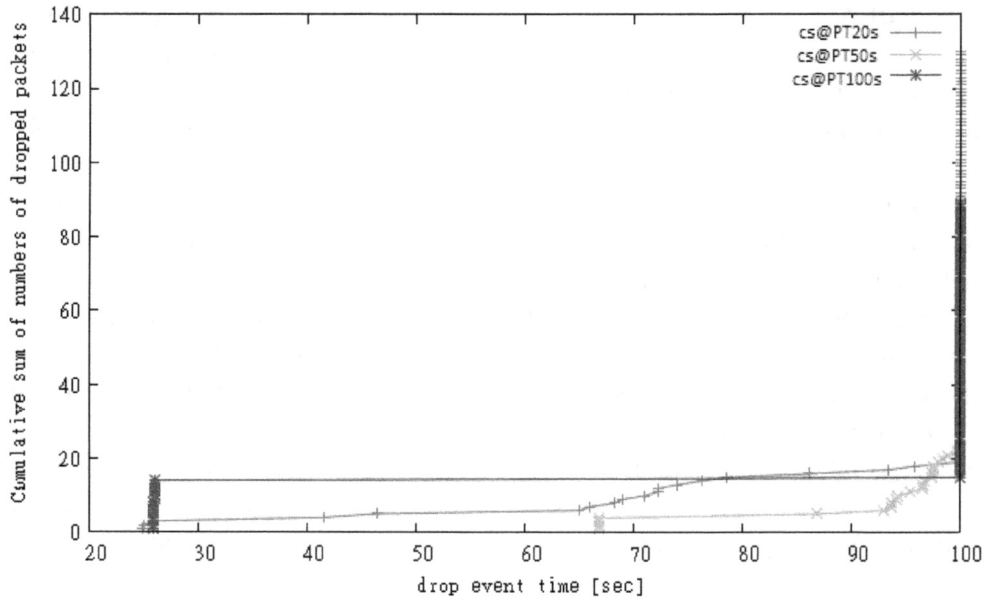

Figure 7. Cumulative sum of numbers of packets dropped using DSR routing protocol at pause time 20s, 50s, 100s

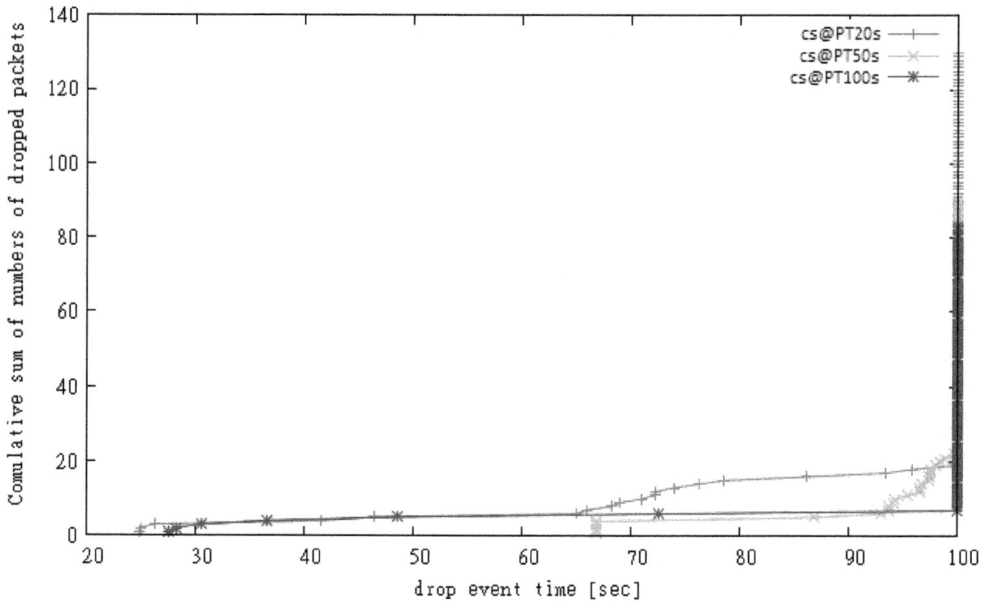

The maximum number of connections setup between the nodes was 20, traffic rate of 5 packets per seconds where each packet size was 512 bytes. A set of two traffic generation files corresponding to each routing protocols were used for same values of maximum connections.

Mobility Generation Models

Inclusion of a mobility model is necessary in order to evaluate the performance of a protocol for ad hoc network in a simulated environment. The mobility model describes the movement of nodes within the simulation area. The Random Waypoint mobility model (RWP) is commonly used in most simulations. In RWP mobility model the nodes move from one waypoint to the next. A specific speed and duration is chosen for every transition. After the stipulated transition duration ends the node may pause for a specific duration of time before starting its transition towards the next waypoint. This pause time is optional and may be equivalent to zero in the case of a network comprising of nodes that are under continuous mobility. A pause time that is equivalent to the duration of the simulation indicates a completely stable network that does not undergo any topological changes due to mobility. NS2 comes with a built in tool called 'setdest' that can be used to generate a large number of nodes and their movements. It is available under the ns2.34/indep-utils/cmu-scen-gen/setdest directory. The command is:

```
./setdest [-n num_of_nodes] [-p pau-
setime] [-s maxspeed] [-t simtime] \
[-x maxx] [-y maxy] > [outdir/move-
ment-file]
```

Setdest generates the position of nodes during the start of the simulation, their movement speed and direction throughout the simulation.

For nodes positions and their movement, a mobility generation file is created with the statements which set nodes' positions and nodes movement using CMUs scenario generation tool setdest (Pan, 2007). To study the effect of high node mobility, the simulation is carried out with movement patterns generated for a pause time of 0 seconds using the 'random waypoint algorithm'. The simulation parameters used are:

- Number of nodes- 50
- Simulation Time- 100sec.
- Area -500m*500m
- Max Speed -20 m/s
- Traffic Source- TCP
- Pause Time (sec)- 0 sec
- Packet Size- 512 Bytes
- Packets Rate- 5 Packets/s
- Max Connections-20
- Band Width- 10Mbps
- Mobility model used-- Random way point

The routing protocols are first evaluated at pause time set to 0s, where pause time is the stop time between successive movements of nodes and 0 s denotes high mobility. Then the various metrics were obtained at pause time set to 20s, 50s and 100s respectively, to study the effect of mobility on the performance of the protocols analyzed.

METRICS USED FOR ANALYSIS

The following metrics were used for the comparison of the protocols.

End to End Delays

It represents the delay encountered between the sending and receiving of the packets.

Throughput of Received Packets

Throughput is the measure of how fast we can actually send through network. This represents the number of packets received within a given time interval.

Figure 8. Throughput

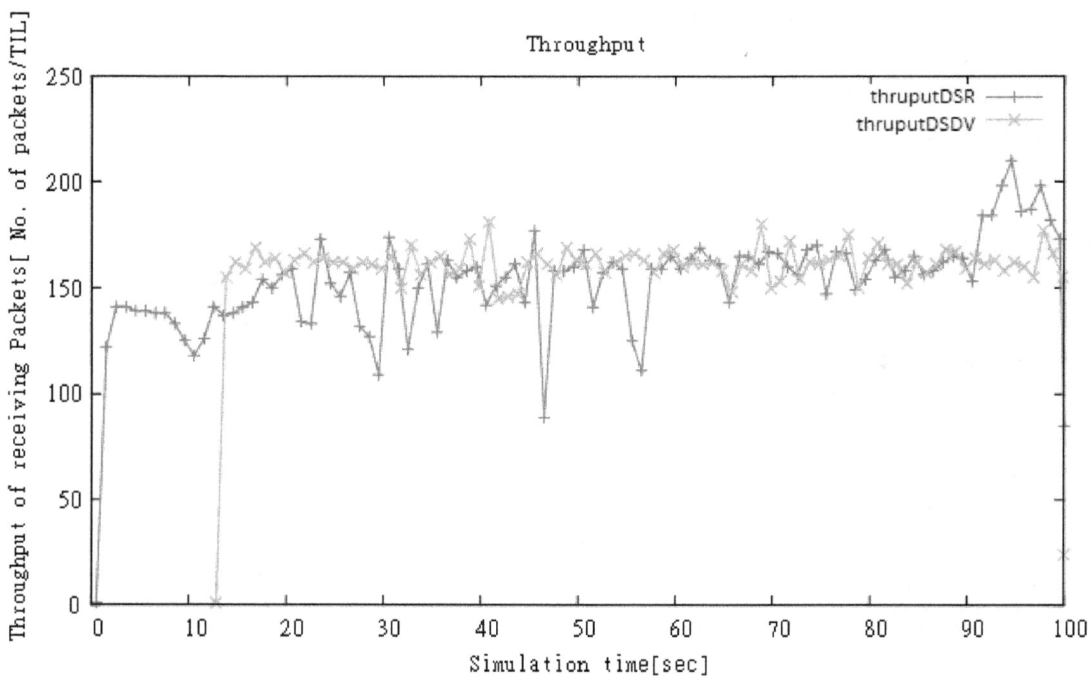

Figure 9. Throughput obtained using DSDV routing protocol at pause time 20s, 50s, 100s

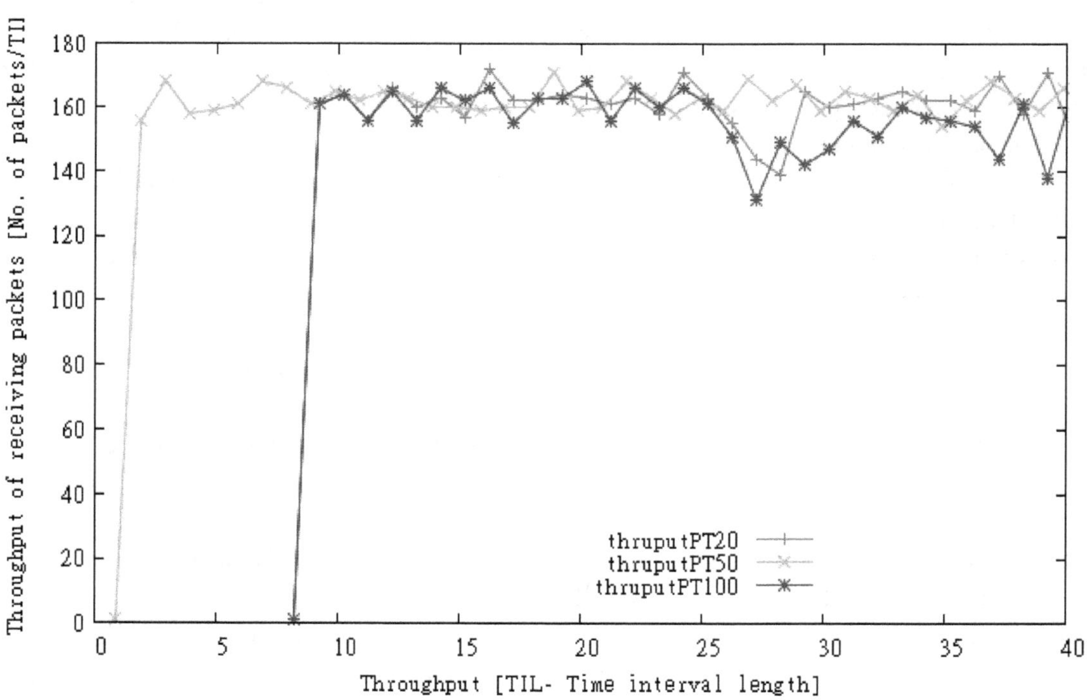

Figure 10. Throughput

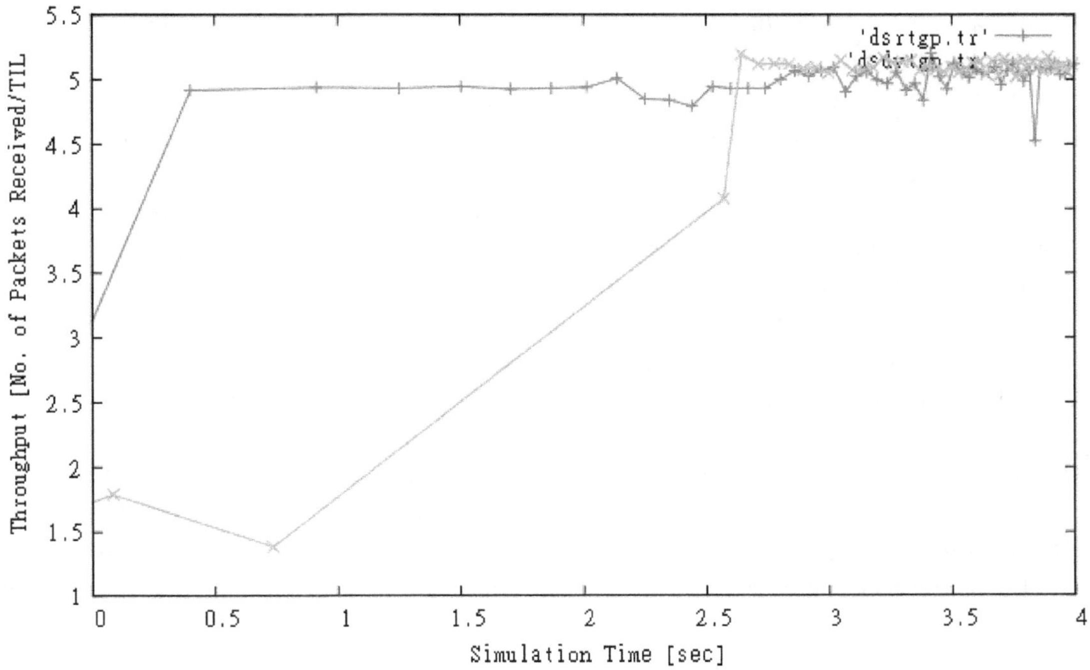

Jitter

It represents any unwanted variation in delays generated during the packet transfer.

RESULTS AND DISCUSSION

The following table (Figure 1) summarizes the result of simulation parameters to evaluate the performance of the two routing protocols.

End-to-End Delay

Average end-to-end delay includes all possible delays caused by buffering during route discovery, latency, queuing at the interface queue, retransmission delays at the MAC, and propagation and transfer times of data packets. Figure 2 shows the end to end delay frequency distribution of the simulated network. DSDV showed less delay in comparison to DSR. However, DSR has a lower delay than DSDV at higher mobility values as the route acquisition procedure in DSR allows more routes to be detected and cached than in DSDV, which obtains a single route per RREQ. Also with DSR, packets wait for lesser time during route acquisition. But DSDV exhibits a low delay because only packets belonging to valid routes at the sending instant get through. A lot of packets are lost until new (valid) route table entries have been propagated through the network by the route update messages in DSDV. DSDV fails to converge at lower pause times hence performance of the protocol decreases as mobility increases. Figure 3 shows the delay frequency distribution of DSDV protocol obtained at pause time set to 20s, 50s and 100s respectively. From the graph it can be seen that the delay increases with increasing mobility. DSR being a source routing protocol, the initial path set up time is significantly higher.

This increases the delay. Figure 4 shows the delay frequency distribution of DSR protocol obtained at pause time set to 20s, 50s and 100s respectively. For DSR, the reactive route acquisition procedures manage to provide new routes with a low packet loss. Figure 5 shows the cumulative sum of numbers of all the packets dropped during transmission via DSR and DSDV ad hoc routing protocols respectively. Figure 6, Figure 7 shows the effect of mobility on the packet loss in DSDV and DSR protocols respectively.

Throughput

Figure 8 shows the throughput of DSDV and DSR protocols for the simulated network. According to the graph the DSR protocol performs better than DSDV. This is due to fact that at high mobility there will be frequent and high degree of changes in the network topology. The proactive nature of DSDV makes it less adaptive to this frequent change and greater number of full dumps needs to be exchanged between the nodes in order to maintain up-to-date routing information at the nodes. This huge volume of control traffic consumes a significant part of the channel bandwidth and lesser channel capacity is left for the data traffic which results in reduced throughput fraction of DSDV at higher mobility. Moreover, in DSDV packets are dropped due to stale routing table entry. DSDV keeps track of only one route per destination. Due to lack of alternate routes, MAC layer drops packets that it is unable to deliver through stale routes. Figure 9 shows the effect of mobility on the throughput in DSDV protocol.

Figure 11. Throughput obtained using DSDV routing protocol at pause time 20s, 50s, 100s

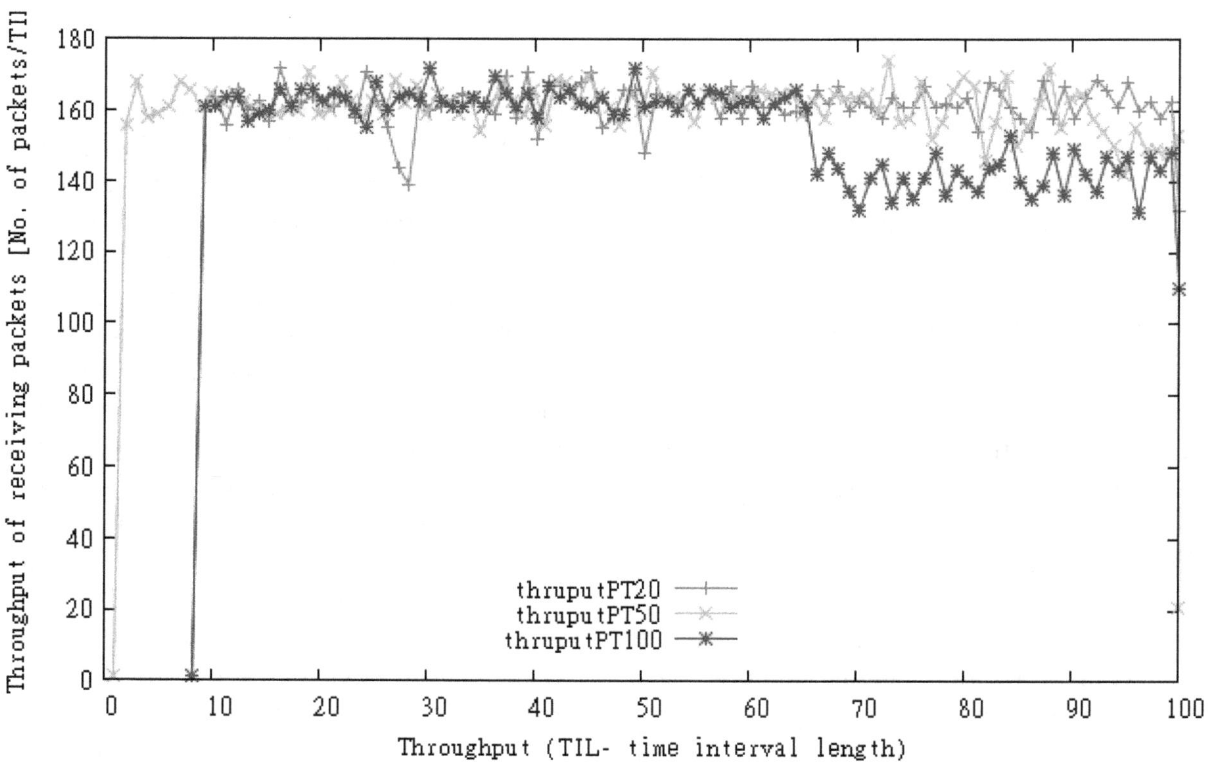

Figure 12. Jitter Comulative Distribution

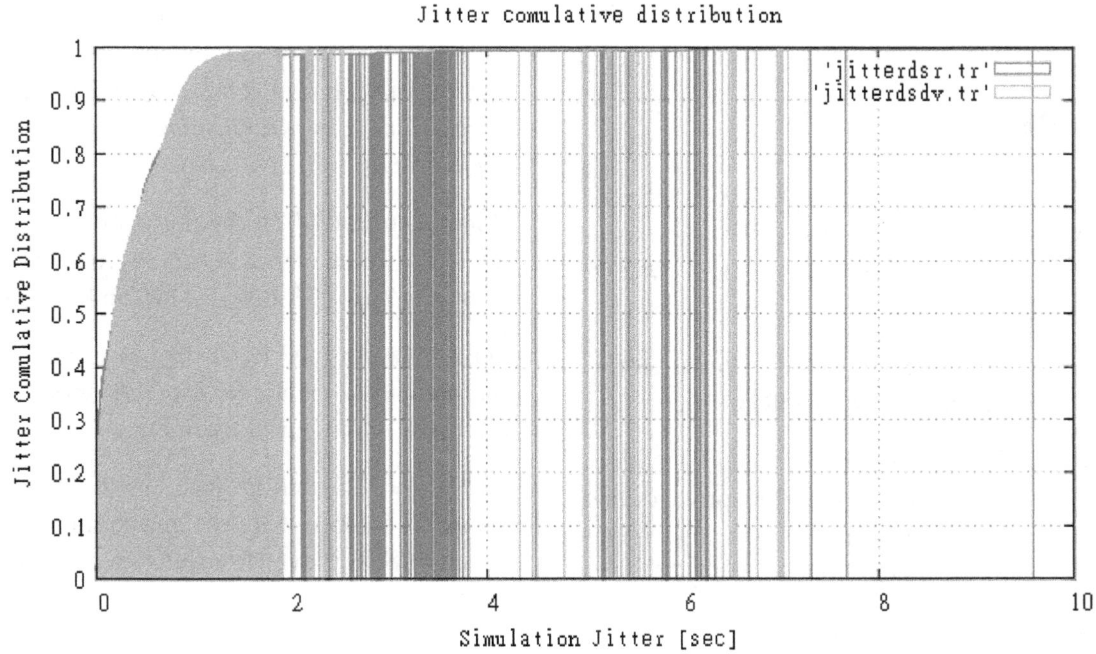

DSR on the contrary, is more adaptive to the frequently changing scenario due to its on-demand routing nature. In case of DSR, multiple routes exist in the cache. Thus, even if a link is broken due to high mobility, alternative routes can be found from the cache. The lower throughput for DSDV during initial time interval (Figure 10) is caused by packets that are sent before routes have converged initially in the network. Note that all simulations are started without any established routes. Figure 11 shows the throughput of DSR protocol at different mobility rates.

Jitter

The Jitter experienced by these protocols cannot to be quantified and they can only be bounded within a delay range. The jitter experienced by DSDV (Figure 12) is the highest than the jitter experienced by DSR.

CONCLUSION

The protocols under study, namely, DSDV and DSR, cover a range of design choices, including periodic advertisements vs. on demand route discovery, hop-by hop routing vs. source routing. This study clearly indicates that in a network of highly dynamic topology, a reactive routing protocol is superior to a proactive one. DSDV have considerable difficulties in maintaining valid routes and drops many packets. With increasing mobility, its' strive to continuously maintain routes to every node increases network load as updates become larger. DSR, on the other hand, is efficient in finding routes in terms of the number of control mobility; alternative routes can be found from the cache. This prevents packet dropping and results in better packet delivery performance and throughput of DSR. Thus we can conclude that if routing delay is of little concern, then DSR shows better performance than DSDV at higher mobility.

REFERENCES

Chen, T. W., & Gerla, M. (1997). Global state routing: A new routing scheme for ad-hoc wireless networks. In . *Proceedings of IEEE SICON, 97*, 197–211.

Clausen, T., & Jacquet, P. (2003). *Optimized link state routing protocol (OLSR)*. RFC-3626.

Dubey, R., Rais, D., Wang, K. Y., & Tripathi, S. K. (1996). Signal stability based adaptive routing (SSA) for ad-hoc mobile networks. *Computer Science Technical Report Series, 3646*, 22.

Johnson, D., Hu, Y., & Maltz, D. (2007). *The dynamic source routing protocol (DSR) for mobile ad hoc networks for IPv4*. RFC-4728.

Liu, W., Chiang, C., Wiu, H., & Gerla, C. (1997). Routing in clustered multihop mobile adhoc networks with fading channel. In . *Proceedings of IEEE SICON, 97*, 197–211.

NS-2. (n.d.). *Network Simulator- ns-2*. Retrieved Febuary 15, 2011, from http://www.isi.edu/nsnam/ns

Pan, Y. (2007). *Design routing protocol performance comparision in NS2: AODV comparing to DSR as example.* Department of CS, SUNY Binghamton, Vestal NY.

Perkins, C., Belding-Royer, E., & Das, S. (2003). *Ad hoc on-demand distance vector (AODV) routing.* RFC-3561 8.

Perkins, C. E., & Bhagwat, P. (1994). Highly dynamic destination-sequenced distance-vector routing (DSDV) for mobile computers. In *Proceedings of the SIGCOMM '94 Conference on Communications Architectures, Protocols and Applications,* (pp. 234–244).

Rajaraman, R. (2002). Topology control and routing in ad hoc networks: A survey. *ACM SIGACT News, 33*(2), 60–73. doi:10.1145/564585.564602

Scott, M., Joseph, P., & Gregory, H. (1999). *Internet based mobile ad hoc networking*. IEEE Internet Computing Magazine.

The CMU Monarch Project. (1998). The CMU Monarch projects wireless and mobility extensions to NS. Retrieved from http://www.monarch.cs.cmu.edu

ADDITIONAL READING

Broch, J., Maltz, D., Johnson, D., Yih-Chun, H., & Jetcheva, J. (1998). A performance comparison of multi-hop wireless ad hoc network routing protocols. *Mobicom '98,* Dallas Texas, (pp. 25–30).

Geetha, J., & Gopinath, G. (2007). Ad-hoc mobile wireless networks routing protocols: A review. *Journal of Computer Science, 14*, 574–582.

Yadav, N., & Yadav, R. (2007). Performance comparison and analysis of table- driven and on- demand routing protocols for mobile adhoc networks. *International Journal of Information Technology, 4*(2), 101–109.

Chapter 10
Overview and Performance Analysis of Ad-Hoc on-Demand Distance Vector Routing Protocol

Jahangir Khan
Sarhad University of Science and Information Technology Peshawar, Pakistan

ABSTRACT

Ad-Hoc on demand distance vector routing protocol (AODV) is a reactive routing protocol in the arena of mobile ad-hoc network. In order to concentrate on the performance in ad-hoc networks, one must pay attention to the technological advancements and applications of MANETS, going on the track of 3G and has the potential to make a tremendous stance in 5G. In evolutionary view, 5G will be capable of supporting Wireless World Wide Web (WWWW), allowing high flexibility in dynamic ad-hoc networks. This intelligent technology is capable of interconnecting the entire world without limitations. AODV can make a very good contribution in 3G-5G technologies in order to prove its correctness and performance evolution in MANETS routing protocols. The objective of this Master's research is to extend and evaluate proposed routing protocol to be suitable for mobile ad hoc networks. In order to improve performance and selecting the packet delivery fraction, end-to-end delay and also route discovery frequency should be compared with global statistics frequencies. The correctness and performance evaluation of the proposed protocol should be done theoretically and implemented through simulation, using OPNET simulator to improve routing performance and end to end data delivery.

DOI: 10.4018/978-1-4666-0321-9.ch010

INTRODUCTION

Mobile Ad Hoc Network (MANET) routing protocols concept as described by Internet Engineering Task Force (IETF) MANET working group and IEEE 802.11 standard. In particular, Ad Hoc On-Demand Distance Vector (AODV) routing, discuses along with a review and simulation of previous research implementation ofAODV Routing protocols. Proposed routing protocol have a strong contribution in performance analysis of on-demand reactive protocol for data delivery ratio and to compare with gobal statistics in AODV routing protocol. For best routing between wireless nodes of ad hoc network a number of qualitative properties is desirable. For best and effective routing loop freedom is very necessary for data packets to avoid collision and waste of time. AODV uses time to live (TTL) to avoid such loops, but more structured approach is required to get best results. The MANET delay effects on routing very badly, as the mobile nodes are battery powered and scarce memory. The routing protocol needs to be routed intelligently and utilize network bandwidth and energy resources in a better way in case of network route delay. One of the important characteristics of MANET is dynamic topology based on node mobility. Mobile nodes in ad hoc network are free to change position frequently, therefore routing protocol needs to quickly adopt topology changes. Due to dynamic mobility and limited energy resources a unique wireless node in MANET cannot be trusted for auto configuration in case of break down. Therefore support of distributed operation is required by routing protocols to solve such type of problems. Due to lack of physical security, MANET protocols are highly exposed to different types of attacks. In MANET it is very easy for attacker to disturb network traffic, corrupt packet header, change addresses of routing messages, and increase traffic to waste bandwidth. To avoid all these threats a sufficient security protection is highly desirable in all MANET routing protocols. Therefore the routing protocol should design to eliminate or minimize control traffic and other periodic messages. AODV is not quite good in route establishment and route maintenance phases but also as good in routing data packets and hop-by-hop routing protocol is good in transmission of data packets. AODV more appropriate in ad hoc networks with a higher mobility and data transfer rate. The simulations results and analysis have shown that mobile ad hoc network definitely needs more precise routing protocols to support increase mobility.

BACKGROUND

Ad hoc on-demand distance vector (AODV) routing algorithm is a routing protocol designed for mobile ad hoc networks and is using on-demand routing approach for establishment of route between nodes. As it uses on-demand routing therefore it built route to transmit data packets when the source node desired and is trying to maintain established route as long as they are needed. AODV protocol has quality to support unicast, multicast, and broadcast routing with loop free, self starting and scalable characteristics. AODV protocol routes data packets between mobile nodes of ad hoc network. This protocol allows mobile nodes to pass data packets to required destination node through neighbor's node which cannot directly communicate. Nodes of network periodically exchange information of distance table to their neighbors and ready for immediate updates. AODV protocol is responsible to select shortest and loop free route from table to transfer data packets. In case of errors or changes in selected route, AODV is able create a new route for the rest of transmission.

Route Management

AODV routing protocol in ad hoc network communicate between mobile nodes through four types of different messages.

- Route Request
- Route Reply
- Route Error
- Hello Message

To establish a route between source and destination node Route Request (RREQ) and Route Reply (RREP) packet query cycle are used. Route Error (REER) and HELLO data packets are used for route maintenance. (Lin)

Route Request (RREQ)

The mobile nodes of AODV routing protocol only contain information about their neighbour nodes. Therefore source node when required a route to destination node in the network, first of all it broadcast a RREQ packet to all the nodes of the network and sets a timer to wait for reply. The RREQ packet of source node contains the following information;

- Source Identifier
- Destination Identifier
- Source Sequence Number
- Destination Sequence Number
- Broadcast Identifier
- Life Span

The RREQ of a source node is uniquely identified by combination of source identifier and broadcast identifier. When a node in network receive RREQ, it performs two tasks either forwards the RREQ to next neighbour node by increasing hop or prepares a RREP if it has a valid route to the destination (Suresh, 2005).

Route Reply (RREP)

In AODV routing protocol the RREP activates forwarding data packets on the selected route. The RREP packet of destination node contains the following information;

- Source Identifier
- Destination Identifier
- Destination Sequence Number
- Hop Count

Route Error (REER)

Route Error (REER) packet is generated when a link breaks down during transmission of data in active path of ad hoc network. Broken node sends REER packet to source node which contains information about failure link. When source node receives REER packets, it removes failed source route form its cache. The REER packet contains IP addresses and sequence numbers of all destinations which are unreachable due to failure of link. Each intermediate node when receive REER packet from broken node delete broken route from its table and forward packet towards source node.

Hello Message

Hello messages are sent periodically by AODV in ad hoc network to ensure symmetric links of active path and to able to detect link failure. Hello message is sent by a node for the purpose to inform its neighbours about his presence. Hello message is broadcast by defining a specific time to live (TTL) operation. When a node receive hello message it updates its routing table by increasing the life time. AODV must utilize small size hello message in order to maintain connectivity. A node can use only that route to its neighbour which is recommended by hello message. When a node receive hello message from a new node in the network, it sends routing table back to that new node.

Routing Table

The AODV routing table consists of different fields to hold information about nodes of the ad hoc network. These fields are;

- Destination Address
- Next Hop Address
- Destination Sequence Number
- Hop Count

In AODV routing packets, sequence number is used to determine freshness of originating node and to guarantee loop freedom. In mobile ad hoc networks, AODV assigns sequence number and broadcast identifier to every node, the sequence number is incremented by node when there is change in its local connectivity information (Lee, 2000).

Route Establishment

Consider the following Figure 1, where source node 1 broadcast flooded RREQ message in network for destination node 9. Node 2, 5, and 6 are neighbour of node 1 therefore receives RREQ packet. When neighbour nodes receive RREQ packet, first of all these nodes create a reverse route entry for source node in their route table. These nodes check their routes to destinations if not available, they rebroadcast RREQ packet to their next own neighbours i.e. 3, 5, 7, and 9 after increasing the hop count. As shown in the Figure 1, an intermediate node 5 can receive multiple RREQ packets from its multiple neighbours, i.e. node 1, 2 and 4. In such situation when a node receives multiple RREQ with same sequence number and broadcast identifier, it drops redundant RREQ and does not forward it. This multicast transmission is continue by the nodes until it reach to the destination node.

Eventually, three different routes are discovered on the arrival of RREQ packet on destination node. This time destination node 9 sends unicast RREP packet which follow reverse route back to source node 1. The RREP packet from destination node includes current sequence number of destination and distance in hops to destination. During reverse routing of RREP packet nodes also set forward route table entry for destination node in their route table. Thus a unicast RREP packet reaches to source node with updated destination sequence number. The source node 1 accepts destination sequence number with shortest route and transmits data packets through forward route. (Feeney, 2005)

During transmission if source node receive RREP packet with greatest destination sequence number or equivalent it already using with less number of hops, it then updates its route table and begins to transmit data packets on new route. In a situation where source node does not receive RREP in required life span. The source node rebroadcast RREQ packet by increasing life span.

Route Maintenance

Link failure is detected by AODV routing protocol in ad hoc network during transmission of data packets from source to destination node on active route as shown in following Figure 2. The AODV routing protocol does not repair a broken link of active route locally, but sends Route Error (REER) packets to both source and destination nodes by incrementing sequence number by one, i.e. n+1. At same time intermediate nodes delete corresponding route entries that contain invalid routes from their routing table. As source node receives REER packet from broken link, it invalidates listed routes. The source node re-initiates rout establishment process by broadcasting RREQ packets to neighbour nodes with new broadcast identifier and same destination sequence number (Ram, 2004).

Properties of AODV

The AODV protocol contained on-demand/reactive functionality to establish a rout to destination and uses destination sequence number to find latest route. Connection setup delay is less as compared to other protocols.

Stale route entries are occurred due to inconsistent routes information by intermediate nodes

Figure 1. Route establishments in AODV

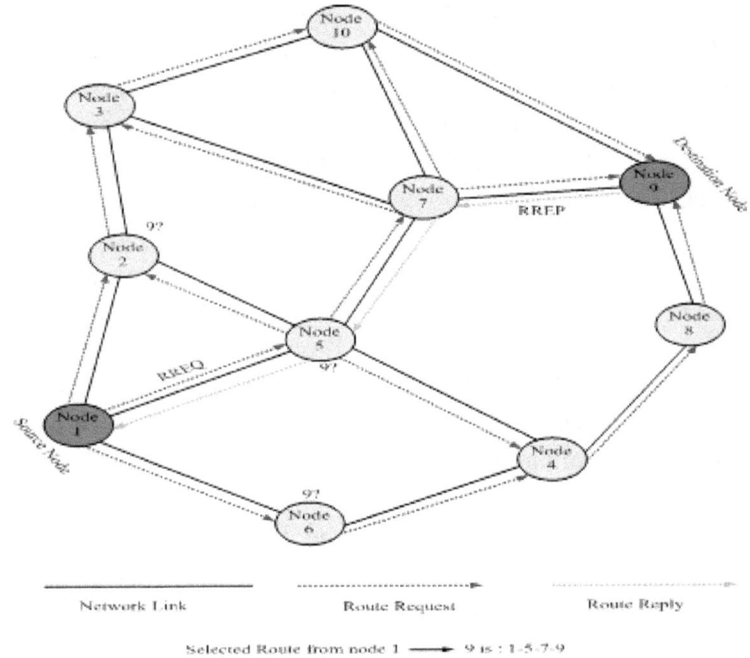

Figure 2. Route maintenance in AODV

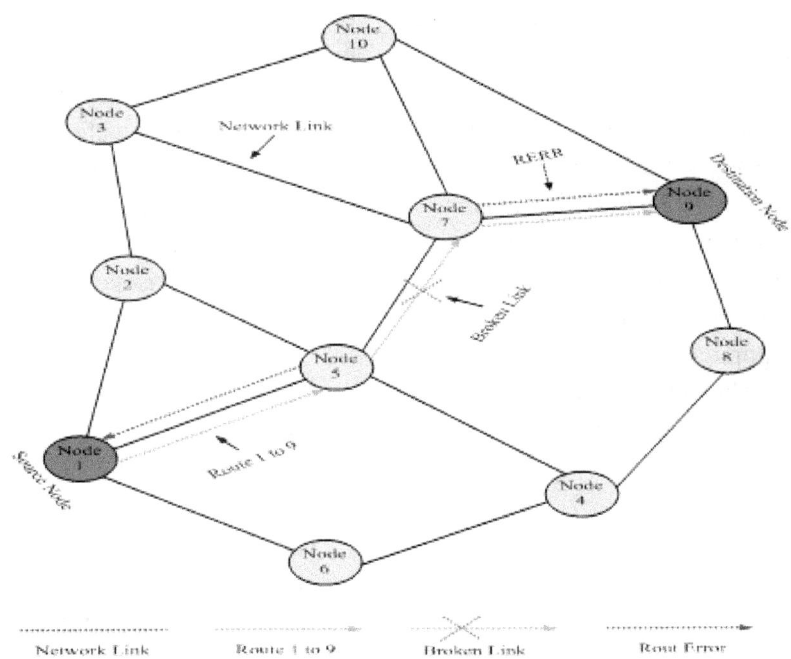

which contain old source sequence number but not latest destination sequence number. Control overhead is high in AODV routing protocol because of multiple replies of RREP packets for a single RREQ packet. The wastage in bandwidth occurs due to periodic beaconing (hello messages) architecture of AODV protocol (Ram, 2004).

SIMULATION STUDY

Conducting real world mobile ad hoc network experiments for researchers are difficult and costly. Therefore MANET research community commonly relies on computer simulation to evaluate and analyze their experiments on different routing protocols for dynamic ad hoc networks. The simulation results are far from perfect real environment but still provide a better understanding. Simulation is the process of designing a model of real system and conducting experiments for purpose of understanding behavior and verification of proposed system control at different stages. To achieve such goals a number of simulation tools are available in market. The selection of good simulation tool helps a lot in better understanding and testing of proposed mechanism in several different possible environments. The emerging MANET routing algorithm studies shows that some sensitive simulation parameters affect simulation results.

In this chapter simulation result of two previously published MANET routing protocol AODV is presented using TCP based application. AODV and DSR are most popular MANET on-demand routing protocols. The analysis of proposed protocols is done using the MANET model in OPNET simulator.

OPNET

Optimized Network Engineering Tools (OPNET) is commercial software developed by OPNET Technologies Inc for simulation, analyzing, and modeling of communication networks, distributed systems, devices, and routing protocols. OPNET is most scalable and efficient simulation tool and friendly to use because of its tremendous features, such as comprehensive graphical user interface and animation. It contains hundreds of protocol and vender device models with great flexibility for experiments and analysis. OPNET provide object oriented modeling and open source code models which give easy understanding of system. OPNET provide ideal simulation environment for wireless ad hoc network including terrain, mobility, and multiple wireless models. OPNET contains 64-bit parallel simulation kernel and provide discrete event, hybrid and analytical simulation environment for researchers. This simulation tool enables its user to optimize performance and maximize availability of communication networks and applications by using 3D visualization and animation. (OPNET, 2007)

The finite state machines of OPNET provide a natural representation of routing protocols and contain an extensive support for communication related applications. OPNET simulation software provides most flexible and scalable wireless mobile ad hoc network performance modeling environment which includes a huge number of powerful advanced technologies for accelerating simulation run time. (Koziniec, 2007)

The OPNET simulation tools are divided into three main categories that communicate in to three different phases of modeling and simulation projects. These three phases are;

- Specification
- Data Collection
- Simulation and Analysis

The three phases of OPNET simulation is performed in a sequence and forms a cycle as shown in the above figure. Specification phase of the tool is actually divided into two parts, i.e. Initial specification and Re-specification, where re-specification phase is part of simulation cycle.

Figure 3. Simulation cycle of OPNET (Suresh, 2005)

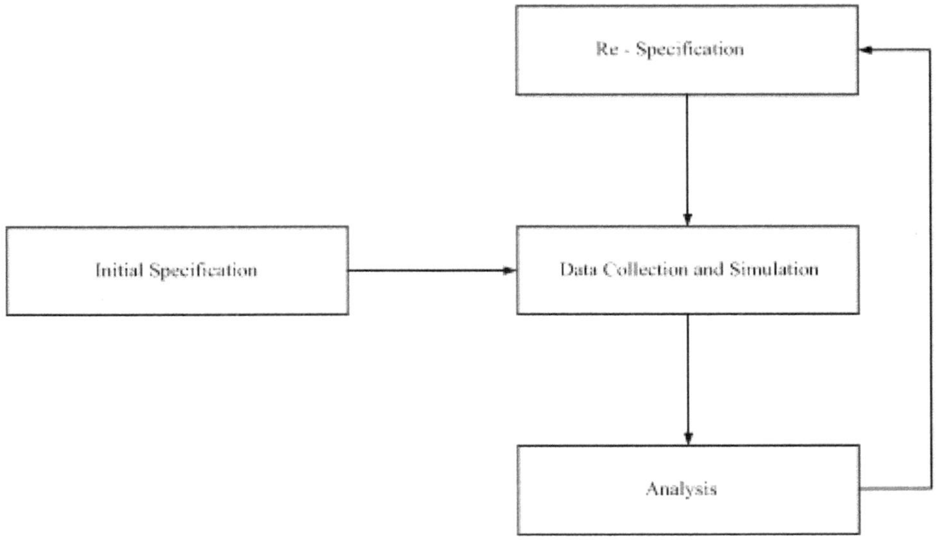

Cost of OPNET

OPNET is the world leading commercial software available for network modeling and simulation. OPNET is available free for academic use in a lot of worldwide universities and more than ten thousand students are using it successfully. Another core product of the OPNET technology is the OPNET IT GURU, which is a free downloadable version contains appropriate modules. The OPNET IT GURU platform is used for enterprise application performance management, network configuration analysis, and IT capacity modeling.

The University program OPNET is used for simulation of two different routing protocols of MANET in this research.

MANET Node Model Structure of OPNET

Routing protocols AODV, DSR, and TORA are implemented at IP layer in MANET model structure as shown in the following figure. Models of AODV, DSR, TORA, and OLSR are available in OPNET version 12. The following figure explains node model architecture of a MANET node in detail.

ip_dispatch is root process for IP in MANET node and creates a child process manet_mgr. The manet_mgr function as manager process and provides a common interface to multiple mobile ad hoc routing protocols in OPNET. The manet_mgr is further child process for required MANET protocol as configured in parameter.

This OPNET ad hoc routing model does not support multiple routing protocols on a single node and redistribution between MANET routing protocols. The model also limits router node interfaces to be running same MANET routing protocol.

Node Model

The MANET object palette contains all MANET capable nodes as shown in the following Figure 5.

To simulate different routing protocols in mobile ad hoc network models some nodes of the above object palette are used. The commonly used nodes in MANET network models are;

Figure 4. MANET node model structure (OPNET model documentation)

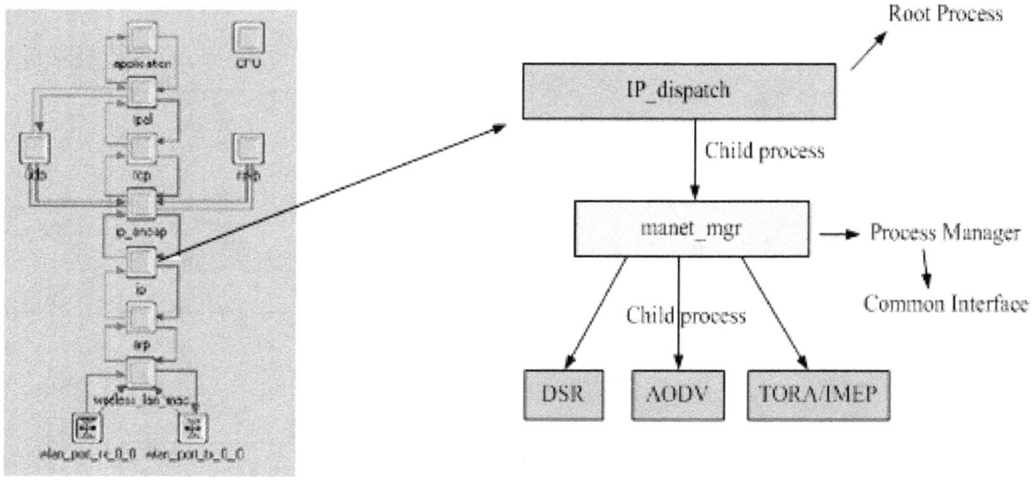

Figure 5. MANET object palette

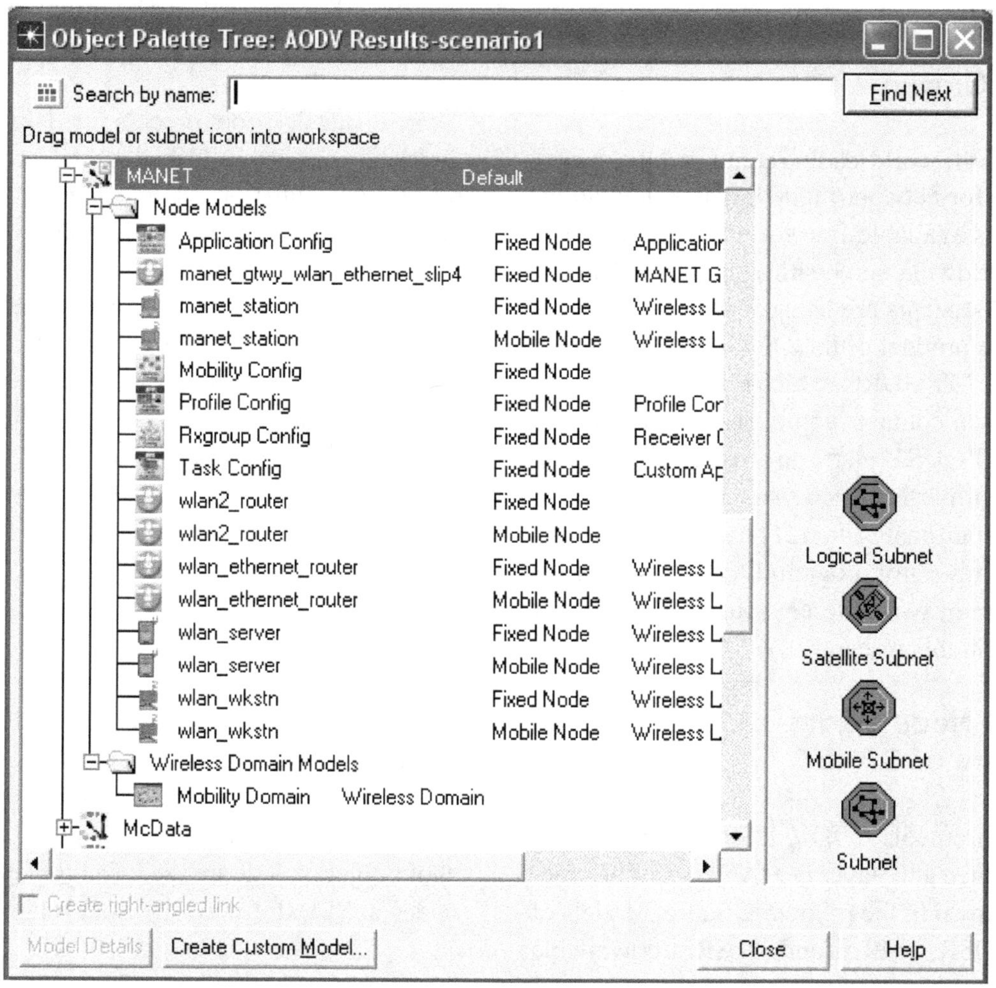

- **Wireless LAN Workstations and Servers**

To generate application traffic such FTP, E-mail, and HTTP over TCP over IP over wireless LAN in a MANET network model these node models of object palette are used. For each MANET routing protocol these nodes can be configured to run in a specific way.

- **MANET Stations**

To generate raw packet over IP over wireless LAN these node models of MANET object palette are used. For each MANET routing protocol these nodes can be configured to function as traffic sources or destination.

- **Wireless LAN routers and MANET gateway**

These node models play very important role in MANET network models by performing function as access point in a mobile ad hoc network. By enabling MANET gateway, these nodes of object palette can also connect mobile nodes of ad hoc network to IP based networks.

Analysis and Results

The following section explains how to model and simulate AODV routing protocol in ad hoc network with different parameters using MANET models in OPNET simulation tool.

Scenario and Simulation

The main parameters that effect mobility in Ad Hoc Network are maximum speed of mobile host nodes and pause time between each moving. In this dissertation we discuss different phases such as design, simulation, data collection, and analysis of mobile ad hoc network in OPNET. The simulation scenario consists of six wireless nodes as shown in the following Figure 6.

In this simple scenario of ad hoc network the following parameters are used for simulation purpose.

Analysis of AODV Routing Protocol

The simulated scenario evaluates the performance of important TCP parameters for AODV based 6-25 nodes network. All nodes in the network are configured to run AODV protocol and FTP sessions. In simulation process different AODV parameters are used as suggested by RFC and WLAN data rate. A single TCP connection is established between wireless nodes of scenario. Two different self mobility trajectories are defined for mobile. The results shows amount of routing traffic generated, route discovery time and the number of hops per route, TCP/IP traffic, no of RREQ and RREP packets, and FTP download and upload time. Simulation of mobile ad hoc network consist of six nodes was run for 300 seconds and the following results were generated.

Global Statistics of AODV Routing Protocol

To establish a route source node initiated route discovery process by sending FTP request to the destination node. In the route establishment

Table 1. Simulation parameters

Routing Protocols	AODV
Number of wireless nodes	6 (static, Dynamic) Intermediate nodes - 25 nodes
Movement space	4000m x 3000m
Maximum speed	2, 10, and 20 m/s
Maximum pause time	0 and 200s
Transmission rates	2, 5, and 10 packets/s
Packet size	512 bytes
Simulation time	300s

Figure 6. Six node campus network for TCP traffic

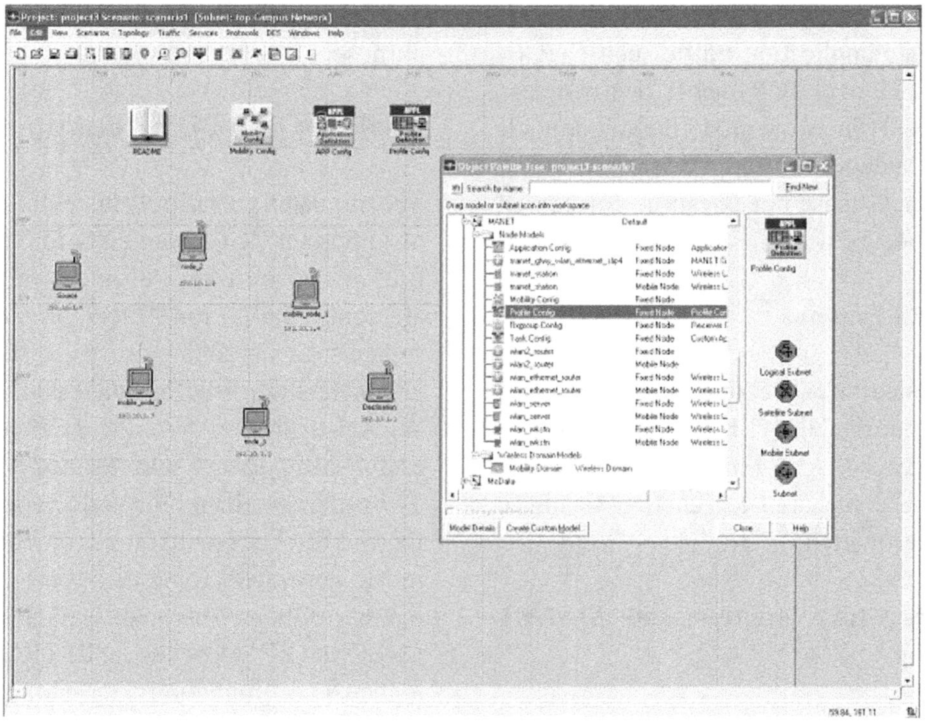

Figure 7. Simulation progress report for AODV routing protocol

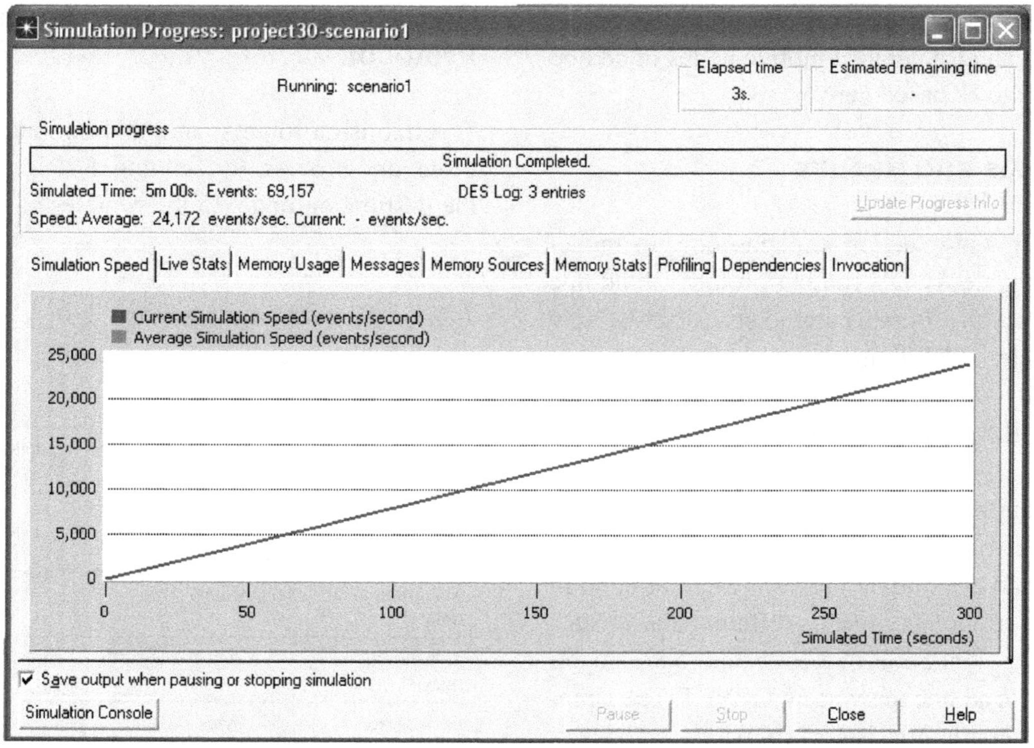

process source node broadcasting RREQ packet and destination node reply by RREP packet. As soon as route is established between source and destination nodes, data transfer is started. The destination node sending data to source node in response of FTP request. Due to delay in Hello messages exchange between neighbor nodes chances of link breakage is higher. In case of link breakage route establishment process is started again. Mobile nodes of define ad hoc network wait until 3 minutes and 20 seconds (i.e. 200seconds) and start moving away towards path defined by trajectory during simulation process as shown in the following Figure 8.

To establish route between source and destination nodes, source node sent three RREQ packets to destination. The first RREQ packet from source node has sent at beginning of application i.e. at 100 seconds, the second RREQ packet has sent at 170 seconds, and third RREQ packet has sent at 210 seconds. The maximum number of times a node will try again to discover a route by broadcasting another route request is keep to 5. Destination node replies with RREP packets in response of RREQ packets received by source node as shown in the following Figure 9.

The figure of AODV route establishment shows that source node request for route to destination three times. First attempt is made at 100 seconds in start of communication. Second and third RREQ is sent in 140 and 270 seconds due to link break down in route respectively.

In the scenario the AODV route from source to destination consist of 3 hops. The discovered route from source to destination consists of the following nodes;

Source ------→ Mobile node 2 ----→ Intermediate node 2--→ Destination

The hops in the established route are shown in the following Figure 10. along with the route discovery time.

The first route from source to destination is established in 100 seconds. The first route discovery takes a long time due to the new destination for which the intermediate nodes do not contains any information and simply forward the RREQ to their neighbour. The second and third route establishments occur very quickly as compared to first one as shown in above figure. During 2nd and 3rd attempt the neighbour nodes route tables have the information about the destination nodes and the establishment occur quickly.

During transmission of data from source to destination nodes in the simulated scenario the total AODV routing traffic sent and received are shown in the following figures X and Y respectively. The routing traffic sent is represented in both packets and bits per second. As the following figure shows that there is sudden rise in the traffic at 100, 140, and 270 seconds due to the RREQ and RREP packets. The traffic signal is also high due to the data packets and a huge number of data bits buffered during establishment process of broken route.

Traffic sent in bits is higher as compared to the packets. Once route is established routing traffic flows in a continuous way along data between wireless nodes of ad hoc network. Each hop in discovered route adds some routing traffic to data.

Traffic received is combination of all hops occurs in route from source to destination. Therefore routing traffic received is very high as compared to traffic sent. The maximum traffic sent in bits per second is between 1300 and 1400, where in packets per second are between 3 and 3.5. most of the time during simulation period the routing traffic is 0 bits/s and packets/s. On the other side maximum traffic received at destination node is much higher which is between 2800 and 3000 bits per second and is 7 packets per second. The figure shows that the traffic signal is very high in first and second time route establishment. The following Figure 12. Shows AODV routing traffic received in both packets and bits per second.

As described earlier that Hello messages are used for maintenance of a route between source and destination. This attribute of AODV is also

Figure 8. Mobile node Trajectory attributes editor

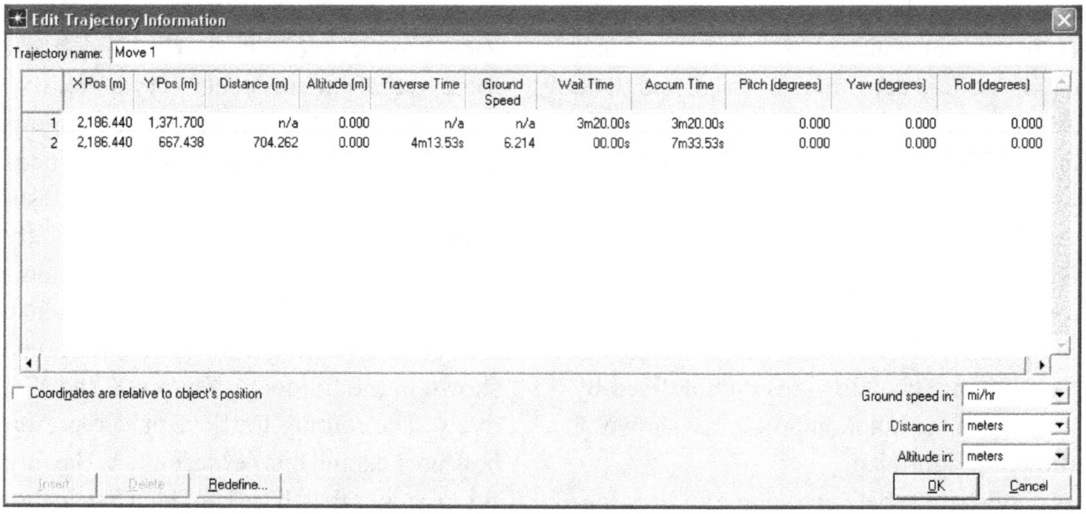

Figure 9. AODV route establishments

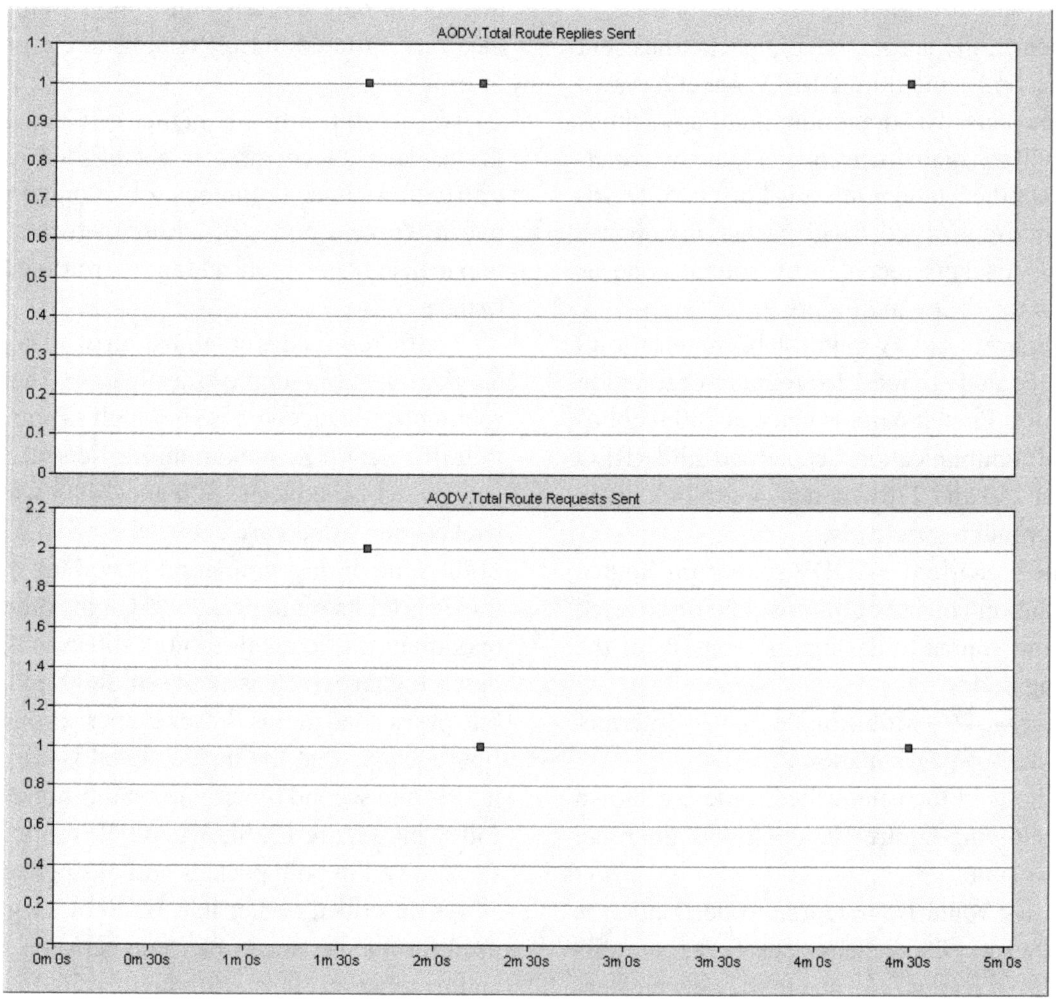

Figure 10. AODV route discovery time and number of hops

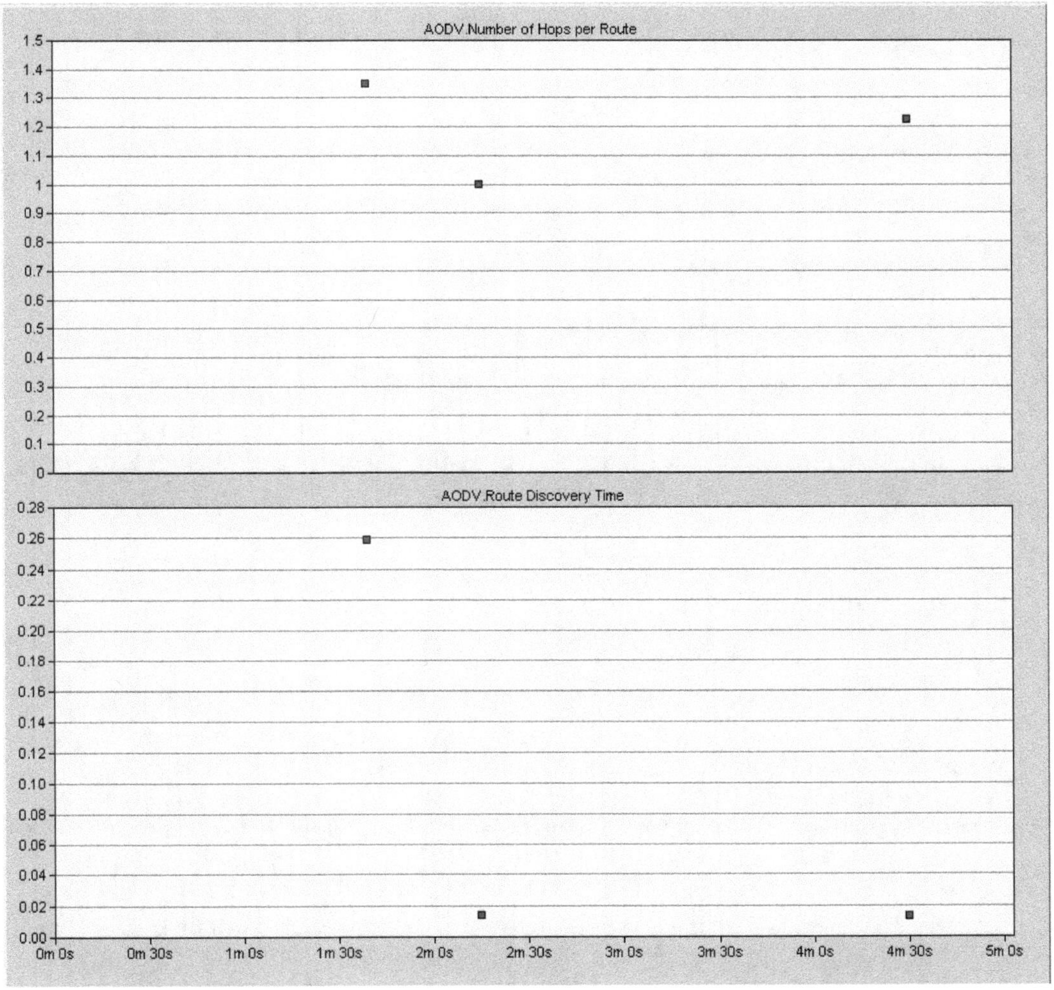

used to determine neighbour nodes connectivity by the following formula,

Allowed Hello Loss * Hello interval

The periodic time interval of hello message transmission for this six node simulated scenario is defined as Uniform (10, 10.1) sec. Allowed Hello loss defines number of packet loss a node can maintain before link break which is defined 3 in this simulation.

Global Statistics of MANET

This characteristic of ad hoc network defines over all statistics of nodes. The following Figure 13. shows total delay in data transmission per second. Delay signal is high in start of the route establishment. The high delay AODV occurs during first attempt for route discovery is very less, which is 0.10 per second. To compare this delay with the routing data discussed above, it is looking very less. The graph also shows that delay occur during second and third time route discovery occurs in ad hoc network. The simulation results shows that during communication there are some times where the delay is zero, such as once after the first route establishment and second in between 230 and 240 seconds. The MANET delay of AODV routing protocol.

Figure 11. AODV routing traffic sent

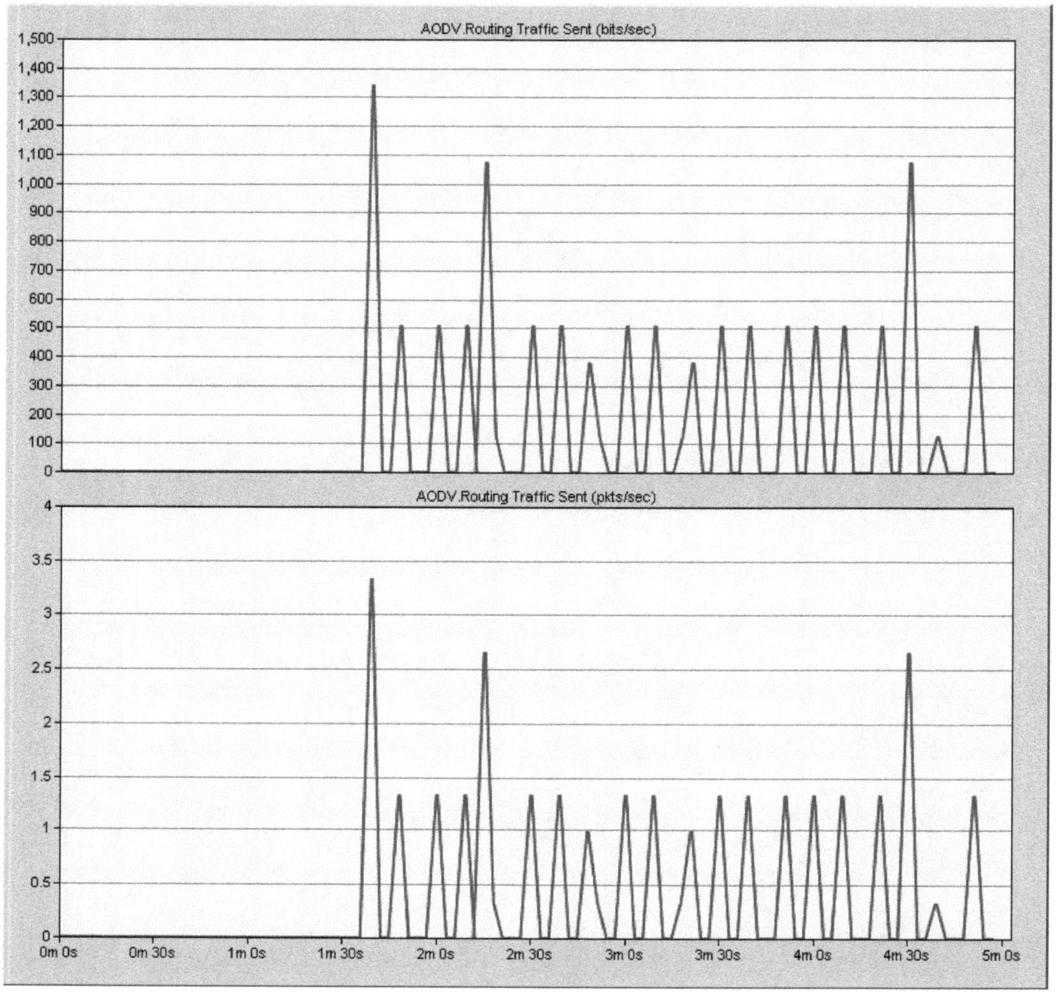

Constant Bit Rate (CBR) packet also affects delay during transmission of data packets. The buffer area in node table becomes full quickly due to transmission of data packets, therefore the packets have to stay for a longer time in buffer area before they are sent to next node. The delay is high even during start of transmission, at this stage buffer contains a lot of packet where route discovery is in progress. Due to high delay there is more chances of packets drops.

Total MANET traffic sent and received during transmission of data in six wireless node ad hoc scenarios is shown in following figure. The Figure 14. shows that total MANET traffic sent and received in both packets and bits are same. As discussed earlier that total AODV routing traffic received is high than total traffic sent due to traffic added by hops occur in route.

Object Statistics of Source and Destination nodes

The lifetime of a routing table entry is initialized to active route timeout and is defined as 5 seconds for this scenario. If a route is not used and refreshed within defined period of time, AODV marks route as invalid and removes it from IP common table. To establish a route for the first time, originating

Figure 12. AODV routing traffic received

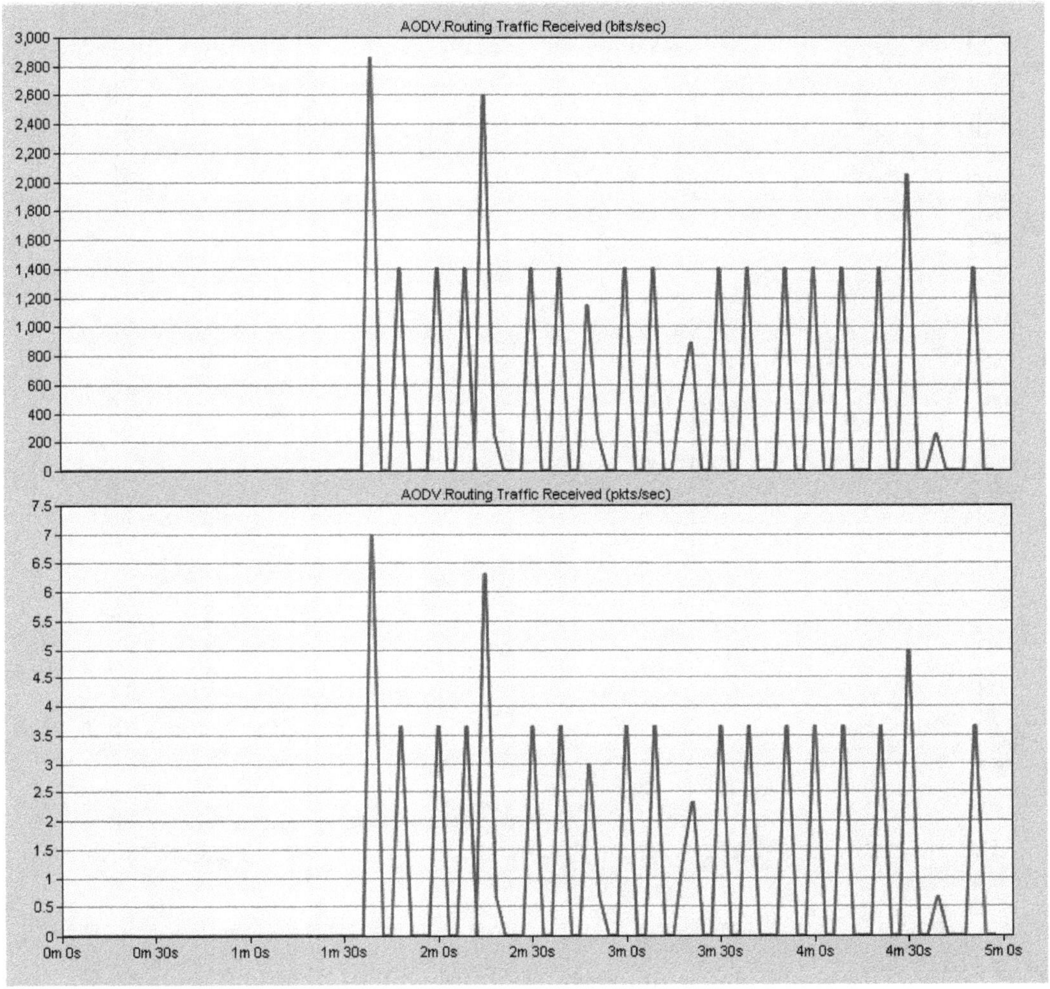

node uses a TTL start value which is 1 for the first RREQ packet. The individual statistics of source and destination nodes in simulated scenario are shown in figure 14. The number of hops covered by source node is 2 where destination node is 1. The AODV packets are queued during route establishment, when route is established then packets are transmitted without any waiting in queue. Both nodes routing table contains same number of nodes information. The routing traffic sent and received by both source and destination nodes are also shown in the Figure 15.

Object Statistics of AODV Mobile Nodes 1 and 2

The individual statistics of both mobile nodes in simulated scenario are shown in Figure 16. The mobile node 2 is used as a hop in selected route for data transmission from source to destination. The AODV routing traffic sent and received in both packets and bits of mobile node 2 is high than mobile node 1. As the graph shows both mobile nodes are used in establishment of routes but OPNET internal mechanism selects the route of mobile node 2 because of shortest way to destination. Mobile node 1 graph shows that it sent

Figure 13. Global statistics of MANET delay

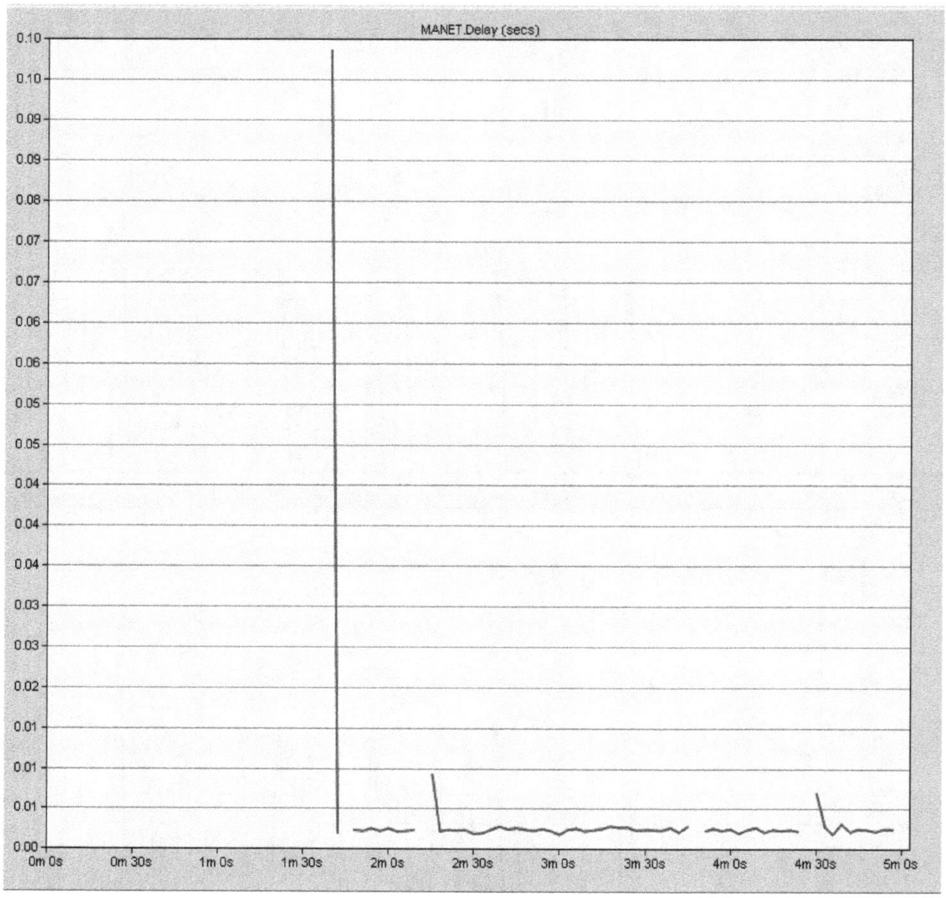

routing traffic only in the establishment of route but is not use any more in the communication of data packets.

Object Statistics of AODV Intermediate Nodes 1 and 2

The individual statistics of both wireless intermediate nodes in simulated scenario are shown in Figure 17. The intermediate node 2 is used as a hop in selected route for data transmission from source to destination. The AODV routing traffic sent and received in both packets and bits of intermediate node 2 is high than intermediate node 1. The following figures shows the same results as discussed above for mobile nodes.

Total AODV Traffic Sent and Received

During transmission of data in simulated ad hoc network routing traffic sent by all wireless nodes is shown in the following figure . As discussed earlier that the selected route consist of four nodes, i.e. source, mobile node 2, intermediate node 2, and destination. The high level of traffic is sent by intermediate node 2 and mobile node 2 during simulation time. The source node sent a bit high routing traffic of 300 bits/sec once during first route discovery process. The intermediate node 2 sends extra high traffic of 290 bits/sec each time the route discovery process occurs. The detail analysis of the graph shows that routing data generated at source node and destination node is nearly the same at. The intermediate nodes show

Figure 14. Global statistics of MANET traffic sent (a) and traffic received (b)

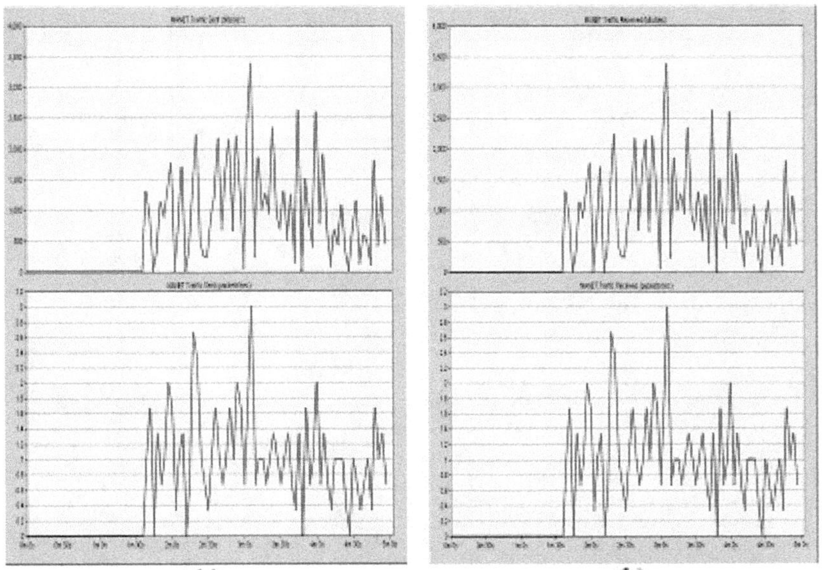

(a) (b)

Figure 15. Object statistics AODV source node (a) and destination node (b)

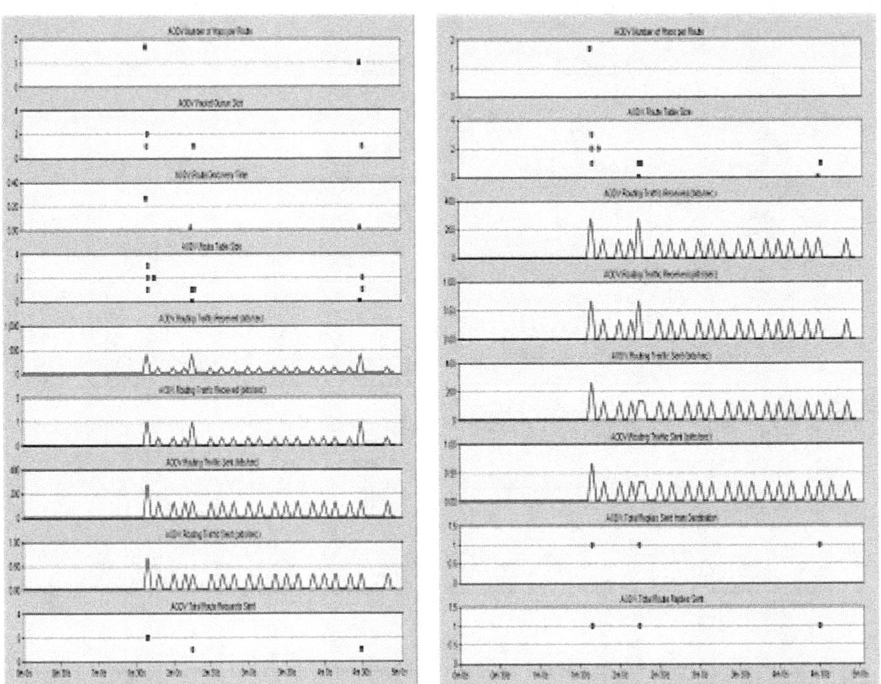

(a) (b)

Figure 16. Object statistics AODV mobile node 1 (a) and mobile node 2 (b)

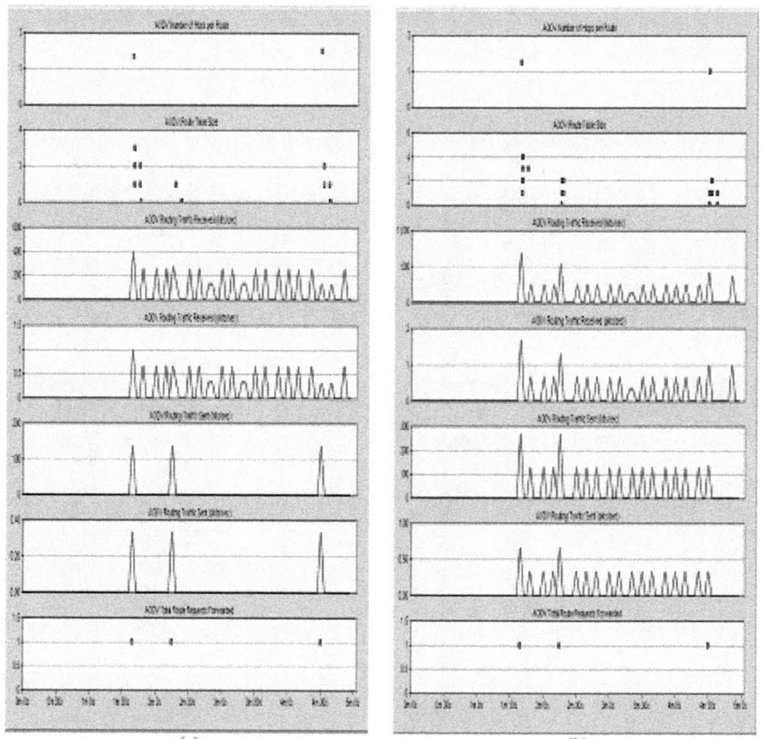

(a) (b)

Figure 17. Object statistics AODV intermediate node 1 (a) and intermediate node 2 (b)

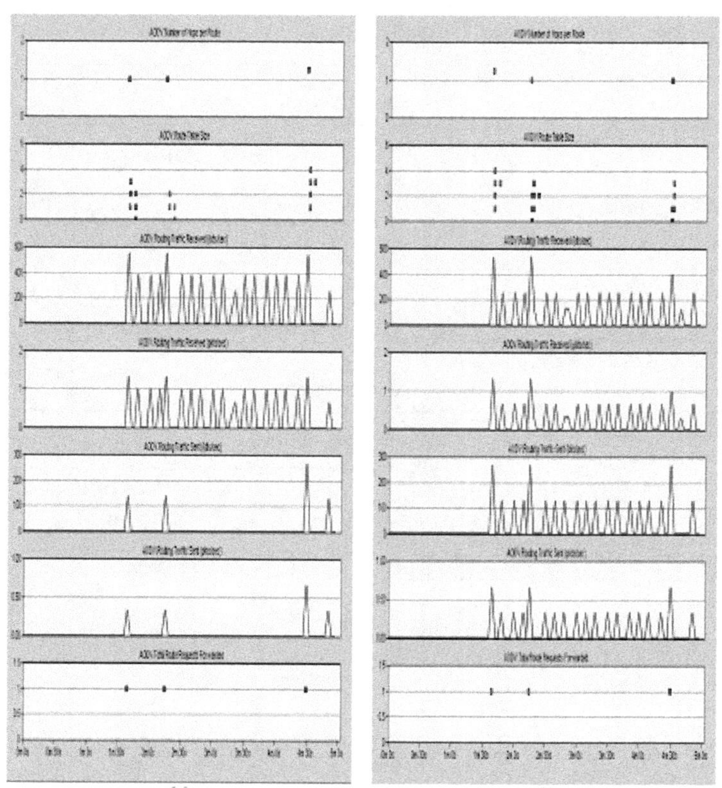

(a) (b)

Figure 18. Total AODV routing traffic sent

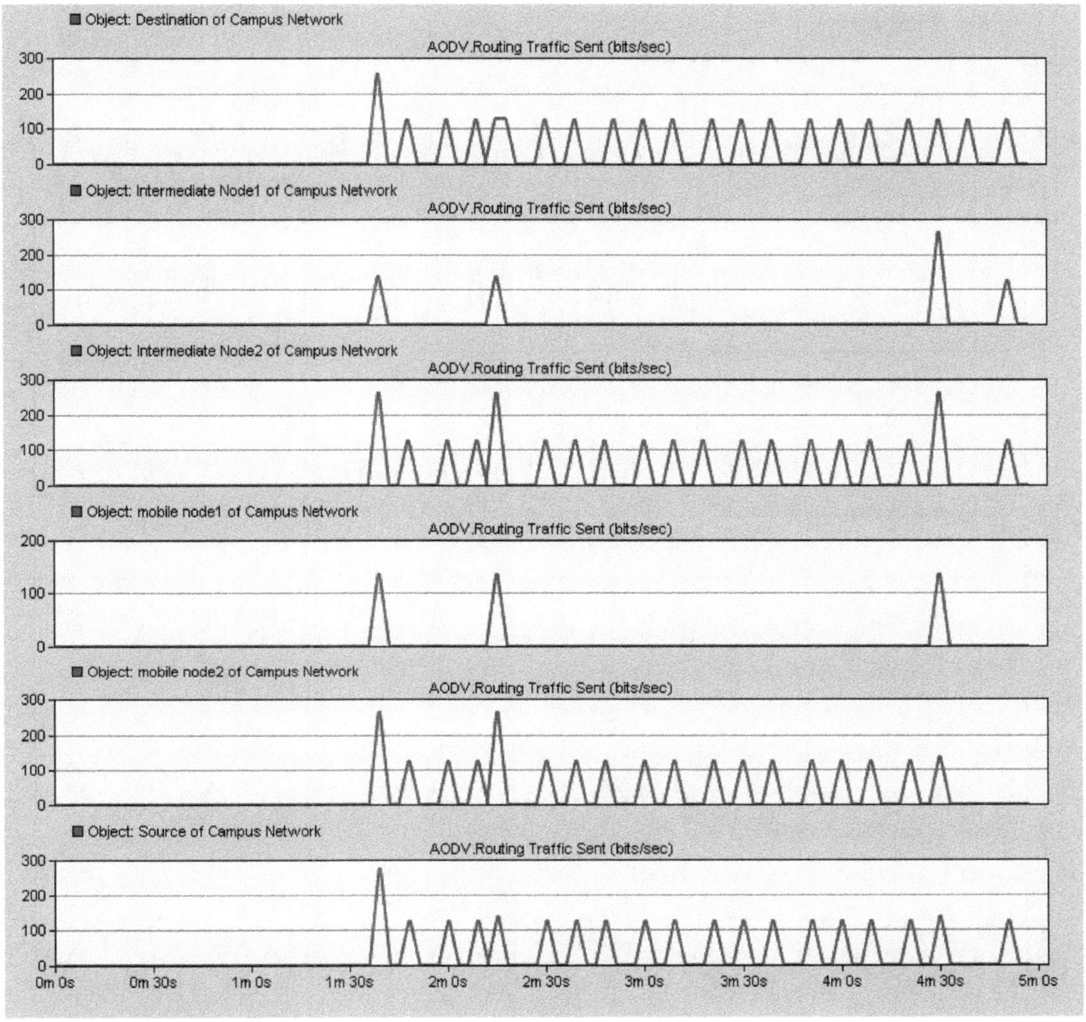

the high traffic signals because of Hello messages to keep alive the next neighbour nodes as shown in the following Figure 18.

The routing traffic received by all wireless nodes of ad hoc network during simulation time is shown in the following Figure 19. The traffic received individually by destination node is less than source node. The high traffic received by nodes occur in selected route are mobile node 2 and intermediate node 2. As discussed above that intermediate nodes have high routing traffic because of Hello messages. During simulation time these nodes receive more than 600 bits/sec traffic as shown.

PROGRAM CODE

See Table 2.

CONCLUSION

In this chapter we discuss in detail the survey of selected routing protocol having varying routing strategies and the simulation scenario and also comparison of the statistics of AODV routing protocol. We also proposed some performance matrices for routing protocols to get best effective results to illustrate; that mobility model changes

Figure 19. Total AODV routing traffic received

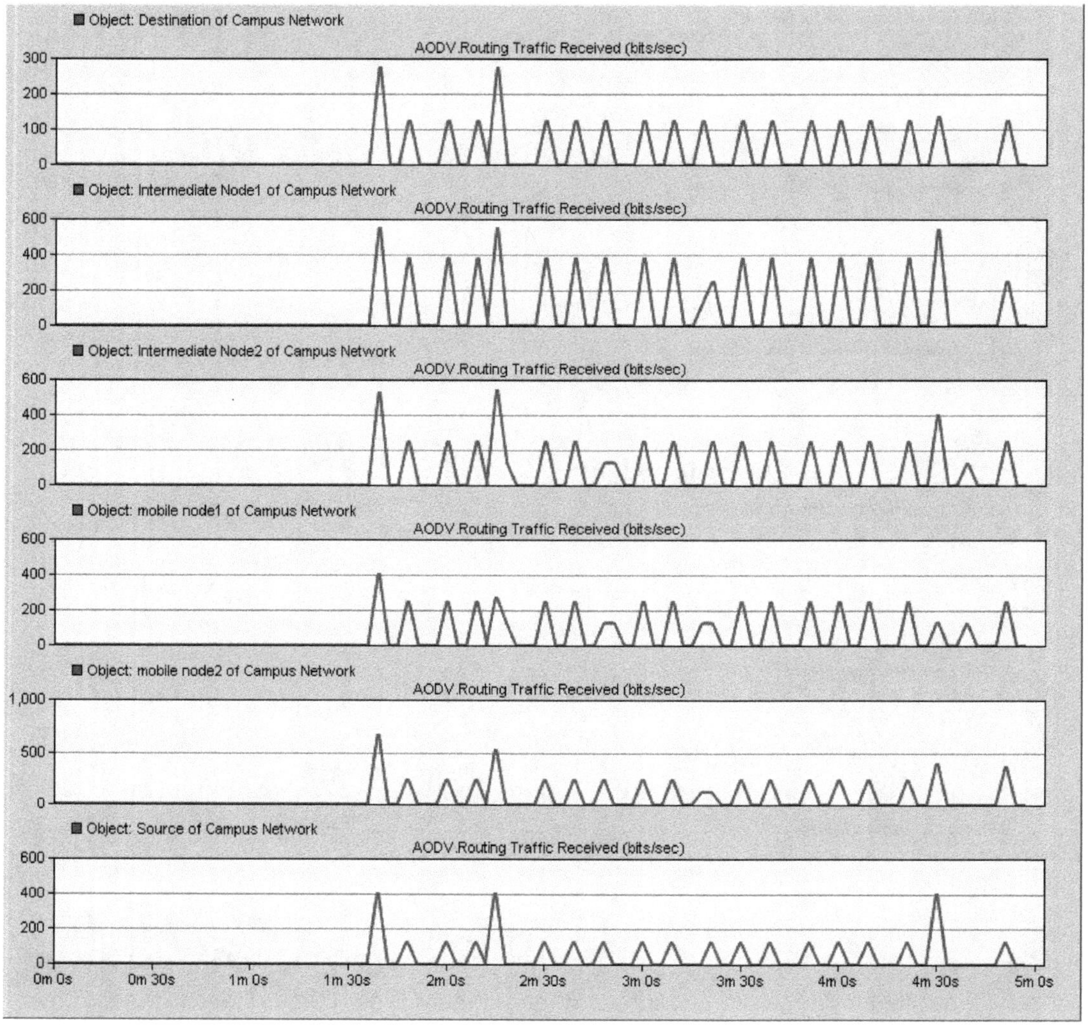

have a significant impact on their performance statically and dynamically. The simulated scenario shows that routing overhead for the routing protocols heavily depends on node mobility. AODV sends many small routing control packets which is more appropriate in ad hoc networks with a higher mobility and data transfer rate. The simulations results and analysis have shown that mobile ad hoc network definitely needs more precise routing protocols to support increase mobility. Study shows that not only fundamental performance properties but also some essential network parameters change performances of various routing protocols. With the increase of number of hops in route throughput degrades due to the higher round trip time delay. The throughput also affect badly by traffic added during the transmission of data through hidden or exposed nodes. Time defined for Hello Loss value also very important in AODV routing protocol to keep neighbor nodes active. The network parameters like network size, number of nodes, number of hops per route, traffic patterns (bits/packets), link capacity, mobility trajectory, and frequency of sleeping nodes do have great results on the performance of routing protocols in the mobile ad hoc networks. The simulations have

Table 2. C File for AODV Data Packet Support

```
/* aodv_pkt_support.ex.c */
#include <opnet.h>
#include <aodv.h>
#include <aodv_pkt_support.h>
#include <aodv_ptypes.h>
#include <ip_addr_v4.h>
staticAODV_Packet*aodv_pkt_memcopy (AODV_Packet*option_ptr);
static void aodv_pkt_memfree (AODV_Packet*option_ptr);
static AODV_Packet*         aodv_mem (void);
static AodvT_Rreq*aodv_rreq_mem (void);
static AodvT_Rrep*aodv_rrep_mem (void);
static AodvT_Rerr*aodv_rerr_mem (void);
static void aodv_rreq_memfr (AodvT_Rreq* rreq_ptr);
static void aodv_rrep_memfr (AodvT_Rrep* rrep_ptr);
static void aodv_rerr_memfr (AodvT_Rerr* rerr_ptr);
Packet*aodv_pkt_create (AODV_Packet* pkt_option_ptr, int option_size)
{
static int index_x = OPC_FIELD_INDEX_INVALID;
Packet* aodv_pkptr;
#ifndef BIN
#define BIN
#endif
#ifndef BOUT
#define BOUT
#endif
 FIN (aodv_pkt_create (<args>));
 aodv_pkptr = op_pk_create_fmt ("aodv");       /* Create the AODV packet*/
        if (index_x == OPC_FIELD_INDEX_INVALID)
index_x = op_pk_nfd_name_to_index (aodv_pkptr, "Options");
op_pk_fd_set(aodv_pkptr, index_x, OPC_FIELD_TYPE_STRUCT, pkt_option_ptr,option_
size,aodv_pkt_memcopy, aodv_pkt_memfree, sizeof (AODV_Packet));
FRET (aodv_pkptr);
}
AODV_Packet*aodv_rreq_create (Boolean join, Boolean rap, Boolean grat_rrep,
Boolean dest_only, Boolean unknown_seq_num, int counthop, int rreq_id, InetT_
Address dest_addr, int dest_seq_num, InetT_Address src_addr, int src_seq_num)
 {
    AodvT_Rreq* rreq_ptr;
    AODV_Packet*aodv_pkt;
FIN (aodv_rreq_create (<args>));  /** Creates the RREQ**/
rreq_ptr = aodv_rreq_mem ();
/* Set the variables of the option*/
```

shown that sudden change in delay, traffic, and link breakage occurs due to nodes mobility as well as the size of the network, traffic load, and delay affects AODV routing protocol. In such situation AODV routing protocol is ideal because of its hop-by-hop routing, so why for the performance of our parameters implemented in AODV routing protocol using OPNET, could be best solution for routing statically and dynamically in ad-hoc networks for efficiency and optimization in such scenarios. AODV typically minimizes the number of required broadcasts by creating routes on an on-demand basis, as opposed to maintaining a complete list of routes as in DSDV algorithm. This chapter contributes in a way form source, mobile node 2, intermediate node 2, and destination in a hop to hop fashion. The high level of traffic is sent by intermediate node 2 and mobile node 2 during simulation time. This research attempts to present some of the analytical comparison of hop-by-hop AODV on- demand routing protocol for better performance. The objective of this master research is to extend and evaluate proposed routing protocol to be suitable for mobile ad hoc networks. In order to prove performance and to selecting the packet delivery fraction, end-to-end delay and also route discovery frequency comparing with global statistics frequencies. The correctness and performance evaluation of the proposed protocol should be done theoretically and implemented through simulation using OPNET simulator to improve routing performance and end to end data delivery.

REFERENCES

Bani Yassein, M., Ould Khaoua, M., Mackenzie, L. M., Papanastasiou, S., & Jamal, A. (2006). *Improving route discovery in on-demand routing protocols using local topology information in MANETs. PM2HW2N'06, October 2, 2006, Torremolinos*. Malaga, Spain: ACM.

De Renesse, R., & Aghvami, A. H. (2004). *Formal verification of ad-hoc routing protocols using SPIN model checker*. Centre for Telecommunication Research, King's College London, UK.

Feeney, M. L. (2005). *Introduction to MANET routing*. Swedish Institute of Computer Science. Retrieved from http://www.nada.kth.se/kurser/kth/ 2D1490/05/lectures/ feeney_mobile_adhoc_routing.pdf

Forouzan, A. B. (2005). *TCP/IP protocol suite* (3rd ed.). McGraw Hill.

Halgamuge, K. S., & Wang, P. L. (2005). *Classification and clustering for knowledge discovery* (1st ed.). Springer.

He, C. (2003). *Throughput and delay in wireless ad hoc networks. Final report of EE359 Class project*. Stanford University.

Hu, Y. C., & Johnson, D. B. (2002). *Ensuring cache freshness in on-demand ad hoc network routing protocols*. POMC'02, October 30–31, 2002, Toulouse, France. ACM.

Khan, J., & Hayder, S. I. (2010). *A comprehensive performance comparison of on-demand routing protocols in mobile ad-hoc network*. FGCN2010 Jeju Island South Korea 13-15 December 2010.

Khan, J., Shah, N., & Jan, H. (2011). A step of mobile ad-hoc on-demand routing protocols towards 4G cellular networks. *International Journal of Computers and Applications, 16*(6). Retrieved from http://www.ijcaonline.org/archives/ volume16/number6/2017-272

Lang, D. (2006). *On the evaluation and classification of routing protocols for mobile ad hoc networks*. Retrieved from http://home.leo.org/~dl/dissertation.pdf

Lee, S. (2000). *Routing and multicasting strategies in wireless mobile ad hoc networks*. Retrieved from http://www.sigmobile.org/phd/ 2000/theses/sjlee.pdf

Lin, C. (n.d.). *AODV routing implementation for scalable wireless ad hoc network simulation (SWANS)*. Retrieved from http://jist.ece.cornell.edu/docs/ 040421-swans-aodv.pdf

Mukija, A. (2001). *Reactive routing protocols for mobile ad hoc network.* Retrieved from http://www.ifi.unizh.ch/~mukhija/ papers/rrp_thesis.pdf

Ram, S. C., & Manog, S. B. (2004). *Ad hoc wireless networks: Architecture and protocol* (1st ed.). Prentice Hall.

Suresh, A. (2005). *Performance analysis of ad hoc on-demand distance vector routing (AODV) using OPNET simulator*. University of Bremen. Retrieved from http://www.comnets.uni-bremen.de/ ~koo/OPNET-AODV-asn.pdf

Chapter 11
Environment Design Architecture of MANET:
Analyzing Parameters, Deviations and Considerations

Krunal D. Trivedi
A.M.Patel Institute of Computer Studies, Ganpat University, India

Narendra J. Patel
Ganpat University, India

Palak. R. Shah
A.M.Patel Institute of computer Studies, Ganpat University, India

ABSTRACT

Mobile Ad-hoc Network (MANET) is a new standard of wireless wearable devices enabling instantaneous person-to-person, person-to-machine, or machine-to-person communications immediately and easily. MANET is a compilation of wireless mobile hosts forming a temporary network without the aid of any stand-alone infrastructure or centralized administration or self organization network. MANET has the aspects such as wireless connection, constantly changing topology, scattered operation, and ease of deployment. These capabilities make MANETs very attractive to applications ranging from simple wireless network coverage in offices and warehouses to emergency rescue/response and military operations, law enforcement, and smart environments (including intelligent transport systems). In this chapter, the authors discuss MANET environment and network architecture, which should be kept in mind at design phase of MANET. The authors also discuss: MANET parameters, environment deviations, routing protocol performance considerations, and MANET security challenges.

DOI: 10.4018/978-1-4666-0321-9.ch011

1. INTRODUCTION TO MOBILE NETWORK AND MOBILE AD HOC NETWORK

- Mobile Network
  - Mobile networks can be classified into infrastructure networks.
  - In an infrastructure mobile network, mobile nodes have wired access points (or base stations) surrounded by their transmission range or area.
- **Mobile Ad Hoc Network**
  - Mobile Ad Hoc Networks (MANET) can be classified according to their reliance on fixed infrastructures network.
  - In a contrast, MANET are autonomously self-organized networks without support of infrastructure.

1.1 Mobile Ad Hoc Network Prologue

In a Mobile Ad Hoc Network, nodes move arbitrarily or randomly, hence the network may familiar with rapid and unpredictable topology changes. Routing paths in Mobile Ad Hoc Network are potentially holding multiple hops, and every node in Mobile Ad Hoc Network has the responsibility to act as a router. Routing in MANET has been an exigent task ever for the wireless networks came into existence. The key reason for this is the invariable change in network topology as a result of high degree of node mobility. A number of protocols have been developed to carry out such kind of task (David B. Johnson, 1994).

An ad-hoc wireless network is a compilation of wireless nodes that self organize into a network without the help of an existing infrastructure. Some or probably all of these nodes are mobile (Jain et al., 2005). Since the network can be deployed swiftly and flexibly, it is attractive to numerous potential applications. A Mobile ad hoc network (MANET) is an autonomous collection of mobile users (nodes) that communicate over bandwidth constrained wireless links as shown in Figure 1.

Because of nodal mobility, the network topology might be change rapidly and impulsively over time to time. The network is decentralized, where network association, connection and message delivery must be executed by the nodes themselves. Message routing could be a problem in a decentralized environment where the topology

Figure 1. Ad hoc wireless network

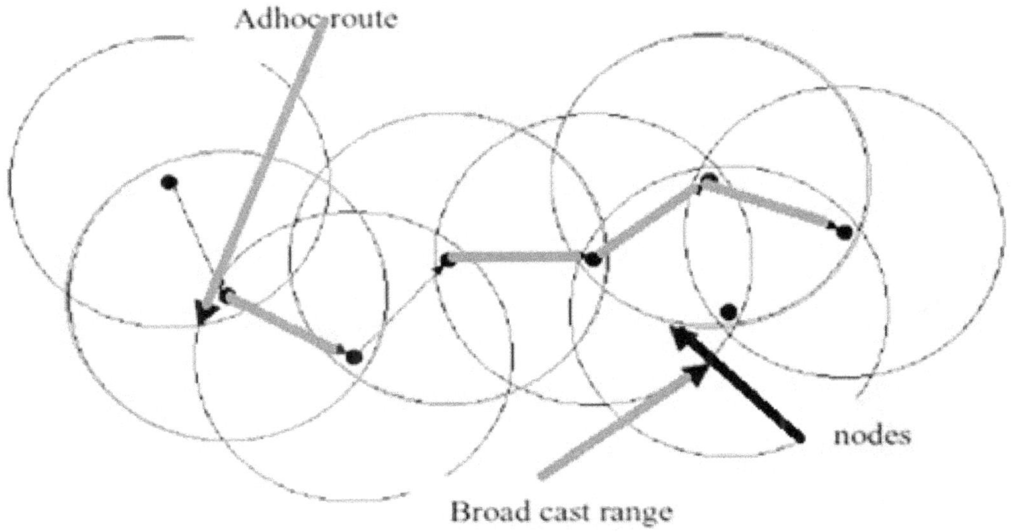

fluctuates and changes. While the shortest path from a source to a destination based on a given cost function in a static network is usually the optimal route, this concept is difficult to extend in Mobile Ad hoc Network. MANET ad hoc nodes present in range of different areas. Factors such as power expended, variable wireless link quality, circulation path loss, fading, multi-user interference, and topological changes, limited bandwidth become appropriate issues. The network should be able to adaptively modify and adjust routing paths to improve any of these properties (Meghanathan, 2011).

E.g.: Possible commercial applications of Mobile Ad hoc Network include business links sharing information during a conference and meeting, students using laptop computers to participate in an interactive sessions, and emergency disaster relief employees coordinating efforts in natural disasters. Mobile ad hoc networks also a good substitute in rural areas or third world countries where basic communication infrastructure is not well recognized.

1.2 Homogenous and Heterogeneous Mobile Ad hoc Network

Most of current routing protocols suppose homogeneous network environment where all nodes have the same capabilities and resources. Although homogenous network are simple and straight forward to model and analysis, they exhibits poor scalability compared with heterogeneous networks that consist of different nodes with different resources. Heterogeneous Mobile Ad Hoc Network comprise of mobile devices as Figure.2 that have different communications potential capacity such as radio range, battery life, data transmission rate, etc (Tamilselvan & Shanmugam, 2010).

2. BACKGROUND

Routing is an expensive operation in Mobile Ad hoc Network. The lack of dedicated and concern equipment to perform routing functions, and the dynamics of the network that cause routes to be unstable, makes routing a resource hungry operation. Furthermore, the more distance the corresponding node and more the more expensive

Figure 2. Heterogenous mobile network

the communication. So, at least which protocol types has been suitable for routing and for node communication that have to be identified before going to use routing Tamilselvan, & Shanmugam, 2010).

A form of wireless network where each node communicates with other node using multi-hop links without motionless and inactive infrastructure is called Ad-hoc network. According to an Ad hoc network are team and group of wireless mobile nodes that creates a network without any support of centralized administrator. It uses multi-hope point-to-point (P2P) routing as an alternative or substitute of inactive network communication to offer network connectivity. In such circumstances, because of partial range of mobile host in wireless transmission, each node required to join up other node in order to communicate with each other and to reach to the destination if located remote places. This communication contains the mechanism of finding paths from one end node to other through which data can be transferred (Murthy & Garcia-Luna-Aceves, 1996).

Routing in ad-hoc networks has been a not only complex but also a very difficult task ever since the wireless networks came into existence. The major reason for this is the environment of ad-hoc networks where network topologies cannot be static or fixed. The non-static nature of Ad-hoc networks raises various performance challenges and issues for routing protocols.

The conceptual framework or skeleton of routing includes decision as to what appropriate optimal or best routing paths should be chosen for transferring the data or packets through an internetwork. The first concept, i.e. determining an optimal or best path, is a very complex and difficult activity while the later one, i.e. forwarding data from end to end taken path, is a straight forward activity. In order to exchange information between two or more nodes, routing needs to be done by using different routing protocols. Therefore efficient routing protocols are key components of successful, reliable and proficient communications. Efficient routing protocol means that an optimal and best route selection is done by the protocol in different scenarios to alleviate the overall network performance and act (Jain et al., 2005).

In order to evaluate the performance of protocols in terms of effectiveness, different performance metrics can be used for more detail.

3. MANET ENVIRONMENT ARCHITECTURE

This Mobile Ad hoc Environment Architecture definitely help those researcher and incumbent who are going to design and implement MANET. Actually, we have classified and managed MANET Environment architecture into four sections like Environment deviations, Routing Protocol performance issues, various Network Parameters that required setting up and few security considerations.

3.1 Environment Deviations

The various and different Mobile Ad Hoc Network environment deviations are listed as follows taking its dynamic topology in to consideration (Mamatha & Sharma, 2010):

1. No Centralized Control: In Mobile Ad Hoc Network all nodes have identical capabilities as well as responsibilities, which are named as fully symmetric environment. A mobile ad hoc network is a network comprising wireless mobile nodes that exchange the information with each other without centralized control or established infrastructure. These nodes which are within each other's radio range can share the information and communicate directly, while distance nodes rely on their neighboring nodes to forward packets. In Mobile Ad Hoc Network every node can be a host or router (Das & Ram

Figure 3. MANET architecture

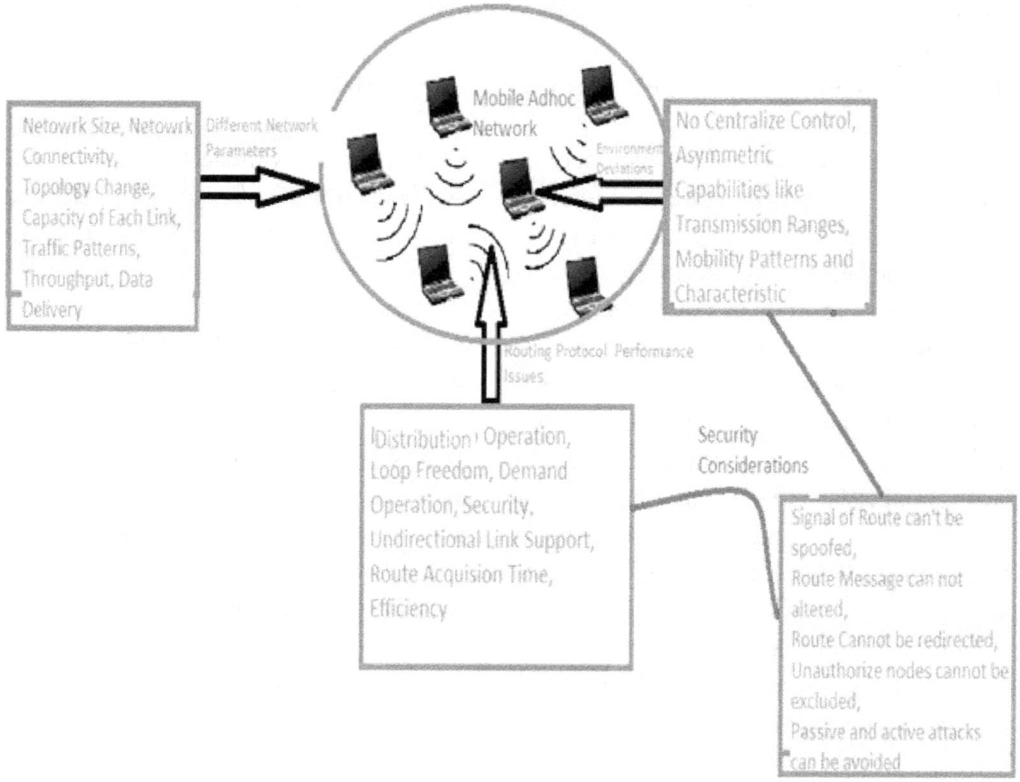

Murthy, 2002). In MANET environment, nodes are free to unite, link or leave the network at any point of time, resulting in a highly dynamic network environment compared to wired network.

2. MANET'S Asymmetric Capabilities and Responsibilities: The Asymmetric Capabilities in Mobile Ad Hoc Network include transmission ranges and radios ranges which may differ. Battery life, speed of movement and processing capacity will be different at different nodes. Asymmetric Responsibilities include that only some nodes may route packets in the network or some nodes may act as leaders for nearby nodes like cluster head (Ioannis et al, 2001).
3. Traffic Characteristics: Traffic characteristics may differ in different ad hoc networks like bit rate, timeliness constraints, reliability requirements, unicast or multicast or geo-cast, host-based addressing or content-based addressing or capability-based addressing. MANETS may co-exist and also co-operate with an infrastructure based network (Thaler, 1996).
4. Mobility patterns: It may be different like people sitting at an airport lounge, citywide taxi cabs, military movements and personal area networks. The performance of a mobile ad hoc network is dependent on the node mobility pattern as well as topology, data traffic patterns, and radio interference (Mamatha & Sharma, 2010).
5. Mobility characteristics: It can include speed, predictability, direction of movement, pattern of movement, uniformity of mobility characteristics among different nodes (Ioannis et al, 2001).

3.2 Protocol Description and Performance Related Issues

Wireless mobile ad-hoc networks are characterized as networks without any substantial connections. In these networks there is no fixed topology due to the mobility of nodes, interference, multipath propagation and path loss. Hence a suitable routing protocol is needed for networks to function properly (Perkins & Royer, 1999).

3.2.1 Multicast Routing Protocol

Many multicast protocols have been implemented as DVMRP, MOSPF, CBT, and PIM, which were designed for static hosts. These protocols do not match the characteristics of wireless mobile network. Therefore some protocols were proposed to solve such problems, such as Mobile IP, IP Multicast, and Mobile Multicast (Das & Ram Murthy, 2002). These protocols are suitable only for infrastructure based networks, which maintain a fixed topology, but for arbitrary network topology such as MANET, other MRP are proposed. Although there are many protocols that have been proposed, the following problems are still not solved completely (Baryun & Al-Begain, 2008):

1. Non-optimal delivery path.
2. Datagram duplication.
3. Overhead resulting from frequent reconstruction of a multicast tree.

In the literature classification of MPs may be based on the algorithm centralization or distribution, or based on type of services and application they support. Moreover, routing protocols may be categorized as reactive and proactive protocols (Kant & Awasthi, 2010). Reactive protocols set up routes on-demand and proactive protocols updates routing tables periodically. In MANET it is more likely to have a reactive routing because of the mobility problem. From we provide a classification of MRP aiming for simplicity and based on how dynamic routes are created and maintained to the member of group as follows (Baryun & Al-Begain, 2008).

3.2.2 Unicast Routing Protocol

Most applications in the MANET are based upon unicast communication. Thus, the most basic operation in the IP layer of the MANET is to successfully transmit data packets from one source to one destination. The forwarding procedure is

Figure 4. Unicast and multicast routing protocol

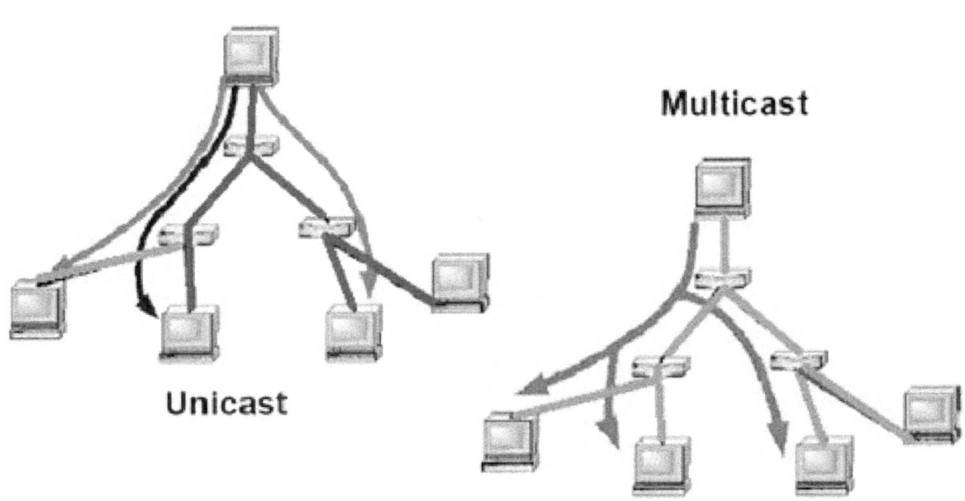

very easy in itself: with the routing table, the relay node just uses the destination address in the data packet to look it up in the routing table. If the longest matching destination address is found in the table, the packet is sent to the corresponding next hop (Ammar & El-Rewini, 2004). The problem that arises is how the routing table is built in the nodes in the MANET. In the unicast routing one separate copy sends to each receiver from the source node. Data packet is replicated at the sender node and then delivered to each destination node. By this process we can easily see that bandwidth is consumed by the redundant data packets (Das, Castaneda, Yan, & Sengupta, 1998),. Many applications use the unicast routing protocol depending upon the need of the application. There are proactive, reactive and hybrid routing protocol in unicast routing for Mobile distributed networks (Kant & Awasthi, 2010).

1. Distributed operation: This is an essential property, but it should be stated nonetheless.
2. Loop-freedom: Not required per se in light of certain quantitative measures (i.e. performance criteria), but generally desirable to avoid problems such as worst-case phenomena, e.g. a small fraction of packets spinning around in the network for arbitrary time periods. Ad hoc solutions such as TTL values can bind the problem, but a more structured and well-formed approach is generally desirable as it usually leads to better overall performance (Mamatha & Sharma, 2010).
3. Demand-based operation: Instead of assuming a uniform traffic distribution within the network (and maintaining routing between all nodes at all times), let the routing algorithm adapt to the traffic pattern on a demand or need basis. If this is done intelligently, it

Figure 5. Unicast and multicast protocol hierarchy

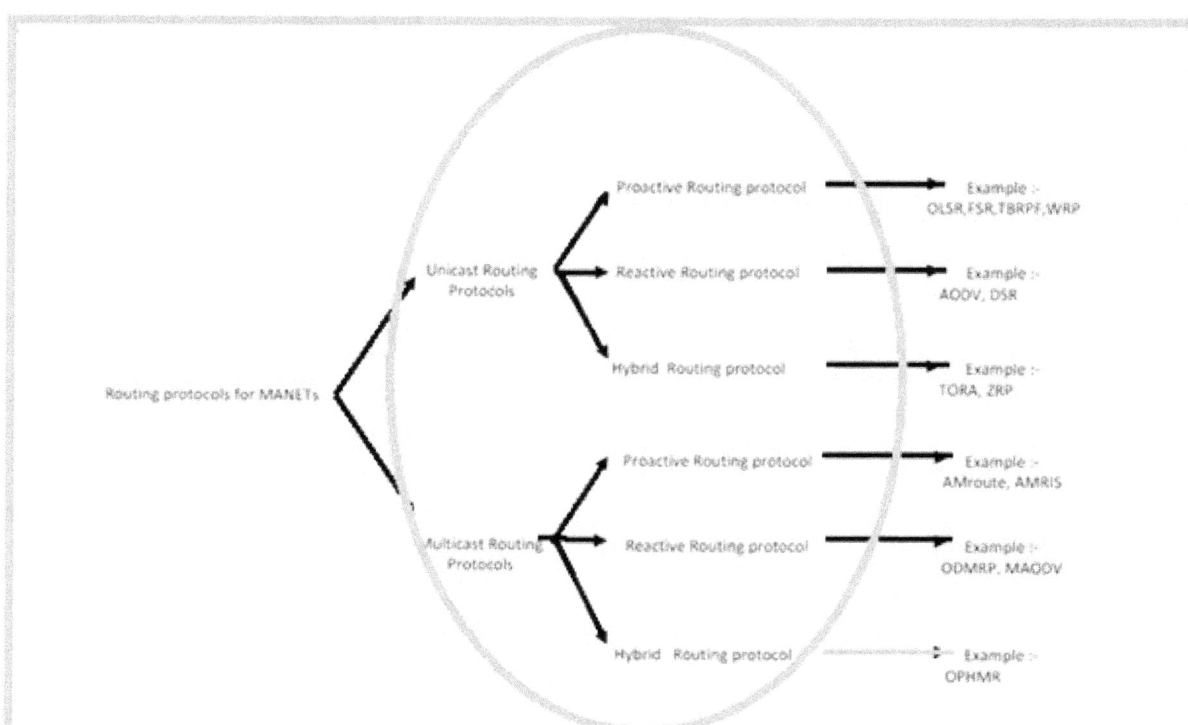

can utilize network energy and bandwidth resources more efficiently, at the cost of increased route discovery delay.

4. Proactive operation: The flip-side of demand-based operation. In certain contexts, the additional latency demand-based operation incurs may be unacceptable. If bandwidth and energy resources permit, proactive operation is desirable in these contexts (Jain et al., 2005).

3.2.3. Composite Headers

One key modification required by our interoperability mechanisms is the addition of a composite header for all routing layer messages. The composite header ensures that nodes in different routing domains running different routing mechanisms have some minimal knowledge about cross-domain routing messages. The information carried by the composite header includes (Ji & Corson, 2001).

Protocol Type: Specifies the type of the protocol e.g. MAODV, ODMRP, AMRIS etc (Wu & Tay, 1999).

Message Type: Specifies the type of message e.g. request, reply, data Source Address
Source Address: Address of node initiating the message Destination Address
Destination Address: Address of destination node or multicast group
Next Hop Address: Next hop towards destination
Sequence Number: Required to detect duplicate messages

The use of the different fields will become clear in the following sections as we illustrate the operation of the interoperability mechanisms through examples.

3.3 Network Parameters

1. *Network Size:* Network size is measured in the number of nodes in infrastructure of MANET.
2. *Network connectivity:* Network connectivity is an average degree of node means average numbers of neighbor node.
3. *Topological rate of change:* the speed with which a network's topology is changing.
   Link capacity: effective link speed measured in bits/second, after accounting for losses due to multiple accesses, coding, framing etc.
4. *Fraction of unidirectional links:* how effectively does a protocol perform as a function of the presence of unidirectional links?
5. *Traffic patterns:* how effective is a protocol in adapting to non-uniform or burst traffic patterns?
6. *Mobility:* when, and under what circumstances, is temporal and spatial topological correlation relevant to the performance of a routing protocol? In these cases, what is the most appropriate model for simulating node mobility in a MANET? (Mamatha & Sharma, 2010)

3.4 Security Considerations

Security is a critical as well as complex issue of ad hoc networks that is still a largely unexplored area. Data should be particularly horizontal to malicious attacks, such as denial of service (DoS). Lack of any centralized network management or certification authority makes the dynamically changing wireless structure.

Traditional methods of protecting the data with cryptographic and encryption / Decryption methods face a challenging task of key distribution and refresh. The most serious problem is probably the risk of a node being captured and compromised.

This node would then have access to structural information on the network, relayed data, but it can also send false routing information which would paralyze the entire network every quickly. One of the current approaches to the security problems is const a self-organized public-key infrastructure for ad hoc networks cryptography (Jain et al., 2005).

FUTURE WORK

Future work could progress along a number of lines. First, one could explore the various protocol parameters such as Multicast routing and Unicast routing. Preferably, appropriate parameters could be derived automatically, based on observed network traffic and rate of neighborhood change. Second, we can explore some ways to increase the network capacity by modifying the MAC protocol. Since this is the more promising approach, we focus on this work. As more traffic can be carried, packet delivery ratios should increase. Even if, you can chose a various protocols from Multicast and Unicast as per your efficiency and requirement. But you should think about the various throughputs, performances, and response time of each and every protocol such as OLSR, DSR, TORA, AMRIS, and ODMRP etc (Thaler, Alaettinoglu, Estrin, & Handley, 1998).

For the future work, we identify the need to develop a light-weighted but reliable multicast protocol for small groups. It can be applied to the upper level multicast in the routing hierarchy to achieve better reliability in packet delivery (Park & Corson, 1997).

CONCLUSION

This section summarizes our efforts towards represents a design approach for MANET architecture.

Mobile Ad-Hoc Networks (MANETs) are comprised of mobile nodes (MNs) that are self-organizing and cooperative to ensure efficient and accurate packet routing between nodes (and, potentially, base stations). Routing is an essential component of information sharing protocols in mobile ad hoc networks. Routing protocols typically fall under two classifications; first one is unicast routing protocol, second one is multicast routing protocol (Royer & Perkins, 1999). The design of the protocols are driven by specific goals and requirements based on respective assumptions about the network properties or application area. In this chapter we studied unicast and multicast routing protocols overview and vital roles for MANETs (Athanasios et al., 2006). In this paper, we apply various network and security issues also for the design of MANET. We somehow categorize the current multicast routing protocols by their reactive and proactive manner. We also study various performance issue of routing protocol (Biradar & Manvi, 2010).

REFERENCES

Ammar, H., & El-Rewini, H. (2004). *A multicast protocol for mobile ad hoc networks using location information*. Boston, MA: Boston University.

Athanasios, B., et al. (2006). A mobility sensitive approach for efficient routing in ad hoc mobile networks. In *Proceedings of the 9th ACM International Symposium on Modeling Analysis and Simulation of Wireless and Mobile Systems*. Terromolinos, Spain: ACM Press.

Badarneh, O. S., & Kadoch, M. (2009). Multicast routing protocols in mobile ad hoc networks: A comparative survey and taxonomy. *EURASIP Journal on Wireless Communications and Networking*, Article ID 764047, 42 pages.

Baryun, A. N., & Al-Begain, K. (2008). *A design approach for MANET multicast protocols*. PGNet. ISBN: 978-1-902560-19-9

Biradar, R., Manvi, S., & Reddy, M. (2010, April). Mesh based multicast routing in MANET: Stable link based approach. *International Journal of Computer and Electrical Engineering, 2*(2), 1793–8163.

Cordeiro, C., Gossain, H., & Agrawal, D. (2003, January). Multicast over wireless mobile ad hoc networks: Present and future directions. *IEEE Network, 17*(1).

Das, B. S. M. S. K., & Ram Murthy, C. S. (2002). A dynamic core based multicast routing protocol for ad hoc wireless networks. *Proceedings of the 3rd ACM International Symposium on Mobile Ad Hoc Networking and Computing, Switzerland* (pp. 24-35).

Das, S., Castaneda, R., Yan, J., & Sengupta, R. (1998). Comparative performance evaluation of routing protocols for mobile ad hoc networks. *Proceedings of IEEE 7th International Conference on Computer Communications and Networks* (I3CN), Lafayette, LA, October, (pp. 153-161).

Ioannis, C., et al. (2001). An experimental study of basic communication protocols in ad-hoc mobile networks. In *Proceedings of the 5th International Workshop on Algorithm Engineering*. Springer-Verlag.

Jain, J., Fatima, M., Gupta, R., & Bandhopadhyay, K. (2005). Overview and challenges of routing protocol and MAC layer in mobile ad-hoc network. *Journal of Theoretical and Applied Information Technology, 8*(1).

Ji, L., & Scott Corson, M. (2001). *Differential destination multicast– A MANET multicast routing protocol for small groups*. IEEE INFOCOM.

Johnson, D. B. (1994, December). Routing in ad hoc networks of mobile hosts. *Proceedings of the IEEE workshop on Mobile Computing Systems and Applications*.

Kant, K., & Awasthi, L. K. (2010). Unicast and multicast routing protocols for MANETs: A comparative survey. *International Journal of IT & Knowledge Management, Special Issue*. ISSN: 0973-4414.

Mamatha, G. S., & Sharma, S. C. (2010). –August). Analyzing the MANET variations, challenges, capacity and protocol issues. *International Journal of Computer Science & Engineering Survey, 1*(1). doi:10.5121/ijcses.2010.1102

Meghanathan, N. (2011). Performance comparison study of multicast routing protocols for mobile ad hoc networks under default flooding and density and mobility aware energy-efficient (DMEF) broadcast strategies. *Informatica, 35*, 165–184.

Murthy, S., & Garcia-Luna-Aceves, J. J. (1996). An efficient routing protocol for wireless networks. *ACM Mobile Networks and Applications Journal, Special Issue on Routing in Mobile Communication Networks*, 183-97.

Park, V. D., & Corson, M. S. (1997). A highly adaptive distributed routing algorithm for mobile wireless networks. *Proceedings Sixteenth Annual Joint Conference of the IEEE Computer and Communications Societies- Driving the Information Revolution, INFOCOM '97,* Vol. 3, (pp. 1405-1413).

Perkins, C. E., & Royer, E. M. (1999). Ad hoc on demand distance vector routing. *Proceedings Second IEEE Workshop on Mobile Computing Systems and Applications, WMCSA '99,* (pp. 90-100).

Royer, E. M., & Perkins, C. E. (1999). Multicast operation of the ad hoc on-demand distance vector routing protocol. *Proceedings of IEEE MOBICOM'99, Seattle, WA,* August 1999, (pp. 207-218).

Tamilselvan, G. M., & Shanmugam, A. (2010, April). Performance analysis of coexistent heterogeneous networks for various routing protocols using Qualnet simulation. *International Journal of Computer Theory and Engineering, 2*(2), 1793–1820.

Thaler, D. (1996, November). *Interoperability rules for multicast routing protocols.* Internet draft draft-thaler-interop.

Thaler, D., Alaettinoglu, C., Estrin, D., & Handley, M. (1998, September). The MASC/BGMP architecture for inter-domain multicast routing. In SIGCOMM, (pp. 93-104). Vancouver, British Columbia.

Wu, C. W., & Tay, Y. C. (1999). AMRIS: A multicast protocol for ad hoc wireless networks. *Proceedings IEEE MILCOM'99, Atlantic City.*

KEY TERMS AND DEFINITIONS

Bandwidth: In computer networks, bandwidth is often used as a synonym for data transfer rate the sum of data that can be carried from one point to another in a given time period (usually a second). This kind of bandwidth is usually expressed in bits (of data) per second (bps). Occasionally, it's expressed as bytes per second (Bps).

Cryptography: Cryptography is the knowledge of information security. The word is derived from the Greek kryptos, meaning hidden. Cryptography includes techniques such as microdots, merging words with images, and other ways to hide information in storage or transit. However, in today's computer-centric world, cryptography is most often associated with scrambling plaintext it's also called ordinary text, sometimes referred to as clear text into cipher text a process called encryption, then back again known as decryption. Individuals who practice this field are known as cryptographers.

Heterogeneous Network: A heterogeneous network is a network connecting computers and other devices with different operating systems and/or protocols. For example, local area networks (LANs) that connect Microsoft Windows and Linux based personal computers with Apple Macintosh computers are heterogeneous. The word heterogeneous network is also used in wireless networks using different access technologies. For example, a wireless network which provides a service through a wireless LAN and is able to maintain the service when switching to a cellular network is called a wireless heterogeneous network.

Homogenous Network: A homogeneous network is a computer network of computers using similar configuration and multiple network layer protocols. For example, a network using Microsoft Windows over TCP/IP is a homogeneous network.

Network Topology: Network topology is the layout pattern of interconnections of the various elements (links, nodes, etc.) of a computer or biological network. Network topologies may be physical or logical.

Routing Protocol: A routing protocol is a protocol that specifies how routers communicate with each other, disseminating information that enables them to select routes between any two nodes on a computer network, the choice of the route being done by routing algorithms.

Section 3
MANET Threat and Security

Chapter 12
Key Generation for Wireless Sensor Networks Using Symmetric Balanced Incomplete Block Design

K. G. Srinivasa
MS Ramaiah Institute of Technology, India

V. Archana
MS Ramaiah Institute of Technology, India

V. Poornima
MS Ramaiah Institute of Technology, India

C. Reshma
MS Ramaiah Institute of Technology, India

ABSTRACT

Wireless sensor networks consist of many tiny sensor nodes deployed at a high density over region, requiring surveillance and monitoring. The sensor nodes typically consist of one or more sensing elements, battery, low powered radio transmitter/receiver, microprocessor, and limited memory. Sensor networks deployed in a hostile environment are prone to malicious attacks like eavesdropping, masquerading, traffic analysis, et cetera. Hence, security is more important in sensor networks than in traditional networks. An important challenge in sensor network security is the design of effective bootstrapping protocols for the nodes, which are pre-initialized with some secret information and have had no prior direct contact with each other. Each node has a set of keys, called a key chain, rather than a single shared key. The keys in the key chain are picked randomly from a key pool. Two neighboring nodes either share a common key or should be able to establish a key path such that every pair in the path shares a key. One problem in such a solution is to select a proper key chain so that the network remains connected with a high probability.

DOI: 10.4018/978-1-4666-0321-9.ch012

INTRODUCTION

This paper brings in the idea of deterministic approach to key generation. There is no need to assume a probability for any two nodes to be neighbors. We use combinatorial based approach for key generation using a block design technique called symmetric Balanced Incomplete Block Design (BIBD). The main contribution to this work is the use of multiple key spaces to increase resilience and decrease memory utilization and still retain connectivity. It eliminates the dependency between the number of keys in the key-chain and number of nodes in the network. OMNeT++ with Mobility Framework is used for simulating the sensor network. The simulation shows each node broadcasting its key-chain, comparing its key-chain with that of its neighbor and then disconnecting the connection with those nodes which do not have a common key with its key-chain. It also generates an optimal number of key-chains to support birth of nodes after the network is deployed.

Sensor Networks are mostly ad-hoc and mobile networks deployed for a wide variety of applications including military sensing and tracking. They are regarded as distributed autonomous systems for information gathering, performing data-intensive tasks such as environment monitoring, seismic monitoring etc. They have the ability to withstand adverse environmental conditions, communication and node failures. They can adapt to dynamic network topology, heterogeneity of nodes and mobility of nodes. They consist of one or more computationally expensive nodes called base stations and a large amount of inexpensive, low capacity nodes called sensor nodes.

The base stations are distinguishable components of wireless sensor networks with high computational capability, energy and communication resources. Sensor nodes can be imagined as small computers, extremely basic in terms of their interfaces and their components. They typically consist of a processing unit with limited computational power and limited memory, one or more sensing elements, a communication device (usually a low powered radio transmitter/receiver), and a power source usually in the form of a battery. The microcontroller is the processing unit which processes data and controls the functionality of other components in the sensor node. Power is used for sensing, communication and data processing. Radio communication allows each node to communicate with nodes only within a limited radius.

As sensor networks are deployed in a hostile environment they are prone to malicious attacks like eavesdropping, masquerading, traffic analysis etc. Hence, security is more important in sensor networks than in traditional networks. To provide security, communication between nodes should be encrypted and authenticated. Encryption and authentication algorithms require that the communicating nodes share a secret key. The problem here is how to set up secret keys between communicating nodes? This problem is known as the key agreement problem. There are basically three types of general key agreement schemes: trusted server scheme, self-enforcing scheme and key pre-distribution scheme. The trusted server scheme depends on a trusted server for key agreement between nodes (B. C. Neumann, T. Tso, 1994). There is no trusted infrastructure in sensor networks. Hence this scheme is not suitable. The self-enforcing scheme depends on asymmetric cryptography, such as key agreement using public key certificates. However, limited computation and energy resources of sensor nodes often make it undesirable to use public key algorithms, such as (R. L. Rivest, A. Shamir and L. M. Adleman, 1978) or (W. Diffie and M. E. Hellman, 1976). The third type of key agreement scheme is key pre-distribution which is nothing but symmetric key cryptography. This scheme is most widely used and suitable for sensor networks.

The secret keys are stored in the ROMs of sensor nodes prior to deployment. The keys stored must be carefully selected so as to increase the probability that two neighboring nodes have at least one key in common. Nodes that do not share a key directly must be able to establish a key path where each pair of nodes in the path shares a key.

There are a number of approaches to key pre-distribution. One naive approach is to store a master key in each of the nodes. The nodes use this master key to obtain a new pair wise key. Though it is simple to implement, it has a major disadvantage of single point of failure. If an intruder compromises a node, he is in possession of the master key with which he can compromise the whole network. Hence, this scheme is not resilient. Another approach is to store a set of $N - 1$ keys in each node where one key is known only to one of the $N - 1$ nodes (Assuming there are N nodes in the network). This scheme overcomes the resilience problem discussed in the previous scheme, since the capture of a single node will not reveal the secret keys between any pair of nodes. However, if the number of nodes increases, the set of keys to be stored in each node increases proportionally. Since the nodes have limited memory this scheme becomes impractical because N could be large. Moreover, if there are more nodes to be added to the network, the set of keys stored in the nodes have to be updated to include the keys for the new nodes.

BACKGROUND

The definitions of some of the terms used in this paper are listed below.

A. Key Pool

A pool from which keys are picked is called a key pool. It is analogous to the universal set from which sets are formed.

Example 1: Let key pool, P = {1,2,3,4,5,6,7,8,9,10,11,12,13}

B. Key-Space

A subset of the key pool which is used to construct key-chains for the sensor nodes is called as a key-space. We know that 2^n subsets exist for a set containing n elements. Hence, for a key pool containing n elements 2^n key-spaces can be constructed.

Example 2: For the key pool P having 13 elements, 2^{13} key-spaces can be constructed. Each Key space is constructed using some criteria. Let's say, the key-spaces, KS_1 = {1,2,3,4,5,6,7}, KS_2 = {1,3,5,7,9,11,13} and KS_3 = {8,9,10,11,12,13} are 3 key-spaces used by a designer.

C. Key-Chain

A set of keys assigned to each sensor node is called as key-chain. Key-chains are subsets of key-spaces.

Example 3: Two key-chains formed from KS_1 are k_1 = {1, 2, 3}, k_2 = {1, 4, 5} and are assigned to sensor nodes numbered N_1, N_2. k_3 = {1, 7, 9}, k_4 = {9, 11, 13} and k_5 = {3, 5, 7} are key-chains formed from KS_2 and are assigned to sensor nodes N_3, N_4, N_5.

D. Key-Path

Sensor nodes that do not share a key directly may use a path where each pair of nodes on the path shares a key. Such a path is called as a key-path.

Example 4: If sensor nodes N_1 and N_2 have to communicate, they find 1 as the common key in their key-chains. Hence they use 1 as the secret key for their communication. Suppose N_1 and N_4 has to communicate, there is no common key in their key-chains. In such a case we say that

Figure 1. Sensor network with 5 nodes

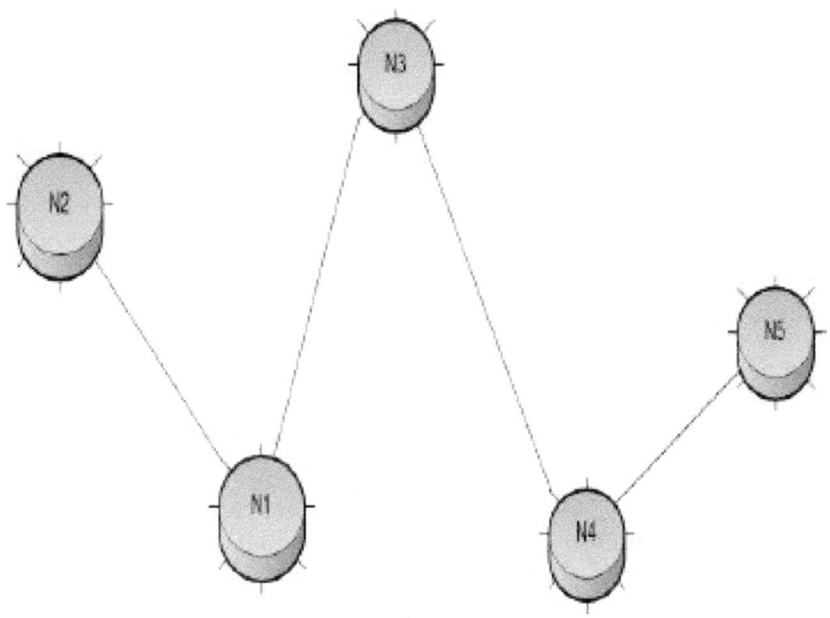

there exists a key-path from N_1 to N_4 via N_3. This is because N_1 and N_3 share 1 as common key and N_3 and N_4 share 9 as common key. So, when N_1 wants to communicate with N_4, N_1 sends the data to N_3 encrypted using the key shared between N_1 and N_3. N3 forwards this data to N_4 by encrypting it using the key that is shared between N_3 and N_4. The network formed with these five nodes is as shown in Figure 1.

E. BIBD

A *Balanced Incomplete Block Design* (BIBD) (I. Anderson, 1990) is an arrangement of v distinct objects into b blocks such that each block contains exactly k distinct objects, each object occurs in exactly r different blocks, and every pair of distinct objects occurs together in exactly ë blocks. The design can be expressed as (v,k, ë), or equivalently (v,b,r,k, ë), where:

ë(v-1)=r(k-1)and bk = vr

A BIBD is called *Symmetric* BIBD or *Symmetric Design* when b = v and therefore r = k. A Symmetric Design has four properties: every block contains k = r elements, every element occurs in r = k blocks, every pair of elements occur in ë blocks and every pair of blocks intersects in ë elements.

Example 5: Consider (v,k,ë) = (7, 3, 1), or equivalently (v,b,r,k,ë) = (7, 7, 3, 3, 1), Symmetric Design. Let's consider key-space, KS_1 = {1, 2, 3, 4, 5, 6, 7} having v = 7 objects. There are b = 7 blocks and each block contains k = 3 objects. Every object occurs in r = 3 blocks. Every pair of distinct objects occurs in ë=1 blocks and every pair of blocks intersects in ë=1 objects. The blocks of the Symmetric Design are: {1, 2, 3}, {1, 4, 5}, {1, 6, 7}, {2, 4, 6}, {2, 5, 7}, {3, 4, 7}, {3, 5, 6}.

F. Latin Square

A Latin Square on n symbols is an *n x n* arrangement such that each of the n symbols occurs exactly once in each row and in each column where n is the order of the square.

Example 6: A Latin square on 4 symbols, {1, 2, 3, 4}, is a 4 × 4 array shown below:

$$\begin{pmatrix} 1 & 2 & 3 & 4 \\ 2 & 1 & 4 & 3 \\ 3 & 4 & 1 & 2 \\ 4 & 3 & 2 & 1 \end{pmatrix}$$

G. Orthogonal Latin Squares

The Latin Squares A = (a_{ij}) and B = (b_{ij}), each of order n are *orthogonal* if all entries of the join of A and B are distinct.

H. MOLS

Latin Squares $A_1, A_2, \ldots, A_r$ are *Mutually Orthogonal Latin Squares* (MOLS) if they are orthogonal in pairs.

I. Complete Set

For prime power n, a set of (n – 1) MOLS of order n is called a Complete Set [10, 11].

Example 7: For n = 4, there exists three MOLS each of order 4 that forms a complete set.

$$\begin{pmatrix} 1 & 2 & 3 & 4 \\ 2 & 1 & 4 & 3 \\ 3 & 4 & 1 & 2 \\ 4 & 3 & 2 & 1 \end{pmatrix}$$

$$\begin{pmatrix} 1 & 2 & 3 & 4 \\ 4 & 3 & 2 & 1 \\ 2 & 1 & 4 & 3 \\ 3 & 4 & 1 & 2 \end{pmatrix}$$

$$\begin{pmatrix} 1 & 2 & 3 & 4 \\ 3 & 4 & 1 & 2 \\ 4 & 3 & 2 & 1 \\ 2 & 1 & 4 & 3 \end{pmatrix}$$

J. Projective Plane

A Projective Plane is an abstract mathematical concept and can be defined as a set of lines and a set of points, the relation between the lines and points being called as incidence and satisfying the following properties:

- Given any two distinct points, there is exactly one line incident with both of them.
- Given any two distinct lines, there is exactly one point incident with both of them.
- There are four points such that no line is incident with more than two of them.

In other words, we can say that a projective plane of order q > 1 is such that;

- Any line is incident with $q + 1$ points.
- Any point is incident with $q + 1$ lines.
- There exists $q^2 + q + 1$ points and the same number of lines.

The smallest possible Projective plane is the Fano plane as shown in Figure 2. It has seven lines and seven points. It is a projective plane of order q = 2. Any line is incident with 2 + 1 = 3 points. Any point is incident with 2 + 1 = 3 lines. There exist $2^2 + 2 + 1 = 7$ points and an equal number of lines.

Figure 2. Fano plane

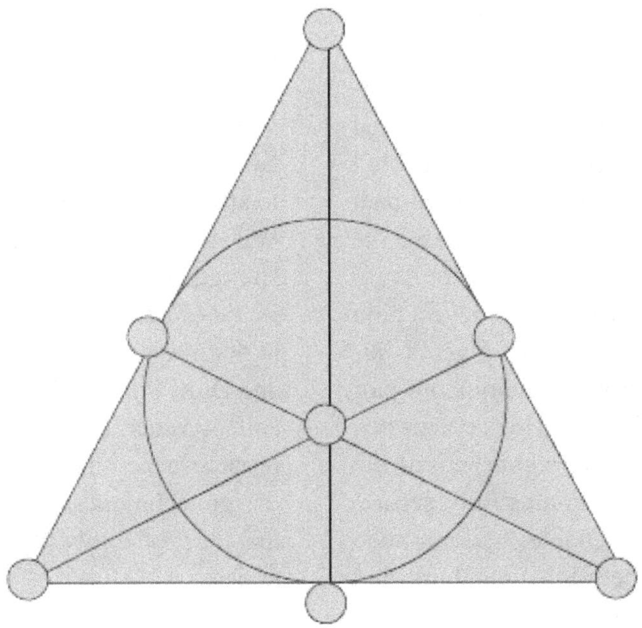

RELATED WORK

(L. Eschenauer, V. D. Gligor, 2002) proposed a random key pre-distribution scheme. In this scheme hundreds of keys are picked from a key pool and are preloaded in the sensor nodes. The keys along with their identities form the key-chains. The nodes exchange their key identities. They propose to employ (R. Merkle, 1978) to secure the key identities. However, the processing at each node increases. They show that for a key pool size of 10000, a key chain of length 75 is enough to obtain connection probability of 0.5.

(L. Eschenauer, V. D. Gligor, 2002) spoke about an interesting feature namely *Revocation* in which the entire key ring of the node which has been compromised will be revoked. To do this there exists a controller node which broadcasts a single revocation message containing the signed list of k key identifiers indicating the key ring to be revoked. After receiving the message from the controller node all the nodes verifies the list of key identifiers, locates those identifiers in its key ring and removes those entries (if any). Once the keys are removed there may be break down of few links in the network so once again shared-key discovery phase takes place and if needed even path-key establishment phase is also repeated.

The disadvantage in this is, if the key-chain length for that revoked node is k, k keys will removed from the key pool (P) for each node revocation. If the number of nodes that is revoked is more, then the key pool size reduces which isn't feasible. The key-discovery and/or path-key establishment phase that follows will take place under this condition and utilizes lots of memory.

One more feature what (L. Eschenauer, V. D. Gligor, 2002) spoke about is *Re-Keying*. This occurs when the lifetime of the key is lesser than that of the node and so re-keying should take place. Re-keying can be considered as self-revocation of a key by a node. As in revocation there is no broadcast of the message by the controller node but a shared-key discovery phase and if needed path-key establishment phase same as in the revocation after the expired-key removal happens.

An enhancement to this scheme was proposed by (H. Chan, A. Perrig and D. Song, 2003).They proposed three new mechanisms to random key pre-distribution. First, they proposed the q-composite random key pre-distribution scheme, which achieves greatly strengthened security under small scale attack while trading off increased vulnerability of a large scale physical attack on network nodes. Second, they present the multi-path key reinforcement scheme, which increases the security of key setup such that an attacker has to compromise many more nodes to achieve a high probability of compromising any given communication. They also propose the random-pair wise keys scheme, which assures that, even when some number of nodes has been compromised, the remainder of the network remains fully secure. Furthermore, this scheme enables node-to-node mutual authentication between neighbors and quorum-based node revocation without involving a base station. However this scheme requires larger key chains and smaller key pool compared to the previous scheme.

Random-pairwise key scheme is a modification over pairwise key scheme. Each node stores a random set of *np* pairwise keys instead of *n − 1* where n is the number of sensor nodes in the network and p is the probability that two nodes are connected. Blom proposed a scheme that allows any pair of nodes to find a pairwise secret key between them. Compared to *(N − 1)*-pairwise key distribution scheme, Blom's scheme uses λ+1 memory space where ë is much smaller than *N*.

Unlike (N − 1)-pairwise key distribution scheme, Blom's scheme is not perfectly resilient against node capture. Instead, it has ë secure property: as long as an adversary captures less than or equal to ë nodes; uncompromised nodes are perfectly secure; when an adversary captures more than ë nodes all the nodes in the network are compromised. Blom's scheme uses one key pool to pick the keys.

An extension to this scheme was proposed by (W. Du, J. Deng, Y. S. Han and P. Varshney, 2003) where multiple key spaces are incorporated. This scheme exhibits a threshold property: when the number of compromised nodes is less than the threshold, the probability that any node other than these compromised nodes is affected is close to zero. This property lowers the initial payoff of smaller scale network breaches to an adversary, and makes it necessary for the adversary to attack a significant proportion of the network. With this scheme the resilience of the network is strengthened.

One disadvantage of these random approaches is that the probability of success is low. (J. Lee and D. R. Stinson, 2005) and (Seyit A. Camtepe, Bulent Yener, 2007) proposed deterministic approach to key distribution using combinatorial design techniques to allocate keys to nodes such that the probability of key share between any two nodes is 1.

Campete and Yener's work brings in a novel construction methodolgy from *combinatorial design theory* to address the key distribution problem. This scheme depicts that determisitic approach has strong advantages over the randomized one since it *i* increases the probability that two nodes share a key, and *ii*decrease the key-path length. Their algorithm is based on symmetric Balanced Incomplete Block Design, Finite Generalized Quadrangle and a hybrid approach. The drawback of this scheme is that the same keys are shared between many pair of nodes leading to weaker resilience to node capture.

1. COMBINATORIAL BASED KEY GENERATION

In the following sections, we describe the use of combinatorics in generation of key-chains for sensor nodes in sensor networks.

A. Mapping

We are interested in constructing Finite Projective Plane of order m which is nothing but a (m^2 + m + 1, m + 1, 1) symmetric design.

Figure 3. List of acronyms and symbols

Acronyms/Symbols	Description
BIBD	Balanced Incomplete Block Design
MOLS	Mutually Orthogonal Latin Squares
N	Total number of sensor nodes in the sensor network
$N_1, N_2, \ldots$	Sensor nodes
P	Key pool
KS_n	Total number of key-spaces
$KS_1, KS_2, \ldots$	Key-spaces
$k_1, k_2, \ldots$	Key-chains
n	Prime power
m	Prime power less than $n/2$
$m+1$	Length of the key-chain

We assume that the sensor nodes are deployed in large numbers randomly and the key-chains are stored in the ROMs of the nodes prior to deployment. Let N be the number of nodes in the network. Each sensor node has K keys in its key-chain. The K keys are selected from a Key-space KS which is a subset of the Key Pool P. Each node has to share ë keys to communicate securely.

If the sensor network contains N nodes, N number of key-chains has to be generated, i.e., a symmetric design with b≥n blocks needs to be generated from an object set S. We now partition the nodes into a k_n number of groups. Each node belonging to one group will pick the key-chain from the corresponding key-space. For example, if a node N_1 belong to group k_3, it will be deployed with a key-chain picked randomly from the 3rd key-space KS_3. Hence k_n is also the number of key spaces generated for the network of size N.

Let m be a prime power such that there are $m^2 + m + 1$ nodes present in each group. Hence we construct a symmetric design with $v = b = m^2 + m + 1$ objects and blocks for the corresponding key-space. We can map the object set S of the symmetric design to the key pool P. A subset of the object set forms the key-space KS. The blocks b can be mapped to the key-chains K. Hence the number of blocks is equal to the number of key-chains, also equal to number of sensor nodes corresponding to a key-space. The objects in the block can be mapped to the keys in the key-chain. This provides b≥n key-chains each having $K = k = m^2+m+1$ keys in its keychain. From the property of $(m^2 + m + 1, m + 1, 1)$ design, each pair of blocks can share ë = 1 object. This means, each pair which picked its key-chain from the same key-space has ë = 1 keys in common. The mapping is as shown in the Figure 4

B. Construction

To generate a $(m^2 + m + 1, m + 1, 1)$ design we use the following approach. First, we generate complete set of (m− 1) MOLS of order m. This can be used to construct an Affine plane of order m which makes it a (m2, m, 1) design. This can be converted to a projective plane which is a $(m^2+m+1, m+1, 1)$ design. The construction process can be summarized as follows:

1. Given N, the number of nodes in the network, find a prime power n where $n^2 + n + 1 \geq N$.
2. Find a suitable value for prime power m and the corresponding value of k_n.
3. Generate a complete set of (m− 1) MOLS or order m [10, 13, Appendix A].

Figure 4. Mapping from symmetric design to key generation

Symmetric Design	Key Generation
Object Set (S)	Key Pool (P)
Subset of Object Set (KS)	Key-spaces (KS)
Subset size $(\|KS\| = v = m^2 + m + 1)$	Key-space size $\|KS\|$
Number of subsets (k_n)	Number of key-spaces (k_n)
Blocks	Key-chains
Number of blocks $(b = m^2 + m + 1)$	Number of Key-chains
Number of blocks $(b = m^2 + m + 1)$	Number of sensor nodes under each key-space
Number of objects in a block $(k = m + 1)$	Number of keys in key-chain
Number of blocks that an object is in $(r = m + 1)$	Number of key-chains that a key is in
Two blocks belonging to the same subset share $\lambda = 1$ objects	Two key-chains belonging to the same key-space share $\lambda = 1$ keys

4. Construct a Affine plane of order m from the MOLS [10, 13]
5. Construct a Projective plane of order m by adding a point at infinity to the affine plane [10, 13, Appendix B]
6. Repeat the steps 4 and 5 k_n times, i.e., for each keyspace.

There are a number of ways in deciding the value of m and k_n. The approach we adopted was to find the value iteratively. We begin with the value of m = n/2 and compute the number of key-spaces necessary to form the required number of key-chains. That is, find k_n such that $k_n(m^2 + m + 1) - N$. Repeat the above process for prime powers less than m such that $k_n(m^2 + m + 1) - N$ is a feasible value. The value $k_n(m^2 + m + 1) - N$ is the number of extra key-chains generated by the design. Stop the iteration when this value becomes reasonably small. Since the iterations cannot be carried out indefinitely, we stop at four iterations and choose the value for m and k_n which gave the least value for $k_n(m^2 + m+ 1) - N$.

The key-spaces are constructed such that any pair of keyspaces do not form a disjoint set. A pair of nodes which picked its key-chain from different key-spaces may or may not contain a key in common. Since, the key-spaces are nondisjoint, there exists a key-path of length greater than 1, if not a direct path. Thus formed projective planes for each of the values k_n becomes the required key-chains.

2. DESIGN AND IMPLEMENTATION

The overall system design is as shown in the Figure 5. The different components of the system implemented in OMNET++ are explained below:

- *Key Generation Component*: This component generates the key-chains which will be deployed in the sensor nodes.
- *Network Formation Component*: This component uses the features of OMNeT++ to create the nodes and establish a network

Figure 5. Comparison of key-chain length

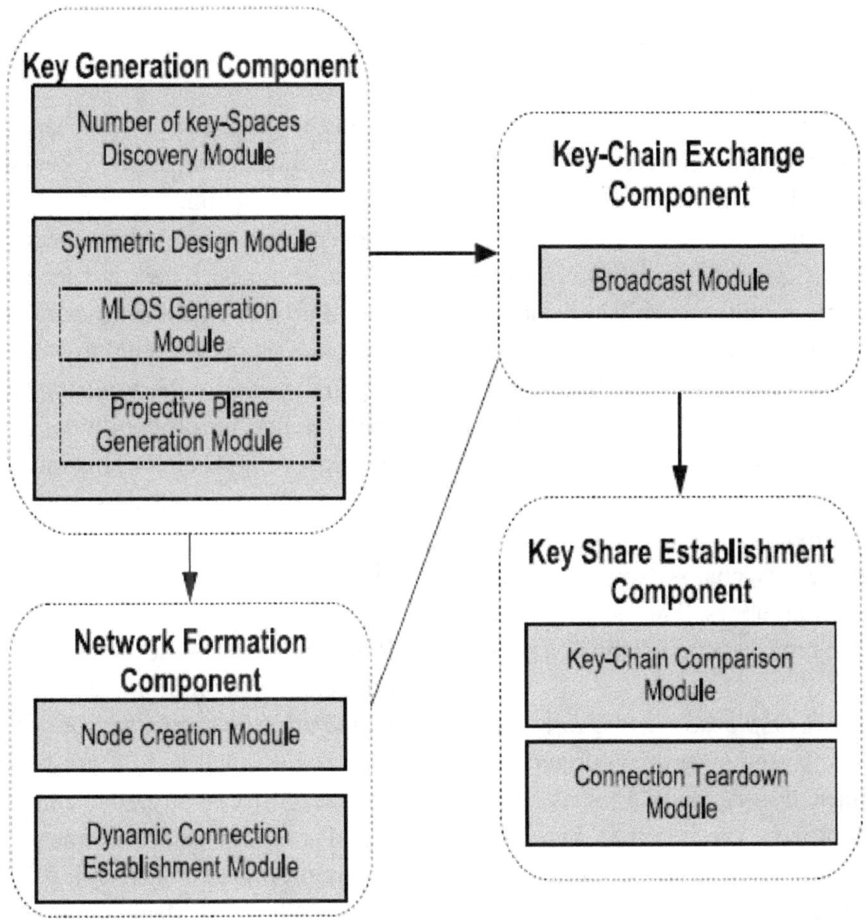

by interconnecting the nodes that are in the interference range.

- *Key Key-chain exchange Component*: This component is responsible for the broadcast of the key-chains to all the nodes that are its neighbors.
- *Key Share Establishment Component*: This component is responsible for checking if there is any key in common to the key-chain of the neighbor and tear down the connection if there is no key share.

Different modules under these components perform the following tasks:

- *Number of Key-Spaces Discovery Module*: This module computes the number of key-spaces required for the given number of sensor nodes.
- *Symmetric Design Module*: This module is used to generate the key-chain based on a combinatorial design technique called Balanced Incomplete Block Design (BIBD). This module is composed of two sub-modules:
- *MOLS Generation Module*: This module generates a complete set of (n-1) Mutually Orthogonal Latin Squares (MOLS) for a given prime power n.
- *Projective Plane Generation Module*: This module generates a projective plane from

the given set of MOLS. The projective plane is what is used as the key-chains.

- *Node Creation Module*: This module creates the sensor nodes over OMNeT++ simulator. Each node comprises a protocol stack of physical layer, MAC layer, Network layer, Transport layer and Application layer.
- *Dynamic Connection Establishment Module*: This module establishes connections between the nodes that are in the interference range dynamically. It automatically creates gates for new connections and deletes the gates when they are not required.
- *Broadcast Module*: This module is used to broadcast the key-chains from each node to all of its neighbors. The neighbors use the broadcast message to establish a secret key.
- *Key-Chain Comparison Module*: This module is activated once the node receives the broadcasted key-chains from its neighbors. It compares the received key-chain and its key-chain to find a common key. This common key can be used as a secret key for encryption of the communication taking places between the two nodes. If there is no common key in the key-chain, the nodes cannot communicate securely.
- *Connection Tear-down Module*: This module is used to disconnect two nodes within the interference range if they do not share a common key in its key-chain. Hence, these two nodes cannot communicate directly.

The flow of the sequence of steps followed in the entire process is illustrated in Figure 6.

3. ANALYSIS

A. Connectivity

In the modified symmetric design, any pair of blocks chosen from a single key-space has at least one object in common, i.e., any pair of key-chains chosen from a single key-space has at least one key in common. The key-spaces are constructed such that two neighboring nodes whose key-chains are taken from different key-spaces have either a key in common or there exists a key-path between them. Hence, the probability of key share between any pair of nodes is 1. This implies that any two nodes can communicate securely.

B. Resilience

Resilience can be defined as resistance against node capture. Compromise of security credentials, which are stored on a sensor node or exchanged over radio links, should not reveal information about security of any other links in the network. Usually higher resilience means lower number of compromised links (Seyit A. Camtepe, Bulent Yener, 2005). An attacker may be able to monitor the network and capture the nodes wisely or he may capture the nodes randomly. Say, an attacker tries to capture the nodes which have the same specific key in their key-chains. From the properties of symmetric design, for a single key space, we know that there are $m + 1$ keys in the key-chain. Every pair of keys can occur in exactly one key-chain within that key space. Hence every $k_n(m^2 + m)$ keys must be pairing with the specific key in these $k_n(m + 1)$ key-chains. So, a wise attacker needs to capture $k_n(m+1)$ key-chains to compromise the entire network. But, if the attacker is unlucky and selects the nodes randomly, he might be capturing those $k_n (m^2)$ key- chains which do not pair with the specific key. Hence, he has to capture one more

Figure 6. Comparison of key-chain length

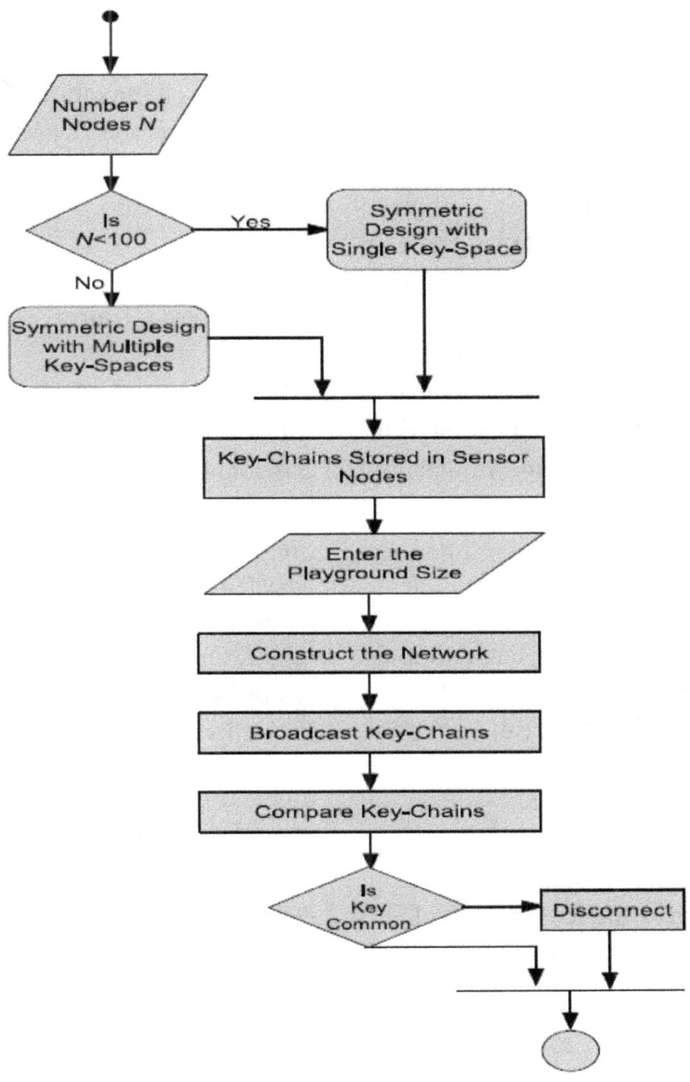

key-chain in each key-space to compromise the whole network. Therefore, an unlucky attacker will need to capture $k_n (m^2 +1)$ key-chains to be able to recover all the keys or in other words, to compromise the entire network.

C. Birth of a Node

In certain applications, new nodes may be added to the network after the complete deployment of the network. This is called birth of a node. The new nodes need to be assigned key-chains before they go into the network. Running the key generation algorithm each time a new node is added to the network makes the process expensive. Hence, the algorithm should compute some extra key-chains the first time it is run and store these key-chains. In our algorithm, the number of key-chains generated is always more than the number of nodes currently deployed in the network. When a new node is to be added to the network it has to get registered at the base station. During this process a new key-chain is assigned to this node from the extra key-chains already generated. It is the responsibility of the base station to maintain a list of unused key-chains out of the entire list of generated key-chains.

D. Temporary Inactivity of a Node

There are instances when a node can go down temporarily, for example, when a running process hangs and the node reboots. We know that, the key-chains are etched on the ROMs of the nodes. Since it is a non-volatile memory, the ROM is not disturbed. When the node is up again its keychain remains intact. It can resume communication with its neighboring nodes using the same key-chain.

4. RESULTS

The performance of the algorithm using multiple key spaces was compared with symmetric design using single key (Seyit A. Camtepe, Bulent Yener, 2007). The following results were obtained.

A. Key-Chain Length

With multiple key spaces the length of the key chain is m + 1, which would have been n + 1 in symmetric design which uses single key-space. Since m < n/2, the key chain length is very much reduced. This is a main advantage since the sensor nodes have limited memory. Even though the key chain length is reduced, the network remains connected with a high probability. The comparison of key-chain lengths for different number of nodes in the sensor network in case of both symmetric design with single key-space and symmetric design with multiple key-spaces is shown in Figure 7.

Resilience

With the use of multiple key spaces, resilience of the network is not hampered. When a node is compromised the intruder has keys taken from one particular key-space. This information is not sufficient to comprise other parts of the network. Hence, he has to compromise more number of nodes to get enough keys to capture the entire network. The graph in Figure 8 shows that the minimum number of nodes to be compromised is almost the same in symmetric design with single key-space and in symmetric design with multiple key spaces.

Extra Key-Chains

The combinatorial design generates more number of key chains than the number of nodes in the sensor network. But, generating extra key-chains will increase the computations. Hence, the number of extra key-chains generated should not be too large. The number of extra key-chains generated should not be too small. Therefore an efficient algorithm should generate an optimal number of extra key-chains. The comparison of the extra key-chains generated is shown in Figure 9.

FUTURE RESEARCH DIRECTIONS

The algorithm can be extended in the following ways to make it more efficient:

- The key-spaces used to construct the key-chains can be designed using a number of ways. If the intruder decodes the way the key-spaces are constructed, it could become easy for him to guess which key-space the corresponding key-chain belongs to. Hence, it becomes easy for him to compromise the network. A more efficient method could be used for the design of key-spaces.
- This algorithm runs in a single base station. It can be extended to work on multiple base stations hence, eliminating single point of failure.

Figure 7. Comparison of key-chain length

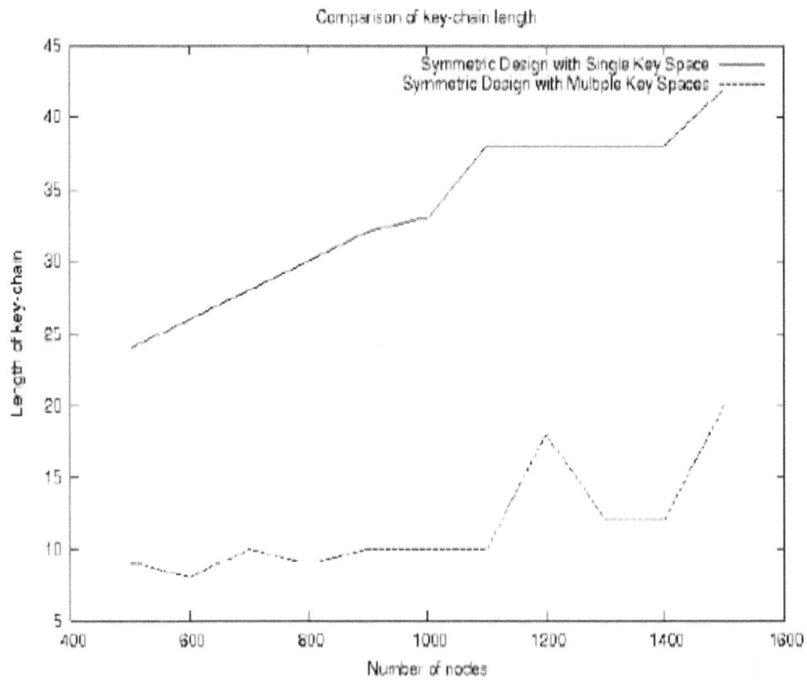

Figure 8. Comparison of resilience

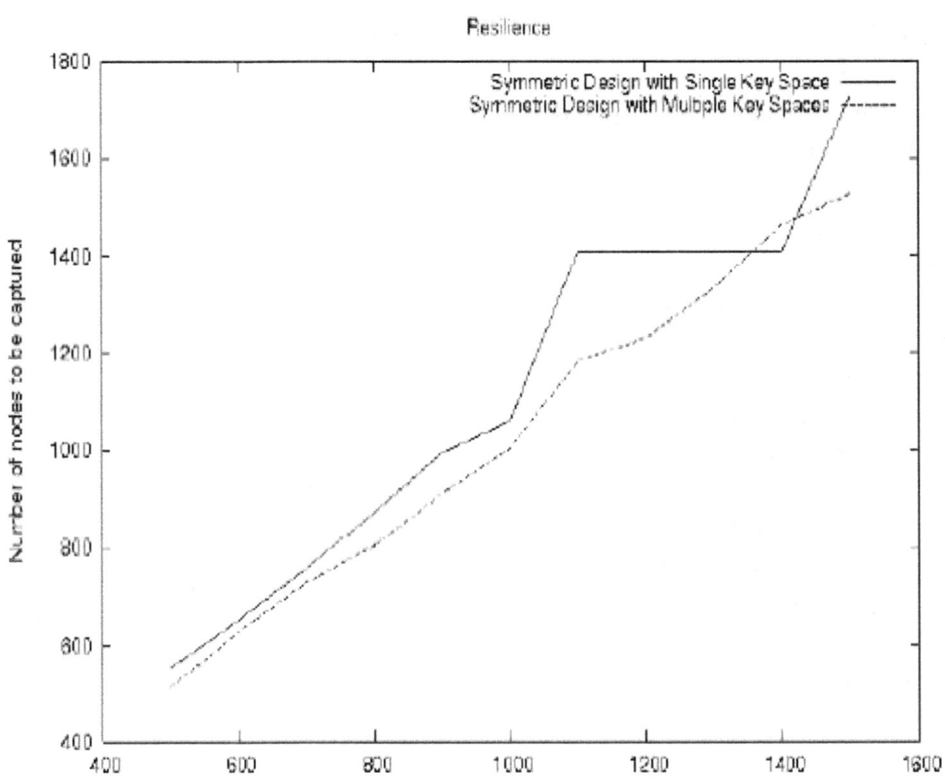

Figure 9. Number of extra key-chains generated

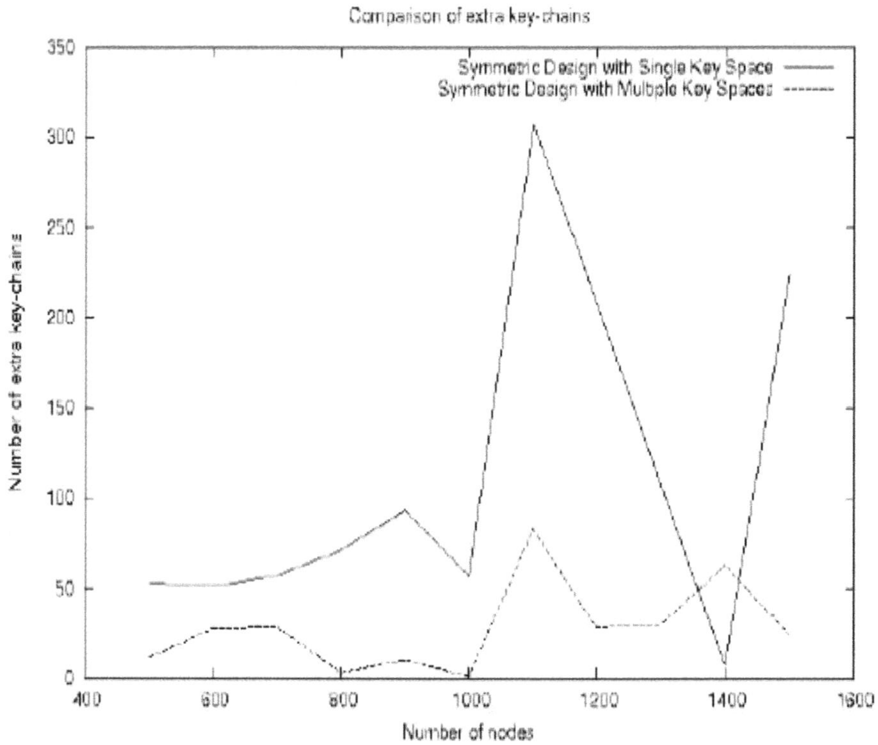

CONCLUSION

The novel idea brought in by this paper is the use of multiple key-spaces instead of a single key-space in the construction of key-chains. Adding to the advantage of using a deterministic approach, we have a few more results derived which prove that this method is better than using a single key-space. Generation of each key-space is left as a part of implementation. The more is the complexity involved in the key-space generation from the key-pool, the better this method proves out to be. The concept of kn(> 1) key-spaces itself proves lesser memory utilization, connectivity in the network and yet have the same network resilience. There exists no dependency between the number of keys in the key-chain and the number of nodes in the network.

REFERENCES

Anderson, I. (1990). *Combinatorial designs: Construction methods*. Ellis Horwood Limited.

Camtepe, S. A., & Yener, B. (2005). *Key distribution mechanisms for wireless sensor networks: A survey*. Technical Report TR-05-07 Rensselaer Polytechnic Institute, Computer Science Department.

Camtepe, S. A., & Yener, B. (2007). Combinatorial design of key distribution mechanisms for wireless sensor networks. *IEEE/ACM Transactions on Networking*, *15*(2). doi:10.1109/TNET.2007.892879

Chan, H., Perrig, A., & Song, D. (2003). *Random key predistribution scheme for sensor networks*. IEEE Symposium on Research in Security and Privacy.

Diffie, W., & Hellman, M. E. (1976). New directions in cryptography. *IEEE Transactions on Information Theory, 22*, 644–654. doi:10.1109/TIT.1976.1055638

Du, W., Deng, J., Han, Y. S., & Varshney, P. (2003). A pairwise key pre-distribution scheme for wireless sensor networks. *Proceedings of the 10th ACM Conference on Computer and Communications Security.*

Eschenauer, V., & Gligor, D. (2002). A key-management scheme for distributed sensor networks. *Proceedings of the 9th ACM Conference on Computer and Communications Security.*

Lee, J., & Stinson, D. R. (2005). Deterministic key predistribution schemes for distributed sensor networks. *Proceedings of ACM Symposium on Applied Computing.*

Merkle, R. (1978). Secure communication over insecure channels. *Communications of the ACM, 21*(4). doi:10.1145/359460.359473

Neumann, B. C., & Tso, T. (1994). Kerberos: An authentication service for computer networks. *IEEE Communications, 32*(9), 33–39. doi:10.1109/35.312841

Rivest, R. L., Shamir, A., & Adleman, L. M. (1978). A method for obtaining digital signatures and public key cryptosystems. *Communications of the ACM, 21*(2), 120–126. doi:10.1145/359340.359342

KEY TERMS AND DEFINITIONS

BIBD: A *Balanced Incomplete Block Design* (BIBD) (I. Anderson, 1990) is an arrangement of v distinct objects into b blocks such that each block contains exactly k distinct objects, each object occurs in exactly r different blocks, and every pair of distinct objects occurs together in exactly ë blocks.

Key pool: A pool from which keys are picked is called a key pool. It is analogous to the universal set from which sets are formed.

Key-chain: A set of keys assigned to each sensor node is called as key-chain. Key-chains are subsets of key-spaces.

Key-path: Sensor nodes that do not share a key directly may use a path where each pair of nodes on the path shares a key. Such a path is called as a key-path.

Key-space: A subset of the key pool which is used to construct key-chains for the sensor nodes is called as a key-space. We know that 2^n subsets exist for a set containing n elements. Hence, for a key pool containing n elements 2^n key-spaces can be constructed.

Latin square: A Latin Square on n symbols is an *n x n* arrangement such that each of the n symbols occurs exactly once in each row and in each column where n is the order of the square.

Resilience: Resilience can be defined as resistance against node capture. Compromise of security credentials, which are stored on a sensor node or exchanged over radio links, should not reveal information about security of any other links in the network. Usually higher resilience means lower number of compromised links.

Sensor Networks: Sensor Networks are mostly ad-hoc and mobile networks deployed for a wide variety of applications including military sensing and tracking. They are regarded as distributed autonomous systems for information gathering, performing data-intensive tasks such as environment monitoring, seismic monitoring etc.

APPENDIX

- **Theorem 1:** A complete set of (n − 1) MOLS exists whenever n is prime.
  Proof: We define square arrays A_1, A2, ..., An_{-1} as follows. For the (i, j) th entry of A_k, take $a_{ij}^{(k)}$ = ki + j, reduced to modulo n to lie in the set 1,...,*n*
  a. We first check that A_k is a Latin square. First, the entries in the ith row are all different; for if $a_{ij}^{(k)} = a_{iJ}^{(k)}$ then ki + j ≡ ki + Jmod(n) whence j = J. Secondly, entries in the jth column are all distinct; for if $a_{ij}^{(k)} = a_{iJ}^{(k)}$ then ki+ jkI + jmod(n) whence k(i− I) ≡0mod(p). Thus p divides k(i−I), whence p divides i−I, so that i ≡ Imod(p) and i must equal I.
  b. To check if A_k and Ah are orthogonal whenever k ≠ h. Suppose $a_{ij}^{(k)} = a_{IJ}^{(k)}$ and $a_{ij}^{(h)} = a_{IJ}^{(h)}$. Then ki + j ≡ kI + J and hi + j ≡ hI + Jmod(n). Subtracting one from the other gives (h − k) i ≡ (h − k)Imod(n) and, as above this gives i = I. Substituting back now, this gives j ≡ Jmod(n) whence j = J.
- To construct a projective plane of order n, given a complete set of MOLS of order n:
  Steps: Let there be a given a complete set of MOLS of order n. Then we can create a $n^2 \times (n + 1)$ matrix (a_{ij}) as follows: The first two columns of the matrix form a succession of ordered pairs, namely, (0,0), (0,1),...,(0,n-1), (1,0), (1,1),..., (1,n-1),...,(n-1,0), (n-1,1),...,(n-1,n-1). The third column repeats the entries of the Latin square A_1 by reading it row by row from left to right. The fourth column does the same for Latin square A_2 and so on.

Label the n^2 row-vectors of (a_{ij}) as 'points' of our projective plane. Create (n + 1) 'ideal points' and label them z_1, z_2, ..., z_{n+1}. Altogether we now have $n^2 + n + 1$ points.

The 'lines' of the plane are stipulated as follows. First, there is an 'ideal line', called L, defined as the set {z_1, z_2, z_3, ..., n+1}. The remaining $n^2 + n$ lines (called 'ordinary lines') are labeled ij, where i = 0, 1, 2,..,(n − 1), j = 1, 2,...,(n + 1). Each such line ij is the set {z_j, those 'points' (= row-vectors) which have i in their j^{th} column}.

Chapter 13
Security Issues and Models in Mobile ad hoc Networks

Revathi Venkataraman
SRM University, India

T. Rama Rao
SRM University, India

ABSTRACT

This chapter covers issues related to security in mobile ad hoc networks. It acts as a comprehensive survey material covering the cryptographic schemes and trust modeling techniques traditionally found in mobile ad hoc network (MANET) survey articles. The need for hybrid security techniques, involving both cryptographic approach and trust based model, in a resource constrained ad hoc network, is also emphasized in this work. Additionally, the lack of realism in the research works related to ad hoc network security is also pointed out. The state of the art in ad hoc security should employ hybrid techniques that can be easily implemented in an ad hoc network. This will lead to the large scale deployment of mobile ad hoc networks in various context-aware applications.

INTRODUCTION

A revolution in ubiquitous computing has arrived due to the technological advances made in the development of wireless computer networks. Wireless networks in which computing nodes are capable of self-organization have become a reality thus enabling deployment of ad hoc networks. The ad hoc capability can potentially lead to the extension of Internet services to places with geographically rough terrain where a communication infrastructure is nonexistent. This capability, however, does not come without a price. Due to unpredictable availability and performance of communication links and dynamically changing topologies of nodes, a lot more design constraints and complexities are to be taken care of.

MANETS are characterized by peer-to-peer communicating nodes on ad hoc basis over an infrastructure-less network. The memory and

DOI: 10.4018/978-1-4666-0321-9.ch013

computational power of these mobile ad hoc devices are limited. Due to this reason, it is not possible to employ any central monitoring audit software over these networks. The communication in a mobile ad hoc network is over a wireless link and a number of security attacks are possible. Some of these links may be lossy and dynamic due to environmental factors and hence, they are unreliable.

MANETs have significant applications in the defence arena. With the move towards the Network Centric Warfare and Global Information Grid (GIG), a variety of services on demand to the various echelons of the defence forces can be easily provided. With these advancements in the MANET field, the security has become a raising concern in MANET deployment in field areas. They are also used for a wide variety of applications like battlefield surveillance, industrial process, machine monitoring, control, medical, healthcare and traffic control systems (Glisic, 2006). The wireless channel of communication is an open channel where any adversary node can easily overhear the transmission between the communicating parties. Since, these networks are infrastructure-less, many of the security approaches suitable for web services cannot be directly applied to these ad hoc devices. This is also due to fact that their resources are limited in terms of energy, bandwidth, computation and memory.

The MANET security issues and challenges are addresses in two broad ways, namely cryptographic approaches and trust-based approaches. The cryptographic approaches extensively used in ad hoc security (Buttayan & Hubaux, 2007) use one or more symmetric and asymmetric encryption algorithms and hashing techniques to achieve confidentiality, authentication and access control. These techniques are considered as hard security approaches. While these techniques are time-tested approaches, they cannot identify 'soft security threats'. Here, the user may be an authenticated and authorized user in the network. But, due to compromise or sabotage, they may behave maliciously thereby causing maximum damage to the network resources. Trust based security approaches act as an extended arm by addressing these challenges. But, they cannot stand independent and provide complete security to a system. The five security requirementsare to be satisfied by any security framework

A comprehensive survey on the traditional cryptographic and trust based approaches is made in this work. The issues that are not addressed by each of these approaches are pointed out and the need for hybrid security schemes employing both cryptographic and trust-based approaches is emphasized. Additionally, this article also addresses the voids in MANET security research where more realistic approaches are required, other than simulation models, for the deployment of these devices in practical applications.

BACKGROUND

Security Issues and Threats

Security attacks in MANETs are classified into two broad categories: *passive* and *active* attacks.

- In a passive attack, the attacker *eavesdrops*the channel used by the victim node. By promiscuously listening to the victim node, an attacking node can gather the unencrypted data that is actually being transmitted. Even otherwise, if the data is encrypted, the attacker analyses the traffic patterns and the frequency of transmissions between the communicating victims. This information can be later used for causing further damage to the system.
- In an active attack, the attacker may *not cooperate* in the network. In a non-cooperating state, the attacker may not involve in the routing process and drop the data packets meant for other nodes in the network. This behavior is *selfish* as these attacking

nodes want to conserve their battery resources. These nodes may be captured or compromised nodes by the enemy forces in a tactical environment.

The attacking nodes may also act malicious by the following ways

- *Content Modification attacks*: They may change the content of the data that are to be forwarded in the network. This leads to loss of data integrity.
- *Impersonation Attacks*: An attacking node may impersonate as another legal entity in the network. It may take control over the victim thereby depriving it of all its resources so as to gain unauthorized access in some of the protected network resources.
- *Wormhole and Replay attacks*: Two cooperating attackers may collude to form a tunnel or *wormhole*, so that data from one side of the tunnel reaches the other side by out-of-band communication. This may be possible by establishing high speed links between two colluding nodes in the network. After establishing the channel, these nodes may replay obsolete routing traffic, thereby fooling other nodes and hijacking the data traffic in the network.
- The communication channel may be *jammed* by the attacker node thereby making the medium impossible for transmitting information. These attacks are dealt with in the physical layer by a variety of techniques like spread spectrum and frequency hopping.

The cryptographic security primitives that should be offered by any security system offering its services are the following:

- *Authentication*: Two peer-to-peer entities communicating with each other should be convinced that they are indeed 'talking' to each other.
- *Access Control*: As an authenticated user of the network, a node should be capable of accessing the network resources like communication channel, data, etc. On the other hand, the system should be secure enough to prevent unauthorized access to privileged resources.
- *Data Integrity*: The destination node should be convinced that the data it has received has not undergone any modifications in the communication channel. Modifications like inserting additional data, changing the current value or deletion of data will lead to loss of data integrity.
- *Confidentiality*: The communication channel should be secure so that the message is received by the receiver only and no other third party intercepts the communication and gains information regarding the data exchange. This is one of the major challenges to be addressed as the wireless channel being prone to eavesdropping,
- *Non-repudiation*: This security requirement ensures that the communicating entities have indeed participated in the communication. They cannot deny their participation, at a later time.

Cryptographic Security Techniques in MANET

In a symmetric cryptographic scheme, both the sender and the receiver share a same secret key. The sender encrypts the message using the key and sends the encrypted data in the communication channel. An adversary can still listen to the transmission, but cannot determine the content of the message being transmitted. In addition, depending on how the input data is handled, there are block ciphers and stream ciphers. Examples of symmetric techniques available in the literature are DES, AES, Triple DES, Blowfish, RC5, etc.

In an asymmetric scheme, two keys are used. If node A wants to send some data to node B, it encrypts the data using B's public key. The node B on receiving the encrypted data deciphers the message by using its private key. This public key is known and distributed to all authorized communicating parties of A.

The message digests are used to provide some of the security requirements. The input message is hashed with or without keys to produce a fixed length digests which are used for message integrity, authentication and digital signatures. Secure hash algorithm (SHA) is a popularly used hash algorithm without keys. It produces a 160-bit hash value for the input message which is processed in 512- bit chunks. Similarly, the MAC algorithms take the message input and a secret key, produces a message digest of fixed size.

In a symmetric encryption, it can be assumed that, when a sender encrypts the message and sends to a receiver, it is authenticated as well since the sender alone knows the secret encryption key. In asymmetric encryption, to authenticate a message, sender A will encrypt the message with its private key. The key distribution mechanisms while using cryptographic primitives is another major issue to be considered in wireless networks. The choice of using these techniques depends on the key size, computational complexity and memory requirements of the participating devices. Hence, these constraints pose a major issue on the usage of these algorithms in mobile ad hoc networks.

Ariadne (Hu, Perrig, & Johnson, 2002) is a secure on-demand routing protocol which uses hash chains to authenticate messages and sign the route request messages. It is based on TESLA which is an efficient broadcast authentication scheme requiring loose time synchronization. The hash values are computed using the one-way hash function and they are recomputed at each hop. A variant of Ariadne is endairA (Buttyan & Vajda, 2004) which signs the route reply messages. It requires less cryptographic computations but it is prone to route request flooding attacks.

SAODV (Secure Adhoc On demand Distance Vector) is another reactive protocol over AODV proposed in (Guerrero Zapata, 2002) . It uses the non-mutable and mutable parts of the message and uses hash chains to ensure that the hop count is never increased by an adversary node. ARAN (Authenticated routing for ad hoc networks) (Sanzgiri et al., 2002) is another on-demand protocol to ensure the integrity of routing messages using public key cryptography. It requires an extensive signature generation and verification in route request flooding.

SEAD (Secure Efficient Ad hoc Distance vector routing) (Hu, Johnson, & Perrig, 2003) is a proactive protocol over DSDV (Destination Sequenced Distance Vector) routing. Similar to SAODV, it secures the destination sequence numbers and hop counts by computing the hash chains. An adversary can never increase these values, but can find loopholes to decrease these values. SMT (Secure Message Transmission) is a secure data communication protocol (Papadimitratos & Haas, 2006) where the source breaks the message into many pieces and sends them using different disjoint alternate routes. MACs are computed for every piece with a key that is shared between the source and the destination. The integrity and authenticity of these messages are ensured by this method.

Trust Based Models in MANETs

In the recent years, trust is a topic of importance in many disciplines like computer science, wireless communications and cognitive science. Even though the abovementioned cryptographic security schemes offer security services like authenticity, confidentiality and integrity of messages between the communicating parties, the extent of trustworthiness of the communicating parties are not known. Hence, the cryptographic schemes offer a 'hard' security solution. On the other hand, trust based solutions offer 'soft' security wherein the communicating parties can judge the trustworthiness of each other.

The concept of trust is a subjective measure and it varies from person to person. It also depends on the functionality of the entity that is being tested. For example, I can trust my car driver for a good car service but not for piloting a plane! Trust is defined as the belief over another person that he will behave according to an established pattern of rules. The following are the different categories of trust

- Centralized vs Decentralized trust

Each of the participating entities in MANET will evaluate the trustworthiness of an entity and send their evaluations to a central authority which will aggregate the values and compute the overall trust of a subject entity. Such centralized evaluations can be performed in a base station or a switching centre where there is no shortage of resources. In a decentralized trust evaluation, every node is its own best judge for evaluating its peer. Each entity makes its own individual decision regarding the behavior of its neighbor.

- Direct vs Indirect trust

The first-hand evidences obtained about the neighbor's behavioral patterns is collected and aggregated to form a direct trust. The indirect trust or recommendation trust refers to the propagation of the direct trust to other participating entities in the network. To what percentage, this recommendation trust should be considered is left to each entity.

Hence, some level of trust is always needed to establish proper communication between the wireless systems, in spite of using the cryptographic primitives. There is a strong need for hybrid schemes employing both cryptographic and trust based techniques to ensure good security in a wireless system.

Some of the early trust evaluation mechanisms use the Watchdog and Pathrater (Marti et al., 2000) to detect and mitigate the greyhole attacks launched by the neighbors. The neighbors are rated by listening in promiscuous mode. The Pathrater identifies the reliable path by aggregating the ratings of the individual nodes in the network.

CONFIDANT (Buchegger & Boudec, 2002) is a secure on demand routing protocol for detecting malicious nodes and making them unattractive for the other nodes. Based on its own observations and the with the recommendations received from the neighbors, the trust and routing decisions are made. Entropy-based trust model and probability-based trust model are proposed in (Sun et al., 2006), which is a distributed approach to acquire, maintain and update the neighbor's trust values. Some of the vulnerabilities in trust management and their defense are proposed in (Sun et al., 2008) where the trust relationship is defined as {subject, agent, making correct recommendations}. Both direct and recommendation trust are taken into account and a probability-based trust evaluation is formulated. The objective trust management framework (OTMF) is proposed in (Li & Kato, 2008), where a modified Bayesian-approach is used. The recommendation trust are weighted differently based on their arrival time and context providers and they are used as second-hand information while performing trust computations. A trust model over Dynamic Source Routing (DSR) protocol is also incorporated for reliable routing (Pirzada, Datta, & McDonald, 2006) in ad hoc communication where the neighbor's behavior is monitored by promiscuous listening.

THE PROPOSED MODEL

To provide complete security to a MANET system, it is evident that a versatile model is needed to thwart all possible security attacks. The security mechanisms can be implemented at any level of the protocol stack. Most of the current research

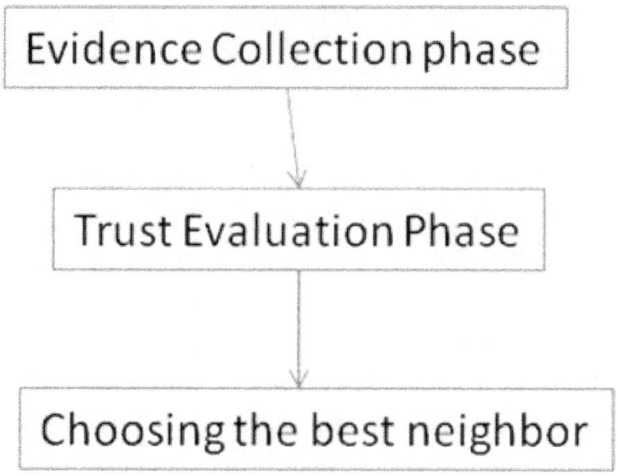

Figure 1. Neighbor evaluation process

in MANET security targets either route discovery or data communication phase. The security primitives should be added in all layers of the protocol stack to provide a total security solution in a wireless network.The underlying model assumes that the cryptographic primitives concerned with authentication, confidentiality and integrity are incorporated in each participating entity. In addition, a hybrid trust model is needed to ensure complete security in an ad hoc network. The evaluation process can be divided into three phases as shown in Figure 1.

Evidence Collection Phase

Each node in an ad hoc network collects as much information about their neighbor's behavior by promiscuous listening. The following are the attacks under consideration

Blackhole and *Greyhole* attack: A selfish neighbor may not cooperate in routing or data forwarding by dropping the RREQ packets in reactive routing and dropping data packets in the communication interval. If they do not forward any packets, they are said to involve in blackhole attack. Else, if they selectively forward the packets, they involve in greyhole attack. The motive behind these attacks is to primarily save their resources which are limited in an ad hoc scenario.

Flooding Attack: A malicious neighbor may indulge in RREQ or DATA flooding by sending route requests to destinations which are not available in the network or by sending junk data packets to exhaust their neighbor's energy resource.

Content Modification Attack: This is a type of active attack where the neighbor changes the data contents and forwards the modified information. This will lead to loss of data integrity.

Misrouting Attack: The misbehaving neighbor node may maliciously change the header information thereby rerouting the packets to some other colluding node in the network. Later on, these nodes may launch wormhole attacks.

Rushing Attack: These malicious nodes may quickly respond to a RREQ message, so that they position themselves as the best neighbors to forward the data. They may later on indulge in blackhole, greyhole or wormhole attacks.

A typical intrusion detection system running in each node gathers the following type of information

- Successful forwarding of data packets (blackhole and greyhole)
- Number of data packets received (flooding attack)
- Number of data packets forwarded without modification (loss of data integrity)
- Number of data packets forwarded without header modification (misrouting)
- Routing parameters (reactive protocols)
    - RREQs forwarded (cooperation in routing)
    - RREPs received (Cooperation in routing, flooding neighbor)
    - RREQs received (flooding neighbor)
    - Time taken to respond to a RREQ (rushing attack)

The information gathered highlights the following behavior of the neighboring nodes like cooperation in routing and sincerity in data forwarding without loss of data integrity. Additionally, by constantly evaluating time taken to respond to a RREQ, it can be assured that the neighbor is not involved in rushing attack. The evidences collectedare helpful in detecting the attacks like blackhole, greyhole, flooding, rushing attacks.

Trust Evaluation Phase

There are plenty of trust evaluation techniques (Sun et al., 2008; Li & Kato, 2008; Pirzata, Datta, & McDonald, 2006; Theodorakopoulous & Baras, 2006; Chang & Kuo, 2009) contributed in MANET security research. The concept of path semiring and distance semiring is introduced in (Theodorakopoulos & Baras, 2006) which effectively finds the most trusted path to transfer the data in a peer-to-peer environment. A Markov-chain trust model and an efficient key management technique is proposed in (Chang & Kuo, 2009), which ensures reliable authentication in multicast MANETs. These techniques invariably have two components: *direct* trust and the *recommendation* trust. A fraction of recommendation trust is taken for the trust evaluation. This fraction varies from each neighbor depending on their direct trust values. The trust evaluation process is as shown in Figure 2.

The trust values are aggregated and normalized into a given range from most trusted to least trusted or malicious. Some of the discrete trust ranges presented in literature are (Pirzada, Datta, & McDonald, 2006; Abdul-Rahman & Hailes, 1998). Along with the trust, confidence plays an important role in deciding a neighbor's behavior. This parameter reflects the quality of trust evaluation done and it generally depends on the duration in which the entities interact. It specifies whether they have been trustworthy for a long period of time. An aging factor (Sun et al., 2008) is introduced in the computation of confidence parameter calculation so that the older behaviors fade away gradually. This type of computation aptly captures the behavior of trust as in social networks.

Choosing an Optimal Path

The neighbor's trustworthiness is evaluated and represented as (T_i, C_i) where i is a neighbor node in an ad hoc network, T_i and C_i are the aggregated trust and confidence values respectively. The following algorithm chooses the best neighbor

- Eliminate all untrustworthy neighbors from the neighbor list
- Choose neighbor i such that

$$\max\left(\bar{T}_i + \sqrt{\frac{2\ln(n)}{n_i}}\right)$$

where $\bar{T}_i$ – trust of neighbor i

n_i: number of times the neighbor i is chosen for a service

n:first specified number of services

The additive log function ensures that the same trustworthy neighbor is not always chosen as a best neighbor which may lead to the quick depletion of its energy. Additionally, a fair chance is also given to other neighbors to prove their trustworthiness.

To detect rushing attacks in a network, if a node detects that it is receiving quick responses from its neighbor against a RREQ, it will initiate the challenge response procedure to distance bound the neighbor (Halfslund et al., 2004). The node generates k random bits α_i. A public string is generated and sent to the neighbor. The sender sends a bit α_i to the neighbor. Immediately after receiving the bit α_i, the neighbor sends $\beta i = \alpha_i \oplus m_i$ to the sender. This process is repeated k times. After this process, the sender verifies $(\alpha_1 \oplus m_1) | (\alpha_2 \oplus m_2) | \ldots(\alpha_k \oplus m_k)$ equals the public bit string. If it is equal, it computes an upper-bound on distance to

Figure 2. Trust establishment procedure

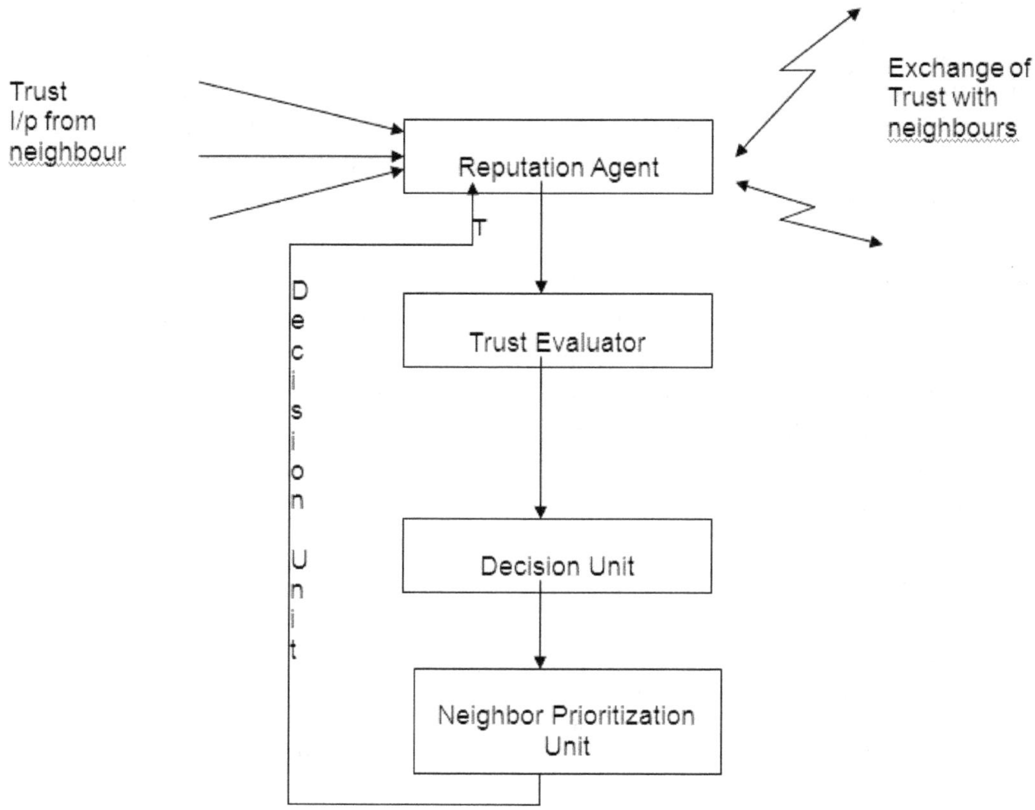

neighbor to be not greater than $\frac{\Delta_{max}}{2} * c$ where Δ_{max} is the maximum time taken for a single challenge response and c is the speed of light. With the current transmission power a node can easily verify its neighbor's distance range and check whether it is indulging in rushing attack.

CONCLUSION

This chapter proposes a trust model for evaluating the neighbors in a MANET and brings out the joint role of cryptographic and trust based schemes in ensuring total security in a wireless system. An analysis highlighting the merits and demerits of the existing security schemes is done. The future work involves coverage of wide range of security attacks and comprehensive models for real time behavior in a mobile ad hoc testbed.

REFERENCES

Abdul-Rahman, A., & Hailes, S. (1998). A distributed trust model. *NSPW '97: Proceedings of the 1997 Workshop on New Security Paradigms,* (pp. 48-60). New York, NY: ACM.

Buchegger, S., & Boudec, J. (2002). Performance analysis of the confidant protocol: Cooperation of nodes - fairness in distributed ad-hoc networks. *Proceedings of the IEEE/ACM Workshop on Mobile Ad Hoc Networking and Computing (MobiHOC)*, (pp. 226–236).

Buttayan, L., & Hubaux, J.-P. (2007). *Security and co-operation in wireless networks*. Cambridge University Press.

Buttyan, L., & Vajda, I. (2004). Towards provable security for ad hoc routing protocols. In *Proceedings of the 2nd ACM Workshop on Security of Ad hoc and Sensor Networks*, 2004, (pp. 94–105).

Chang, B. J., & Kuo, S. L. (2009). Markov chain trust model for trust-value analysis and key management in distributed multicast MANETs. *IEEE Transactions on Vehicular Technology, 58*(4).

Glisic, S. G. (2006). *Advanced wireless networks: 4G technologies*. John Wily. doi:10.1002/9780470035405

Guerrero Zapata, M. (2002). *Secure ad hoc on-demand distance vector (SAODV) routing*. Retrieved from http://ietfreport.isoc.org/idref/draft-guerrero-manet-saodv/

Halfslund, A., Tonnesen, A., Rotvik, R. B., Andersson, J., & Kure, O. (2004). *Secure extension to the OLSR protocol*. OLSR Interop and Workshop, 2004.

Hu, Y.-C., Johnson, D. B., & Perrig, A. (2003). SEAD: Secure efficient distance vector routing for mobile wireless ad hoc networks. *Ad Hoc Networks, 1*(1), 175–192. doi:10.1016/S1570-8705(03)00019-2

Hu, Y.-C., Perrig, A., & Johnson, D. B. (2002). Ariadne: A secure on-demand routing protocol for ad hoc networks. In *Proceedings of the ACM Conference on Mobile Computing and Networking (Mobicom)*, 2002, (pp. 12–23).

Li, J., & Kato, J. (2008). Future trust management framework for mobile ad hoc networks. *IEEE Communications Magazine, April*.

Marti, S., Giuli, T., Lai, K., & Baker, M. (2000). Mitigating routing misbehavior in mobile ad hoc networks. *Proceedings of the Sixth Annual International Conference on Mobile Computing and Networking (MobiCom)*, (pp. 255–265).

Papadimitratos, P., & Haas, Z. J. (2006). Secure data communication in mobile ad hoc networks. *IEEE Journal on Selected Areas in Communication, 24*(2).

Pirzada, A. A., Datta, A., & McDonald, C. (2006). Incorporating trust and reputation in the DSR protocol for dependable routing. *Computer Communications* [Elsevier Press.]. *Special Issue on Internet Communications Security, 29*, 2806–2821.

Sanzgiri, K., Dahill, B., Levine, B. N., Shields, C., & Belding-Royer, E. M. (2002). *A secure routing protocol for ad hoc networks*. In International Conference on Network Protocols (ICNP), 2002.

Sun, Y. (2006). Information theoretic framework of trust modeling and evaluation for ad hoc networks. *IEEE Journal on Selected Areas in Communications, 24*(2).

Sun, Y., *et al.* (2008). Defense of trust management vulnerabilities in distributed networks. *IEEE Communications Magazine, February*.

Theodorakopoulos, G., & Baras, G. S. (2006). On trust models and trust evaluation metrics for ad hoc networks. *IEEE Journal on Selected Areas in Communications, 24*(2). doi:10.1109/JSAC.2005.861390

Chapter 14

Incorporating Security and Energy Efficiency for Multimedia Communications in WANets: A Cross-Layer Design

Lamia Kaddar
University of Versailles/PRiSM, France

ABSTRACT

The security of multimedia data in Wireless Ad hoc Networks (WANet) is commonly provided by encryption, which consists in transforming a plain text message into an unintelligible ciphertext. Nevertheless, the classical and modern ciphers have all been developed without taking into consideration the intrinsic characteristics of multimedia flows. In this chapter, the author proposes to reduce the computational requirements for the multimedia flow's encryption when the energy is a limited resource, as it is the case of WANets. Thus, the chapter defines a new profile of Secure Real Time Protocol (SRTP), named EE_SRTP. EE_SRTP provides energy efficiency to SRTP, which provides essentially confidentiality, message authentication, and replay protection to the RTP traffic. In EE_SRTP, the author exploits the dependency existing between the frames created by the inter-frame coding to further improve the energy-efficiency of the encryption process. In contrast with the existing approaches, this chapter presents an experimental model that integrates EE_SRTP to Video LAN Coding (VLC). Therefore, to validate EE_SRTP, the author implements a secure version of VLC (SecVLC). The performance evaluation demonstrates clearly that using this new scheme in the context of wireless ad hoc nodes allows saving energy while ensuring a high level of content confidentiality, without adding an overhead to the wireless network.

DOI: 10.4018/978-1-4666-0321-9.ch014

1. INTRODUCTION

Nowadays, Wireless Ad hoc Networks (WANet) have a large area of applications especially in the real time communication space. However, the success of such applications depends strongly on the user's trust in the provided degree of security.

Wireless Ad hoc Networks (WANET) can be classified by their application:

- Mobile Ad hoc Networks (MANET)
- Wireless Mesh Networks (WMN)
- Wireless Sensor Networks (WSN)

MANET: Mobile Ad hoc Networks

- A mobile ad hoc network (MANET), is a self-configuring infrastructureless network of mobile devices connected by wireless links. Ad hoc is Latin and means "for this purpose".
- Each device in a MANET is free to move independently in any direction, and will therefore change its links to other devices frequently. Each must forward traffic unrelated to its own use, and therefore be a router. The primary challenge in building a MANET is equipping each device to continuously maintain the information required to properly route traffic. Such networks may operate by them selves or may be connected to the larger Internet.
- MANETs are a kind of wireless ad hoc networks that usually has a routeable networking environment on top of a Link Layer ad hoc network.
- The growths of laptops and 802.11/Wi-Fi wireless networking have made MANETs a popular research topic since the mid 1990s. Many academic papers evaluate protocols and their abilities, assuming varying degrees of mobility within a bounded space, usually with all nodes within a few hops of each other. Different protocols are then evaluated based on measure such as the packet drop rate, the overhead introduced by the routing protocol, end-to-end packet delays, network throughput etc.

WMN: Wireless Mesh Networks

Wireless Mesh networks is becoming a major paradigm for constructing user access networks that provide community or city-wide Internet connectivity. The network coding, as a new technique which intermediate nodes in the network not only forward but also process the incoming information flows, will have a good application in wireless Mesh networks. Firstly, comparing traditional routing and network coding under unicast introduce the basic idea of network coding.

WSN: Wireless Sensor Networks

Recent advances in micro-electro-mechanical systems (MEMS) technology, wireless communications, and digital electronics have enabled the development of low-cost, low-power, multifunctional sensor nodes that are small in size and communicate untethered in short distances. These tiny sensor nodes, which consist of sensing, data processing, and communicating components, leverage the idea of sensor networks based on collaborative effort of a large number of nodes. Sensor networks represent a significant improvement over traditional sensors, which are deployed in the following two ways:

- Sensors can be positioned far from the actual phenomenon, i.e., something known by sense perception. In this approach, large sensors that use some complex techniques to distinguish the targets from environmental noise are required.
- Several sensors that perform only sensing can be deployed. The positions of the sensors and communications topology are carefully engineered. They transmit time

series of the sensed phenomenon to the central nodes where computations are performed and data are fused.

MWANet: Multimedia WANet

The development of application-layer protocols for the real-time distribution of audio and video over IP has been under study for several years. These researches gave birth to the Multimedia WANet (MWANet). One of the most prominent protocols is the Real-Time Transport Protocol (RTP), described in RFC 1889 (Crow et al., 1997), which is dedicated to multicast and unicast delivery of multimedia flows over IP. A secure profile of RTP named Secure RTP (SRTP) was defined in the RFC 3711 (Network Working Group, 2004). SRTP provides confidentiality, message authentication, and replay protection to the RTP traffic and to the RTP's control traffic.

Providing security in MWANet is a very important and challenging issue to the success of such networks. In fact, tackling the WANet's constraints augmented with the huge communication requirements of multimedia flows is sometimes incompatible with their intrinsic nature. Indeed, the computationally intensive processes of video images encryption and decryption introduce important delays which may degrades the quality of real-time video streams. Moreover, all the services in WANets are created and configured on the fly without any pre-existing or centralized infrastructure. They are characterized by low computation capabilities, nodes' heterogeneity, dynamic topology, wireless links with variable quality and limited energy resources. All these constraints add additional burden on the multimedia communication, and narrows the margin to deploy processing-hungry applications.

In this work, we mainly focus on saving energy at the data encryption phase using selective encryption algorithm. While achieving the same level of security, selective encryption is a technique to reduce computational complexity and energy consumption by encrypting only portion of the compressed bitstream.

Reducing the encrypted fraction can help us to:

- Reduce the total complexity of implementation in both encoders and decoders, reducing size, weight, power consumption, and allowing new task partitioning within devices (e.g., moving encryption and decryption from a hardware to a software implementation).
- Enable different system level functionality. For instance, selective encryption allows to overlay multiple security systems on a common multimedia stream with little bandwidth overhead by sharing the bulk of the stream in-the-clear and encrypting a small portion separately under each security system.

In this paper, we introduce Secure Video Lan Coding (SecVLC) for secure multimedia streaming in WANets for this we used the open source VLC and we define a new profile of SRTP named Energy Efficiency SRTP (EE_SRTP) which is compliant with MWANet's exigencies. This Profile is an extension that provides energy efficiency to the SRTP profile. Our solution doesn't increase the overhead. In fact, the EE_SRTP uses the existing header fields and unused fields in SRTP to implement an in-bound signaling channel necessary to our selective encryption scheme. In order to evaluate the performance of our scheme, we have carried out several sets of simulation experiments. The validation of our scheme was done in two parts, in a first time with test by implementing SecVLC; to validate that the security is preserved when using EE_SRTP.

The remainder of the chapter is organized as follows. In section 2 we introduce the wireless ad hoc networks and their problems, with a short discussion about security in the WANet, then we present in section 3 about the different applications of WANets. In section 4 we introduce the video coding and the different compression standards. In section 5 discuss the choice of SRTP protocol and describe the encryption algorithm used within SRTP, and then we, followed by providing a description of our solution SecVLC in section 6. We conclude in section 7.

2. WIRELESS AD HOC NETWORKS AND SECURITY

Ad hoc networks are a new paradigm of wireless communication for mobile hosts (which we call nodes). In an ad hoc network, there is no fixed infrastructure such as base stations or mobile switching centers. Mobile nodes that are within each other's radio range communicate directly via wireless links, while those that are far apart rely on other nodes to relay messages as routers. Node mobility in an ad hoc network causes frequent changes of the network topology.

Challenges in the area of WANets include security, dynamic network topology, routing, quality of service, and power efficiency.

There are a number of problems that affect packet video streaming. Video streaming over IP is a challenging task since IP networks are shared environments that provide a rather best effort service. These networks do not offer any quality of service nor guarantees of resources in terms of (i) bandwidth, (ii) packet losses, (iii) delay variation (jitter), and (iv) transfer delay. Therefore, a key goal of video streaming is to design a system that efficiently handles the video over IP and deals with possible changes in network conditions.

Security, as well, is an important challenge for ad hoc networks, especially for those security-sensitive applications. To secure an ad hoc network, we consider the following attributes: availability, confidentiality, integrity, authentication, and non-repudiation.

Availability ensures the survivability of network services despite denial of service attacks. A denial of service attack could bon e launched at any layer of an ad hoc network. On the physical and media access control layers, an adversary could employ jamming to interfere with communication on physical channels. On the network layer, an adversary could disrupt the routing protocol and disconnect the network. On the higher layers, an adversary could bring down high-level services. One such target is the key management service, an essential service for any security framework.

Confidentiality ensures that certain information is never disclosed to unauthorized entities. Network transmission of sensitive information, such as strategic or tactical military information, requires confidentiality. Leakage of such information to enemies could have devastating consequences. Routing information must also remain confidential in certain cases, because the information might be valuable for enemies to identify and to locate their targets in a battlefield.

Integrity guarantees that a message being transferred is never corrupted. A message could be corrupted because of benign failures, such as radio propagation impairment, or because of malicious attacks on the network.

Authentication enables a node to ensure the identity of the peer node it is communicating with. Without authentication, an adversary could masquerade a node, thus gaining unauthorized access to resource and sensitive information and interfering with the operation of other nodes.

Finally, *non-repudiation* ensures that the origin of a message cannot deny having sent the message. Non- repudiation is useful for detection and isolation of compromised nodes. When a node A receives an erroneous message from a node B, non-repudiation allows A to accuse B using this message and to convince other nodes that B is compromised.

The major design constraints for a secure protocol are listed as follows:

- *Autonomous* No centralized administration entity is available to manage the operation of the different mobile nodes.

- *Dynamic topology* Nodes are mobile and can be connected dynamically in an arbitrary manner. Links of the network vary timely and are based on the proximity of one node to another node.
- *Device discovery* Identifying relevant newly moved in nodes and informing about their existence need dynamic update to facilitate automatic optimal route selection.
- *Bandwidth optimization* Wireless links have significantly lower capacity than the wired links.
- *Limited resources* Mobile nodes rely on battery power, which is a scarce resource. Also storage capacity and processing power are severely limited.
- *Scalability* Scalability can be broadly defined as whether the network is able to provide an acceptable level of service even in the presence of a large number of nodes.
- *Limited physical security* Mobility implies higher security risks such as peer-to- peer network architecture or a shared wireless medium accessible to both legitimate network users and malicious attackers. Eavesdropping, spoofing and denial-of-service attacks should be considered.
- *Infrastructure-less and self operated* self-healing feature demands MANET should realign itself to blanket any node moving out of its range.
- *Poor Transmission Quality* This is an inherent problem of wireless communication caused by several error sources that result in degradation of the received signal.
- *Ad hoc addressing* Challenges in standard addressing scheme to be implemented.
- *Network configuration* The whole MANET infrastructure is dynamic and is the reason for dynamic connection and disconnection of the variable links.
- *Topology maintenance* Updating information of dynamic links among nodes in MANETs is a major challenge

3. APPLICATIONS OF WIRELESS MULTIMEDIA NETWORKS

The field of wireless networking emerges from the integration of personal computing, cellular technology, and the Internet. This is due to the increasing interactions between communication and computing, which is changing information access from "anytime anywhere" into "all the time, everywhere." At present, a large variety of networks exist, ranging from the well-known infrastructure of cellular networks to non-infrastructure wireless ad-hoc networks.

Unlike a fixed wireless network, wireless ad-hoc or on-the-fly networks are characterized by the lack of infrastructure. Nodes in a mobile ad-hoc network are free to move and organize themselves in an arbitrary fashion. Each user is free to roam about while communicating with others. The path between each pair of the users may have multiple links, and the radio between them can be heterogeneous. This allows an association of various links to be a part of the same network. Wireless ad hoc networks can operate in a standalone fashion or could possibly be connected to a larger network such as the Internet.

WANets are suited for use in situations where an infrastructure is unavailable or when to deploy one is not cost effective. One of many possible uses of WANets are in some business environments, where the need for collaborative computing might be more important outside the office environment than inside, such as in a business meeting outside the office to brief clients on a given assignment.

A WANet can also be used to provide crisis management services applications, such as in disaster recovery, where the entire communication infrastructure is destroyed and resorting communication quickly is crucial. By using a WANet, an infrastructure could be set up in hours instead of weeks, as is required in the case of wired line communication. Another application example of a WANet is Bluetooth, which is designed to support a personal area network by eliminating the need of wires between various devices, such as printers and personal digital assistants. The famous IEEE 802.11 protocol also supports an ad-hoc network system in the absence of a wireless access point.

In conclusion, WANets allow the construction of flexible and adaptive networks with no fixed infrastructure. These networks are expected to play an important role in the future wireless generation. Future wireless technology will require highly-adaptive mobile networking technology to effectively manage multi-hop WANets clusters, which will not only operate autonomously but also will be able to attach at some point to the fixed networks.

4. VIDEO COMPRESSION CHARACTERISTICS

Audio and video coding algorithms are highly correlated to the growth of computation speed of systems. Nowadays, the most important compression standards for streaming video (such as H.261, H.263, MPEG-1, MPEG-2, MPEG-4 and recently H.264) are proposed by ITU (International Telecommunication Union) and ISO (International Organization for Standardization) standardization bodies. The majority of available video codecs do not perform well when used to streaming media in wireless IP environments, because they require a high scalability, lower computational complexity, high resiliency to network losses and lower encoding / decoding latency. Building a video codec that respond to all these requirements is not sufficient. The codec must be aware of network conditions in order to switch to its best operation point for any particular network condition, producing the highest possible user perceived quality. Current research is investigating a new scalable and flexible coding algorithm, and ways to adapt existing codecs to heterogeneous environment such as Internet and WLAN. We present in this subsection a non-exhaustive list of video codecs that are widely used in packet video applications.

Basically, most video codecs are able to generate either Constant Bit Rate (CBR) or Variable Bit Rate (VBR) traffic. However, coded video sequence typically has time-varying complexity which cannot be generally achieved by a CBR coding. For instance, during scene changing, the predictive (Inter) coding involves high data rate. Therefore, VBR coding produces a constant visual quality while CBR coding produces a rather time-varying quality. In variable network channel condition, it is important for streaming system to match the available bandwidth.

4.1. Overview of Video Compression

Video compression is achieved by exploiting similarities or redundancies that exists in typical video signal. A video sequence consists of a sequence of video frames or images. Each frame may be coded as a separate image. A frame to be coded is divided into a set of blocks, which can be compressed using two main techniques. Block Discrete Cosing Transform (DCT) and Discrete Wavelet Transform (DWT). Coding each frame separately is not efficient since neighboring video frames are typically very similar. Much higher compression can be achieved by exploiting similarity between frames. A given frame can be predicted from a previous one. The difference between the two frames is considered rather than each time independently coding the last frame. Consecutive video frames typically contain the same imagery with possible different special locations because of motion. To improve the prediction, it is important to estimate the motion between the frames and to form an appropriate prediction that compensates this motion. This process is known as Motion Estimation. The process of forming a prediction in motion is called Motion Compensation. It represents the difference or error between the best-match frame block and the current (to-be-coded) frame block when the motion compensation is achieved in block basis.

Three types of coded frames are generally generated by the codec. (1) Intra-coded frame

Figure 1. Example of the prediction dependencies between frames

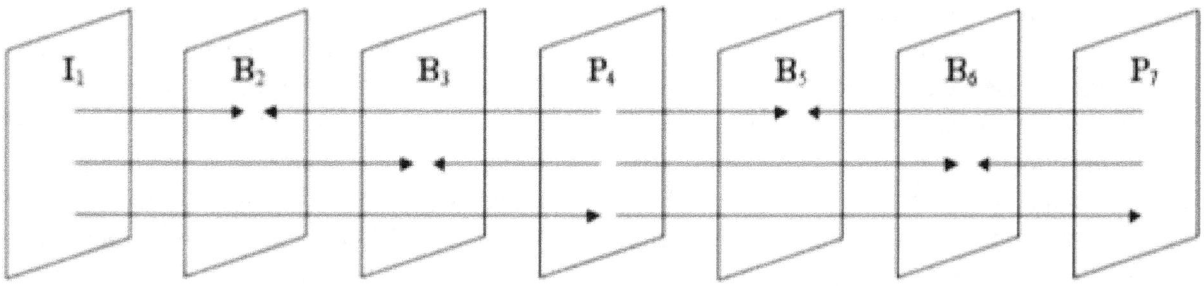

(I-Frame) where the frame is coded independently of other frames; (2) Predicatively coded frame or (P-Frame) where the frame is coded based on a previously coded frame; and (3) Bi-directionally predicted frame (B-Frame) where the frame is coded using both previous and future coded frames. Figure 1 illustrates an MPEG video Group of Picture (GOP) coded using predictions dependencies between frames.

Current video codecs achieve video compression by applying combination of the previously described techniques, namely, color space conversion, motion compensation, motion estimation, and signal transformation.

Moving Picture Experts Group (MPEG) was created to develop video packing standards. MPEG-4 standard (ISO, 1998)(ISO, 1998b) (ISO, 1998c) is an emerging digital multimedia standard with associated protocols for representing, manipulating and transporting natural and synthetic multimedia content (i.e. audio, video and data) over a broad range of communication infrastructures including IP and ATM. The original characteristic of MPEG-4 resides in its ability to provide an integrated object-oriented representation of multimedia content for the support of new ways of communication, access, and interaction with digital audiovisual data, and offering a common technical solution to various telecommunications, broadcast, and interactive services. MPEG-4 addressed a broad range of existing and emerging multimedia applications such as multimedia broadcasting, content-based audiovisual database access, games, audiovisual home/studio editing, advanced audiovisual communications, and video over mobile networks.

The MPEG-4 standard introduces a new technique for multimedia scenes coding called "object-based video coding". This technique allows an independent encoding of different audio-visual objects in the scene. An MPEG-4 scene consists of one or more Audio Visual Object (AVO), each of them is characterized by temporal and spatial information. The hierarchical composition of an MPEG-4 scene is depicted in Figure 2. Each audio visual object may be encoded by separate encoders in a scalable (multi-layer) or non scalable (single layer) manners. A layer is composed of a sequence of a Group of Video-Object-Plane (GOV). A Video Object Plane (VOP) is similar to the MPEG-2 frame. VOP supports intra coding (I-VOP), uni-directional temporally predicted (P-VOP), and bi-directional temporally predicted (B-VOP).

5. RELATED WORKS

The real-time constraints for many multimedia applications cannot be met by an increasing number of smaller client devices (such as PDAs and videophones) with limited processing and communication power. Hence, there is a need to develop techniques that exploit the structure of multimedia data by maximizing the network efficiency.

Figure 2. Hierarchical organization of MPEG-4 data stream

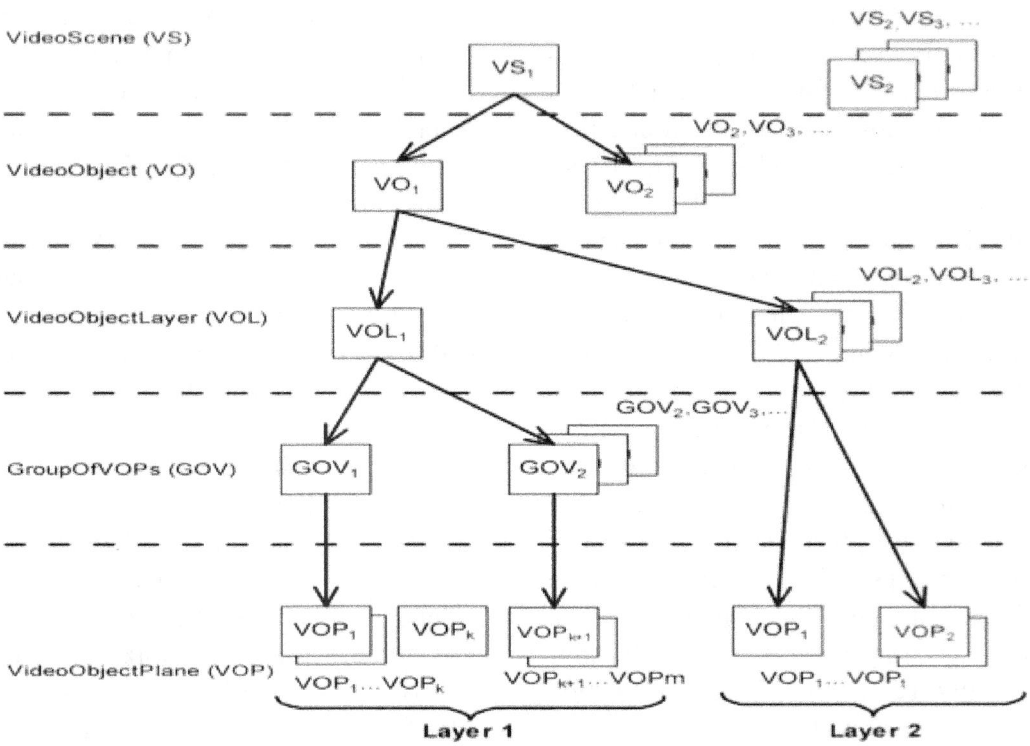

Saving energy in WMANet can be done on (1) multimedia encoding techniques and on (2) used security protocol, more specifically on data encryption. For the first one, the main objectives to design a coder for a MWANet are:

Low complexity: It is mandatory to achieve a high ratio of compression to efficiently limit both bandwidth and energy consumption.

High compression: Multimedia encoders are embedded in sensor devices. These must be low-complexity and low-power encoders to extend the lifetime of sensor nodes.

Error resiliency: The source coder should provide robust and error-resilient coding of source data.

The idea of the selective encryption is to encrypt chosen units of message while leaving the remaining units unencrypted. This concept was introduced by Spanos and Maples in (Spanos & Maples, 1995). Meyer and Gadegast (Meyer & Gadegast, 1995) propose to reduce the amount of encrypted MPEG data in a video sequence while at the same time providing an acceptable level of security. However, these works do not take into account the constraints of Multimedia Wireless ad hoc Networks. In addition, the main works on selective encryption focus on performance issues and not on security issues. That is why, Lundin and Lindskog have evaluated the security impact of the selective encryption by using the entropy and guesswork (Lundin & Lindskog, 2010).

Wu & Kuo, (Lima et al., 2007): Base on a number of observations, the authors argue that selective encryption (i.e., the selection of the most important coefficients from a compression system and their encryption with a conventional cipher) may not be effective in two situations: (1) The media compression system is based on an orthogonal transform followed by quantization, and (2) The media compression system contains entropy coding at the last stage. Using multiple statistical models, they investigate a different encryption methodology that turns entropy coders into ciphers. This methodology allows the construction of two selective encryption schemes by application to the Huffman coder and the QM coder, two of the most popular entropy coders in multimedia compression. Both coders have very simple statistical models; the model of the Huffman coder is usually a fixed-size non-adaptive binary tree, and the initial state of the QM coder includes only three integer numbers. Since hiding the Huffman coding table or the initial state of the QM coder does not provide sufficient secrecy, it is argued that the problem of a limited key/model space can be overcome by using m statistical models instead of only one. The first of the proposed encryption schemes makes use of multiple Huffman tables and the second multiple indices in the QM coder estimation state machine.

Restricting the encryption to I frames (Spanos & Maples, 1995) does not make the information in the P and B frames useless if their base frames are correlated (Agi & Gong, 1996). The Zig-Zag Permutation Algorithm (Tang, 1997) is vulnerable to known-plaintext and ciphertext-only attacks (Qiao & Nahrstedt, 1998) Encryption of the sign bit (and also more significant bits) of every DCT coefficient does not work because useful image content can be recovered by assigning all DC coefficients to 128 and all AC coefficients to positive (Lima et al., 2007). To create visible degradation, a minimum number of 4 or 5 least significant bitplanes should be encrypted in uncompressed images (Van Droogenbroeck & Benedett, 2002). Two types of ciphertext only attacks on bitplane encryption (replacement attack and reconstruction attack) show that encryption of the most significant bitplane is not secure enough; at least 4 bitplanes need to be protected (Podesser, Scmidt, & Uhl, 2002). In (Pommer & Uhl, 2003), the amount of data to be encrypted for a given image is extremely small as no image data (e.g., transform coefficients) needs to be protected. Simulation results in (Alattar, Al-Regib, & Al-Semari, 1999) show that it is sufficient to encrypt every other I-macroblock and the header of every other predicted macroblock to disguise the video completely. When this simulation is repeated for n=3, there is a clear degradation in the security level. The motion becomes easily recognizable when the encrypted MPEG video sequences are played back.

Spanos and Maples (Spanos & Maples, 1995) were the first proponents of a selective encryption style algorithm; with a new security mechanism called Aegis. Aegis encrypts the I frames of all frames groups in an MPEG video stream. The encryption of the I frames is justified by the fact that they have great significance in the decompression of an MPEG stream whereas B and P frames represent only translations of the picture information found in adjacent I frames. The proposed solution is not efficient because compare to the P and B-frames the I-frame's size is too large. This is why in our proposition of saving energy with EE-SRTP we propose a selective encryption for the packets constituting the different type of frames.

Those are the most prominent works done in the selective encryption; many works are done in this area (Cheng & Li, 2000) (Pommer & Uhl, 2003), to further describe these aspects; you can refer to the above reference for more information.

6. SECURE VLC (SECVLC)

The aim of SecVLC (Kaddar & Mehaoua, 2009) is to improve the lifetime of a secure transmission in MWNet. We save energy on data encryption by taking advantage of the dependencies between frames made by the MPEG video coding described in section 4. SecVLC transmit in both modes, with or without confidentiality. When the sender wants to send a secure video, SecVLC use the SRTP mode on UDP with EE_SRTP.

In a SRTP packet, the payload with media data is encrypted and this payload and the packet header are authenticated. Sender and receiver must keep a cryptographic context with a set of parameters about security. Some of this information is obtained from a key exchange protocol or manually achieved and some are extracted from the RTP packet itself.

Cryptographic Context

The main parameters in a cryptographic context are the master key and a secure session-derived key. The construction of a SRTP packet has the following steps:

1. Determining a cryptographic context.
2. Determining the master key and derivation of encryption and authentication session keys.
3. Encryption of the RTP payload to produce an encrypted portion of the packet.
4. Computation of an authentication tag to produce an authenticated portion of the packet.

Extraction of a SRTP Packet

The extraction of a SRTP packet has the following steps:

1. Determining a cryptographic context.
2. Determining the master key and derivation of encryption and authentication session keys.
3. Confirmation if the packet has been replayed and verification of the authentication tag.
4. Decryption of the SRTP payload.

In contrast with Datagram Transport Layer Security (DTLS), SRTP supports both unicast and multicast applications. On the other hand, SRTP has some limitations compared to DTLS. First, it cannot be used to protect traffic other than RTP. Second, it does not include key management and requires an external (out-of-bound) signalling protocol/channel such as SIP to set up the keying material.

In extremely bandwidth-constrained applications SRTP has advantages over DTLS due to its integration within RTP, which allows lower network overhead.

The SRTP uses Advanced Encryption Standard (AES) (NIST, 2001) as default cipher algorithm. AES (previously called Rijndael) is a block cipher that has a variable key length of 128, 192, or 256 bits to encrypt data blocks of 128 bits (192 and 256 bit blocks are also possible with Rijndael). Both block and key lengths are extensible to multiples of 32 bits. To produce a block of ciphertext, a plaintext is first manipulated with the key (called Add Round Key process), then it then goes through several operational rounds are, the more the operational rounds, the more is the strength of the AES algorithm. The Add Round Key requires creation of round keys that is a computational overhead for AES encryption. AES encryption is fast and flexible. It can be implemented on various platforms especially in small devices and smart cards. AES has been rigorously tested for security loopholes for a few years before the National Institute of Standards and Technology (NIST) standardized it. It is considered to be very secure for more than seven rounds. In terms of choice of algorithms in WLANs, both RC4 and AES have different tradeoffs. AES for instance, needs to be used in the counter mode to generate a pseudorandom stream in WLANs. Additionally, a study of their energy consumption is needed to decide on their use in security protocols. A preliminary study of energy consumption of RC4 and AES algorithms for WLAN is presented in (Prasithsangaree & Krishnamurthy, 2004).

AES encryption algorithm presents lower computational cost (Prasithsangaree & Krishnamurthy, 2004). Thus, for multimedia streaming over ad hoc networks that have scarce bandwidth and energy resources is much suitable to use SRTP protocol.

Since SRTP requires a mechanism to exchange security keys, the Multimedia Internet KEYing (MIKEY) (IETF, 2004) protocol that was developed to be used in such real-time applications, can be used for the key management purpose.

Energy Efficiency SRTP (EE_SRTP)

We proposed the Energy Efficiency SRTP (EE_SRTP) a new profile of SRTP that offers, in addition to the security, an energy saving model for streaming in ad hoc networks using the selective encryption. For EE_SRTP we defined three levels of security (high, medium and low), where it corresponds to each one a packet encryption frequency f (f_{high}, f_{medium}, f_{low}). Also, we define two thresholds (Th_{HIGH}, Th_{LOW}) for the node's residual energy (R_{ENERGY}). The Level of security use depends on the node's residual energy. The following equation defines the switching between different security levels:

$$\begin{cases} if(R_{ENERGY} >= Th_{HIGH}) - then - f = f_{high} \\ if(Th_{LOW} =< R_{ENERGY} < Th_{HIGH}) - then - f = f_{medium} \\ if(R_{ENERGY} < Th_{LOW}) - then - f = f_{low} \end{cases}$$

With EE_SRTP we exploit dependency between frames created by the inter-frame coding. Inter-frame coding uses motion estimation and compensation between successive video frames. In an MPEG-coded video stream intra coded frames (I-frames) are compressed without reference to any other frame in the sequence, they are intra coded. The P-frames are forward predicted from the last I-frame or P-frame, i.e. it is impossible to rebuild them without the data of another frame (I or P). The B-frames are both forward predicted and backward predicted from the last/next I-frame or P-frame, i.e. there are two other frames necessary to reconstruct them. P-frames and B-frames are referred to as inter coded frames. For better energy saving, we propose to not encrypt every RTP packet to be sent. The strategy is achieved twofold: (1) we select the frequency of encryption depending on the residual energy thresholds that are defined by the user.

We implemented EE_SRTP approach by modifying the popular VIDEOLAN VLC player in bump-in-the stack fashion, to be Secure VLC (SecVLC). We used a field of the SRTP packet, to notify the receiver if this packet is encrypted or not. In the SRTP packet we have some fields already defined but not used. We used the field M, Marker, in the SRTP packet; therefore we will not add any overhead. In the definition of our profile EE_SRTP, the sender after choosing the frequency of encryption *f*, it study the frames to detected the different type of frames (I, P or B) then return the bit M to 1 for the encrypted packets. The receiver checks the field M to know if the packet is encrypted or not.

SecVLC framework can be summarized in two flowcharts illustrated in Figures 5 and 6. Figure 5 shows the set of operations realized by the sender multimedia wireless node where figure 6 shows the set of operations by the receiver multimedia wireless node.

In addition, with EE_SRTP we are at the same time performing the Quality of Service (QoS) in the WNs, in fact, with reducing the time of encryption we improve the WN'QoS (Lima et al., 2007).

For the selection of the packet to encrypt, we use EE_SRTP for the encryption frequency. Each packet encrypted is marked by the field M=1, M=0 in the other case, then the packets are sent. The operation at the receiver side is described in Figure 4.

7. PERFORMANCE EVALUATION

The performance evaluation of our model is done in two phases. First, with the implementation of EE_SRTP on VLC we have evaluated the security offered by EE_SRTP. As result we had sequences of video encrypt with different encryption fre-

Figure 5. View capture from the original video

quency. We can see the results in Figure 5, Figure 6 and Figure 7.

Figure 5. shows a screen shot of the video we used in the SecVLC test. The video is 25fps, for better results we choose a video with lot of movements to make the experiment more realistic.

Figure 6 and Figure 7 are view capture of the receive video in a node who is not allowed to see the video (it hasn't the decryption key), when using an encryption frequency = 3 and *f*=6 respectively. We can clearly see that this node without the encryption key can't know the content of the video.

In the experiment we notice that with a frequency *f*> 6 we can see some sequences of the video, because of the high motion level and difference between each sequence.

Using the abovementioned videos we objectively evaluated the quality and the level of the security. For more concrete evaluations, we analysed the different videos using the MSU Video Quality Measurement Tool. MSU VQMT is a free software tool for video quality measurements. This application allows one to create objective comparison of video codec and perform filters video analysis. Video quality is a characteristic of a video passed through a video transmission/processing system, i.e., a formal or informal measure of perceived video degradation (typically, compared to the original video). Video processing systems may introduce some amounts of distortion or artifacts in the video signal, so video quality evaluation is an important problem. Using this tool we performed extensive tests on the videos using EE_SRTP.

Figure 8 and Figure 9 are respectively the Peak Signal-to-Noise Ratio (PSNR) and Mean Squared Error (MSE) measurements.

PSNR is an engineering term for the ratio between the maximum possible power of a signal and the power of corrupting noise that affects the fidelity of its representation. Because many signals have a very wide dynamic range, PSNR is usually expressed in terms of the logarithmic decibel scale.

Figure 8 represents the videos PSNR variation with different frequency encryption values. Typical values for the PSNR in lossy image and video compression are between 30 and 50 dB, where higher is better. Acceptable values for wireless transmission quality loss are considered to be about 20 dB to 25 dB (Boulgouris, Thomos, & Strintzis, 2006). The average PSNR for f=2 is 6.68db, for f= 6 is 14.08 db and for f=8 is 17.99 db which is considered high.

MSE (Spanos & Maples, 1995) of an estimator is one of many ways to quantify the amount by which an estimator differs from the true value of the quantity being estimated. As a loss function, MSE is called squared error loss. MSE measures the average of the square of the "error". The error is the amount by which the estimator differs from

Figure 6. View capture from an encrypted video with f= 3

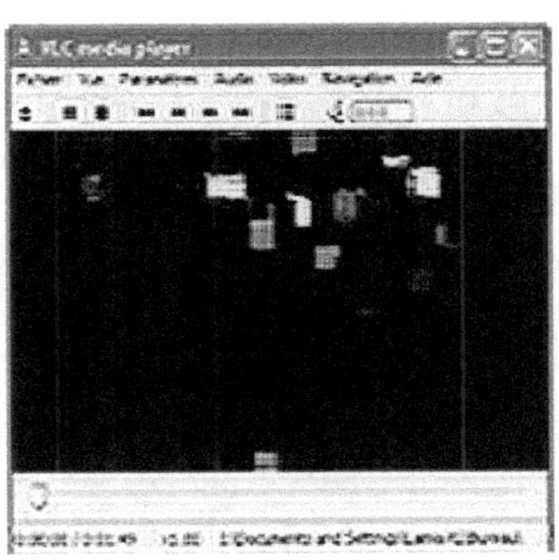

Figure 3. SecVLC datagram at sender

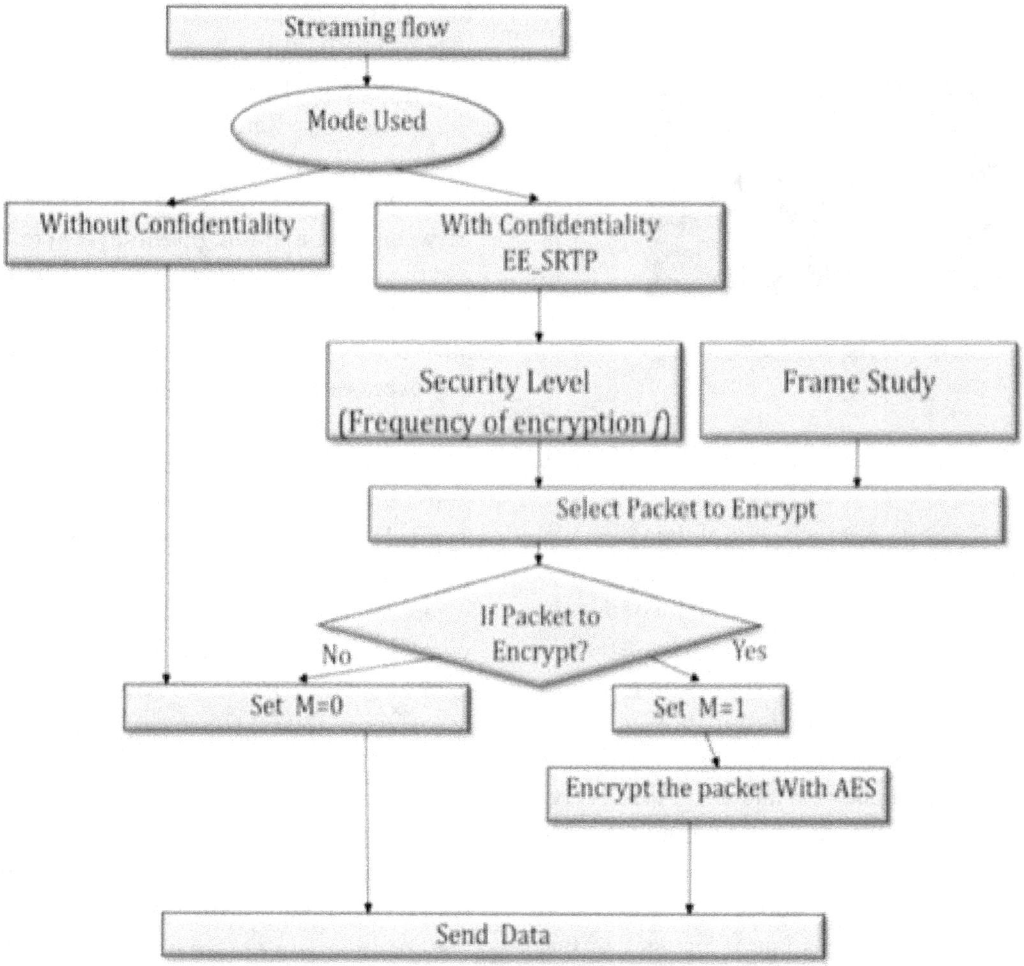

the quantity to be estimated. The difference occurs because of randomness or because the estimator doesn't account for information that could produce a more accurate estimate.

Figure 9 shows clearly that the MSE increase significantly between two frames encrypted with a frequency of encryption f = 8. The MSE is relatively stable with the other frequencies (with f<6).

To evaluate the energy performances, we used trace file of the video. Trace file gives the statistics of the video [frame number, frame type, time (ms), frame length (byte)]. The Frame type can be I, P or B. The time indicates the presentation time of the frame in the video sequence. The frame length is different from a frame type to another, and is the size of the frame to be encrypted or not.

On the other hand, it is commonly known that encrypting each byte will consume 0.02 μJ on average, when using a key of 128 bits and by breaking it up into 1000 byte packets (Prasithsangaree & Krishnamurthy, 2004) With those data we calculate the estimated energy consumption using the equation Eq.1.

$$E=0.02 \times PacketLength \quad (1)$$

Figure 4. SecVLC datagram at receiver

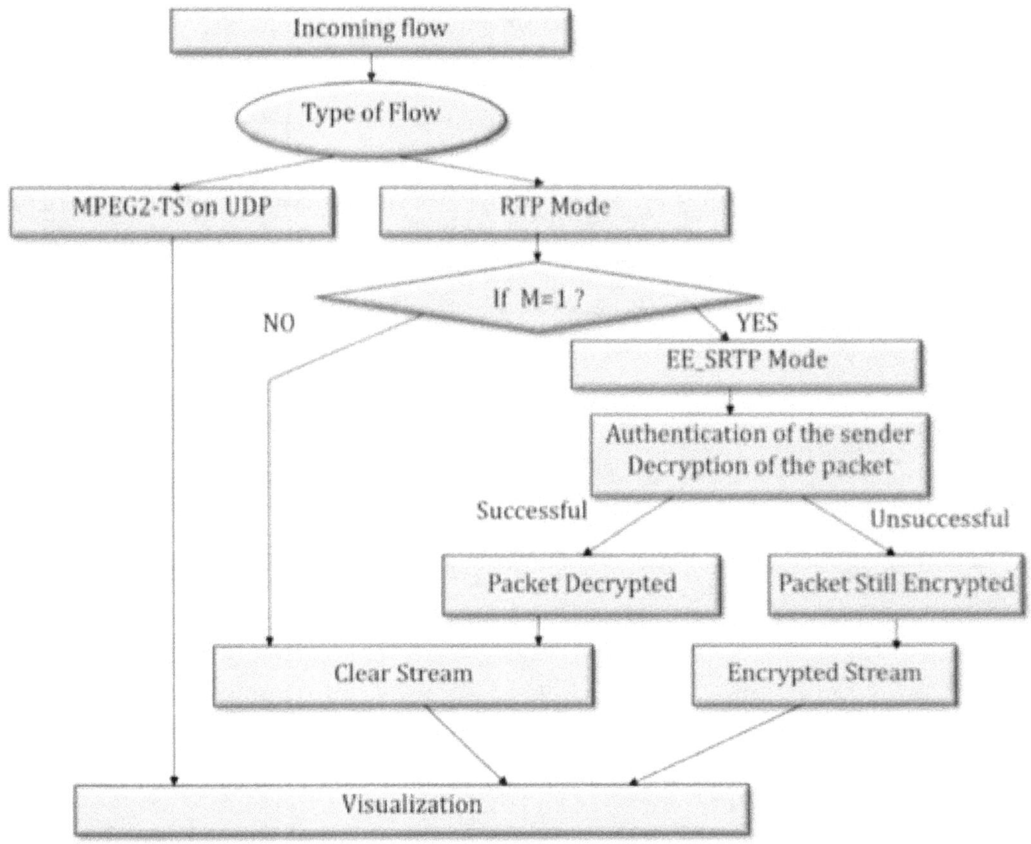

Figure 10 represents the differences between the lengths of the three types of frames (I/P/B). It shows that I frames have the larger size of data at the most of the time, only in the time t= 6400ms P frame have more data than I. The moments, like t, are interpreted in the encryption video, when encrypting only I frame, by clear part of a scene. Since, this case is really not frequent; there is no loss of security. Figure 11 shows the difference of energy consumption when using confidentiality with the standard SRTP and with EE_SRTP.

Figure 12. represents the energy consumption of SRTP and EE_SRTP. This shows that the difference between them and that EE_SRTP secure the MWANet and allows a gain of energy, then improve the lifetime of MWANet.

CONCLUSION

In this chapter, we proposed a new framework called Efficient Energy Secure Real Time Protocol

Figure 7. View capture from an encrypted video with *f*= 6

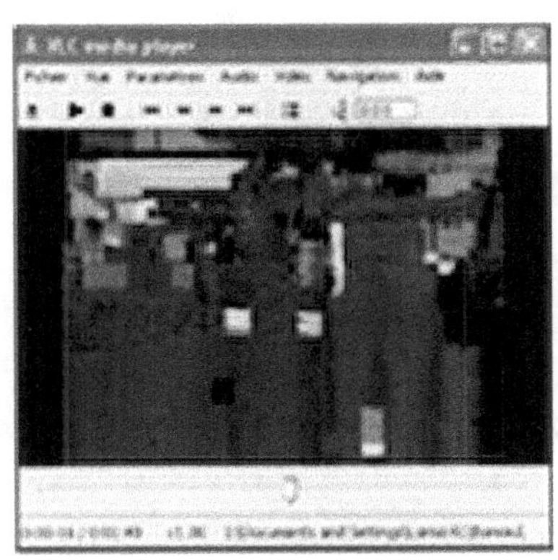

Figure 8. PSNR measure of the encrypted videos

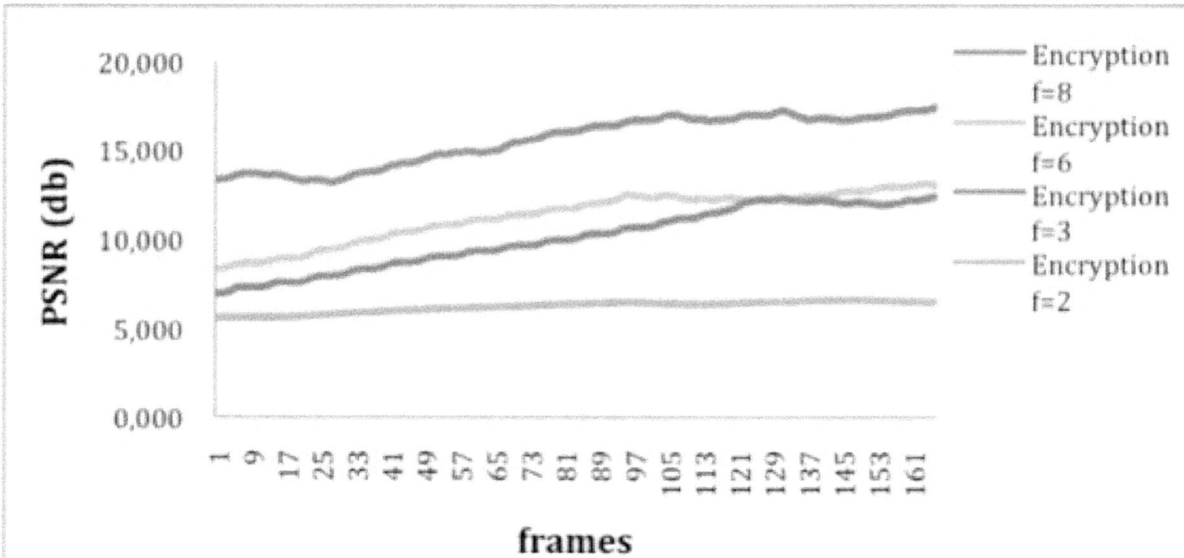

Figure 9. MSE measure of the encrypted videos

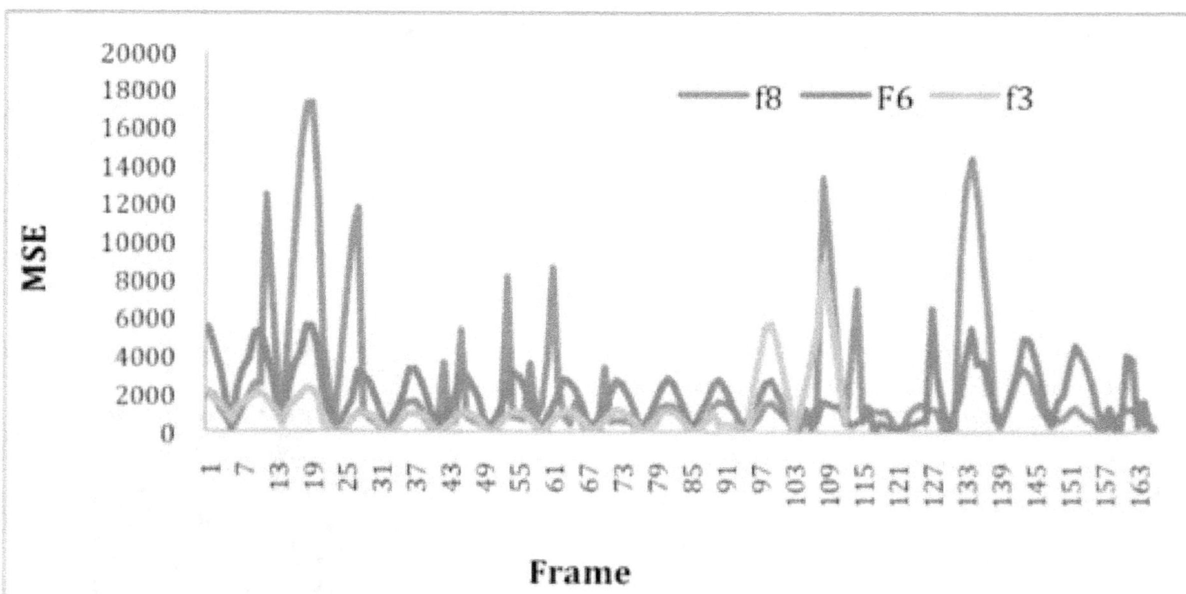

Figure 10. Statistics of I/P/B frames length

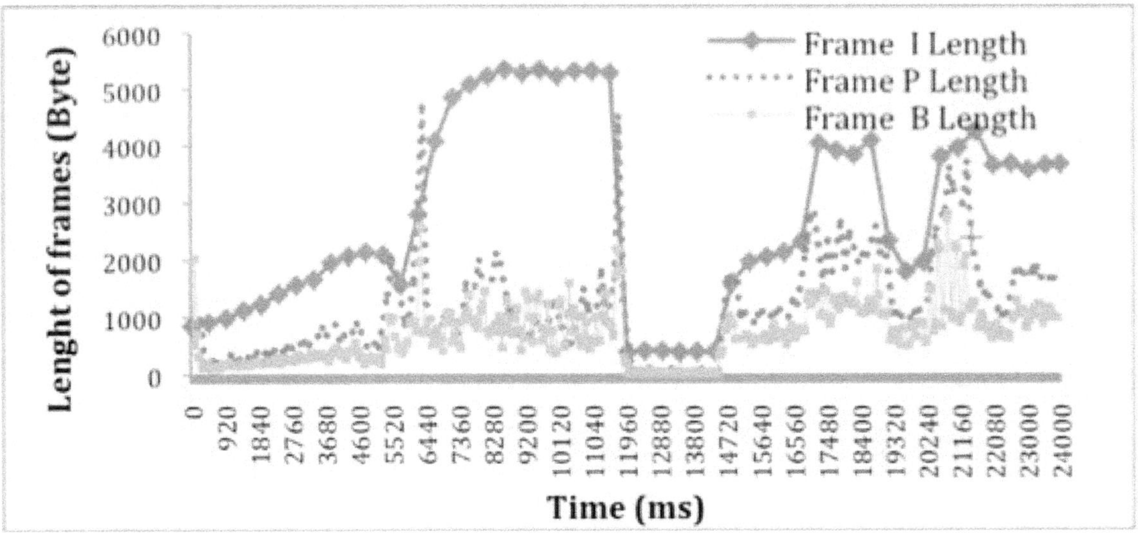

Figure 11. Energy consumption for encryption: EE_SRTP versus SRTP

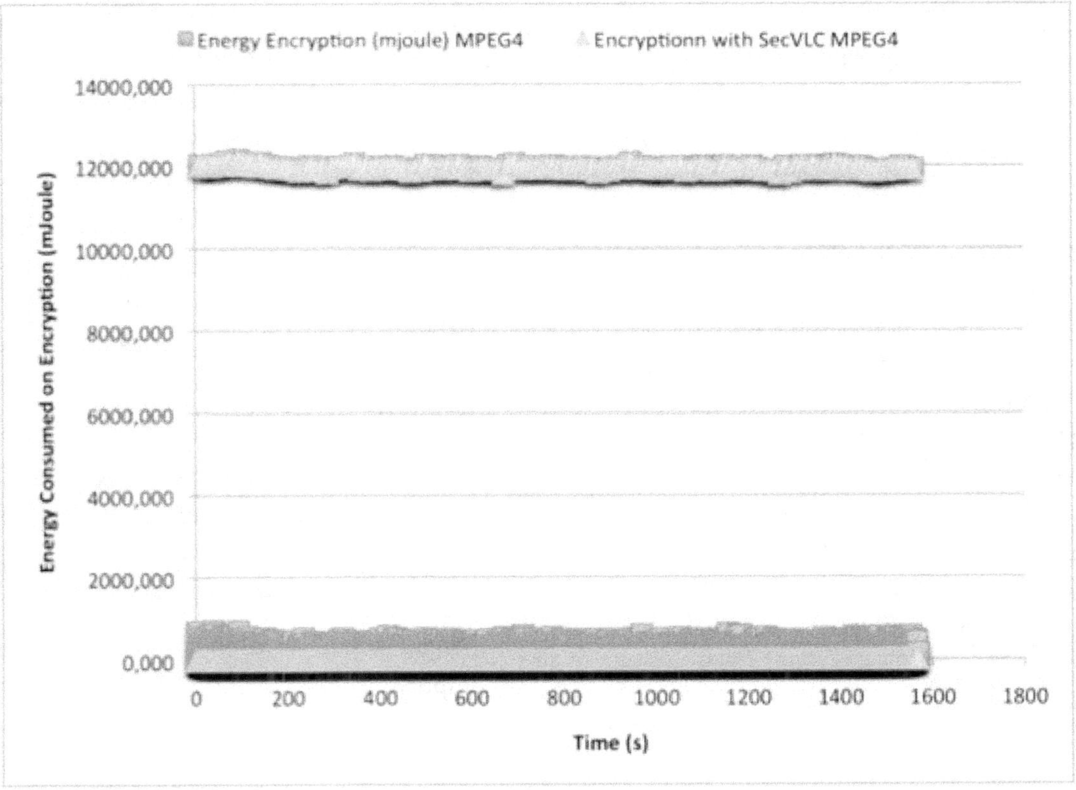

Figure 12. Lifetime for a sender: with SecVLC versus standard SRTP

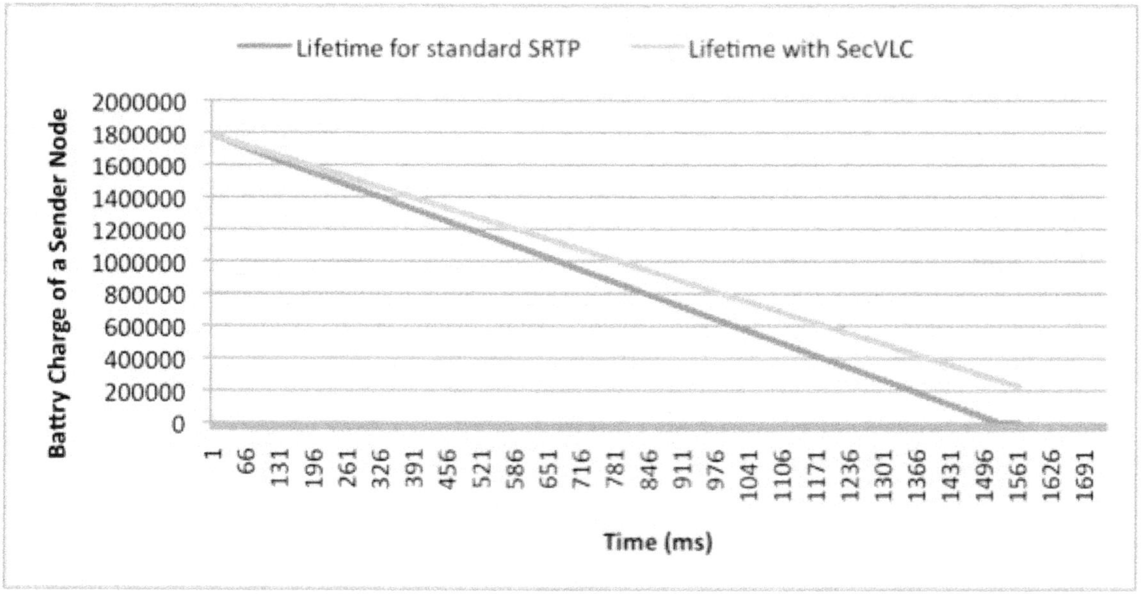

(EE_SRTP) implemented on SecVLC in order to ensure three different security levels while taking into account the energy consumption. The innovation of the proposed EE_SRTP consists in a frequency selective encryption in the packet level and the implementation of our solution in VLC. SecVLC proposes three security levels (high, medium and low) with the ability to ensure the authentication, and the confidentiality. The selection of each level depends on the node's residual energy. EE_SRTP increases the lifetime duration of nodes by around 50% compared to classical encryption algorithm. In future works, we plan to adapt EE_SRTP to ESTREL (Kaddar & Mehaoua, 2007) and EVAN Kaddar, Aoul, & Mehaoua, 2011) a model of energy saving on transmission using enhancement layers of MPEG4. Another future research direction is to use probabilistic model for the selection of the encrypted frame.

REFERENCES

Agi, I., & Gong, L. (February 1996). An empirical study of secure MPEG video transmission. *Proceedings of the Internet Society Symposium on Network and Distributed System Security*, San Diego, CA, (pp. 137-144).

Alattar, A. M., Al-Regib, G. I., & Al-Semari, S. A. (1999). Improved selective encryption techniques for secure transmission of MPEG video bit-streams. *Proceedings of the 1999 International Conference on Image Processing (ICIP '99)*, Vol. 4, (pp. 256-260). Kobe, Japan, October 24-28.

Boulgouris, N. V., Thomos, N., & Strintzis, M. G. (2006, January). Optimized transmission of JPEG2000 streams over wireless channels. *IEEE Transactions on Image Processing*, *15*(1).

Cheng, H., & Li, X. (2000). Partial encryption of compressed images and video. *IEEE Transactions on Signal Processing*, *48*(8), 2439–2451. doi:10.1109/78.852023

Crow, B., Widjaja, I., Kim, J. G., & Sakai, P. (1997). Investigation of the IEEE 802.11 medium access control (MAC) sublayer functions. In . *Proceedings of INFOCOM, 1*, 26–33.

IETF. (2004). *MIKEY: Multimedia internet keying*. The Internet Society 2004. Retrieved from http://tools.ietf.org/html/rfc3830 ISO/IEC 14496-1. (May 1998). *Coding of audio-visual objects, part 1: Systems*. Final committee draft.

ISO/IEC 14496-2. (May 1998). *Coding of audio-visual objects, part 2: Visual*. Final committee draft.

ISO/IEC 14496-3. (May 1998). *Coding of audio-visual objects, part 3: Audio*. Final committee draft.

Kaddar, L., Hadjadj Aoul, Y., & Mehaoua, A. (2011). *EVAN: Energy-aware SVC video streaming over wireless ad hoc networks*. IEEE 73rd Vehicular Technology Conference: VTC2011-Spring.

Kaddar, L., & Mehaoua, A. (2007). ESTREL: Transmission and reception energy saving model for wireless ad hoc networks. In *Proceedings 32nd IEEE Conference on Local Computer Networks (LCN)* (Short paper), (pp. 784-785).

Kaddar, L., & Mehaoua, A. (2009). *SecVLC: Secure transmission over multimedia wireless ad hoc networks with energy-awareness*. In Global Information Infrastructure Symposium 2009 (GIIS 2009).

Lima, A., Carvalho, L., Souza, J., & Mota, E. (2007). VoIP management: Speech quality monitoring architecture based on statistical analysis. *International Journal of Network Management, 17*(4).

Lundin, R., & Lindskog, S. (2010). Security implications of selective encryption. In *Proceedings of Metric2010*, Bilzano-Bozen, Italy, 2010.

Meyer, J., & Gadegast, F. (1995). *Security mechanisms for multimedia data with example MPEG-I video*.

National Institute for Standards and Technology (NIST). (2001). *Advanced encryption standard (AES)*. FIPS 197. Retrieved from http://csrc.nist.gov/encryption/aes/

Network Working Group. (2004). *The secure real-time transport protocol (STRP)*. Retrieved from http://www.ietf.org/rfc/rfc3711.txt

Podesser, M., Schmidt, H.-P., & Uhl, A. (2002). *Selective bitplane encryption for secure transmission of image data in mobile environments*. 5th Nordic Signal Processing Symposium, on board Hurtigruten, Norway, October 4-7.

Pommer, A., & Uhl, A. (2003). Selective encryption of wavelet-packet encoded image data. *ACM Multimedia Systems Journal, Special Issue on Multimedia Security, 9*(3).

Prasithsangaree, P., & Krishnamurthy, P. (2004). On a framework for energy-efficient security protocols in wireless networks. *Computer Communications, 27*, 1716–1729. doi:10.1016/j.comcom.2004.05.022

Qiao, L., & Nahrstedt, K. (1998). Comparison of MPEG encryption algorithms. *International Journal on Computer and Graphics, Special Issue on Data Security in Image Communication and Network, 22*(3).

Spanos, G. A., & Maples, T. B. (1995). Performance study of a selective encryption scheme for security of networked, real-time video. In *Proceedings of the 4th International Conference on Computer Communications and Networks* (ICCCN'95), (pp. 72-78). Las Vegas, Nevada, USA.

Tang, L. (2007). Methods for encrypting and decrypting MPEG video data efficiently. *Proceedings of the 4th ACM International Multimedia Conference,* Boston, MA, November 18-22, (pp. 219-230).

Van Droogenbroeck, M., & Benedett, R. (2002). Techniques for a selective encryption of uncompressed and compressed images. *Proceedings of Advanced Concepts for Intelligent Vision Systems (ACIVS) 2002*, Ghent, Belgium, September 9-11.

Chapter 15
Security Threats and Issues with MANET

Sudha Singh
Bengal College of Engineering and Technology, India

ABSTRACT

In recent years, mobile ad hoc networks have received tremendous attention because of their self configuration capabilities. Although security has long been an active research topic in wireless networks, the unique characteristics of MANETs present a new set of non trivial challenges to security design. These challenges include open network architecture, shared wireless medium, stringent resource constraints, and highly dynamic network topology. Consequently, the existing security solutions for wired networks do not directly apply to the MANET domain. The ultimate goal of the security solution for MANET is to provide security services, such as authentication, confidentiality, integrity, anonymity, and availability to the mobile users. One distinguishing characteristics of MANETs from the security design perspective is the lack of a clear line of defense. The wireless channel is accessible to both legitimate network users and malicious attackers. There is no well defined place where traffic monitoring or access control mechanisms can be deployed. As a result, the boundary that separates the inside network from the outside world becomes blurred.

I. INTRODUCTION

The research on MANET security is still in its early stage. The existing proposals are typically attack-oriented in that they first identify several security threats and then enhance the existing protocol or propose a new protocol to thwart such threats. Because the solutions are designed explicitly with certain attack models in mind, they work well in the presence of designated attacks but may collapse under unanticipated attacks. Therefore, a more ambitious goal for ad hoc network security is to develop a multi-fence security solution that

DOI: 10.4018/978-1-4666-0321-9.ch015

is embedded into possibly every component in the network, resulting in in-depth protection that offers multiple lines of defense against many both known and unknown security threats.

A MANET is an autonomous collection of mobile users that communicate over relatively bandwidth constrained wireless links. Since the nodes are mobile, the network topology may change rapidly and unpredictably over time. The network is decentralized, where all network activity including discovering the topology and delivering messages must be executed by the nodes themselves, i.e., routing functionality will be incorporated into mobile nodes. Anonymity is another important issue. A MANET environment is vulnerable to eavesdropping, where an adversary can track and analyze the traffic flow through a network, and identify or locate a certain sender/receiver, even if it cannot crack the encrypted data (Cordasco and Wetzel,2009).

Ad hoc have several salient characteristics (Toh,2007; Ilyas,2003): *(1) Distributed Operation:* The ad-hoc concept differs from other wireless systems, such as cellular system in terms of network operation. An ad-hoc network comprises stations that have the same capabilities and responsibilities. No centralized entry that controls the network exists. In an ad-hoc network there are no BSS or MSCs thus all network protocols operate in a distributed manner. *(2) Dynamic topologies:* Nodes are free to move arbitrary, thus the network topology which is typically multihop may change randomly and rapidly at unpredictable times, and may consist of both bi-directional and unidirectional links. *(3) Bandwidth-constrained, variable capacity links:* Wireless links will continue to have significantly lower capacity than their hardwired counterparts. In addition, the realized throughput of wireless communication after accounting for the effects of multiple access, fading, noise and interference condition, etc. is often less than a radios maximum transmission rate. One effect of the relatively low moderate link capacities is that congestion is typically the norm rather than the exception i.e. often much less than a radios maximum transmission rate. *(4) Energy constrained operation:* Some or all of the nodes in a MANET is a may rely on batteries or other exhaustible means for their energy. For these nodes, the most important system design criteria for optimizing may be energy conservation. *(5) Multihop communication:* In case where a station A needs to send data to another station B out of range, the transmission needs to relayed through other nodes. Such networks are known as multihop wireless ad-hoc network. Example: HIPERLAN,WLAN. *(6) Limited physical security:* mobile networks than fixed networks (wired and wireless) are more vulnerable to security threats. Need to overcome the security weaknesses in wireless links and mobile topologies arising from the new security risks.

Literature Survey

In wireless ad hoc networks, existing security routing protocols are evaluated in the traditional threat model, such as *active-n-m* attacker model (Hu and Perrig,2004). It is assumed that the attacker has the same power and capability of a non- malicious node in the traditional threat model. This limitation which doesn't allow the attacker use out-of-band resources or other more powerful methods to corrupt a given routing protocol is not reasonable. In the traditional threat model, authors tend to claim protocol security based on a fixed environment. However, the security results are just only applicable if the protocol is analyzed under the author's assumptions of the attacker capabilities. These secure routing protocols may not be secure out of the respective assumptions (Djenouri,Khelladi and Badache,2005 ; Argvroudis and Mahony,2005). Additionally, it is infeasible to compare multiple protocols without common assumptions or security definitions. Because of its tractability, the Dolev-Yao attacker model (Doley and Yao,1983) is the most prevalent model for security analysis of protocol such as authentication protocol. The Dolev-Yao attacker model can be disassembled into the following three assumptions: (1)Perfect cryptography, this assumption states that cipher-text can only be decrypted if the decryption key is present, and generated using encryption with the appropriate encryption key and plaintext. (2) Finite set of capabilities, an idealized attacker is defined which have a finite set of capabilities at his disposal. It is assumed that this attacker is equivalent to one which has an infinite set of capabilities which is playing according to the rules of perfect cryptography. (3) Omnipresence, the view that in order to prove a protocol correct, an attacker controlling all communication paths has to be considered. Following the first two assumptions is correct in ad hoc network. However, it is impossible to establish a view that an attacker in wireless ad hoc networks which would be "safe" in sense of the Dolev-Yao attacker, since the omnipresent attacker is imaginable but cannot be defended. To address the above issues, new threat models are proposed in (Cordasco and Wetzel,2009 ; Todd and Alec,2008). Inspired by these works, xian et al (Xian,Tao,Ting and Fing, 2010) develop the hierarchical threat model of routing security for ad hoc networks. Another contribution is to analysis security of two routings such as endairA (Gergely, Buttyan and Vajda,2006) and ARAN (Sanzgiri,Laflamme,Dahill,Levine, Shields and Belding-Royer,2005) in their model. Both of them are provable security routing in extended active-*n*-*m* model.

Currently, more sophisticated uses of ad hoc networks are under study in virtual class- rooms (Vijayaragayan et. al.,2009), vehicular ad hoc networks (VANET) (Valery et. al.,2006), and ambient intelligent environments (Cabri et. al.,2008). Communication features exhibited by ad hoc networks greatly depend on their internal routing protocols. These protocols are in charge of collecting information about network nodes in order to establish routes for those willing to communicate. In most cases, the evaluation of ad hoc networks simply relies on assessing their performance under regular (non-faulty) conditions (Ahmed and Alam,2006). This seems contradictory to the widely accepted idea that the existence of wireless links, dynamic topologies, and implicit trust-your-neighbor policies in routing protocols, make ad hoc networks very sensitive to malicious faults (attacks) (Yang,Luo,Ye,Lu and Zhang,2004). As a result, the confident use of these networks, and their success in real-life products, strongly relies not only in our ability to produce secure ad hoc routing protocols, but also on our capacity to evaluate them in the presence of attacks.

It is the ambitious chapter of this book in terms of security. This chapter is to provide a unified view of the broad field of security in mobile ad hoc network. In this chapter the security requirements for the MANET and problems related to them are also discussed. Section II will explain security requirement in MANET. "How security in MANET is different from security in wireless network?" is also explained in this section. Section III is to describe the attack model and layered security issues in MANET. Section IV will identify the challenges in MANET security design. This Section will also describe overview security issues in MANET. Section V gives recent development in security issues. Section VI is to predict future direction of security issues in MANET. Section VII will present some open problems in this area.

II. HOW IS SECURITY IN MANET DIFFERENT?

Unlike Wired networks, MANET pose a number of challenges to security solutions due to their unpredictable topology, wireless shared medium, heterogeneous resources and stringent resource constraint etc. Security is not a single layer issue but a multilayered issue over here. MANET requires a multifence security solution that provides complete security spanning over the entire protocol stack.

The mobile nodes in MANETs are highly susceptible to malicious damage. Security issues are important in MANETs to prevent potential attacks, threats and system vulnerabilities. The fundamental aspects of computer security (Fisher,1999)–confidentiality, integrity, authentication and non-repudiation are valid when the protection of routing in MANET network is discussed.

- Confidentiality ensures the survivability of network services in case of denial of service attacks.
- Integrity guarantees that a message being transferred is never corrupted. A message could be corrupted because of beginning failure because of malicious attacks.
- Authentication guarantees every piece of routing information, which can always be confirmed to be valid and to have originated from the correct sender.
- Non-repudiation means that the node cannot deny having sent or handled certain piece of routing information in the past. Authentication mechanisms allow partial non-repudiation.

Finally, it should be noted that authentication gets often confused with authorization issues that involve more or less the control of policy and are typically implemented only in the application layer and not in the network layer (Sheikh,Singh and Mishra,2010; Yang and Zheng,2001).

Vulnerabilities in MANET

The protection of security always involves the identification of potential attacks, threats and vulnerabilities of a certain system. The vulnerabilities in MANET are the poor wireless channel security of the mobile nodes and the trust on any centralized resource. Potential attacks against routing can be divided into two groups: passive attacks typically involve only eavesdropping of the routing messages, while active attacks involve actions performed by adversaries such as routing message replication and deletion. External attacks are typically active attacks that may lead to the sending of false routing information, generation of routing loops, network portioning and merging and congestion. External attacks typically be prevented by using standard security mechanisms. Internal attacks are more severe attacks, since malicious nodes sending incorrect routing traffic are already protected with the security mechanism the routing framework offers. Thus such malicious insiders who may even operate in a group may use the standard security means to actually protect the attacks.

Thirdly, due to the frequent changes in its topology and membership, the ad hoc network is highly dynamic. Trust relationship among nodes changes accordingly. Our security mechanism must be able to adapt to such changes.

III. ATTACK MODEL

Routing in wireless ad hoc networks are vulnerable to traffic analysis, spoofing and denial of service attacks due to open wireless medium communications. Layer wise security issues are given in Table 1 (Cordasco and Wetzel,2009).

IV. OVERVIEW OF SECURITY ISSUES AND CHALLENGES IN MANET

Attacks in MANETs are Denial of Service, Impersonation, Eavesdropping and Routing. Based on the security criteria like availability, integrity, confidentiality, authenticity, non-repudiation, authorization and anonymity; following are the security and threat issues in MANET (Gergely,Buttayan and Vajda,2006):

Figure 1. Attack model

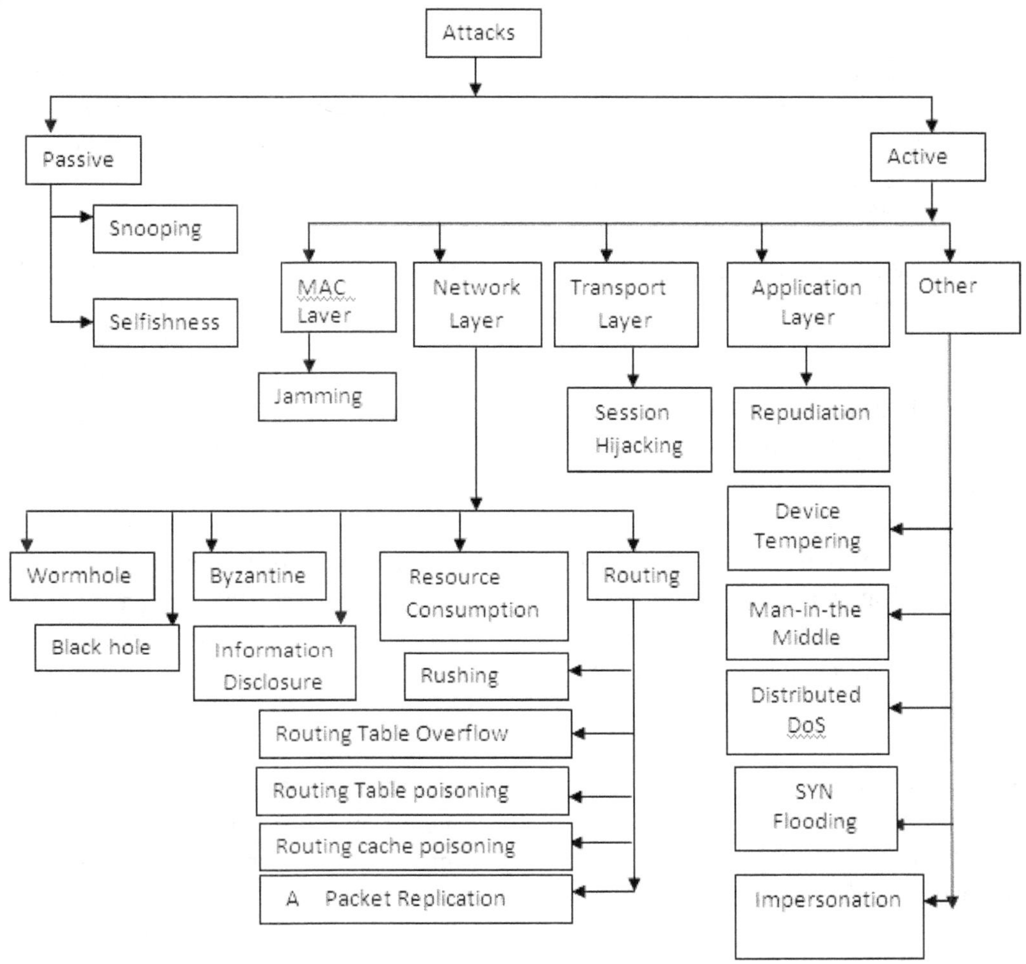

1. *No predefined boundary:* In MANETs, we can not precisely define a physical boundary of the network. The nodes work in a nomadic environment where they are allowed to join and leave the wireless network. As soon as an adversary comes in the radio range of a node it will be able to communicate with that node.
2. *Adversary inside the network:* The nodes within the network can also behave maliciously. This is hard to detect that the behavior of the node is malicious. Thus this type of attack is more dangerous than the external attack.
3. *No centralized control facility:* Since MANETs do not have any centralized control facility, it leads to many security problems. It becomes very difficult to detect ant attack. Due to the lack of security association we can not classify the nodes as trusted node or untreated node.
4. *Limited energy resource:* All nodes in a MANET depend on battery power for their operation. The adversary can sent huge traffic to the target node. The target node may be continuously busy in handeling these packets, this will cause the battery power to be exhausted and so denial of service attack.

Table 1. Layer wise security issues

Application Layer	Detecting and preventing viruses, worms, malicious codes, and application abuses
Transport layer	Authenticating and securing end-to-end communications through data encryption
Network layer	Protecting the ad hoc routing and forwarding protocols
Link layer	Protecting the wireless MAC protocol and providing link-layer security support
Physical layer	Preventing signal jamming denial-of-service attacks

Due to limited energy resource some nodes may behave as selfish nodes.

5. *Changing scale:* The Scalability of the MANET keeps changing all the time. It is very difficult to predict the number of nodes in a MANET at some future time.

Following are the challenges to security in MANET

- Shared radio channel
- Insecure environment
- Lack of central authority
- Lack of permanent association
- Limited resources
- Physical vulnerability
- Limited wireless transmission range
- Broadcast nature of the wireless medium
- Packet losses due to transmission errors
- Mobility-induced route changes
- Mobility-induced packet losses
- Battery constraints
- Potentially frequent network partitions
- Ease of snooping on wireless transmissions (security hazard)

We are briefly describing the challenges to security in MANET. An ad hoc network is a dynamic type of network which is both similar and very different to its parent fixed communication network. In the following we introduce the properties of an ad hoc network as a way of defining its shortcomings and to highlight its security challenges (Dawoud et al.,2011).

Dynamic Network Architecture: Ad hoc networks have no fixed or existing network infrastructure. The network architecture is continuously changing as the network evolves. There is no pre-existing or fixed architecture which handles all network tasks such as: routing security and network management. Instead, the network infrastructure is spontaneously set up in a distributive manner. Each participating node shares the network's responsibilities. Distribution of network functionality avoids single point attacks and allows for the network to survive under harsh network circumstances.

A fixed entity structure, such as a base station or central administration, is crucial for security mechanisms. A trusted third party member, which is expected in traditional networks, is similar to a fixed entity as both define security services; manage and distribute secret keying information (which allows secure communication of data through encryption and decryption techniques). Therefore the absence of such a control entity introduces new opportunities for security attacks on the network.

Self Organized Nature: Wireless ad hoc nodes cannot rely on an off-line trusted third party member. The security functions of the trusted third party member are distributed among the participating nodes. Each node takes responsibility for establishing and maintaining its own security and is, therefore, the centre of its own world and authority. A wireless ad hoc network is therefore referred to as a self organized network.

No Prior relationships: In ad hoc networks, nodes can have no prior relationships with other nodes within the network. Prior acquaintance between nodes can be considered as pre-trust relationships between nodes. However, the ad hoc nature of these networks does not allow for these assumptions, as it cannot be assumed that secrets exist between the respective pair of nodes. If nodes can join and leave the network at random without prior trust relationships with nodes, access control becomes a difficult task for the security mechanism.

Multi-hop communication channel: Wired networks include fixed nodes and fixed wired communication lines. Wireless ad hoc networks have mobile wireless nodes (often in the form of hand held devices) and, as suggested, their communication medium is wireless. This allows for greater network availability and easy network deployment. Each node's transmission range is limited and network communication is realized through multi-hop paths. Co-operation and trust along these paths is a crucial aspect of the security mechanism and ensures successful communication. The shared wireless communication medium means that any user can participate in the network. This creates access control problems for security mechanisms as adversaries are able eavesdrop on communication or launch active attacks to alter message data.

Mobility: Nodes are expected to be mobile within an ad hoc network, creating a dynamic and unpredictable network environment. In certain situations the nodes' mobility is not totally unsystematic and assumptions can be made in the form of mobility patterns . An example of these patterns is evident in a vehicular ad hoc network where vehicles move along fixed paths, or roads, at speeds which have a high probability of being within the local speed limit. However, nodes demonstrate random mobility within these predictions. Connectivity between nodes is sporadic. This is due to the shared, error-prone wireless medium and frequent route failures which caused by the unpredictable mobility of nodes . Increased mobility can result in the multi-hop communication paths being broken and network services becoming unavailable. Security mechanisms must account for the weak connectivity and unavailability. Furthermore, due to mobility and sporadic connectivity, these mechanisms must also aim to be scalable with the changing network density.

Resource Limitations: Wireless nodes allow for the freedom of mobility and easy network establishment and deployment. Wireless nodes are often smaller hand-held devices that do not experience the same resource privileges of traditional wired nodes . Mobile nodes are ideally low cost and small in size as to maximize node availability and mobility. In attempt to achieve these objectives wireless nodes have limited resource, specifically in the following areas:

- Battery life
- Communication range
- Bandwidth
- Computational capacity
- Memory resources

If mobility is to be attained, nodes must be battery powered. Battery powered nodes suffer from the consequences of power failures which break connectivity. They also run a high possibility of failing to be on-line the entire duration of the network. This could hinder network service availability. Cost and power restrictions limit the design features of wireless nodes. Power and transmission range are directly related, resulting in wireless devices having limited transmission ranges and bandwidths. Low powered, low cost CPU's are preferred, as this reduces the computational capacity and memory resources available for routing and security operations. As discussed above, network and security tasks are not performed by a central authority, but rather distributed among all the nodes. This creates a heavy burden upon the nodes to perform their own tasks as well as the network services. If the security mechanisms do not distribute the load fairly, adversaries can act in a selfish manner, forcing other nodes to perform extra tasks. In some instances malicious nodes will flood a single node with service requests in the aim of depleting its limited resources. A well designed security algorithms optimizes computational processing and operation to meet the limited resource requirements of these dynamic networks.

Physical Vulnerability: Another challenge in ad hoc networks is the physical vulnerability of nodes. In a mobile ad hoc network nodes are mobile and often small devices. This contributes to a higher probability of being capture or compromised when compared to traditional wired networks with stationary entities. This means that wireless ad hoc networks are more prone to insider attacks and security mechanisms and must be designed with this in mind. An inside attacker could analyze the node to gain secret keying information or use the node to compromise other nodes. The same threats exist in wired formal networks. Although they may rely on a secure host to detect and recover compromised nodes. Sensitive security information may also be stored on that host, minimizing the consequences upon the network if a single node is captured. In an attempt to enhance security within hybrid ad hoc networks a fixed architecture is combined with a volatile distributive architecture.

Security objectives and services

Securing mobile ad hoc networks requires certain services to be met. A security service is a made available by a protocol which ensures sufficient security for the system or the data transferred. The security objectives for mobile ad hoc networks are similar to that of fixed wired networks. The security objects are described in six categories:

- Authentication
- Access Control
- Data Confidentiality
- Data Integrity
- Non-repudiation
- Availability Services

ROUTING PROTOCOL GOALS

Any routing protocol can protect the privacy of nodes and routes and also ensure the security of discovered routes. Any anonymous routing protocol need to achieve following goals (Jai and Kannammal,2009):

A. Ensure Privacy: The Privacy can be ensured in MANET by the following attributes: identity privacy, location privacy and routing anonymity. Identity privacy ensures that mobile nodes cannot be identified by each other, except the communicating parties (source and destination). Location privacy ensures that each mobile nodes geographic position cannot be located by other mobile nodes. Routing anonymity ensures that all the intermediate nodes cannot identify the source and the destination of a data packet.

B. Ensure Security: The protocol can protect the necessary functionalities such as discover and maintain the route from various type of attacks.

Routing Problems

Routing is an important topic in MANET, because changes in network topology occur frequently. An efficient routing protocol is required to cope with highly fluid network conditions. Before proceeding ahead, let us have some routing problems in MANETs available in the recent literatures:

1. **Distributed Network**: There is no centralized server to maintain the state of the clients.
2. **Dynamic topology**: We have to design routing protocol which can take care of the changes in the topology.
3. **Power awareness**: MANET protocol should have power awareness.
4. **Addressing scheme**: Dynamic topology and distributed network concept of MANET implies problems in addressing scheme.
5. **Cross Layer interactions**: Both at the conceptual level and the implementation level, underlying protocols have to take care of cross layer interactions.

6. **Network Size**: Protocols generally puts a bound on the size of the network.
7. **Error Prone broadcast channel**: Routing protocols have to take care of attenuation, multipath fading etc.
8. **Quality of Service(QoS) limitations**: QoS guarantees is required for all type of data and applications.
9. **Security**: Confidentiality, integrity and authenticity are very difficult to achieve in MANET.

V. RECENT DEVELOPMENTS IN MANET

For comprehensive security we need protocol security, communication security and physical security. To achieve above mentioned securities, many solutions have been proposed and are under research. Major solutions are for protocol security, authentication & key management schemes and trust management. Secure protocols available in the literatures are SLSP(Secure Link State Routing Protocol), ARAN(Authentication routing for ad hoc networks), SAR(security aware ad hoc routing protocol), SEAD(Security efficient ad hoc distance vector routing protocol),SAODV(Secure on demand distance vector)(Perkins et al., 1999 and 2003), SRP(secure routing protocol) etc. The primary goal of key management is to share a secret among a specified set of participants. Authentication and key management available in the literatures are password-based group system, threshold cryptography, self organized public key management for MANETs etc. In Ad hoc network every node work as a router to forward packets of other nodes. So we need authentic users.

Different solutions are available in literatures like (1) KPI(Key distribution infrastructure)-trusting the device instead trusting the user, (2) Mobility helps security, (3) SAFE(Secure packet forwarding in ad hoc networks)- Here every node is responsible to monitor its neighbor and reputation values are assigned based on the successful forwarding transactions(+1,0,-1), (4) Currency based solutions(Nuglets)-Network is a market where services are exchanged. Based on virtual economy where node pay for service. It introduces Currency "Nuglets". Only data packets require "nuglets". It works with many protocol. Here misbehavior not forbidden; only discouraged. (5) Token based co-operation enforcement.

VI. FUTURE DIRECTION IN SECURITY ISSUES OF MANET

Future direction of security issues is tabulated in Table 2 (Hu and Perrig,2004).

VII. OPEN PROBLEMS

1. A more ambitious goal for ad hoc network security is to develop a multi-fence security solution that is embedded into possibly every component in the network, resulting in in-depth protection that offers multiple lines of defense against many both known and unknown security threats. This new design perspective is what we call resiliency-oriented security design. Yang et al.(2004) envision the resiliency-oriented security solution as possessing several features.

First, the solution seeks to attack a bigger problem space. It attempts not only to thwart malicious attacks, but also to cope with other network faults due to node misconfiguration, extreme network overload, or operational failures. In some sense, all such faults, whether incurred by attacks or misconfigurations, share some common symptoms from both the network and end-user per spectives, and should be handled by the system. Second, resiliency-oriented design takes a paradigm shift from conventional intrusion prevention to intrusion tolerance.Third, as far as the solution space is concerned, cryptography-based techniques just offer a subset of toolkits in a resiliency-oriented

Table 2. Future directions of security issues of MANETs

Sl. No.	Problems	Details
1	Security Issues • What's New ?	• Wireless medium easy to snoop on • With ad hoc networking, hard to guarantee connectivity • Easier for intruders to insert themselves into network
2	• Authentication	• How to authenticate a node ? • May not have access to a certification authority
3	• Resource Depletion Attack	• Intruders may send data with the objective of congesting a network or depleting batteries
4	Routing Attacks • Intruders may mis-route the data o not delivering it to the destination at all, or o delaying it significantly	• How to detect such attacks ? • How to tolerate such attacks ?
5	Traffic Analysis	• Despite encryption, an eavesdropper can identify traffic patterns • Traffic patterns can divulge information about the operation mode • Traffic analysis can be prevented by presenting "constant" traffic pattern • Insert dummy traffic • How to make this cheaper ?

design.Fourth, the solution should be able to handle unexpected faults to

some extent. One possible approach worth exploring is to strengthen the correct operation mode of the network by enhancing more redundancy at the protocol and system levels. At each step of the protocol operation, the design makes sure what it has done is completely along the right track. Anything deviating from valid operations is treated with caution. Whenever an inconsistent operation is detected, the system can raise a suspicion flag and query the identified source for further verification. This way, the protocol tells right from wrong because it knows right with higher confidence, not necessarily knowing what is exactly wrong. The design strengthens the correct operations and may handle even unanticipated threats in runtime operations. Next, the solution may also take a collaborative security approach, which relies on multiple nodes in a MANET to provide any security primitives. Therefore, no single node is fully trusted. Instead, only a group of nodes will be trusted collectively. The group of nodes can be nodes in a local network neighborhood or all nodes along the forwarding path. Finally, the solution relies on multiple fences, spanning dif- ferent devices, different layers in the protocol stack, and different solution techniques, to guard the entire system. Each fence has all functional elements of prevention, detection/ver- ification, and reaction.

The above mentioned resiliency-oriented MANET security solution poses grand yet exciting research challenges. How to build an efficient fence that accommodates each device's resource constraint poses an interesting challenge. Device heterogeneity is one important concern that has been largely neglected in the current security design process. However, multifence security protection is deployed throughout the network, and each individual fence adopted by a single node may have different security strength due to its resource constraints. A node has to properly select security mechanisms that fit well into its own available resources, deployment cost, and other complexity concerns. The security solution should not Stipulate the minimum requirement a component must have. Instead, it expects best effort from each component. The more powerful a component is, the higher degree of security or resiliency it has. Next, identifying the system principles of how to build such a new-genera-tion of network protocols remains unexplored. The state-of-the-art network protocols are all designed for functionality only. The protocol specification fundamentally assumes a fully trusted and well-behaved network setting for all message exchanges and protocol operations. It does not anticipate any faulty signals or ill- behaved nodes. We need to identify new principles to build the next-generation network protocols that are resilient to faults. There only exist a few piecemeal individual efforts. Finally, evaluating the multifence security design also offers new research opportunities. The effectiveness of each fence and the minimal number of fences the system has to possess to ensure some degree of security assurances should be evaluated through a combination of analysis, simulations, and measurements in principle. However, it is recognized that the current eval- uation for state-of-the-art wireless security solutions is quite ad hoc. The community still lacks effective analytical tools, particularly in a large- scale wireless network setting. The multidimensional trade-offs among security strength, communication overhead, computation com- plexity, energy consumption, and scalability still remain largely unexplored. Developing effective evaluation methodology and toolkits will probably need interdisciplinary efforts from research communities working in wireless networking, mobile systems, and cryptography.

2. To identify different attack techniques and their consequences at different layers of the TCP/IP stack like sen (2010). A plan can be to investigate and determine the optimum number of clusters that maximizes the system performance.
3. Wasef et al.(2010) have presented a number of security mechanisms to complement the PKI security services for privacy, efficient authentication, and revocation. The presented solutions for location privacy can protect the location privacy of vehicles against outsiders. How to protect the location privacy of vehicles against legitimate insiders in traditional certificate-based PKI is still an open research issue. Also, mitigating the impact of DoS attacks launched by insider attackers is an important research issue as insider attackers can generate authentic signatures, and it will be difficult for other vehicles to detect this type of DoS attack. In addition, the problem of efficiently distributing the revocation information (e.g., CRL) to the vehicles in VANETs is a challenging issue as the network scale is very large. Another challenging topic is building a global reputation-based system while supporting privacy preservation because preserving the privacy of users requires frequent identity changes. Consequently, linking the reputation of a user to all its identities may contradict preserving the privacy of that user.
4. A paper by Friginal et al.(2010) has shown results issued from attack injection experiments can be useful with purposes that go beyond the mere assessment of protocol robustness. It has illustrated how experimental results can be exploited for i) the detection of implementation flaws that can lead protocols to exhibit vulnerabilities at runtime, ii) the benchmarking (comparison and selection) of the various protocol candidates to select the most suitable one for a particular solution, and finally iii) the fine tuning of the protocol configuration parameters to fulfill the performance and dependability requirements of each particular ad hoc solution. Other context of use where attack injection can be valuable include, but are not limited to, the validation of intrusion detection/tolerance mechanisms for ad hoc networks, and the collection of actual data to animate, and make more realistic, existing ad hoc network simulations. These are still open challenges requiring further research.

5. In a model, Xian et al (2010) appropriately limit the communication capability of the attacker, and introduce two important parameters such as the number of nodes controlled by the adversary and the knowledge owned by the adversary to evaluate the attack strength. In contrast to the old two models, the hierarchical threat model can analyze the security of routing protocol which runs in asynchronous and concurrent network environment and make it possible to develop a provable security routing for ad hoc network. This hierarchical threat model can also identify the minimum attacker capability which breaks a given protocol and can facilitate the security comparison between different protocols. Extensive future work remains to be done including further exploring integration of our threat model and formal analysis technique of secure routing protocol, analysis of additional existing protocols, and expression of the security properties of these protocols in our model for security analysis purpose. Especially, it is the future main work that developing provable security routing protocol for Ad hoc networks in our treat model.
6. In the paper by Umang et al.(2010), an enhanced intrusion detection algorithm for malicious node detection in ad hoc routing is proposed to protect against attacks. The proposed algorithm is implemented using AODV(Perkins,1999&2003). The simulation results have illustrated that performance of enhanced IDS approach is better than AODV routing in the presence of malicious nodes. Future studies can be extended to implement enhanced IDS for other routing protocols, by considering a lot of the energy consumption.

CONCLUSION

1. One distinguishing characteristics of MANETs from the security design perspective is the lack of a clear line of defense. The wireless channel is accessible to both legitimate network users and malicious attackers.
2. The research on MANET security is still in its early stage. The existing proposals are typically attack-oriented in that they first identify several security threats and then enhance the existing protocol or propose a new protocol to thwart such threats.
3. Unlike Wired networks, MANET pose a number of challenges to security solutions due to their unpredictable topology, wireless shared medium, heterogeneous resources and stringent resource constraint etc.
4. Security is not a single layer issue but a multilayered issue over here.
5. Confidentiality ensures the survivability of network services in case of denial of service attacks.
6. Integrity guarantees that a message being transferred is never corrupted. A message could be corrupted because of beginning failure because of malicious attacks.
7. Authentication guarantees every piece of routing information, which can always be confirmed to be valid and to have originated from the correct sender.
8. Non-repudiation means that the node cannot deny having sent or handled certain piece of routing information in the past. Authentication mechanisms allow partial non-repudiation.
9. Characteristics of MANET are Distributed operation, dynamic topologies, bandwidth constrained, variable capacity links, energy constrained operation, multihop communication, and limited physical security.
10. Routing problems in MANET are Distributed Network, Dynamic topology, Power aware-

ness, Addressing scheme, Cross Layer interactions, Network Size, Error Prone broadcast channel, Quality of Service(QoS) limitations and Security.

11. Routing in wireless ad hoc networks are vulnerable to traffic analysis, spoofing and denial of service attacks due to open wireless medium communications. Layer wise security is required for MANET.
12. Security issues in MANET are No predefined boundary, Adversary inside the network, No centralized control facility, Limited energy resource and Changing scale.
13. The protection of security always involves the identification of potential attacks, threats and vulnerabilities of a certain system. The vulnerabilities in MANET are the poor wireless channel security of the mobile nodes and the trust on any centralized resource.
14. The challenges to security in MANET are Shared radio channel, Insecure environment, Lack of central authority, Lack of permanent association, Limited resources, Physical vulnerability, Limited wireless transmission range, Broadcast nature of the wireless medium, Packet losses due to transmission errors, Mobility-induced route changes, Mobility-induced packet losses, Battery constraints, Potentially frequent network partitions, Ease of snooping on wireless transmissions (security hazard).
15. The security objectivess are described in six categories: Authentication, Access Control, Data Confidentiality, Data Integrity, Non-repudiation, and Availability Services.
16. For comprehensive security we need protocol security, communication security and physical security. To achieve above mentioned securities, many solutions have been proposed and are under research. Major solutions are for protocol security, authentication & key management schemes and trust management.

REFERENCES

Ahmed, S., & Alam, M. S. (2006). Performance evaluation of important ad hoc network protocols. *EURASIP Journal on Wireless Communications and Networking*, *2*, 42–53.

Argyroudis, P. G., & Mahony, D. O. (2005). Secure routing for mobile ad hoc networks. *IEEE Communications Surveys & Tutorials*, *7*(3), 2–21. doi:10.1109/COMST.2005.1610547

Cabri, G. (2008). The LAICA project: An ad-hoc middleware to support ambient intelligence. *Multiagent and Grid Systems Journal*, *2*(3), 235–247.

Cordasco, J., & Wetzel, S. (2009). An attacker model for MANET routing security. In *Proceeding of the Second ACM Conference on Wireless Network Security*, Zurich, Switzerland, (pp. 87-94).

Dawoud, D. S., Richard, L. G., Suliman, A., & Raja, S. V. K. (2011). *Trust establishment in mobile ad hoc networks: Key management*. Intech Publication.

Djenouri, D., Khelladi, L., & Badache, A. N. (2005). A survey of security issues in mobile ad hoc and sensor networks. *IEEE Communications Surveys & Tutorials*, *7*(4), 2–28. doi:10.1109/COMST.2005.1593277

Dolev, D., & Yao, A. (1983). On the security of public key protocols. *IEEE Transactions on Information Theory*, *29*(2), 198–208. doi:10.1109/TIT.1983.1056650

Dow, C. R., Lin, P. J., Chen, S. C., Lin, J. H., & Hwang, S. F. (2005). A study of recent research trends and experimental guidelines in mobile ad-hoc networks. *Proceedings of the IEEE 19th International Conference on Advanced Information Networking and Applications (AINA'05).*

Fisher, S. (1999, February 5). *Policy management: An introduction.* Gartner Group/Datapro Technical report.

Friginal, J., David, A., Ruiz, J., & Gil, P. (2010). Attack injection to support the evaluation of ad hoc networks. *29th IEEE International Symposium on Reliable Distributed Systems*, (pp. 21-29).

Gergely, A., Buttyan, L., & Vajda, I. (2006). Provably secure on-demand source routing in mobile ad hoc networks. *IEEE Transactions on Mobile Computing*, *5*(11), 1533–1546. doi:10.1109/TMC.2006.170

Hu, Y. C., & Perrig, A. (2004). A survey of secure wireless ad hoc routing. *IEEE Security & Privacy*, *2*(3), 28–39. doi:10.1109/MSP.2004.1

Ilyas, M. (2003). *The handbook of ad hoc wireless networks*. New York, NY: CRC Press.

Jai, K. E. H., & Kannammal, A. (2009). Privacy and security on anonymous routing protocols in MANET. *Proceedings of IEEE Second International Conference on Computer and Electrical Engineering*, (pp. 431-435).

Perkins, C. E., Belding-Royer, E. M., & Das, S. R. (2003, February 17). *Ad hoc on-demand distance vector (AODV) routing*. IETF Internet Draft- draft-ietf-manet-aodv-13.txt.

Perkins, C. E., & Royer, E. M. (1999). Ad hoc on-demand distance vector routing. *Proceedings of 2nd IEEE Workshop on Mobile Computing Systems and Applications* (pp. 90-100).

Sanzgiri, K., Laflamme, D., Dahill, B., Levine, B. N., Shields, C., & Belding-Royer, E. M. (2005). Authenticated routing for ad hoc networks. *Selected Areas in Communications*, *23*(3), 598–610. doi:10.1109/JSAC.2004.842547

Sen, J. (2010). An intrusion detection architecture for clustered wireless ad hoc networks. *Proceedings of IEEE Second International Conference on Computational Intelligence, Communication Systems and Networks*, (pp. 202-207).

Sheikh, R., Singh, C., & Mishra, D. K. (2010). Security issues in MANET: A review. *Proceedings of IEEE Seventh International Conference on Wireless and Optical Communications Networks* (WOCN), (pp. 1-5).

Todd, R. A., & Alec, Y. (2008). Adaptive threat modeling for secure ad hoc routing protocols. [ENTCS]. *Electronic Notes in Theoretical Computer Science*, *197*(2), 3–14. doi:10.1016/j.entcs.2007.12.013

Toh, C. K. (2007). *Ad hoc mobile wireless networks: Protocols and systems*. Prentice Hall.

Umang, S., Reddy, B. V. R., & Hoda, M. N. (2010). Enhanced intrusion detection system for malicious node detection in ad hoc routing protocols using minimal energy consumption. *IET Communications*, *4*(17), 2084–2094. doi:10.1049/iet-com.2009.0616

Valery, N., et al. (2006). An evaluation of inter-vehicle ad hoc networks based on realistic vehicular traces. In *International Symposium on Mobile Ad Hoc Networking & Computing*, (pp. 108–119).

Vijayaragavan, S. (2009). A performance study of reactive multicast routing protocols in virtual class room using mobile ad hoc network. *Journal of Computer Science*, *5*(11), 788–793. doi:10.3844/jcssp.2009.788.793

Wasef, A., Lin, X., & Shen, X. (2010). *Complementing public key infrastructure to secure vehicular ad hoc network* (pp. 22–28). IEEE Wireless Communication.

Xian, G., Tao, F., Ting, Y. Z., & Feng, M. J. (2010). The hierarchical threat model of routing security for wireless ad hoc networks. *Proceedings of IEEE First International Conference on Networking and Distributed Computing* (pp. 36-40).

Yang, H., Luo, H., Ye, F., Lu, S., & Zhang, L. (2004). Security in MANET: Challenges and solutions. *IEEE Wireless Communications*, *11*(1), 38–47. doi:10.1109/MWC.2004.1269716

Yang, P., & Zheng, S. (2001). Security management in hierachical ad hoc network. *Proceedings of IEEE International Conference on Info-tech and Info-net*, (pp. 642-649).

ADDITIONAL READING SECTION

Aggelou, G. (2005). *Mobile ad hoc networks from wireless LAN to 4G networks*. McGraw Hill Companies.

Azzedine, B. (2005). *Handbooks of algorithm for wireless networking and mobile computing*, 1st ed. Chapman and Hall/CRC Computer and Information Science series, Chapman and Hall/CRC.

Azzedine, B. (2008). *Algorithms and protocols for wireless mobile ad hoc network. Wiley series on Parallel and Distributed Computing*. Wiley-IEEE Press.

Basagni, S., Conti, M., Giordano, S., & Stojmenović, I. (2004). Mobile ad hoc networking. IEEE Press, Wiley-Interscience Publication.

Boban, M., & Vinhoza, T. V. (2011). *Modeling and simulation of vehicular networks: Towards realistic and efficient models*. Intech Publication.

Bulent, T., & Wendi, H. (2006). *Mobile ad hoc networks: Energy efficient real time data communications* (1st ed.). Springer.

Caballero-Gil, P. (2011). *Security issues in vehicular ad hoc networks*. Intech Publication.

De Rango, F., & Socievole, A. (2011). *Meta-heuristics techniques and swarm intelligence in mobile ad hoc networks*. Intech Publication.

Della Rosa, F., Leppäkoski, H., Ghalib, A., Ghazanfari, L., Garcia, O., Frattasi, S., & Nurmi, J. (2011). *Ad hoc networks for cooperative mobile positioning*. Intech Publication.

Forouzan, B. A. (2008). *Data communication and networking* (4th ed.). Tata McGraw Hill.

Galaviz Mosqueda, G. A., Santos, R. A., & Villaseñor González, L. A. (2011). *Survey on multi-hop vehicular ad hoc networks under IEEE 802.16 technology*. Intech publication.

Gonzalez-Cañete, F. J., & Casilari, E. (2011). *Impact of the mobility model on a cooperative caching scheme for mobile ad hoc networks*. Intech Publication.

Hekmat, R. (2002). *Ad hoc networks: Fundamental properties and network topologies*. Springer.

Hutchison, D., Mariani, J. A., & Shepherd, W. D. (1985). Local area networks: An advanced course. *Lecture Notes series in Computer Science, 184*.

Ilyas, M. (2003). *The handbook of ad hoc wireless networks*. New York, NY: CRC Press.

Ivan, S. (2002). *Handbook of wireless network and mobile computing* (1st ed.). Wiley-Interscience Publication.

Mishra, S., Issac, W., & Misra, S. C. (2010). *Guide to wireless ad hoc networks*. Springer Computer Communications and Network.

Rawashdeh, Z. Y., & Mahmud, S. M. (2011). *Communications in vehicular ad hoc networks*. Intech Publication.

Sharieh, S. (2011). *Ad-hoc networks as an enabler of brain spectroscopy*. Intech Publication.

Toh, C. K. (2007). *Ad hoc mobile wireless networks: Protocols and systems*. Prentice Hall.

Wang, X. (2011). *Mobile ad hoc network: Applications*. Intech Publication.

KEY TERMS AND DEFINITIONS

Asymmetric Link: A link with transmission characteristics which are different depending upon the relative position or design characteristics of the transmitter and the receiver of data on the link. For instance, the range of one transmitter may be much higher than the range of another transmitter on the same medium.

Bandwidth: The total capacity of a link to carry information.

Bandwidth Utilization: The actual amount of information delivered over a link,expressed as a percent of the available bandwidth on that link.

Base Station: A centralized node coordinating the channel access of a population of mobile nodes within its transmission range.

Beacon: A control message issued by a node (especially, a base station) informing all the other nodes in its neighborhood of the continuing presence of the node, possibly along with additional status information.

Channel: A subdivision of the physical medium allowing possibly shared independent uses of the medium. Channels may be made available by subdividing the medium into distinct time slots, or distinct spectral bands, or decorrelated coding sequences.

Channel Access Protocol: A protocol for mediating access to, and possibly allocation of, the various channels available within the physical communications medium. Nodes participating in the channel access protocol can communicate only when they have uncontested access to the medium, so that there will be no interference.

Cluster: A group of nodes located within close physical proximity, typically all within range of one another, which can be grouped together for the purpose of limiting the production and propogation of routing information.

Control Message: Information passed between two or more network nodes for maintaining protocol state which is not associated to any specific application.

Convergence: The process of approaching a state of equilibrium in which all nodes in the network agree on a consistent collection of state about the topology of the network, and in which no further control messages are needed to establish the consistency of the network topology.

Convergence Time: The time which is required for a network to reach convergence after an event (typically, the movement of a mobile node) which changes the network topology.

Distance Vector: A style of routing protocol in which, for each desired destination, a node maintains information about the distance to that destination, and a vector (next hop) towards that destination.

Fairness: A property of channel access protocols whereby a medium is made fairly equal to all eligible nodes on the link. Fairness does not strictly imply equality, especially in cases where nodes are given link access according to unequal priority or classification.

Flooding: It is the process of delivering data or control messages to every node within the ad hoc network.

Forwarding Node: It is a node within an ad hoc network which performs the function of forwarding datagram from one of its neighbors to another.

Goodput: The total bandwidth used, less the volume of control messages and protocol overhead from the data packets.

Hidden-Terminal Problem: The problem whereby a transmitting node can fail in its attempt to transmit data because of destructive interference which is only detectable at the receiving node, not the transmitting node.

Laydown: The relative physical location of the nodes within the ad hoc network.

Link: A physical medium which can sustain data communications between multiple network nodes.

Link State: A style of routing protocol in which every node within the network is expected to maintain information about every link within the network topology.

Link-level Acknowledgement: A protocol strategy, typically employed over wireless media, requiring neighbors to acknowledge receipt of packets (typically unicast only) from the transmitter. Such strategies aim to avoid packet loss or delay resulting from lack of, or unwanted characteristics of, higher level protocols.

Local Broadcast: The delivery of data to every node on a link (i.e., within range of the transmitter).

Loop-free: A property of routing protocols whereby the path taken by a data packet from source to destination never transits the same intermediate node twice before arrival at the destination.

MAC-layer Address: An address (sometimes called the link address) associated with the link interface of a node on a physical link.

Mobility Factor: The relative frequency of node movement, compared to the convergence time of the routing protocols used in the ad hoc network.

Neighborhood: All the nodes which can receive data on the same link from one node whenever it transmits data.

Next Hop: A neighbor which has been designated to forward packets along the way to a particular destination.

Pathloss: A reduction in signal strength caused by traversing the physical medium constituting the link.

Pathloss Matrix: A matrix of coefficients describing the pathloss between any two nodes in an ad hoc network. When the links are asymmetric, the matrix is also asymmetric.

Payload: The actual data within a packet, not including network protocol headers which were not inserted by an application.

Scalability: Wide applicability of a protocol to large as well as small populations of nodes participating in the protocol.

Scenario: The tuple <laydown, pathloss matrix, mobility factor, traffic> characterizing a class of ad hoc networks.

Signal Strength:The detectable power of the signal carrying the data bits: as seen by the receiver of the signal.

Source Route: A route to a source (i.e., a path with indications of intermediate forwarding nodes) made available to a receiver by the source of the data arriving at the receiver.

Spatial Re-use: Simultaneous use of channels with identical or close physical characteristics, but located spatially far enough apart to avoid interference (i.e., co-channel interference)

System-wide Broadcast: Same as flooding, but used in contrast to local broadcast.

Throughput: The amount of data from a source to a destination processed by the protocol for which throughput is to be measured for instance, IP, TCP, or the MAC protocol.

Chapter 16
Security Aware Routing Protocols for Mobile Ad hoc Networks

M.S. Ali
Prof Ram Meghe College of Engineering & Management, India

P.M. Jawandhiya
Jawaharlal Darda Institute of Engineering & Technology, India

ABSTRACT

An ad hoc network is the cooperative engagement of a collection of mobile nodes without the required intervention of any centralized access point or existing infrastructure. There is an increasing trend to adopt ad hoc networking for commercial user; however, their main applications lie in military, tactical, and other security-sensitive operations. In these and other applications of ad hoc networking, secure routing is an important issue. Designing a foolproof security protocol for ad hoc network is a challenging task due to its unique characteristics, such as lack of central authority, frequent topology changes, rapid node mobility, shared radio channel, and limited availability of resources. A number of protocols have been proposed for secure routing. However, most of these protocols are either proactive or reactive in approach. Both the approaches have their own limitations; for example, the proactive protocols use excess bandwidth in maintaining the routing information, while the reactive ones have long route request delay.

1.1 INTRODUCTION

We quote the definition of a mobile ad hoc network from the charter of the corresponding Internet Engineering Task Force (IETF) (Internet Engineering Task Force, www.ietf.org):

DOI: 10.4018/978-1-4666-0321-9.ch016

"A 'mobile ad hoc network' (MANET) is an autonomous system of mobile routers (and associated hosts) connected by wireless links-the union of which forms an arbitrary graph. The routers are free to move randomly and organize themselves arbitrarily; thus, the network's wireless topology may change rapidly and unpredictably. Such a network may operate in a standalone fashion, or may be connected to the larger Internet".

Mobile computing (Stefano Basagni, Marco Conti, Silvia Giordano, Ivan Stojmenovic, 2004) is proliferating as devices are becoming smaller, cheaper, and more powerful. By combining mobile devices with wireless communication abilities, the vision of being connected *anytime, anywhere, anyhow* will soon be a reality. New applications arise from mobile entities interacting and collaborating towards a common goal. With cellular phones being widely employed (Bin Xie, Anup Kumar, Dharma Agrawal and S. Srinivasan, 2006) and the mobile Internet emerging into the market place, concepts of dynamic wireless networks that do not depend on expensive infrastructure draws attention to the area of ad hoc networks. Mobile devices constitute a mobile ad hoc network when they directly and wirelessly communicate with other devices nearby without any fixed infrastructure. The mobile nodes move and thus the network topology changes dynamically and frequently. The absence of any hierarchy, established infrastructure, or centralized administration forces the nodes to control the network on their own.

Wireless mobile ad hoc nature of Mobile Ad hoc Network (MANET) brings new security challenges to network design. Mobile ad hoc networks, due to their unique characteristics, are generally more vulnerable to information and physical security threats than wired networks or infrastructure-based wireless networks. In this chapter, we explore the various security requirements (goals) for wireless ad hoc network and the different types of threats an ad hoc network faces. We identify the new challenges and opportunities posed by this new networking environment and explore new approaches to secure its communication. The ultimate goal of MANETs is to provide secure routing of data resources to mobile users at anytime and from anywhere. In conjunction with the existing routing protocols, providing security for MANETs gives rise to significant challenges and performance opportunities (P. Michiardi and R. Molva, 2002).

1.2 ROUTING IN MANETS

Routing in mobile ad hoc networks faces additional problems and challenges when compared to routing in traditional wired networks with fixed infrastructure. There are several well known protocols (A. K. Verma, Mayank Dave and R C Joshi,IJSCI) in the literature that have been specifically developed to cope with the limitations imposed by ad hoc networking environments. The problem of routing in such environments is aggravated by limiting factors such as rapidly changing topologies, high power consumption, low bandwidth and high error rates (E. M Royer, C. K Toh, 2007). Most of the existing routing protocols follow two different design approaches to confront the inherent characteristics of ad hoc networks, namely the table-driven and the source-initiated on-demand approaches.

The following sections analyze in more detail these two design approaches, and briefly present example protocols that are based on them.

1.2.1 Table-driven Ad hoc Routing Protocols

Table-driven ad hoc routing protocols maintain at all times routing information regarding the connectivity of every node to all other nodes that participate in the network. Also known as proactive, these protocols allow every node to have a clear and consistent view of the network topology by propagating periodic updates (E. M Royer et.al., 2007). Therefore, all nodes are able to make immediate decisions regarding the forwarding of a specific packet. On the other hand, the use of periodic routing messages has the effect of having a constant amount of signaling traffic in the network, totally independent of the actual data traffic and the topology changes. As an example of two protocols that follow the table-driven design approach we will briefly present the Destination-Sequenced Distance-Vector (DSDV) protocol (C. Perkin and P. Bhagwat, 1994) and the Optimized Link State Routing(OLSR) protocol (T. Clausen, G. Hansen, L. Christensen, and G. Behrmann, 2001).

1.2.1.1 Destination-Sequenced Distance-Vector Routing (DSDV)

DSDV is a table-driven routing protocol based on the Bellman-Ford algorithm (Dimitri P. Bertsekas and Robert G. Gallager, 1987). The DSDV protocol can be used in mobile ad hoc networking environments by assuming that each participating node acts as a router. Each node must maintain a table that consists of all the possible destinations. In more detail, an entry of the table contains the address identifier of a destination, the shortest known distance metric to that destination measured in hop counts and the address identifier of the node that is the first hop on the shortest path to the destination (C. Perkin et.al. 1994). Furthermore, the DSDV protocol adds a sequence number to each table entry assigned by the destination node, preventing the formation of routing loops caused by stale routes. The routing tables are maintained by periodically transmitted updates by each router to all the neighboring routers.

1.2.1.2 Optimized Link State Routing (OLSR)

The Optimized Link State Routing (OLSR) protocol is a proactive link state routing protocol based on the Open Shortest Path First (OSPF) protocol (J. Moy, 1998). OLSR has been specifically developed to support mobile ad hoc networks and the constraints they impose on routing. The OLSR protocol can be conceptually divided into three different operations, namely neighbor sensing, distribution of signaling traffic and distribution of topological information (T. Clausen et.al. 2001). Neighbor sensing in OLSR is accomplished by transmitting periodic hello messages that contain the generating node's address identifier, a list of its neighboring nodes and the type of the link it has with each neighbor (e.g.: symmetric or asymmetric). For the distribution of signaling traffic OLSR adopts a flooding mechanism where every node forwards a flooded message that it has not forwarded previously. Finally, the distribution of topological information function is realized with the use of periodic topology control messages that result in each node knowing a partial topology graph of the network which is then used for the computation of optimal routes (Rai Tirthraj, Verma A K, 2009 and E. M Royer et.al.,2007).

The Table 1 and Table 2 outline the basic characteristics and complexity of two routing protocols discussed above.

1.2.2 Source-Initiated On-Demand Ad hoc Routing Protocols

An alternative approach to the one followed by table-driven protocols is the source initiated on-demand routing. According to this approach a route is created only when the source node requires one to a specific destination. A route is acquired by the initiation of a route discovery function by the source node. The data packets transmitted while a route discovery is in process are buffered and are sent when the path is established. An established route is maintained as long as it is required through a route maintenance procedure. The Ad hoc On demand Distance Vector (AODV) routing protocol (C. E. Perkin and E. M. Royer, 2000) and the Dynamic Source Routing protocol (D. Johnson and D. Maltz, 1996) are examples of this category of protocols also known as reactive.

1.2.2.1 Ad hoc On-demand Distance Vector Routing (AODV)

The AODV protocol uses route request (RREQ) messages flooded through the network in order to

Table 1. Basic characteristics of DSDV and OLSR

Protocol	Routing Structure	Number of Table	Critical Nodes	Characteristics Feature
DSDV	Flat	2	No	Loop Free
OLSR	Flat	4	No	Loop Free

discover the paths required by a source node. An intermediate node that receives a RREQ replies to it using a route reply message only if it has a route to the destination whose corresponding destination sequence number is greater or equal to the one contained in the RREQ (C.E. Perkin et.al. 2000). This effectively means that an intermediate node replies to a RREQ only if it has a fresh enough route to the destination. Otherwise, an intermediate node broadcasts the RREQ packet to its neighbors until it reaches the destination. The destination unicast a RREP,back to the node that initiated the route discovery by transmitting it, to the neighbor from which it received the RREQ.

1.2.2.2 Dynamic Source Routing (DSR)

The Dynamic Source Routing (DSR) protocol is based on a method known as source routing (D. Johnson et.al., 1996). The route discovery process in DSR is similar to the one used by AODV, except that each intermediate node that broadcasts a route request packet adds its own address identifier to a list carried in the packet. The destination node generates a route reply message that includes the list of addresses received in the route request and transmits it back along this path to the source. Route maintenance in DSR is accomplished through the confirmations that nodes generate when they can verify that the next node successfully received a packet. These confirmations can be link-layer acknowledgements, passive acknowledgements or network-layer acknowledgements specified by the DSR protocol. When a node is not able to verify the successful reception of a packet it tries to retransmit it. When a finite number of retransmissions fail, the node generates a route error message that specifies the problematic link, transmitting it to the source node.

Tables 3 and 4 outline the basic characteristics and complexity of the two routing protocols discussed in this section.

1.3 SECURITY PROBLEMS WITH EXISTING AD HOC ROUTING PROTOCOLS

The main assumption of the previously presented ad hoc routing protocols is that all participating nodes do so in good faith and without maliciously disrupting the operation of the protocol. However, the existence of malicious entities cannot be disregarded in any system, especially in open ones like ad hoc networks. In ad hoc network the routing function can be disrupted by internal or external attackers. An internal attacker can be any legitimate participant of the routing protocol. An external attacker is defined as any other entity. Cryptographic solutions can be employed to pre-

Table 2. Complexity comparison of DSDV and OLSR

Protocol Convergence	Time	Memory Overhead	Control Overhead	Advantage/Disadvantage
DSDV	O(Diameter of network-1)	O(Number of node in the network)	O(Number of node in the network)	Loop Free/Higher overhead
OLSR	O(Greater than the diameter of routing tree)	O(Number of node in the network)	O(Number of node in the network)	Loop Free/ Low overhead than DSDV

Table 3. Basic characteristics of AODV and DSR

Protocol	Multiple Route	Route Metric Method	Route Maintenance In	Route Reconfiguration Strategy
AODV	No	Freshest and shortest path	Route Table	Erase Route then Source Notification or Local
DSR	Yes	Shortest path or next path available	Route Cache	Erase Route then Source Notification.

vent the impact of external attackers by mutual authentication of the participating nodes through digital signature schemes.

However, the underlying protocols should also be considered since an attacker could manipulate a lower level protocol to interrupt a security mechanism in a higher level. Internal attackers having capability to complete access the communication link they are able to advertise false routing information at will and force arbitrary routing decisions on their peers.

1.4 SECURITY GOALS

In providing a secure networking environment some or all of the following service may be required (M. G. Zapata and N. Asokan, 2002 and Y. C. Hu, A. Perrig, and D. B. Johnson, 2003).

1.4.1 Authentication

This service verifies the identity of node or a user, and to be able to prevent impersonation. In wired networks and infrastructure-based wireless networks, it is possible to implement a central authority at a point such as a router, base station, or access point. But there is no central authority in MANET, and it is much more difficult to authenticate an entity. Authentication can be providing using encryption along with cryptographic hash function, digital signature and certificates.

1.4.2 Confidentially

Keep the information sent unreadable to unauthorized users or nodes. MANET uses an open medium, so usually all nodes within the direct transmission range can obtain the data. One way to keep information confidential is to encrypt the

Table 4. Complexity comparison of AODV and DSR

Protocol	Time Complexity for Route Discovery	Time Complexity for Route Maintenance	Advantage	Disadvantage
AODV	O(2*Diameter of network)	O(2*Diameter of network)	Adaptable to highly dynamic topologies.	Scalability Problems and large Delays
DSR	O(2*Diameter of network)	O(2*Diameter of network)	Multiple Routes, Promiscuous overhearing	Scalability problem due to source routing and flooding.

data, and another technique is to use directional antennas. It also ensures that the transmitted data can only be accessed by the intended receivers.

1.4.3 Integrity

Ensure that the data has been not altered during transmission. The integrity service can be provided using cryptography hash function along with some form of encryption. When dealing with network security the integrity service is often provided implicitly by the authentication service.

1.4.4 Availability

Ensure that the intended network security services listed above are available to the intended parties when required. The availability is typically endure by redundancy, physical protection and other non-cryptographic means, e.g. use of robust protocol.

1.4.5 Non-repudiation

Ensure that parties can prove the transmission or reception of information by another party, i.e. a party cannot falsely deny having received or sent certain data. By producing a signature for the message, the entity cannot later deny the message. In public key cryptography, a node A signs the message using its private key. All other nodes can verify the signed message by using A's public key, and A cannot deny that its signature is attached to the message.

1.4.6 Access Control

To prevent unauthorized use of network services and system resources, access control is tied to authentication attributes. In general, access control is the most commonly thought of service in both network communications and individual computer systems.

1.5 ISSUES AND CHALLENGES IN SECURITY PROVISIONING

Designing a foolproof security protocol for ad hoc routing is a very challenging task due its unique characteristics such as, shared radio channel, insecure operational environment, lack of central authority and association rules among nodes and limited availability of resources (C. Siva Ram Murthy and B. S Manoj, 2004). A brief discussion on how each of the above mentioned characteristics causes difficulty in providing security in ad hoc wireless network is given below.

1.5.1 Shared Radio Channel

Unlike the wired networks where a separate dedicated transmission line can be provided between a pair of end users, the radio channel used for communication in ad hoc networks is broadcast in nature and shared by all nodes in the network. Data transmitted by a node is received by all the nodes within its direct transmission range. So a malicious node can easily obtain data being transmitted in the network.

1.5.2 Insecure operational environment

The operational environment in which MANETs are generally used may not be always secure, for example, a battle field. In such environment, nodes may move in and out of hostile and insecure enemy territory, where they would be highly vulnerable to security attacks.

1.5.3 Lack of Central Authority

In wired networks or infrastructure based wireless networks it would be possible to monitor the network traffic through routers or base stations and implement security mechanisms at those points. Since MANETs don't have any such central points, these mechanisms can't be applicable to them.

1.5.4 Lack of Association Rules

In MANET, since nodes can leave or join the network at any point of time, if no proper authentication mechanism is used for associating nodes with the network intruders can easily join the network and carry out attacks.

1.5.5 Limited Availability of Resources

Resources such as bandwidth, battery power and computational power are scare in ad hoc networks. Hence, it is difficult to implement complex cryptography-based security mechanisms in such networks.

1.6 SECURITY ATTACKS ON AD HOC ROUTING PROTOCOLS

1.6.1 Classification of Attacks

Malicious and selfish nodes are the ones that fabricate attacks against physical, link, network, and application-layer functionality. Current routing protocols are exposed to two types of attacks.

External attacks, in which the attacker aims to cause congestion, propagate fake routing information or disturb nodes from providing services. Internal attacks, in which the adversary wants to gain the normal access to the network and participate the network activities, either by some malicious impersonation to get the access to the network as a new node, or by directly compromising a current node and using it as a basis to conduct its malicious behaviors. The security attacks in MANET can be roughly classified into two major categories, namely passive attacks and active attacks are as described in the Figure 1 The active attacks further divided according to the layers.

1.6.1.1 Passive Attacks

A passive attack does not disrupt the normal operation of the network; the attacker snoops the data exchanged in the network without altering it. Here the requirement of confidentiality gets violated. Detection of passive attack is very difficult since the operation of the network itself doesn't get affected. One of the solutions to the problem is to use powerful encryption mechanism to encrypt the data being transmitted, there by making it impossible for the attacker to get useful information from the data overhead.

Eavesdropping

Eavesdropping is another kind of attack that usually happens in the mobile ad hoc networks. It aims to obtain some confidential information that should be kept secret during the communication. The information may include the location, public key, private key or even passwords of the nodes. Because such data are very important to the security state of the nodes, they should be kept away from the unauthorized access.

Traffic Analysis & Monitoring

Traffic analysis attack adversaries monitor packet transmission to infer important information such as a source, destination, and source-destination pair.

1.6.1.2 Active Attacks

An active attack attempts to alter or destroy the data being exchanged in the network there by disrupting the normal functioning of the network. Active attacks can be internal or external. External attacks are carried out by nodes that do not belong to the network. Internal attacks are from compromised nodes that are part of the network. Since the attacker is already part of the network, internal attacks are more severe and hard to detect than external attacks. Active attacks, whether carried out by an external advisory or an internal

Figure 1. Different types of security attacks

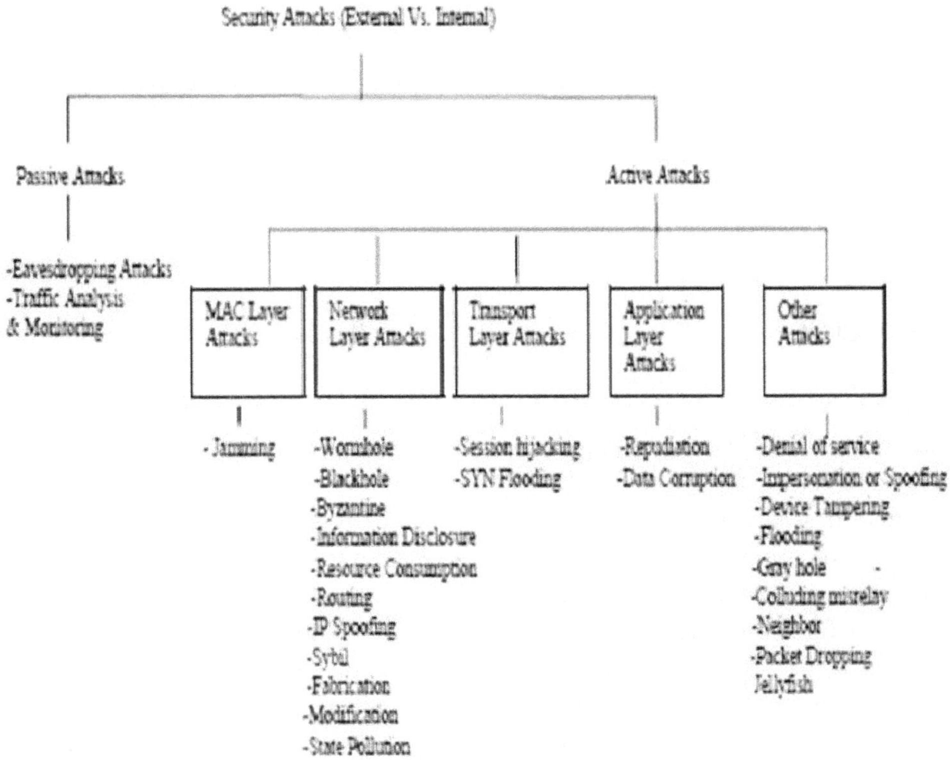

compromised node involves actions such as impersonation, modification, fabrication and replication. As shown in Figure 1.5 the active attacks.

MAC LAYER ATTACKS

Jamming Attack

Jamming is the particular class of DoS attacks. The objective of a jammer is to interfere with legitimate wireless communications. A jammer can achieve this goal by either preventing a real traffic source from sending out a packet, or by preventing the reception of legitimate packets.

NETWORK LAYER ATTACKS

Wormhole Attack

An attacker records packets at one location in the network and tunnels them to another location. Routing can be disrupted when routing control messages are tunneled. This tunnel between two colluding attackers is referred as a wormhole. Wormhole attacks are severe threats to MANET routing protocols. For example, when a wormhole attack is used against an on-demand routing protocol such as DSR or AODV, the attack could prevent the discovery of any routes other than through the wormhole.

Figure 2. Wormhole attack

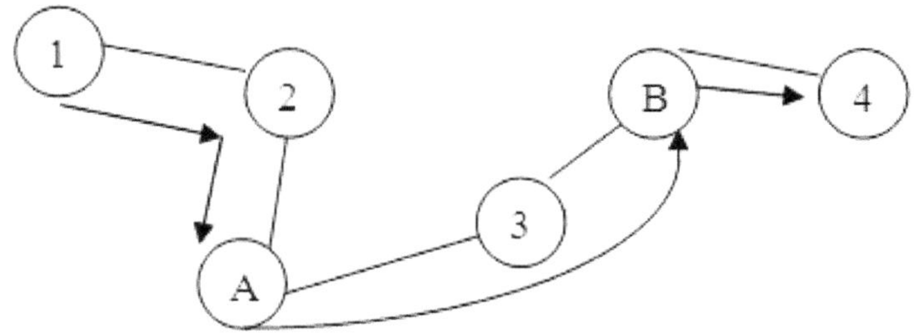

Black Hole Attack

The Black hole attack has two properties. First, the node exploits the mobile ad hoc routing protocol, such as AODV, to advertise itself as having a valid route to a destination node, even though the route is spurious, with the intention of intercepting packets. Second, the attacker consumes the intercepted packets without any forwarding. However, the attacker runs the risk that neighboring nodes will monitor and expose the ongoing attacks. There is a more subtle form of these attacks when an attacker selectively forwards packets. An attacker suppresses or modifies packets originating from some nodes, while leaving the data from the other nodes unaffected, which limits the suspicion of its wrongdoing.

Byzantine Attack

A compromised intermediate node works alone, or a set of compromised intermediate nodes works in collusion and carry out attacks such as creating routing loops, forwarding packets through non-optimal paths, or selectively dropping packets, which results in disruption or degradation of the routing services.

Routing Attacks

There are several types of attacks mounted on the routing protocol which are aimed at disrupting the operation of the network. Various attacks on the routing protocol are described briefly below:

1. *Routing Table Overflow:* In this attack, the attacker attempts to create routes to nonexistent nodes. The goal is to create enough routes to prevent new routes from being created or to overwhelm the protocol implementation. Proactive routing algorithms attempt to discover routing information even before it is needed, while a reactive algorithm creates a route only once it is needed. An attacker can simply send excessive route advertisements to the routers in a network. Reactive protocols, on the other hand, do not collect routing data in advance.
2. *Routing Table Poisoning:* Here, the compromised nodes in the networks send fictitious routing updates or modify genuine route update packets sent to other uncompromised nodes. Routing table poisoning may result in sub-optimal routing, congestion in portions of the network, or even make some parts of the network inaccessible.
3. *Packet Replication:* In this attack, an adversary node replicates stale packets. This consumes additional bandwidth and battery power resources available to the nodes and also causes unnecessary confusion in the routing process.
4. *Route Cache Poisoning*: In the case of on-demand routing protocols (such as the AODV protocol), each node maintains a route cache which holds information regarding routes that have become known to the node in the recent past. Similar to routing table poison-

ing, an adversary can also poison the route cache to achieve similar objectives.

5. *Rushing Attack:* On-demand routing protocols that use duplicate suppression during the route discovery process are vulnerable to this attack. An adversary node which receives a Route Request packet from the source node floods the packet quickly throughout the network before other nodes which also receive the same Route Request packet can react. Nodes that receive the legitimate Route Request packets assume those packets to be duplicates of the packet already received through the adversary node and hence discard those packets. Any route discovered by the source node would contain the adversary node as one of the intermediate nodes. Hence, the source node would not be able to find secure routes, that is, routes that do not include the adversary node. It is extremely difficult to detect such attacks in ad hoc wireless networks.

Resource Consumption Attack

This is also known as the sleep deprivation attack. An attacker or a compromised node can attempt to consume battery life by requesting excessive route discovery, or by forwarding unnecessary packets to the victim node.

IP Spoofing Attack

In conflict-detection allocation, the new node chooses a random address (say y) and broadcast a conflict detection packet throughout the MANET. Any veto from a node will prevent it from using this address. If the malicious node always impersonates a member that has occupied the same IP address and keeps replying with vetoes, it is called an IP Spoofing attack as illustrated in Figure 4.

In figure 1.5.3, N represents the new node, and M represents a malicious node. Node P is a neighbor of node M. Although node P may be aware that it has no direct neighbor with the address of y by means of a neighbor detection mechanism, it still thinks that the veto message is forwarded by node M from another node N'.

State Pollution Attack

If a malicious node gives incorrect parameters in reply, it is called the state pollution attack. For example, in best effort allocation, a malicious allocator can always give the new node an occupied address, which leads to repeated broadcast of Duplication Address Detection messages throughout the MANET and the rejection of new node.

Sybil Attack

If a malicious node impersonates some nonexistent nodes, it will appear as several malicious nodes conspiring together, which is called a Sybil attack. This attacks aims at network services when cooperation is necessary, and affects all the auto configuration schemes and secure allocation schemes based on trust model as well. However, there is no effective way to defeat Sybil attacks.

Fabrication

Instead of modifying or interrupting the existing routing packets in the networks, malicious nodes also could fabricate their own packets to cause chaos in the network operations. They could launch the message fabrication attacks by injecting huge packets into the networks such as in the sleep deprivation attacks. However, message fabrication attacks are not only launch by the malicious nodes. Such attacks also might come from the internal misbehaving nodes such as in the route salvaging attacks.

Modification

In a message modification attack, adversaries make some changes to the routing messages, and

Figure 4. IP spoofing attack

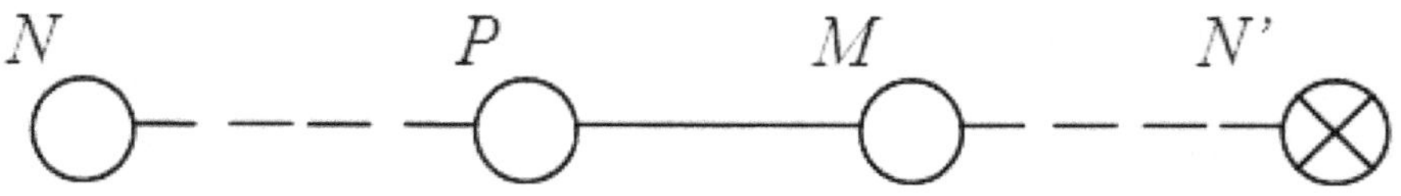

thus endanger the integrity of the packets in the networks. Since nodes in the ad hoc networks are free to move and self-organize, relationships among nodes at some times might include the malicious nodes. These malicious nodes might exploit the sporadic relationships in the network to participate in the packet forwarding process, and later launch the message modification attacks. Examples of attacks that can be classified under the message modification attacks are packet misrouting and impersonation attacks.

TRANSPORT LAYER ATTACKS

Session Hijacking Attack

Session hijacking takes advantage of the fact that most communications are protected (by providing credentials) at session setup, but not thereafter. In the TCP session hijacking attack, the attacker spoofs the victim's IP address, determines the correct sequence number that is expected by the target, and then performs a DoS attack on the victim. Thus the attacker impersonates the victim node and continues the session with the target.

SYN Flooding Attack

The SYN flooding attack is a denial-of-service attack. The attacker creates a large number of half-opened TCP connections with a victim node, but never completes the handshake to fully open the connection.

APPLICATION LAYER ATTACKS

Repudiation Attack

In the network layer, firewalls can be installed to keep packets in or keep packets out. In the transport layer, entire connections can be encrypted, end-to-end. But these solutions do not solve the authentication or non-repudiation problems in general. Repudiation refers to a denial of participation in all or part of the communications. For example, a selfish person could deny conducting an operation on a credit card purchase, or deny any on-line bank transaction, which is the prototypical repudiation attack on a commercial system.

OTHER ATTACKS

Denial of Service Attack

Denial of service (DoS) is another type of attack, where the attacker injects a large amount of junk packets into the network. These packets overspend a significant portion of network resources, and introduce wireless channel contention and network contention in the MANET. A routing table overflow attack and sleep deprivation attack are two other types of the DoS attacks. In the routing table overflow attack, an attacker attempts to create routes to non-existent nodes. Meanwhile the sleep deprivation attack aims to consume the batteries of a victim node. For example, consider the following Figure 3. Assume a shortest path exists from *S* to *X* and *C* and *X* cannot hear each other, that nodes *B* and *C* cannot hear each other, and that *M* is a malicious node attempting a denial of service attack. Suppose S wishes to communicate with X and that S has an unexpired route to X in its route cache. S transmits a data packet toward X with the source route S --> A --> B --> M --> C --> D --> X contained in the packet's header. When M receives the packet, it can alter the source route in the packet's header, such as deleting D from the source route. Consequently, when C receives the altered packet, it attempts to forward the packet to X. Since X cannot hear C, the transmission is unsuccessful.

Location Disclosure Attack

An attacker reveals information regarding the location of nodes or the structure of the network. It gathers the node location information, such as a route map, and then plans further attack scenarios. Traffic analysis, one of the subtlest security attacks against MANET, is unsolved. Adversaries try to figure out the identities of communication parties and analyze traffic to learn the network traffic pattern and track changes in the traffic pattern. The leakage of such information is devastating in security sensitive scenarios.

Flooding Attack

In flooding attack, attacker exhausts the network resources, such as bandwidth and to consume a node's resources, such as computational and battery power or to disrupt the routing operation to cause severe degradation in network performance. For example, in AODV protocol, a malicious node can send a large number of RREQs in a short period to a destination node that does not exist in the network. Because no one will reply to the RREQs, these RREQs will flood the whole network. As a result, all of the node battery power, as well as network bandwidth will be consumed and could lead to denial-of-service.

Impersonation or Spoofing Attack

Spoofing is a special case of integrity attacks whereby a compromised node impersonates a legitimate one due to the lack of authentication in the current ad hoc routing protocols. The main result of the spoofing attack is the misrepresentation of the network topology that may cause network loops or partitioning. Lack of integrity and authentication in routing protocols creates fabrication attacks that result in erroneous and bogus routing messages.

Colluding Misrelay Attack

In colluding misrelay attack, multiple attackers work in collusion to modify or drop routing packets to disrupt routing operation in a MANET. This attack is difficult to detect by using the conventional methods such as watchdog and pathrater. Figure 6 shows an example of this attack. Consider the case where node A1 forwards routing packets for

Figure 3. Black hole attack

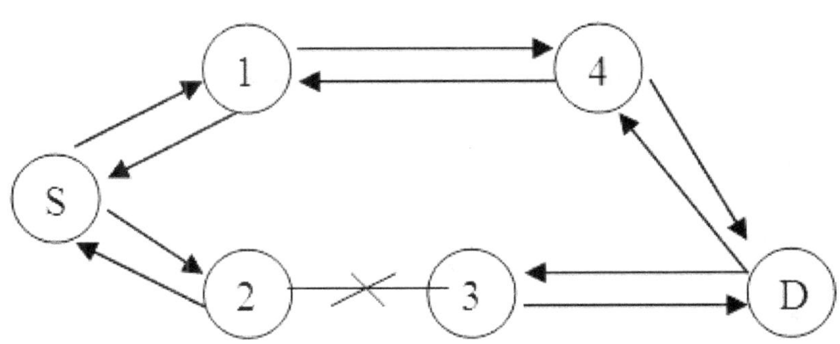

Figure 5. Denial of service attack

S ↔A↔ B↔ M ↔C↔ D↔ X

node T. In the figure, the first attacker A1 forwards routing packets as usual to avoid being detected by node T. However, the second attacker A2 drops or modifies these routing packets. This type of attack in OLSR protocol and show that a pair of malicious nodes can disrupt up to 100 percent of data packets in the OLSR MANET.

Device Tampering Attack

Unlike nodes in a wired network, nodes in ad hoc wireless networks are usually compact, soft, and hand-held in nature. They could get damaged or stolen easily. In the process of route discovery, control messages created by a node must be signed and validated by a receiving node. Thus the route discovery prevents anti-authenticating attacks, such as creating routing loop, fabrication because no node can create and sign a packet in the name of a spoofed or invented node. In the absence of centralized administration it is easy for MN's to change their identities.

Gray Hole Attack

We now describe the gray hole attack on MANETS. The gray hole attack has two phases. In the first phase, a malicious node exploits the AODV protocol to advertise itself as having a valid route to a destination node, with the intention of intercepting packets, even though the route is spurious. In the second phase, the node drops the intercepted packets with a certain probability. This attack is more difficult to detect than the black hole attack where the malicious node drops the received data packets with certainly. A gray hole may exhibit its malicious behavior in different ways. It may drop packets coming from (or destined to) certain specific node(s) in the network while forwarding all the packets for other nodes. Another type of gray hole node may behave maliciously for some time duration by dropping packets but may switch to normal behavior later. A gray hole may also exhibit a behavior which is a combination of the above two, thereby making its detection even more difficult.

Link Spoofing Attack

In a link spoofing attack, a malicious node advertises fake links with non-neighbors to disrupt routing operations. For example, in the OLSR protocol, an attacker can advertise a fake link with a target's two-hop neighbors. This causes

Figure 6. Colluding misrelay attack

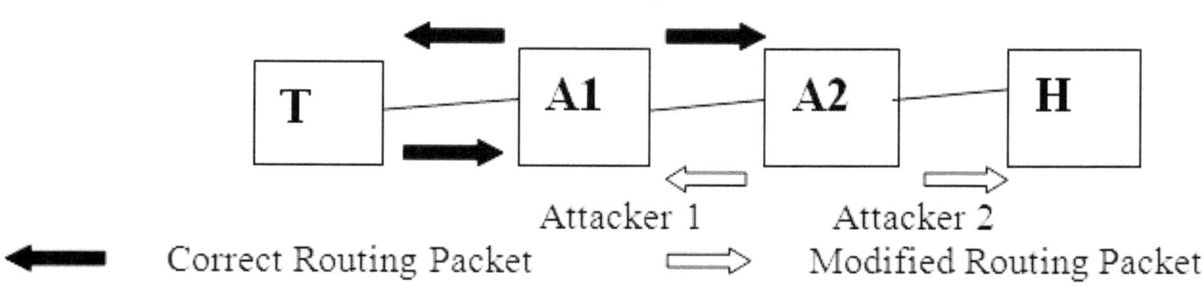

the target node to select the malicious node to be its MPR. As an MPR node, a malicious node can then manipulate data or routing traffic, for example, modifying or dropping the routing traffic or performing other types of DoS attacks.

Neighbor Attack

Upon receiving a packet, an intermediate node records its ID in the packet before forwarding the packet to the next node. However, if an attacker simply forwards the packet without recording its ID in the packet, it makes two nodes that are not within the communication range of each other believe that they are neighbors (i. e. one-hop away from each other), resulting in a disrupted route.

Jellyfish Attack

Similar to the black hole attack, a jellyfish attacker first needs to intrude into the forwarding group and then it delay data packets unnecessarily for some amount of time before forwarding them. This result in significantly high end-to-end delay and delay jitter, and thus degrades the performance of real time applications.

Packet Dropping Attacks

Direct interruption to the routing messages could be done by using the packet dropping attacks. In a standard packet dropping attack, an adversary collaborates as usual in the route discovery process and launches the constant packet dropping attacks if it is included as one of the intermediate nodes. In addition, instead of constantly dropping all the packets, adversaries might vary their techniques using random, selective, or periodic packet dropping attacks to help their interrupting behavior remain concealed.

Sleep Deprivation Torture

These kinds of attacks are most specific to wireless ad hoc networks, but may be encountered in conventional or wired networks as well. The idea behind this attack is to request the services a certain node offers, over and over again, so it can not go into an idle or power preserving state, thus depriving it of its sleep (hence the name). This can be very devastating to networks with nodes that have limited resources, for example battery power. It can also lead to constant business of the component, hindering other nodes to (legitimately) request services, data or information from the targeted entity.

1.7 SECURITY MECHANISM AND SOLUTIONS

Having seen the various kinds of attacks possible on ad hoc routing, we now look at various techniques employed to overcome these attacks.

Figure 7. Link spoofing attack

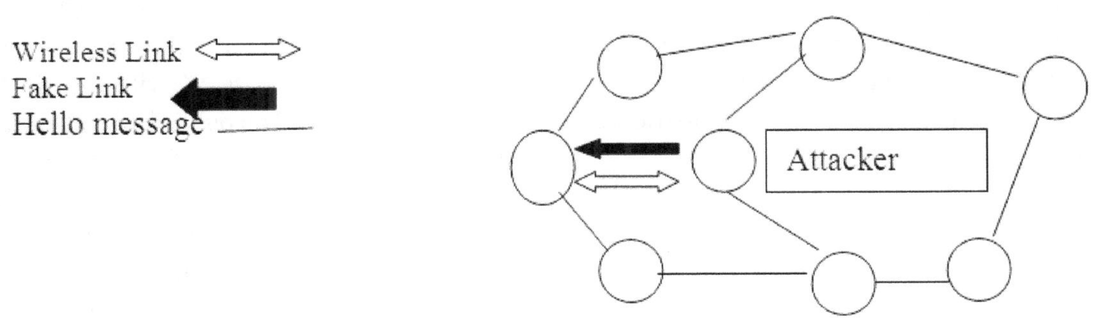

There can be two types of security mechanisms: preventive and detective. Preventive mechanisms are typically based on message encryption techniques, while detective mechanisms include the application of digital signature and cryptographic hash functions (Stefano Basagni, Macro Conti, Silvia Giordano and Ivan Stojmenovic, 2003).

1.7.1 Message Encryption

Encipherment or message encryption is the science and art of transforming a message into a disguised version which no unauthorized person can read, but which can be recovered in its original form by an intended recipient. In the parlance of cryptography, the original message is called *plaintext* and the secret version of the message is called *cipher text*. The plaintext is converted into cipher text by the process of *encryption*, that is, by the use of certain algorithms or functions. The reverse process is called *decryption*. The process of encryption and decryption are governed by *keys*, which are small amount of information used by the cryptographic algorithms.

There are two types of encryption techniques: symmetric key and asymmetric key (or public key). Symmetric key cryptosystem uses the same key (the secret key) for encryption and decryption of a message, where as asymmetric key cryptosystems use one key (the public key) to encrypt a message and another key (the private key) to decrypt it. Public and private keys are related in such a way that only the public key can be used to encrypt messages and only the corresponding private key can be used for decryption purpose. Even if attacker comprises a public key, it is virtually impossible to deduce the private key. Symmetric key algorithms are usually faster to execute electronically than the asymmetric key algorithms.

1.7.2 Digital Signature and Hashing

The process of encryption only ensures the confidentiality of the message being sent. Digital signature is a technique by which one can achieve the other security goals like message integrity, authentication and non-repudiation. In this the sender uses a *signing algorithm* and its *private key to sign* the message. The message and the signature are sent to the receiver. The receiver receives the message and the signature and applies the *verifying algorithm* on the message-signature pair. The verification algorithm requires a *verification key,* which is a public key provided by the signer, to verify the document. After verification if the result is true, the message is accepted; otherwise, it is rejected. Hashing can be used for the digital signature process especially when the message is long. In this the message is passed through an algorithm called *cryptographic hash function* or *one-way hash function* before signing. It is an algorithm which creates a compressed image of the message in the form of a hash value (or message digest) which is usually much smaller than the message and unique to it. Any change to the message will produce a different hash result even when the same hash function is used.

1.7.3 Key Management Approaches

Both digital signature and encryption mechanisms are key-based approaches. Key distribution and management is therefore at the center of these mechanisms. There are several methods given in (C. Siva Ram Murthy et.al., 2004) that can be employed to perform this operation, all requiring varying amounts of initial configuration, communication and computation. We will however, focus on the method based on public key certificates (J. J. Tardo and K. Algappan, 1991), as we have used this approach in the proposed protocol.

According to this approach, the key management responsibility is shared among a set of trusted certification servers called the certification authorities (CAs). Each CA has a public/private key pair, with its public key known to every node, and signs certificates binding public keys to nodes after verifying their authenticity secretly. The trusted CA has to stay on-line to reflect the current bindings, because the bindings could change over time: a public key should be revoked if the owner node is no longer trusted or is out of the network; a node may refresh its key pair periodically to reduce the chance of a successful brute-force attack on its private key.

1.8 SECURE AD HOC ROUTING

1.8.1 Symmetric Key Cryptography Solutions

SEAD

Hu et al. presented a proactive secure routing protocol based on the Destination- Sequenced Distance Vector protocol (DSDV)(C. Perkin et.al. 1994). In a proactive (or periodic) routing protocol nodes periodically exchange routing information with other nodes in attempt to have each node always know a current route to all destinations.

I. Brief Operation

SEAD (Y-C Hu, D. B. Johnson, and A. Perrig) authenticates the sequence number and metric of a routing table update message using hash chains elements. In addition, the receiver of SEAD routing information also authenticates the sender, ensuring that the routing information originates form the correct node. The source of each routing update message in SEAD must also be authenticated, since otherwise, an attacker may be able to create routing loops through the impersonation attack.

II. Features

- SEAD deals with attackers that modify routing information broadcasted during the update phase of the DSDV-SQ protocol: in particular, routing can be disrupted if the attacker modifies the sequence number and the metric field of a routing table update message.
- SEAD makes use of efficient one-way hash chains rather than relaying on expensive asymmetric cryptography operations.
- SEAD assumes some mechanism for a node to distribute an authentic element of the hash chain that can be used to authenticate all the other elements of the chain.
- SEAD does not cope with wormhole attacks.

SRP

The Secure Routing Protocol (SRP) (A. K. Verma, Mayank Dave and R C Joshi, IJSCI and P. Papadimitratos and Z. Haas,2002) was designed as an extension compatible with a variety of existing reactive routing protocols. SRP combats attacks that disrupt the route discovery process and guarantees the acquisition of correct topological information.

I. Brief Operation

SRP allows the initiator of a route discovery to detect and discard bogus replies. SRP relies on the availability of a security association (SA) between the source node (S) and the destination node (T). The SA could be established using a hybrid key distribution based on the public keys of the communicating parties. S and T can exchange a secret symmetric key (KS, T) using the public keys of one another to establish a secure channel. S and T can then further proceed to mutual authentication of one another and the authentication of routing messages.

II. Features

- SRP copes with non-colluding malicious nodes that are able to modify (corrupt), replay and fabricate routing packets.
- Assuming that the neighbor discovery mechanism maintains information on the binding of the medium access control and the IP addresses of nodes, SRP is proven to be essentially immune to IP spoofing.
- In case of the Dynamic Source Routing (DSR) protocol (A. Perrig, R. Canetti, J. D. Tygar, and D. Song, 2000), SRP requires including a 6-word header containing unique identifiers that tag the discovery process and a message authentication code (MAC) computed using a keyed hash algorithm.
- The basic version of SRP suffers from the route cache poisoning attack.
- SRP suffers from the lack of a validation mechanism for route maintenance messages
- SRP is not immune to the wormhole attack: two colluding malicious nodes can misroute the routing packets on a private network connection and alter the perception of the network topology by legitimate nodes.

ARIADNE

ARIADNE (Y.-C. Hu, A. Perrig, and D.B. Johnson, 2002) is an on-demand secure ad hoc routing protocol based on DSR that withstands node compromise and relies only on highly efficient symmetric cryptography. ARIADNE guarantees that the target node of a route discovery process can authenticate the initiator, that the initiator can authenticate each intermediate node on the path to the destination present in the RREP message and that no intermediate node can remove a previous node in the node list in the RREQ or RREP messages.

I. Brief Operation

As for the SRP (Asad Amir Prizada and Chris McDonald, 2003) protocol, ARIADNE needs some mechanism to bootstrap authentic keys required by the protocol. In particular, each node needs a shared secret key (KS, D) is the shared key between a source S and a destination D) with each node it communicates with at a higher layer, an authentic TESLA (A. Perrig, R. Canetti, 2000) key for each node in the network and an authentic "Route Discovery chain" element for each node for which this node will forward RREQ messages. Various terminologies used are depicted in Figure 8.

II. Features

i. ARIADNE provides point-to-point authentication of a routing message using a message authentication code (MAC) (A. Perrig, Y-C Hu, and D. B. Johnson) and a shared key between the two parties.
ii. For authentication of a broadcast packet such as RREQ, ARIADNE uses the TESLA broadcast authentication protocol.
iii. Selfish nodes are not taken into account.
iv. ARIADNE copes with attacks performed by malicious nodes that modify and fabricate routing information, with attacks using impersonation and, in an advanced version, with the wormhole attack
v. ARIADNE is protected also from a flood of RREQ packets that could lead to the cache poisoning attack (Wensheng Zhang, Guohong Cao, 2007).
vi. ARIADNE is immune to the wormhole attack (Y-C Hu, D. B. Johnson, and A.

Perrig) only in its advanced version: using an extension called TIK (TESLA with Instant Key disclosure) that requires tight clock synchronization between the nodes, it is possible to detect anomalies caused by a wormhole based on timing discrepancies.

1.8.2 Asymmetric Key Cryptography Solutions

ARAN

The ARAN (B. Dahill, B. N. Levine, E. Royer, and C. Shields, 2002) secure routing protocol is an on-demand routing protocol that detects and protects against malicious actions carried out by third parties and peers in the ad hoc environment. ARAN introduces authentication, message integrity and non-repudiation as part of minimal security policy for the ad hoc environment and consists of a preliminary certification process, a mandatory end-to-end authentication stage and an optional second stage that provides secure shortest paths.

I. Brief Operation

ARAN requires the use of a trusted certificate server (T): before entering in the ad hoc network, each node has to request a certificate signed by T. The certificate contains the IP address of the node, its public key, a timestamp of when the certificate was created and a time at which the certificate expires along with the signature by T. All nodes are supposed to maintain fresh certificates with the trusted server and must know T's public key.

The goal of the first stage of the ARAN protocol is for the source to verify that the intended destination was reached. As with any secure system based on cryptographic certificates, the key revocation issue has to be addressed in order to make sure that expired or revoked certificates do not allow the holder to access the network. In ARAN, when a certificate needs to be revoked, the trusted certificate server T sends a broadcast message to the ad hoc group that announces the revocation. Any node receiving this message re-broadcasts it to its neighbors. Revocation notices need to be stored until the revoked certificate would have expired normally. Any neighbor of the node with the revoked certificate needs to reform routing as necessary to avoid transmission through the now un-trusted node. This method is not failsafe.

In some cases, the un-trusted node that is having its certificate revoked may be the sole connection between two parts of the ad hoc network. In this case, the non trusted node might not forward the notice of revocation for its certificate, resulting in a partition of the network, as nodes that have received the revocation notice will no longer forward messages through the un-trusted node, while all other nodes depend on it to reach the rest of the network. This only lasts as long as the un trusted node's certificate would have otherwise been valid, or until the un-trusted node is no longer the sole connection between the two partitions. At the time that the revoked certificate should have expired, the un-trusted node is unable to renew the certificate, and routing across that node ceases. Additionally, to detect this situation and to hasten the propagation of revocation notices, when a node meets a new neighbor, it can exchange a summary of its revocation notices with that neighbor; if these summaries do not match, the actual signed notices can be forwarded and re-broadcasted to restart propagation of the notice.

II. Features

- The ARAN protocol protects against exploits using modification, fabrication and impersonation
- The ARAN protocol uses of asymmetric cryptography makes it a very costly pro-

Figure 8. Route discovery in ARIADNE, along with terminologies used

Term	Meaning	Term	Meaning
Brdcast	*Broadcast*	K_{AB} **and** K_{BA}	*MAC keys shared between A and B*
$MAC_{KAB}(M)$	*computation of the message authentication code (MAC) of message M with the MAC key KAB*	***REQ***	*REQuest*
REP	*REPly*	H_n	*One way hash function executed n times.*
id	*Unique id*	t_i	*time interval*
S	*Initiator node*	***D***	*Target node*
		M_A	*MAC address of node A*

$S: \quad h_0 = MAC_{KSD}(REQ, S, D, id, t_i)$

$S \rightarrow * : < REQ, S, D, id, ti, h0, (), () >$

$A: \quad h_1 = H[A, h_0]$

$M_A = MAC_{KAti}(REQ, S, D, id, t_i, h_1, (A), ())$

$A \rightarrow * : < REQ, S, D, id, ti, \underline{h_1}, (\underline{A}), (\underline{M_A}) >$

$B: \quad h_2 = H[B, h_1]$

$M_B = MAC_{KBti}(REQ, S, D, id, t_i, h_2, (A,B), (M_A))$

$B \rightarrow * : < REQ, S, D, id, ti, \underline{h_2}, (A, \underline{B}), (M_A, \underline{M_B}) >$

$C: \quad h_3 = H[C, h_2]$

$M_C = MAC_{KCti}(REQ, S, D, id, t_i, h_3, (A,B,C), (M_A, M_B))$

$C \rightarrow * : < REQ, S, D, id, ti, \underline{h_3}, (A, B, \underline{C}), (M_A, M_B, \underline{M_c}) >$

$D: \quad MD = MAC_{KDS}(REP, D, S, ti, (A,B,C), (M_A, M_B, M_c))$

$D \rightarrow C : < REP, D, S, ti, (A,B,C), (M_A, M_B, M_c), \underline{M_D}, () >$

$C \rightarrow B : < REP, D, S, ti, (A,B,C), (M_A, M_B, M_c), M_D, (\underline{K_{Cti}}) >$

$B \rightarrow A : < REP, D, S, ti, (A,B,C), (M_A, M_B, M_c), M_D, (K_{Cti}, \underline{K_{Bti}}) >$

$A \rightarrow S : < REP, D, S, ti, (A,B,C), (M_A, M_B, M_c), M_D, (K_{Cti}\ K_{Bti}, \underline{K_{Ati}}) >$

tocol to use in terms of CPU and energy usage.
- ARAN is not immune to the wormhole attack

SAR

Security-Aware Ad-Hoc Routing (SAR) (C. E. Perkins, 2000 and *S. Yi, P. Naldurg, R. Kravets*,2002) is the generalized framework for any on-demand ad-hoc routing protocol. SAR requires that nodes having same trust level must share a secret key. SAR augments the routing process using hash digests and symmetric encryption mechanisms. The signed hash digests provide message integrity while the encryption of packets ensures their confidentiality.

I. Brief Operation

SAR when implemented on AODV protocol adds two additional fields to the ROUTE REQUEST packet and one additional to the ROUTE REPLY packet. The first field added to the ROUTE REQUESTPACKET is the security requirement field and is set by the sender .It indicates the preferred level of trust for the path to the destination. The Second field added to is the security guarantee that signifies the maximum level of security provided by the discovered paths. If the security requirement field has an integer representation then the security guarantee field will be minimum of all security levels of the participating nodes in the path. If the security requirement field is represented in vectors then the security guarantee field value id computed by ANDing the security requirement values of the participating nodes in the path. The value thus computed is copied into additional security guarantee field of the ROUTE REPLY packet and sent back to the sender. This value is also copied into the routing table of nodes in the reverse path, to preserve the security information with reference to cashed paths

II. Features

- SAR uses security information to dynamically control the choice of routes installed in the routing table.
- SAR enables applications to selectively implement a subset of security services based on the cost-benefit analysis.
- The routes discovered by SAR may not always be the shortest between any two communicating entities in terms of hop-count. However these routes have quantifiable guarantee of the security.
- SAR will find the optimal route if all the nodes on the shortest path satisfy the security requirements.
- SAR may fail to find the route if the ad hoc network does not have a path on which all nodes on the path satisfy the security requirements in spite of being connected.

1.8.3 Hybrid Solutions

SAODV

The Secure Ad hoc On Demand distance Vector (SAODV) (M. Zapata, 2002 and Zapata, M. G., 2001) protocol is an extension of the AODV protocol. The Secure AODV scheme is based on the assumption that each node possesses certified public keys of all network nodes.

I. Brief Operation

The originator of the routing control packet appends its RSA signature and the last element of a hash chain to the routing packets. A packet transverse the network, intermediate nodes cryptographically authenticates the signature and the hash value. The intermediate nodes generate the kth element of the hash chain, with k being the number of transverse hops, and place it in packet. The SAODV protocol gives two alternatives for ROUTE REQUEST and ROUTE REPLY mes-

Figure 9. Route discovery in ARAN, along with terminologies used

K_{A+}	*Public-key of node A.*	N_A	*Nonce issued by node A.*
K_{A-}	*Private-key of node A.*	IP_A	*IP address of node A.*
$\{d\}K_{A+}$	*Encryption of data d with key KA+.*	**RDP**	*Route Discovery Packet identifier.*
$\{d\}K_{A-}$	*Data d digitally signed by node A.*	**REP**	*REPly packet identifier.*
$cert_A$	*Certificate belonging to node A.*	**RREQ**	*Route REQuest*
T	*Timestamp.*	**RREP**	*Route REPly*
E	*Certificate expiration time.*	**ERR**	*ERRor packet identifier.*
brdcast	*Broadcast*		

$A \rightarrow brdcast$: $[RDP, IP_x, cert_A, N_A, t]\ K_{A-}$

$B \rightarrow brdcast$: $[[RDP, IP_x, cert_A, N_A, t]\ K_{A-}]\ K_{B-}, cert_B$

$C \rightarrow brdcast$: $[[RDP, IP_x, cert_A, N_A, t]\ K_{A-}]\ K_{c-}, cert_c$

$X \rightarrow D$: $[REP, IP_a, cert_x, N_A, t]\ K_{x-}$

$D \rightarrow C$: $[[REP, IP_a, cert_x, N_A, t]\ K_{x-}]\ K_{D-}, cert_D$

$C \rightarrow B$: $[[REP, IP_a, cert_x, N_A, t]\ K_{x-}]\ K_{c-}, cert_c$

sages. In the first case when a ROUTE REQUEST is sent, the sender creates a signature and appends it to packet. Intermediate nodes authenticate the signature before creating or updating the reverse route to the host. The reverse rout is stored only when the signature is verified. When the node reaches the destination, the node signs the ROUTE REPLY with its private key and sends it back. The intermediate nodes again verify the signature. The signature of the sender is again stored with the along with the route entry.

II. Features

- Ownership of certified public keys enables intermediate enable intermediate nodes to authenticate all in-transit routing packets.
- The protocol operates mainly by using the new extension message with the AODV protocol.
- The SAODV can be used to protect the route discovery mechanism of the AODV by providing security features like integrity, authentication and non-repudiation.

Figure 10. Route discovery in SAODV, along with terminologies used

RREQ	*Route Request*	***brdcast***	*Broadcast*
Seq_s	*Sequence No issued by s*	K_{A-}	*Private-key of node A*
$\{d\}K_{A-}$	*Data d digitally signed by node A*	Seq_A	*Sequence number issued by A.*
S	*Initiator node*	***D***	*Target node*
h_n	*Hash chain function operated n times.*	***brdcast***	*Broadcast*
RREP	*Route Reply*	$oldseq_A$	*Old sequence number issued by A*
Id	*Unique Id for the packet*	***N***	*Nonce*

$S \rightarrow brdcast$: $((RREQ, id, S, seq_s, D, oldseq_D, h_0, N)_{KS-}, 0, h_N)$
$A \rightarrow brdcast$: $((RREQ, id, S, seq_s, D, oldseq_D, h_0, N)_{KS-}, 1, h_{N-1})$
$B \rightarrow brdcast$: $((RREQ, id, S, seq_s, D, oldseq_D, h_0, N)_{KS-}, 2, h_{N-2})$
$C \rightarrow brdcast$: $((RREQ, id, S, seq_s, D, oldseq_D, h_0, N)_{KS-}, 3, h_{N-3})$
$D \rightarrow C$: $((RREP, D, seq_D, S, lifetime, h'_0, N)_{KD-}, 0, h'_N)$
$C \rightarrow B$: $((RREP, D, seq_D, S, lifetime, h'_0, N)_{KD-}, 1, h'_{N-1})$
$B \rightarrow A$: $((RREP, D, seq_D, S, lifetime, h'_0, N)_{KD-}, 2, h'_{N-2})$
$A \rightarrow S$: $((RREP, D, seq_D, S, lifetime, h'_0, N)_{KD-}, 3, h'_{N-3})$

1.9 FUTURE RESEARCH DIRECTIONS

There are many research problems that must be solved to support the Mobile Ad Hoc Networks. Solutions to these problems are needed both in terms of proactive and reactive approaches. The solution should comprise of all three components prevention, detection and reaction.

Following security issues are to be handled in Mobile Ad Hoc Networks.

- **Secure Multicasting**
  - Communicating group keys to dynamic subsets/clusters of nodes in MANET's
  - Architectures that off-load heavy computation to more powerful devices are needed.
- **Secure Routing**
  - Most MANET routing protocols are vulnerable to attacks that can freeze the whole network.
  - Need solutions that work even if some nodes have been compromised.
- **Privacy-aware Routing**

- Building routing protocols that prevent intermediate nodes from performing traffic analysis.
- Schemes for minimizing size of crypto-tags signatures, MACs) are needed.

- **Group Membership Control**
  - Secure admission of members to a group while tolerating adversaries from both outside inside.
  - Use of distributed cryptography in MANETs.
- **Key Distribution**
  - Combining key pre-distribution with secret sharing to achieve key distribution in MANETs.
  - Need a secure and efficient key-distribution mechanism allowing simple key establishment for large-scale sensor networks.
- **Intrusion detection and Response Systems**
  - Anomaly detection is difficult in MANETs.
  - Collaborative IDS schemes are needed.
- **Denial of Service Attacks**
  - Needs better solutions than Client Puzzles.

1.10 CONCLUSION

Due to the rapidly changing telecommunication industry and the increasing popularity of wireless networks, there has been a great deal of concern about security in wireless and mobile telecommunication systems.

Of the five areas of network management – configuration, failures, performance, accounting and security – the last area has not received its fair share of attention.

In this chapter, we have introduced MANET, its features and characteristics. We have considered the routing approaches in mobile ad hoc networks from the security point of view. We have analyzed the threats against ad hoc routing protocols and presented the requirements that need to be addressed for secure routing. Existing secure routing protocols for MANET are either proactive or reactive in nature; hence are limited in their approach in terms of providing security across diverse network applications. We explored the advantages of hybrid routing in dealing with these limitations, where the proactive and the reactive behavior is mixed in the amounts that best match these operational conditions.

REFERENCES

Aad, I., Hubaux, J. P., & Knightly, E. W. (2004). Denial of service resilience in ad hoc networks. *Proceedings of ACM MobiCom 2004*, Philadelphia, PA, September 2004, (pp. 202-215).

Banerjee, S. (2008). Detection/removal of cooperative black and gray hole attack in mobile ad-hoc networks. *Proceedings of the World Congress on Engineering and Computer Science 2008 WCECS 2008*, October 22 - 24, 2008, San Francisco, USA.

Basagni, S., Conti, M., Giordano, S., & Stojmenovic, I. (Eds.). (2004). *Mobile ad hoc networking*. IEEE Press, Wiley Interscience. doi:10.1002/0471656895

Basagni, S., Conti, M., Giordano, S., & Stojmenovic, S. (2003). *Mobile ad hoc networks*. IEEE Press, John Wily & Sons, Inc.

Bertsekas, D. P., & Gallager, R. G. (1987). Distributed asynchronous Bellman- Ford algorithm . In *Data networks* (pp. 325–333). Englewood Cliffs, NJ: Prentice Hall.

Buchegger, S., & Le Boudec, J.-Y. (2002). *Cooperative routing in mobile ad hoc networks: Current efforts against malice and selfishness*. In Lecture Notes on Informatics, Mobile Internet Workshop, Germany, October 2002.Springer.

Clausen, T., Hansen, G., Christensen, L., & Behrmann, G. (2001). The optimized link state routing protocol. *Proceedings of the 4th International Symposium on Wireless Personal Multimedia Communications,* Aalborg, Denmark, September 2001.

Dahill, B., Levine, B. N., Royer, E., & Shields, C. (2002). *ARAN: A secure routing protocol for ad hoc networks*. UMass Tech Report 02-32.

Douceur, J. R. (2002). The Sybil attack. In *Proceedings of 1st International Workshop on Peer-to-Peer Systems, LNCS 2429*, (pp. 251-260). March 2002.

Garg, N., & Mahapatra, R. P. (2009). MANET security issues. *International Journal of Computer Science and Network Security, 9*(8).

Hu, Y. C., Johnson, D. B., & Perrig, A. (2003). SEAD: Secure efficient distance vector routing for mobile wireless ad hoc networks. In *Proceedings of 4th IEEE Workshop on Mobile Computing Systems and Applications.*

Hu, Y. C., Perrig, A., & Johnson, D. B. (2002). Ariadne: A secure on-demand routing protocol for ad hoc networks. In *Proceedings of 8th ACM International Conference on Mobile Computing and Networking (Mobicom'02)*, Atlanta, Georgia, (pp. 12-23).

Hu, Y. C., Perrig, A., & Johnson, D. B. (2002). *Wormhole detection in wireless ad hoc networks*. Tech. Rep. TR01-384, Department of Computer Science, Rice University, June 2002.

Hu, Y. C., Perrig, A., & Johnson, D. B. (2003). Packet leashes: A defense against wormhole attacks in wireless ad hoc network. *Proceedings of the 22nd Annual Joint Conf. IEEE Computer and Communication Societies,* San Francisco, CA, April 2003.

IETF MANET Working Group. (n.d.). *Mobile ad hoc networks (MANET).* Working Group. Retrieved from http://www.ietf.org/html.charters/manet-charter.html

Ilyas, M. (2003). *The handbook of ad hoc wireless networks*. CRC Press.

Internet Engineering Task Force (IETF). (n.d.). *Website*. Retrieved from www.ietf.org

Jawandhiya, P. M., Ghonge, M. M., Ali, M. S., & Deshpande, J. S. (2010). A survey of mobile ad hoc network attacks. *International Journal of Engineering Science & Technology, 2*(9), 4063-4071. ISSN 0975-5462

Johnson, D., & Maltz, D. (1996). Dynamic source routing in ad hoc wireless network . In Imielinski, T., & Korth, H. (Eds.), *Mobile computing*. Kluwer. doi:10.1007/978-0-585-29603-6_5

Lan, H., & Nguyen, U. T. (2006). Study of different types of attacks on multicast in mobile ad hoc networks. *Proceedings of ICNICONSMCL'06.* ISBN 0-7695-2552-0/06

Lundberg, J. (2000). *Routing security in ad-hoc networks*. Retrieved from http://citeseer.nec.com/400961.html

Madhu Viswanatham, V., & Chari, A. A. (2008). An approach for detecting attacks in mobile ad hoc networks. *Journal of Computer Science, 4*(3), 245-251. ISSN 1549-3636

Michiardi, P., & Molva, R. (2002). *Game theoretic analysis of security in mobile ad hoc networks.* Institut Eurecom Research Report RR-02-070, April 2002.

Misgra, A., & Nadkarni, K. M. (2003). Security in wireless ad hoc networks . In Ilyas, M. (Ed.), *The handbook of ad hoc wireless networks*. CRC Press LLC.

Moy, J. (1998). *Open shortest path first (OSPF) version 2.* RFC 2328, April 1998.

Murphy, S. (2002). *Routing protocol threat analysis.* Internet Draft, draft-murphy-threat-00.txt, October 2002.

Nait-Abdesselam, F., Bensaou, B., & Taleb, T. (2003). Detecting and avoiding wormhole attack in wireless ad hoc networks. *IEEE Communications Magazine*, *46*(4), 127–133. doi:10.1109/MCOM.2008.4481351

Papadimitratos, P., & Haas, Z. (2002). Secure routing for mobile ad hoc networks. In *Proceedings of SCS Communication Networks and Distributed Systems Modeling and Simulation Conference (CNDS)*, 2002.

Papadimitratos, P., & Haas, Z. J. (2002). Securing the routing infrastructure. *IEEE Communications*, *10*(40), 60–68. doi:10.1109/MCOM.2002.1039858

Perkin, C. E., & Bhagwat, P. (1994). *Highly dynamic destination-sequenced distance-vector routing (DSDV) for mobile computers.* ACM SIGCOMM, 1994.

Perkin, C. E., & Royer, E. M. (2000). The ad hoc on-demand distance vector routing protocol . In Perkin, C. E. (Ed.), *Ad hoc networking* (pp. 173–219). Addison Wesley.

Perkins, C. E. (2000). *Ad hoc networking*. Addison-Wesley Longman.

Perrig, A., & Canetti, R. (2000). *The TESLA broadcast authentication protocol*. Internet Draft, 2000.

Perrig, A., Canetti, R., Tygar, J. D., & Song, D. (2000). Efficient authentication and signing of multicast streams over lossy channels. In *Proceedings of IEEE Symposium on Security and Privacy,* 2000.

Perrig, A., Hu, Y.-C., & Johnson, D. B. (2001). *Wormhole protection in wireless ad hoc networks*. Technical Report TR01-384, Dept. of Computer Science, Rice University.

Prizada, A. A., & McDonald, C. (2003). Secure routing protocols for mobile ad hoc wireless networks. In *Proceedings of 2nd Workshop on Internet, Telecommunication and Signal Processing," WITSP 2003*, Australia, (pp. 57-80).

Rai, A. K., Tewari, R. R., & Upadhyay, S. K. (2010). Different types of attacks on integrated MANET-Internet communication. [IJCSS]. *International Journal of Computer Science and Security*, *4*(3).

Royer, E. M., & Toh, C. K. (2007). A review of current routing protocols for ad hoc mobile wireless networks. *IEEE Personal Communications*, *2*(6), 46–55.

Shanthi, N., Lganesan, & Ramar, K. (2009). Study of different attacks on multicast mobile ad hoc network. *Journal of Theoretical and Applied Information Technology*, *10*(1).

Siva Ram Murthy, C., & Manoj, B. S. (2004). *Ad hoc wireless networks, architecture and protocols*. Prentice Hall PTR.

Tardo, J. J., & Algappan, K. (1991). SPX: Global authentication using public key certificates. In *Proceedings of the 1991 IEEE Symposium on Security and Privacy*, (pp. 232–244). Oakland, CA, May 1991. IEEE Computer Society Press.

Tirthraj, R., & Verma, A. K. (2009). Survey and analysis of secure routing protocols for MANETs. In the *Proceedings of National Conference on Cutting Edge Computer and Electronics Technology* (CECT 2009), Pantnager, February 14-16, (pp. 501-06).

Toh, C. K. (2002). *Ad hoc mobile wireless networks: Protocols and systems*. Englewood Cliff, NJ: Prentice Hall.

Verma, A., Dave, M., & Joshi, R. C. (in press). Secure routing in mobile networks: A review. *International Journal of of Systemics, Cybernetics and Informatics (IJSCI), in press*. ISSN 0973-4864

Verma, A. K., Dave, M., & Joshi, R. C. (in press). Secure data sharing in mobile adhoc networks. *International Review on Computers and Software (IRECOS), in press*. ISSN 1828-6003

Wu, B., Chen, J., Wu, J., & Cardei, M. (2006). A survey on attacks and countermeasures in mobile ad hoc networks . In Xiao, Y., Shen, X., & Du, D.-Z. (Eds.), *Wireless/mobile network security*. doi:10.1007/978-0-387-33112-6_5

Xie, B., Kumar, A., Agrawal, D., & Srinivasan, S. (2006). Secured macro/micro-mobility protocol for multi-hop cellular IP. *Journal of Pervasive and Mobile Computing*, *2*(2), 111–136. doi:10.1016/j.pmcj.2005.11.004

Yi, P., Wu, Y., Zou, F., & Liu, N. (2010). A survey on security in wireless mesh networks. *IETE Technical Review*, *27*(1). doi:10.4103/0256-4602.58969

Yi, P., Zhong, Y. P., Zhang, S. Y., & Dai, Z. L. (2006). Flooding attack and defence in ad hoc network. *Journal of Systems Engineering and Electronics*, *17*(2), 410–416. doi:10.1016/S1004-4132(06)60070-4

Yi, S., Naldurg, P., & Kravets, R. (2002). *Security aware ad hoc routing for wireless networks*. Report No. UIUCDCS-R-2002-2290, UIUC, 2002.

Zapata, M. (2002). *Secure ad hoc on-demand distance vector* (SAODV). Internet Draft, draft-guerrero-manet-saodv-01.txt, 2002.

Zapata, M. G. (2001). *Secure ad-hoc on-demand distance vector (SAODV) routing*. IETF MANET, internet draft (Work in progress), draft-guerrero-manet- saodv-00.txt.

Zapata, M. G., & Asokan, N. (2002). Secure ad hoc routing protocols. In *Proceeding of the ACM Workshop on Wireless Security,* Atlanta, GA September, 2002.

Zhang, W., & Cao, W. (2007). Defending against cache consistency attacks in wireless ad hoc networks. *Journal of Ad Hoc Networks*, *6*(3).

Zhou, L., & Haas, Z. J. (1999). Securing ad hoc networks. *IEEE Network Magazine*, *13*(6), 24–30. doi:10.1109/65.806983

Section 4
Emerging Trends of Research in MANET

Chapter 17
Energy Conservation Issues and Challenges in MANETs

Niranjan Kumar Ray
NIT Rourkela, India

Ashok Kumar Turuk
NIT Rourkela, India

ABSTRACT

Energy efficiency is a major issue of concern in wireless ad hoc networks as mobile nodes rely on batteries, which are limited sources of energy, and, in many environments, it is quite a cumbersome task to replace or recharge them. Despite the progress made in battery technology, the lifetime of battery powered devices continues to be a key challenge, requiring additional research on efficient design of platforms, protocols, and systems. Many tangible efforts are made by many researchers to reduce the power consumption at protocol level by designing an energy efficient protocol to prolong the lifetime of the networks. The main focus of this chapter is to present a comprehensive analysis of energy efficient techniques in wireless ad hoc networks, integrating various issues and challenges to provide a big picture in this area. This chapter addresses energy management techniques in wireless ad hoc networks, especially in decentralized ad hoc environments.

1. INTRODUCTION

Energy efficiency is the most crucial design criteria for mobile ad hoc networks (MANETs), as mobile nodes are battery operated. If a node runs out of battery, its ability to route the traffic of other nodes gets affected which in turn affects the life time of the overall network negatively. Life period of a network can be enhanced by maximizing the battery power of nodes, and by minimizing total power consumption in the network. Despites the progress made in battery technologies in the recent times, it has not achieved as much as expected in comparison to semiconductor technology. This difference creates a gap between the amount of energy needed to operate in a wireless environment and a battery capacity that powered the nodes. Hence, energy efficiency appears to be an extremely important property for new protocols in MANETs. In recent years, numerous energy

DOI: 10.4018/978-1-4666-0321-9.ch017

efficient techniques such as, dynamic transmission power adjustment, topology control, multipath routing, adaptive sleeping, directional antennas, etc. have been proposed in literature. However, they suffer from their own pitfalls. For example, a protocol that reduces energy consumption by selecting an energy efficient path in the network may lead to quicker network partition, due to some overloaded node routing heavy traffic of other nodes. This may disturb the whole network leading to its early disconnection.

Therefore MANET protocols must try to choose a path that balances between the energy consumption of all the nodes and maximize the lifetime of network as a whole. It is also important to maintain a tradeoff between the energy consumed and other metrics, such as: *throughput*, *end-to-end delay*, *reliability of the links*, *network capacity*, etc. Researchers have made studies on exclusive layers of the protocol stack as well as on cross layer optimization to conserve energy.

This chapter explores different approaches to reduce energy consumption in wireless ad hoc networks and focused on issues involved in it. The chapter opens with a brief introduction to energy efficiency in wireless network. Second section discusses about energy management in ad hoc network. The third section categorically presents some energy aware protocols with their strength and weakness. Conclusions are made in fourth section.

2. BACKGROUND INFORMATION

In this section we detail on different energy management issues such as: need of energy managements, causes of energy wastes, energy consumption, radio energy model, etc. This helps us in understanding energy issues, challenges and requirements in MANETs.

2.1. Need for Energy Management

Now a day's energy efficiency has become a pervasive issue in all layers of communication protocols, and in a wide range of technological applications spanning from simple microelectronic systems to more complex vehicular ad hoc networks. Until now, research and development in the field of communication networks was mainly targeted at their functionality and performance, but for battery-driven devices, such as mobile and sensor nodes, energy efficiency is a significant consideration. The intensity of its importance increases due to the massive use of such devices in public, professional, and private life, coupled with the increasing cost of energy and with the recognition of the critical need to reduce carbon dioxide(CO_2) emissions. These economic and ecological drivers have induced a new research area with "energy efficient of communication networks" as the main objective. The primary reasons behind the energy management in ad hoc networks are as follows:

- **Limited Battery Power:** Nodes in MANETs are battery operated, and in many applications like disaster management, battle field communication, environment monitoring, etc., it is difficult to replace and recharge them. Since the capacity is limited, some energy management techniques need to be performed to prolong the network lifetime.
- **Increasing Demand of Battery Operated Devices:** Massive use of battery operated devices in communication, personal and professional life, demand more requirements of battery. Devices operated by battery are omnipresent in the form of cell phone to micro sensor nodes.
- **Relay Traffic:** Nodes in ad hoc network act as a source as well as router to relay the traffic from other nodes in the network. These relay nodes play a vital role in defining the life time of the network. If the pro-

portions of relay nodes are large in the network, then it may lead to a faster depletion of node's battery power. For lesser numbers of relay nodes in the network, nodes have to transmit with maximum power. Transmitting with maximum power may increase the interference.

- **Interference Reduction:** Interference is the major cause of throughput degradation which occurs, when nodes transmits with maximum power. Interference not only degrades network throughput and node's life time but also the main cause of bad carrier sensing. Selection of optimal transmission power is a basic requirement of energy constrained networks to reduce interference.

2.2. Cause of Energy Waste

- **Carrier Sensing:** MAC protocols for wired networks such as CSMA/CD ALOHA are not suitable for wireless networks due to the physical characteristics of wireless medium. Carrier sensing is a challenging task in wireless network. A node expend maximum energy in carrier sensing, which is also affected by hidden and exposed terminals.
- **Collisions:** Re-transmission occurs due to packet collision, and this increases per packet energy consumptions. Retransmission also affects other network parameters such as: *end-to-end delay*, *jitter*, *throughput* etc.
- **Idle Channel Sensing:** Continuous monitoring of an idle channel for possible traffic consumes more energy without increase in the network throughput. Experimental results have shown that idle power consumption is nearly equal to receive power consumptions, which can't be ignored from power consumption point of view. Reducing idle power consumptions is a major task of all energy efficient protocols in ad hoc network.
- **Over Hearing:** It occurs when a node hears a message meant for other nodes. Over hearing can't be completely avoided but its effect can be minimized by implementing a suitable power control technique.
- **Higher Bit Rate and Larger Packet Size:** Higher bit rate and larger packet size, consumes more power as compare to lower bit rate and smaller packet size. Larger packet size also increases the probability of retransmission due to collision.
- **Message Flooding:** Flooding is the basic mechanism in most of the routing protocols in MANETs. In a large network, message flooding is a major overhead in terms of bandwidth utilizations, energy consumptions, and interferences.
- **Dynamic Topology:** Mobility makes routing problematic, requiring frequent update of path information. MANETs has dynamic and unpredictable mobility as a result route discovery and route maintenance are most challenging tasks.

2.3. Energy Consumption

We discuss energy consumption behavior of MANETs at network interface and at network traffic level.

- **Network interface level**

Energy consumed by a network interface depends on, operating states of mobile nodes such as: *transmit, receive, idle, and sleep*. Transmit and receive states are used for sending and receiving network packets respectively, their energy consumptions are more as compared to idle and sleep states. A node in idle state neither transmits nor receives any packets but it wait for the same. Stemm and Karz (1999) have shown that the ratio of *idle: receive: transmit* power is *1:1.05:1.4*, while Chen *et al.* (2002) shown the ratio as 0.*83:1:1.4*. From the above two power consumption ratio, it is observed that the idle state power consump-

Table 1. Power consumption of different state

Interface	transmit (watt)	receive (watt)	idle (watt)	sleep (watt)	Mbps
Aironet PC4800	1.4-1.9	1.3-1.4	1.34	0.075	11
Lucent Bronze	1.3	0.97	0.84	0.066	2
Lucent Silver	1.3	0.90	0.74	0.048	11
Cabletron Roamabout	1.4	1.0	0.83	0.13	2

Source: Feeney, L.M. (2004), Energy efficient communication in ad hoc networks

tion cannot be ignored. Mobile nodes stay most of the times in idle state, as compared to other states. Sleep state consumes very less amount of energy as the transceiver is switched off. during that periods. Table 1 provides some experimental power consumption measurements (Feeney, 2004) for IEEE 802.11 (2.4 GHz) interface. From the Table 1, it is concluded that for power conservation purpose the network interface must be put into sleep state for more periods of time.

- **Network traffic level**

Feeney (2001) presented some results of energy consumption measurement at different traffic level. The cost of energy expend by a node to send and receive a network-layer packet varies at different traffic level. There is a fixed cost associated with channel acquisition, and an incremental cost associated with size of the network packets. Energy consumption of a network packet is the sum of costs incurred by source node *S*, and all receivers. Potential receivers are destination node *D*, and nodes in the radio range of *S*. In simplest receiving and sending pattern, the energy cost is expressed as:

$$Cost = m \times size + b$$

where *b* is the fixed cost associated with *channel acquisition*, and *m* is the *incremental pay load cost*. For broadcast and point to point traffic, packet cost is given below.

- *Broadcast traffic*

In broadcast traffic there is one sender but many receivers. Sender access the channel before sending the data. If the channel is free then it sends the packets. Therefore, energy cost for broadcast traffic is given by:

$$Cost = m_{send} \times size + b_{send} + \sum_{n \in S} \left(m_{recv} \times size + b_{recv} \right)$$

where *bsend* and *brecv* are channel acquisition costs for sending and receiving packets respectively, and corresponding payload costs are *msend* and *mrecv*. In broadcast traffic there is one sender but many receivers. The *n* in the equation is the nodes presents within the radio range of source node *S*.

- *Point to point traffic*

In point to point traffic, source sends a *RTS* signal and receiver on receiving that responds with *CTS* signal. After receiving *CTS* source sends *DATA* and that is acknowledged by *ACK* from receiver. RTS, CTS, and ACK are used to avoid collision of data frames in wireless networks but these control signals consume a small amount of energy. So the cost of the packet in point to point traffic at sender and receiver is given by:

At Sender:

$$Cost = b_{send\_con} + b_{recv\_con} + m_{send} \times size + b_{send} + b_{recv\_con}$$

At Receiver:

$$Cost = b_{send\_con} + b_{recv\_con} + m_{recv} \times size + b_{recv} + b_{send\_con}$$

where *bsend_con* and *brecv_con* are the costs for sending and receiving control packets, respectively.

2.4. Radio Energy Model

In wireless network the maximum numbers of bits a node can transmit depends on the total battery energy. Energy consumption by radio interface is discussed in sub-section 2.3. The energy '*E*' consumed by a node to transmit '*k*' bit can be written as:

$$E = P_{active}T_{active} + P_{sleep}T_{sleep} + P_{idle}T_{idle}$$

where, *Pactive, Psleep, Pidle* and *Tactive, Tsleep, Tidle* are the corresponding power consumption and time spent in the active, sleep, and idle mode respectively. The power consumed during switching of states (transient mode) is ignored in the above equation.

Further active power is the combination of transmitted signal power (*Psig*) and circuit power (*Pckt*). So the total energy is given as:

$$E = \left(P_{sig} + P_{ckt}\right)T_{active} + P_{sleep}T_{sleep} + P_{idle}T_{idle}$$

Circuit power *Pckt* is combination of other powers, as given below:

$$P_{ckt} = P_{amp} + P_{mix} + P_{syn} + P_{filt} + P_{DAC} + P_{LNA} + P_{IFA}$$

where, *Pamp, Pmix, Psyn, Pfilt, PDAC, PLNA, PIFA* are corresponding power consumption in a circuit (such as amplifier power, frequency synthesizer power, filter power, etc). Amplifier power consumption *Pamp* can be expressed as:

$$P_{amp} = \alpha P_{sig} \text{ where } \alpha = \frac{\beta}{\mu} - 1$$

β, μ are the *Peak to Average Ratio*(PAR) and drain efficiency of RF power amplifier respectively. The β value depends upon modulation scheme, and μ value depends upon different class of amplifiers. Considering all parameters discussed above the energy consumption of radio interface is expressed as:

$$E = \begin{bmatrix} \left[P_{sig}\left(\frac{\beta}{\mu} - 1\right) + P_{mix} + P_{syn} + P_{filt} + P_{DAC} + P_{LNA} + P_{IFA}\right] \\ T_{active} + P_{sleep}T_{sleep} + P_{idle}T_{idle} \end{bmatrix}$$

2.5. Energy-Aware Communication Issues

Some of the energy-aware issues in wireless ad hoc network protocols are discussed below:

- *Finding an energy efficient path:* It is difficult to choose between minimum energy path and balanced power consumed path. Minimum energy path always conserve more energy but it does not guaranty any increase in network life period. Nodes in this path are over utilized, while balanced power consumed path prevents the nodes from being overloaded but does not necessarily lead to minimized power consumption.
- *Layer specific designing:* Energy optimization can be done in all layers of protocol stack. However, most of the energy aware protocols focused either at routing layer

or at MAC layer. Routing layer protocols does not consider the link layer issues and vice versa. This difference creates serious performance degradation in network throughput. Therefore, designing a protocol for multi-layer is a potential area of energy conservation.

- *Assumption based computation:* Some network protocols assume that the transmission power is controllable and nodes location information is available (e.g, through GPS). With this information in mind, the problem of finding a path with least consumed power becomes a conventional optimization problem on a graph, where the weighted link cost is mapped to transmission power required for transmitting a packet between the two ends of the link. However, acquisition of location information by this way will introduce computation delay, extra message overhead and energy consumption at each node.

3. ENERGY AWARE SCHEMES

We organize energy-aware protocols into three major categories: *power managements*, *power control* and *topology control*. Strength and weakness of each protocol covered under each category is discussed and summarized bellow.

3.1. Power Management Techniques

The basic objective of power management scheme is to save energy by putting the nodes in idle states to sleep state. Few power management protocols for ad hoc networks are discussed below.

Preliminary

Power Save Protocol and Sleep Scheduling

In order to reduce energy consumptions, *power save protocols* switches off the radio transceiver of a node when the node is in idle state. It is observed that a node transmitting and receiving data at two Mbps from four neighbors uses only 1% more power than an idle interface. So power save protocols tries to minimize the idle power consumption, and maximizes energy saving without affecting the network connectivity and throughput.

Sleep scheduling in ad hoc networks can be synchronous or asynchronous. In *synchronous sleep scheduling* the selected nodes will goes to sleep and wake state periodically based on the sleep cycles. IEEE 802.11 PS mode adopts synchronous sleep scheduling. In *asynchronous sleep scheduling*, nodes are not synchronized to a common clock. They have different sleep time as well as sleep durations. Asynchronous sleep scheduling is more challenging as compared to synchronous approach.

3.1.1. Power Management Protocols

- IEEE 802.11 Power Saving

There are two types of power management in IEEE 802.11 DCF (IEEE 802.11 Working Group, 1999). One is known as power save (PS) mode for infrastructure based wireless network and the other type is known as IBSS PS mode, which is relevant to the ad hoc network. Both schemes are briefly discussed below:

- ***PS mode for infrastructure based network***

A node in this scheme may be in one of the two different mode, (i) Active mode (AM), and (ii) Power save (PS) mode. In AM, node receives frames at any time. In PS mode, node listens to selected beacons and sends *PS-Poll* frame to the access point (AP). AP buffers the MAC service data unit (MSDU) and transmits them at designated time by the help of traffic indication map (TIM) and delayed traffic indication map (DTIM). TIM identify the nodes for which traffic are pending.

Figure1. Data transmission in IBSS PS mode

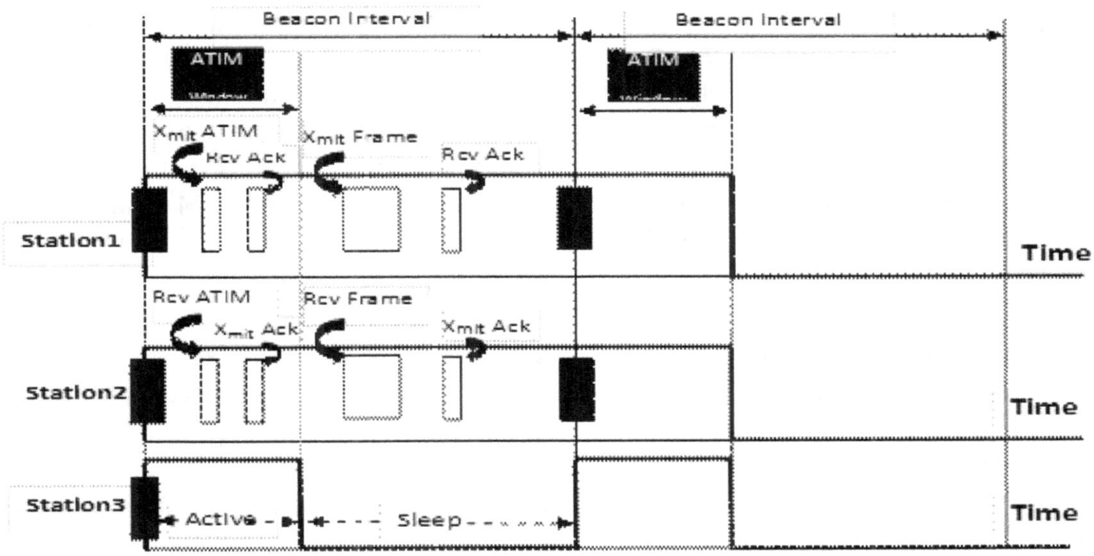

The AP deletes the buffered traffic when it is buffered for an excessive periods of time. Nodes in PS modes shall periodically listen for beacons as determined by the node's *Listen-Interval-Amount.*

- ***IBSS PS mode***

In this mode, synchronized beacon intervals are established by the node which initiates the IBSS and is maintained in a distributed fashion. It defines the fixed size length, *announcement traffic indication message* (ATIM). A node transmitting an ATIM management frame shall remain awake for the entire current beacon interval. All nodes wake at the beginning of the *beacon interval* (BI) and remain wake till the end of the traffic window. A node transmits RTS-CTS control frames and beacon and ATIM management frames during the ATIM window. If a node is unable to transmit an ATIM during the ATIM window due to contention with other nodes, the node shall retain the buffered MSDU and attempt to transmit the ATIM during the next ATIM window. Nodes participating in the traffic announcement remain awake till the end of beacon interval and the non-participator of the traffic goes to sleep at the end of the traffic window. Beacon announcements and acknowledgements are transmitted during the ATIM window to avoid contention with the data traffic. Figure 1 shows the data transmission in IBSS PS mode. A node may discard frames buffered for later transmission, if it has been buffered for an excessive amount of time. The effectiveness of power saving depends on the value selected for the ATIM window and beacon interval. If the ATIM window is short then not enough traffic will be announced and if it is long then wake up time for the nodes will be more, causing more power consumption. Similarly beacon with larger interval causes congestion in the network and shorter beacon interval increases wakeup periods of nodes causing more energy consumptions. Some recent work (Kim, Lin, & Shroff, 2011) focused on *sleep-wake scheduling* and *any-cast forwarding* policies to reduce further *end-to-end delay* and energy consumptions.

- Dominating-Awake-Interval Protocol

IEEE 802.11 PS mode is based on clock synchronization and is applicable to single-hop fully connected networks. Its power saving is not suitable for multi-hop ad hoc networks where, achieving clock synchronization is a major difficult task due to unpredictable mobility and communication delay. Tseng *et al.* (2003) have proposed a power management based protocol called *dominating-awake-interval* for multi-hop ad hoc network by redesigning the IEEE 802.11 PS mode. This protocol is based on multiple beacons and overlapping awake intervals. The beacon interval contains three windows: active window, beacon window and MTIM window. Beacon window and MTIM window are part of active window as shown in the Figure 2(a). During an active window a node switches on its radio transceiver for listening and sending the packets. Beacon window is used for sending its own frames, and a node receives the beacon frame from other nodes through MTIM window. MTIM window is similar to ATIM window of IEEE 802.11 PS mode. Within each beacon, the lengths of all three windows are constant. When a node decides to enter the PS mode, it divides its time axis into fixed length beacon intervals. The sequences of beacon intervals are alternatively labeled as odd and even as shown in Figure 2(b). In odd beacon interval an active window is started with a beacon window followed by MTIM window, and in even beacon interval an active window is always terminated by beacon window.

In the Figure 2(b), it is found that node *A* can respond to node *B* in even beacon interval and vice-versa in odd beacon interval. Beacon intervals are alternatively designed as even and odd, so that both the node can able to access each other. To satisfy the dominating awake property an active window length (AW) is set as: AW ≥ BI/2 + BW, where BI is the beacon interval length, BW is the beacon window length.

Figure 2. Structure of beacon interval (a) active window, beacon and MTIM window (b) even and odd beacon intervals

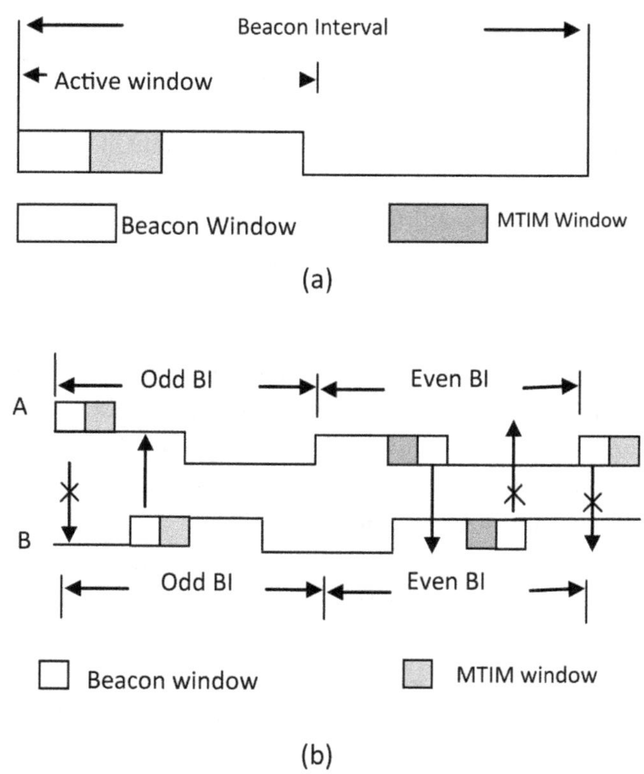

The limitation of this protocol is that a node remains awake for a longer period of time as compared to IEEE 802.11 DCF.

- Dynamic Power Saving Mechanism (DPSM)

DPSM (Jung & Vaidya, INFOCOM 2002) is a variation of IEEE 802.11 DCF PS mode approach. It optimizes power saving mechanism of DCF by dynamically selecting a suitable ATIM window based on network condition. The concept of ATIM window and beacon interval is same as that of DCF but it uses increased and deceased size ATIM window. The major limitation of DCF was at lower traffic loads its energy saving is drastically reduced due to fixed size ATIM window. DPSM in other way allows the sender and receiver to change their ATIM window dynamically. Size of an ATIM window is increased when transmission of few packets are pending after expire of current window. Nodes on overhearing the *marked packets* modify their length of ATIM window. A *marked packet* indicates failure to deliver the corresponding ATIM frame in the last few attempts, due to small ATIM window size of recipient node. When a node transmits an ATIM frame for another node, it will not transmit different frame to that node. If a sender node could not complete all its pending traffic in the current beacon interval, then both sender and receiver node will stay up for next beacon interval. After the delivery of all pending packets, the sender and receiver node may go to sleep mode immediately, if they have no other packet to send or to receive.

DPSM specifies rules under which nodes can dynamically increase or decrease their ATIM window. A node can dynamically adjust ATIM window based on:

a. Pending packets, which are not announced during the current ATIM window
b. Over-hearing of larger ATIM window size
c. Receiving an ATIM frames after expires of ATIM window
d. On receiving a *marked packet* ATIM

Nodes based on the above condition adjust the ATIM window size at the beginning of next beacon interval. DPSM performs better in comparison to IEEE 802.11 DCF in term of power saving. However it is more complex in selecting ATIM window size.

- Power Aware Multi-access with Signaling

Energy efficiency is the primary goal in power aware multi-access with signaling (PAMAS) protocol (Singh & Raghavendra, 1998). This scheme uses *RTS-CTS* mechanism with a separate signaling channel with MACA protocol. PAMAS uses different channels for data and control, RTS/CTS signals are transmitted over control channel, while data are transmitted over data channel. Signaling channel enables nodes to determine when and how long they can power themselves off. The protocol is discussed in two-fold:

1. Node states and their functionality

A node in PAMAS can be in any one of the six states. The states are stated below with their functionality.

a. *Await CTS state*: When a node has a packet to send it transmit the signal and enter to this state.
b. *Binary exponential back-off state (BEB)*: A node enters to this state if it does not receive *CTS* after the *await CTS state;* it hears a busy tone signal on control channel.
c. *Transmit packet state*: A Node enter to this state for packet transmission after receiving *CTS.*
d. *Await packet state*: A node when receives a *RTS*, it replies back by sending *CTS* to the sender and enter to this state, only if no other

nodes exist in the *Await CTS* or *Transmit packet states*.

e. *Receive packet state*: If packets arrive on time at the node. A node enters to this state and start receiving packets.

f. *Idle state*: A node enters to this state, if packets do not arrive on time at the node.

When a node has packet to transmit, it sends *RTS* over the control channel, and waits for *CTS*; if no *CTS* arrives then node enters to *BEB* state. However, if it received *CTS*, node then enters *transmit packet state* to transmits the data packets over the data channel. The receiving node transmits a *busy tone* over the control channel to inform others, that the data channel is busy. *Await CTS*, *BEB* and *transmit packet* states are used for packet transmission while the rest are used for packet reception.

2. Turning of the radios

A node turns off its radio under two situations: (a) When it has no packet to send, and (b) if it sense ongoing transmissions in its neighborhood.

Node determines the power off duration through a probe protocol. When packet transmission begins at the neighborhood of a node, it determines the duration of transmission through control channel and power off itself for that period of time. But in some case when neighbors begin transmission, while the node power off and these transmission are ongoing when node wake up. To find the remaining transmission time node uses probe signal and determines the switch off time. For example, node transmits a probe signal *t-probe(l)* through its control channel where *l* is the maximum packet length duration. Nodes expecting to finish the transmission by period [*l*/2,*l*] in its neighborhood, response the probe signal by transmitting *t-probe-response(t)*, where *t* is the time left for the current transmission. Intended node on receiving this probe signal turn off its radio for *t* period of time. However collisions in signaling and data channel make it difficult to determine the length of the current transmission.

PAMAS achieves power saving, which varies from 10% to 70%, depending on the network type. Network with higher node density and traffic load saves more energy but it is affected in sparsely connected network under light load conditions.

3. BECA and AFECA

Xu *et al.* (2000) proposes two power management based protocols, (i) *Basic Energy-Conserving Algorithm (BECA)* and (ii) *Adaptive Fidelity Energy Conserving Algorithm (AFECA)* for ad hoc networks. These protocols attempt to turn off node's radios as often as possible to improve power consumption. They use information from application layer to decide when the radios should be on or off based on the traffic conditions.

In BECA each node can be in one of three states: *sleeping*, *listening*, or *active*. Each state is associated with a timer to determine how long a node will be in that state. Initially nodes are in the listening state, and listens for traffic from their neighbors. If node does not receive any traffic in the specified listening time, it enters to sleeping state. A node goes to active state, when it has some pending traffic to forward. Maintaining a

Figure 3. Node states in BECA

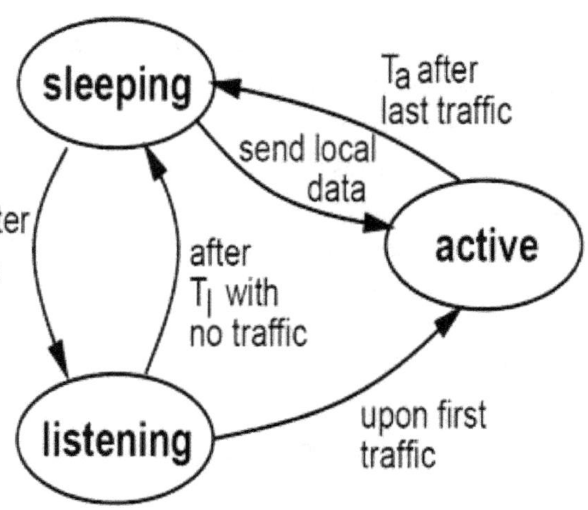

live path to destination is a major issue in BECA, as nodes take their own decisions about the state transition. This issue is handled in a probabilistic way, in which broadcasting of route request messages (RREQ) is repeated for several times with increasing redundancy. Nodes enter to active state on receiving RREQs. Once the route is established, the nodes in the path should remain active and other node will enter to sleep state at the end of their current active state. Figure 3 shows the node states in BECA.

AFECA is an extension of BECA; its working principles remain same as that of BECA. It takes advantage of node density to allow nodes in dense areas to sleep for longer periods. Sleep time varies depending on the number of neighbor a node has. AFECA uses variable sleep time as a result it is more likely that at any given time, one nearby node will wake up to hear RREQ packets while other nodes will be in sleep state for a longer duration. AFECA works well for network having uniform density of nodes. Per node energy-saving in AFECA is not more than BECA, but it enhances the lifetime of the network. Though AFECA and BECA offer substantial energy saving, it comes at the cost of higher packet loss and longer route setup latency.

4. EC-MAC

Energy conservation is the main design goal in *energy conserving MAC* [EC-MAC] (Sivalingam, Chen, & Agrawal, 2000). The protocol is defined for infrastructure based network where a base stations (BT) serving the mobile nodes in its coverage area. This definition can be extended to an ad hoc network by selecting a node to perform the task of BT. Transmission in EC-MAC is organized by BT in term of frames. A frame consists of fixed number of slots divided into different phases. Mainly a frame consists of *request-update phase*, *new-user phase*, *downlink phase* and *uplink phase*. Each frame begins with a *frame synchronization message* (FSM) which contains synchronization informations. New user phase allows a new user to register with the base station and operate in contention mode, using slotted aloha protocol. In *request-update phase* a node transmits new connection requests to BT. Uplink and down link phases are used for data transmission which also reduces turnaround times and collisions.

Energy saving is done by the help of centralize scheduler. A centralize scheduler reduces collisions also optimizes the transmission so that nodes transmit and receive within contiguous transmission slots. Scheduler also considers the remaining battery power level of nodes in addition to packet priority while scheduling packets for transmission. At heavy traffic power saving of EC-MAC is close to IEEE 802.11 DCF.

3.1.2. *Summarization*

Most of the power management protocols for ad hoc networks are based on sleep-wake scheduling and their energy savings depend on the time spent by the nodes in sleep state. Clock synchronization is necessary for synchronous sleep-awake scheduling protocols like *IEEE 802.11 DCF* and is not necessary for asynchronous sleep-wake scheduling protocol like *dominating-awake interval*. Some variance of DCF, optimizes power saving mechanism by dynamically selecting a suitable ATIM window, but their power saving is not measured properly in different type of network and at different traffic conditions. Summary of the power management protocols for ad hoc network discussed in this chapter is given in Table 2.

3.2. Power Control Techniques

Transmitting with higher power may improve the performance of network layer in MANETs by reducing the number of forwarding nodes and signaling overhead to maintain routes. However, high power transmissions always have negative impact on the performance of MAC layer due to their increased interference. Therefore, selecting

Table 2. Summary of power management protocols

Protocol	Approach	Remark
IEEE 802.11 DCF	• Synchronized sleep scheduling • Fixed ATIM and Beacon window	Applicable for fully connected network
Dominating-Awake-Interval	• Asynchronous sleep scheduling • Dominating awake property • Even and odd beacon interval, MTIM window	Applicable for multi-hop network
DPSM	• Variable ATIM window	Better energy saving More complex in implementation
PAMAS	• Addition of separate signaling channel • Probe protocol to decide switch off condition	Most effective in network with high density and higher traffic load
BECA/AFECA	• Adaptive sleep scheduling based on node density and on traffic load	Intended for networks of uniform density
EC-MAC	• Scheduler based energy saving • Good power saving at heavy traffic load	Designed for infrastructure based wireless network

an appropriate transmission power is a major task for protocol designer as it can affects overall performance of the network. We illustrate below the impact of transmission power with an example.

Consider a wireless network as shown in Figure 4(a) where source node *C* is reachable to destination E through its maximum transmission power TR1. Transmitting with maximum power TR1,

Figure 4. (a) Transmission without power control; (b) Transmission with power control

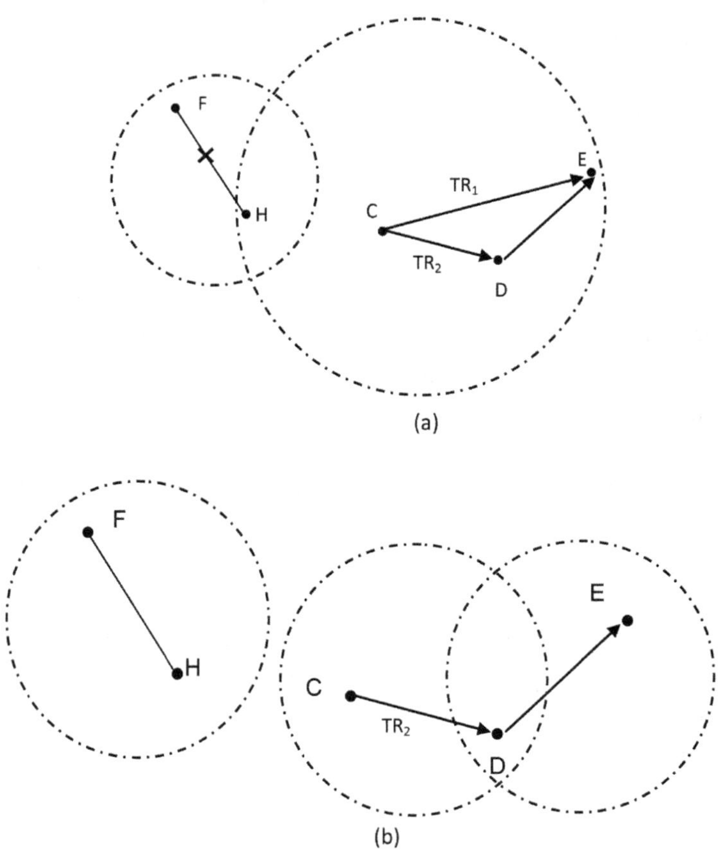

node E is directly reachable from node C but it affects the ongoing transmission within this radio range. In such a situation communication between node *H* and *F* is not possible due to higher radio interference from node *C*. This interference can be reduced by reducing the transmission power of node *C*. If node *C* transmits with a relatively lesser power say *TR2* then communication to node *E* will be possible through relay node *D*. The power required to transmit a packet between two nodes increases with increase in their relative distance between them. Hence, it is always desirable to transmit through short multi-hop communication rather than at maximum.

Power control is a technique to adjust nodes transmission power to a proper range at node level. Figure 4(a) shows the transmission without power control which affects communication between node H and F. Figure 4(b) shows the transmission with power control which allows simultaneous transmission between H and F.

Power control has been successfully implemented in cellular networks, where the base stations manage power control features such as interference control, energy saving, and connectivity, etc. Some of the power control techniques of cellular networks are*: linear and bilinear power control, opportunistic power control technique, Kalman filter based power control, distance based power control* etc. While the power control in ad hoc network has many complexity and challenges due to its self-organizing, self-creating, self-administrating nature and absence of fixed infrastructure. Most of the power control algorithms in ad hoc networks are adapted from cellular network. Though power control algorithms for ad hoc network contribute to network throughput by enabling simultaneous transmissions but it inflicts weaker network connectivity, higher error rate, lower reliability, etc. Balancing these factors is a major design criteria in MANETs. We discussed below several powers control techniques for ad hoc network.

PRELIMINARY

- **Centralized and distributed power control**

In centralized form of power control a central entity such as access point or base station manages the information and controls the transmission power.

Distributed power control approach does not require any central entity and is done at individual node level. Extensive computational work is not required in distributed method. *Signal to interference ratio (SIR)* can be controlled and managed in distributed approach.

- **Distanced and received signal strength based power control**

Transmission power is a function of distance gradient. If the distance between sender and receiver is known then the power required to communicate can be calculated as:

$$p_{uv} = d^{\beta} + C$$

where *Puv* is the power required to communicate from node *u* to node *v*, relative distance between them is expressed as *d*; β is the distance-power gradient ranging between [2,4] and *C* is a constant.

In received signal strength based approach, a node can determine the minimum power from the signal strength of receiving packets. Let *u* and *v* be two communicating nodes. Suppose *v* sends a message to *u* with its maximum transmission power *Pmax*, and *u* receive the message with power *Pr*, and *Pmin* denotes the node's smallest possible receiving power. Then minimum power required for communication between *u* and *v* can be calculated as:

$$P_{uv} = \frac{P_{max} \times P_{min}}{P_r} \times C$$

Pr is given by the free space propagation model as:

$$P_r = P_t (\frac{\lambda}{4\pi d})^n G_t G_r$$

where, *Pt* and *Pr* denotes the signal power at transmitting and receiving antenna respectively, λ denotes the carrier wavelength, *d* denotes the distance between the sender and the receiver, and *Gt* and *Gr* denotes the antenna gains at the sender and receiver respectively, *n* is the path loss coefficient which depends on the environment.

- **Common and variable range transmission control**

When nodes in a network use a common transmission power, Rcom to communicate with peer nodes is called common range transmission control. A minimum common transmission range is the minimum value of Rcom which maintains connectivity in network.

In variable range transmission control nodes transmit with different power level. It is assumed that each node dynamically controls the transmission power it uses independently of other nodes.

3.2.1. Power Control Protocols

- Modified IEEE 802.11

Power control can be achieved in IEEE 802.11 standard by modifying the transmission frame's power level. Rather than transmitting the frames in common power, the idea is to transmit them in different power level. The RTS-CTS and DATA-ACK frames are transmitted at different power level. RTS-CTS frames are transmitted at highest power level such that nodes within this range can be aware about the intended communications, and accordingly differs their communication. DATA-ACK frames are transmitted at lower power to save energy. Figure 5, shows the transmission range of RTS-CTS and DATA-ACK frames of node *u* and *v*. In Figure 5, RTS and CTS frames are transmitted at highest power level such that, neighboring nodes *w* and *x* can set their *network allocation vector* (NAV) for the duration of the DATA-ACK transmission. The above scheme is affected when nodes in the carrier senescing zone (CSZ) do not properly decode the RTS-CTS frame. Nodes present in CSZ can detect the signal but can't decode it correctly. The nodes in this situation set their NAV value equal to *extended inter frame spacing* (EIFS), period for avoiding possible collision with DATA-ACK frames. After expire of EIFS periods these nodes mistakenly consider that the wireless channel is free and transmit their own packets leading to collisions. To avoid this Jung and Vaidya (MOBICOM, 2002,) proposed

Figure 5. Modified IEEE 802.11 schemes where RTS-CTS are transmitted with higher power level

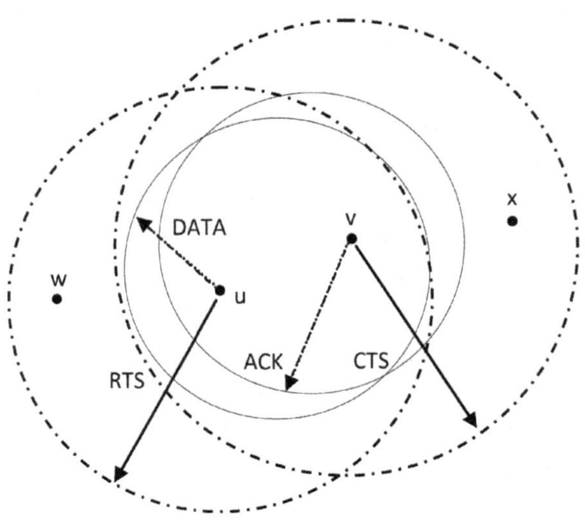

Figure 6. Transmission range and carrier sensing zone of node u

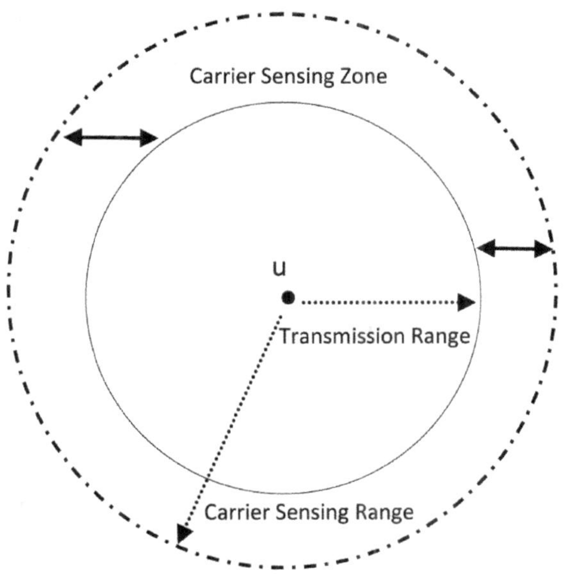

a modification to this technique in which source node periodically transmits DATA at power level *Pmax*, for just enough time so that nodes in the CSZ can sense it. Nodes that can potentially interfere with the reception of ACK at the sender will periodically sense the channel and defer their own transmission. Figure 6 shows the transmission range, carrier sensing range and carrier sensing zone of node u.

In this manner, the wireless channel is first reserved by RTS-CTS frames, and the nodes in the transmission range of the sender and receiver then adjust their NAVs on receiving RTS or CTS so that the probability of packet collision is greatly decreased.

- PCMAC

Lin and Lou (2003) propose PCMAC which is a modification of IEEE 802.11 standard. It eliminates collision of ACK frames at sender by using only RTS-CTS-DATA handshaking. Acknowledgement is done by maintaining two tables: *sent-table*, and *receive-table*. These two tables stores information about packet sent and received respectively. When a sender transmits data packet to receiver it records the *session ID, receiver ID* and *packet sequence number* together in its *sent-table*. In the same ways receiver records the corresponding information in its *receive-table*. For example when *v* receives a RTS form *u*, it replies with CTS, which includes the *session ID, sequence number* of the last received data packet from *u*. When *u* receives these CTS it checks these two fields, comparing them with those values recorded in its *sent-table*. If the two data matches, this indicates that, the last data packet has been successfully received by *v*; otherwise, *u* retransmits the last sent data packet to *v*.

The necessary power level is determined by receiver node based on received signal strength of the packet received. This is computed as:

$$P_{required} = \frac{Rx_{th} \times p_t}{E}$$

where Rxth is the minimal threshold power, *Pt* is the transmitted power level at which packet is sent, and *E* is the signal strength at receiver.

- Power Control by Busy Tone

Wu *et al.* (2000) extended the work of Deng and Hass (1998), in which intelligent power control along with RTS-CTS dialogues and busy tone signals are used. Busy-tones are two narrow band signals *BTt* for transmitting and *BTr* for receiving are placed on the medium at different frequency. A host should refrain from sending if it hears any *BTr* and should not consent to send if it hears any *BTt*. RTS-CTS exchanges are done to find out the appropriate power level between sender and receiver. Power control technique work as follows:

a. RTS is transmitted at a power level determined based on *BTr*

b. CTS are transmitted at the maximum available power level, and
c. DATA packets are transmitted at a power level derived from the power level of the received CTS

Sender turns on the *BTt* while sending the data and receiver turn on *BTr* while transmitting CTS signals. Busy tone avoids collisions both at receiver and sender. Neighboring nodes does not interrupt the transmission till the end of *BTt* or *BTr*.

- Neighbor Detection Power Control

Abasgholi *et al.* (2008) proposed a power control mechanism in which required power is based on number of neighboring nodes. If the number of neighbors increases then transmission power accordingly decreases. There is a tradeoff between physical and network connectivity for selecting transmission power. A node increases its transmission power until it finds some neighbors, and then its power is adjusted based on number of neighbors. The resulting network topology increases network lifetime by reducing transmission power and decreases interference by having low node degrees. The transmission power varied between maximum and minimum and is computed as:

$$p_t = p_c - 5n \log\left(d_t / d_c\right)$$

where *Pt* is the targeted power, *Pc* is the current transmission power, *n* is the propagation loss which depends on environments, *dt* and *dc* are the targeted node degree and current node degree respectively. The targeted node degree *dt* and current node degree *dc* are calculated as:

$$d_t = D\pi r_t^2 \; and \; d_c = D\pi r_c^2$$

- Adaptive Power Control

The basic objective of adaptive power control approach is to adjust the transmission power to a suitable level according to the network condition. APMAC (Choi, Byun & LIM, 2008) is an example of this family. It modifies RTS-CTS frames by adding an extra field of one byte to store current transmission power of RTS and CTS frames respectively. Required power level is calculated by receiver and is appended in CTS frame and send to sender. It assumes that each node has a same signal to interference ratio (SIR) and fixed symmetric channel gain between source and destinations. Nodes maintain power level of neighboring nodes. All the power levels in the list are initialized as *Pmax*. When a node waiting for ACK frames from destination it may receive:

a. RTS/CTS frames from other nodes,
b. Un-decodable frames from some nodes,
c. Transmission fail message, or
d. ACK frames from destination.

If a source receives RTS or CTS packets from other nodes during its transmission then it indicates that the source is in the transmission range of other nodes, but they are not within the transmission range of source. For a transmission power higher than the received frame power level neighboring nodes will be aware of the source node's transmission. When node receives un-decodable frames it indicates source node is in the carrier sensing zone of another node for which it could not decode the frame correctly, in such scenario source requests to destination to adjust transmission power. Whenever the packet transmission succeeds or fails the source updates its transmission power *Ps* depending upon the receive frames as follows:

For case (a), (b) and (d)

$$p_s = \min\left(p_{\max}, \max\left(p_{urc,} \, p_{cs}\right)\right)$$

For case (c)

$$p_s = \min\left(p_{\max}, \max\left(p_{s,} + \Delta\right)\right)$$

where *Purc* is recorded power in new RTS/CTS frame, *Pcs* is received power of un-decodable packets, Δ is small value to avoid the interferences and *Pmax* is the initialized power of nodes.

Zhang *et al.* (2005) proposed, adaptive power control technique, based on the transmission power pair of *RTS-CTS*, *CTS-DATA*, and *DATA-ACK*. They define a relationship between transmission powers *Pt* and receipting power *Pr* as follows:

$$p_r = p_t \times G_t \times G_r \times h_t^2 \times h_r^2 / d^4$$

where *ht* and *hr* are heights of transmitter and receiver antenna respectively and *d* is there separating distance. *Gt* and *Gr* are the corresponding gains of transmitter and receivers antenna respectively. The relationships between transmitting and receiving power of RTS, CTS, and DATA are defined as follows:

$$p_{rts,r} = p_{rts,t} \times G_t \times G_r \times h^4 / d^4$$

$$p_{cts,r} = p_{cts,t} \times G_t \times G_r \times h^4 / d^4$$

$$p_{DATA,r} = p_{DATA,r} \times G_t \times G_r \times h^4 / d^4$$

where *Prts,r* and *Prts, t* are the *Pcts,r* and *Pdata,r* are the corresponding power, *d* is the distance between two nodes, *h* is the antenna height.

3.2.2. Summarization

Power control technique in ad hoc network uses the transmission power effectively by adjusting the transmission power to a desired level. Different power control techniques discussed in this section is summarized in Table 3.

3.3. Topology Control Techniques

We will illustrate the topology control scheme with an example. Consider Figure 7 where node *u* wants to communicate with node *v* which is at a distance *d* from node *u*. Node *w* is in region *C* circumscribed by circle of diameter *d* and is located at a distance of *d1* from *u* and *d2* from *v*. As *d1* and *d2* is less than *d* so communication from *u* to *v* can be possible through node *w*. In

Table 3. Summary of power control techniques

Techniques	Approaches
Modified IEEE 802.11	• Data and control frames are transmitted at different power level • Source node periodically transmits DATA at maximum power level to reduce carrier sensing zone problem
PCMAC	• Uses three way of handshaking to eliminates collision of ACK frames • Acknowledgement is managed by *sent-table*, *received-table*
Power Control by Busy Tone	• Busy tone avoids collisions both at receiver and sender
Power control by neighbor information	• Transmission power adjusted based on number of neighbor node
Adaptive Power Control	• Adaptively adjust the transmission based on power level of neighbor

Figure 7. Path $u \rightarrow w \rightarrow v$ is preferred in place of $u \rightarrow v$ for energy saving.

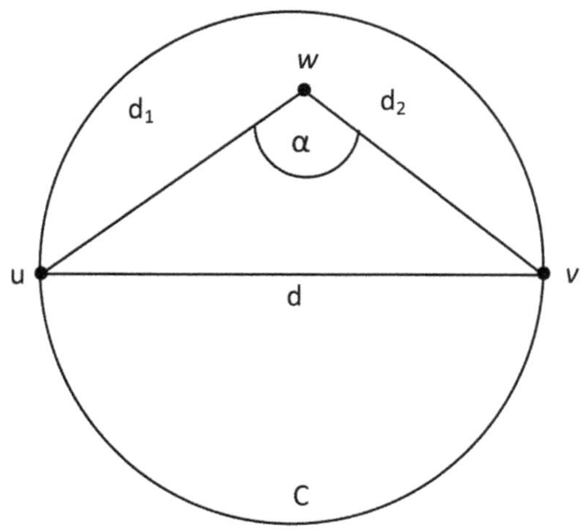

this scenario there exist two paths from *u* to *v*, one is a direct path $u \rightarrow v$ while the other is a relay path through $u \rightarrow w \rightarrow v$. Distance of the path $u \rightarrow v$ is represented as $d(u \rightarrow v) = d$ and thepath $u \rightarrow w \rightarrow v$ as $d(u \rightarrow w \rightarrow v) = d_1 + d_2$. Among these two paths we have to find an energy efficient path.

To compute an energy efficient path we need to calculate the energy consumed in each path. We know that radio signal propagates according to free space model. Power required to transmit a signal in the path $u \rightarrow v$ is proportional to square of the distance between them, which is d2. Power consumption in the path $u \rightarrow w \rightarrow v$ is the sum of power consumption from u to w and w to v, which is equal to. d12+d22. Now consider triangle uwv, where α is the angle between uw and vw.

$$d^2 = d_1^2 + d_2^2 - 2d_1d_2 \cos\alpha$$

as $w \in C$, so it implies that

$$\cos\alpha \leq 1, \text{as } w \in C, d^2 \geq (d_1^2 + d_2^2)$$

when is equal to 90°,

$$d^2 = d_1^2 + d_2^2$$

From the Figure 7 we can conclude that a node expend higher energy in a long direct path in comparison to shorter path. Therefore, it is always better to communicate using short multi-hop path rather than a direct path.

This above observation gives first argument in favor of topology control. So far energy is concerned, it is better to go through short multi hop path rather than a long direct path. In power control technique node *u* adjust its transmission power to a specific level by which it can communicate to node *v* through *w* only. If we map this observation to topology control approach it can be represent as: node *u* reduce its transmission power, such a way that it can reach *v* via *w*. In other way we can say that objective of node *u* is to remove the energy inefficient link *uv* from its edge set. Minimizing the maximum power used by a node is one of the main design goals for most of the topology control protocols.

The second motivation for topology control is based on network capacity enhancement by selecting an appropriate interference model. In ad hoc networking environment, increasing transmit power reduces packet drop due to far and near problem. However, it increases interference in the network. In this conflicting scenario there must be some trade-off between these two approaches. Gupta and Kumar (2000) design a protocol model to derive the upper and lower bounds on the capacity of the ad hoc networks. In this model packets from node *u* can correctly receive node *v* if,

$$d(vw) \geq (1 + \lambda)(d(u, v))$$

This means, distance between receiver and interference node should be more than distance between sender and receiver by a factor $(1 + \lambda)$ for any $\lambda > 0$. Here network capacity refers to the minimum value of $(1 + \lambda)$ such that it satisfies the network constraints. The above two observation motivates to study topology control protocols (Labrador,&Wightman, 2009, Santi, 2005).

Preliminary

The network topology is considered as a graph G= (V, E) where V= ($v1$, $v2$, ..., vn) a set of nodes and E is the set of links. Each node $u \in$ V has a unique id (ui) = i, where $1 \leq i \leq n$. By topology control we have a sub graph G'= (V, E') of G in G' the node has shorter and fewer numbers of edges as compare to G. The problems, assumptions and objectives for topology control approach is stated below:

Problem: Topology control is formulated as the problem of achieving one or more *objectives* under the certain *assumptions* subject to satisfying some *constraints* by employing some *approaches*.

Assumption: Without topology control all the nodes in the network are transmitting with common maximum power via Omni directional antenna.

Objective: Objective can be states as follows:

- Minimize the maximum power used at node level also known as *min-max* objective.
- Minimize the total power used by all the nodes in the network also known as *min-total* objective.

Constraints: To maintain network connectivity.

Physical and Logical Topology Control

Nodes in physical topology control (PTC) apply power control at node level to adjust the transmission power to a certain level to achieve topology control objectives. It reduces interference and energy consumptions in the network.

Logical topology control (LTC) in other way adjusts transmission power like PTC and also considers other parameters such as: reducing the logical node degree, transmission radius etc. In this approach a node carefully selects a set of neighbors to establish links and dynamically adjusts its transmission radius for different links. The PTC and LTC are depicted in Figure 8.

Figure 8. (a) PTC based on the transmission power only, and (b) LTC based on transmission power and neighbor set information

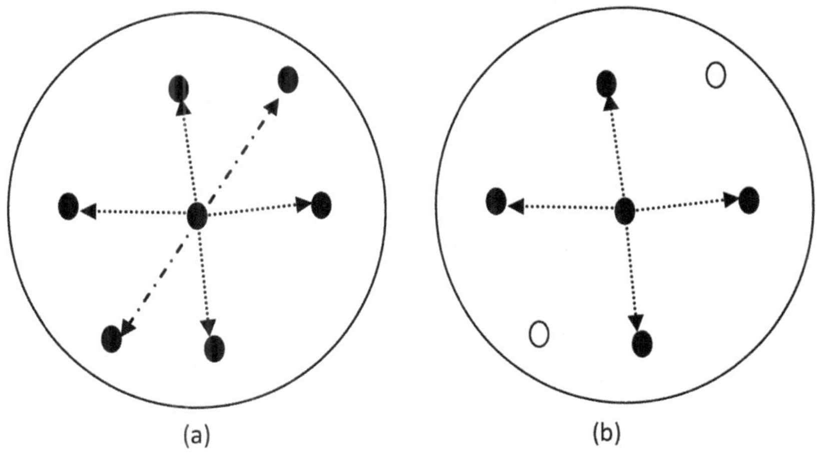

RNG and GG

RNG (Relative Neighborhood Graph) and GG (Gabriel Graph) are two topology control approach based on computational geometry. In RNG the edge between node *u* and *v* can be exists if there is no other node *w* such that:

d(uw)<d(uv) and d(vw)<d(uv)

where d(uv) represents the distance between node *u* and node *v*. It implies that in RNG no node is in the intersection area of two circles centered at node *u* and node *v* respectively.

In GG the edge between node u and v can be exists if and only if there is no node *w* such that

$$d^2(uw) + d^2(vw) \leq d^2(uv)$$

That states no node exists inside the circle where node *u* and node *v* are the two end points in a circle. Both RNG and GG are connected graph. Figure 9 shows the RNG and GG.

Unit Disk Graph and Point Graph

Unit disk graphs (UDG) have been used to model the topology in ad-hoc networks. In geometric graph theory a UDG is defined as the intersection graph of a family of unit circles in a Euclidean plane. A UDG can be formed by connecting vertex for each circle by an edge whenever the corresponding circles cross each other. Major weakness of UDG is that it assumes the coverage area should be perfectly regular which is difficult to achieve in real-life situations. Figure 10 shows an example of a unit disk graph.

Point graph is defined as a set of points distributed according to some probability distribution in a certain region. Points are then connected according to some rule. Most of the location based approach in topology control protocols relies on Euclidian distance to find the energy cost and assume that the topology is a unit disk graph or point graph.

3.3.1. Topology Control Protocols

Topology control protocols for ad hoc network can be classified into two types: (i) centralize control, and (ii) distributed control. In centralized control the central entity (access point or base station) takes the decision to adjust the transmission

Figure 9. (a) RNG: no node is present within the intersection area of two overlapping circle; (b) GG: no node presents inside the circle where u and v are two end points

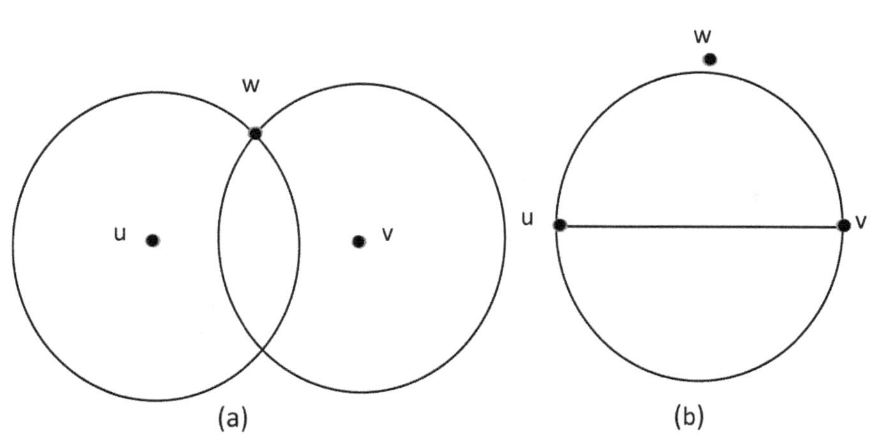

Figure 10. Unit disk graph consisting of unit distance circles

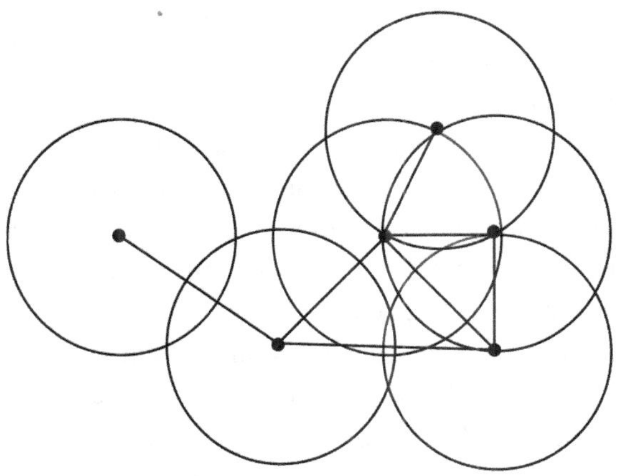

power according to the requirement. However such methods are not suitable for ad hoc network environments for which must of the topology control protocol for ad hoc networks are distributed in nature. Distributed topology control protocols are further classified as: *location based, direction based* and *neighbor based.* Each approach is discussed below briefly.

Location Based Approach

In this approach it is assumed that location information of a node is available by means of global positioning systems (GPS) triangulation-based positioning or any other positioning methods. The information obtained from these technologies will help to construct better topology. From that topological structure other parameters like transmission range, neighborhood information, and energy consumption can be obtained.

In order to reduce hardware cost some protocols assumes that a subset of the nodes are equipped with GPS receiver while other nodes get their node information by exchanging message with these nodes. Two important location based protocols LMST (Li, Hou, & Sha2003) and R&M (Rodoplu, & Meng, 1999) are explained below.

- LMST

In local minimum spanning tree [LMST] protocol, it assumed that the nodes are homogeneous, they are transmitting with common maximum power and their links are symmetric. The protocol has three phases: *information exchange, topology construction*, and *determination of transmission power*. These phases are stated below.

- *Information exchange phase*: Node sends *hello* message to their neighbors, the message contains sender's own ID and location information.
- *Topology construction phase*: A node constructs its local MST based upon the message received from its neighbors using Prim's algorithm. The link cost used to build the local MST is the Euclidean distance to its neighbor. The MST obtained in this way may not be unique. Since uniqueness is needed by the algorithm to preserve connectivity. Therefore protocol defines a link weight function for which after the execution of Prim's algorithm every node obtained their unique local MST.
- *Determination of transmission power*: Transmission power to a neighbor node is derived from the received signal strength of *hello* message. Transmit power required to send a message to any neighbor is generated from the final topology. Final topology is obtained by the intermediate neighbors of the local MST. The node also determines the broadcast power which is the minimum power needed to reach the farthest node in its neighborhood.

Authors have showed that the link obtained by LMST in its final topology may be unidirectional. In order to avoid these links they proposed two techniques. The first technique forces all links to be unidirectional by notifying the neighbors to include reverse path with unidirectional links.

The second technique deletes the unidirectional links from the final topology.

LMST has the following properties.

- It is fully distributed and uses local information to construct topology
- It has maximum node degree equal to six.
- It preserves connectivity in worst case and its message complexity is *O(n)* and computational complexity is *O(m log n),* where *n* is the number of node and *m* is the number of edges.
    - R&M protocol

The protocol is based on *relay region* and *enclosure graph,* and uses *all-to-one communication.* In this approach, communication through relay node is preferred. The protocol tries to find the most energy efficient node in the relay region through which the communication will be more energy efficient.

The protocol has two phases. In the first phase; nodes compute their neighbor set by means of local information. Then they broadcast the beacon message. Beacon message contains *sender node ID, sender location information* and *transmission power level.* On receiving the beacon information of other node, generic node *u* computes the enclosure graph based on the relay region. A relay region (RR) is defined as:

$$RR(u \to w) = \{(x,y) \in \Re^2 : P_{u \to w \to (x,y)} < P_{u \to (x,y)}\}$$

The shape of the relay region depends on the radio propagation model, and depends on the value of distance power gradient. The relay region is used to define the enclosure of a node. The nodes are said to be *dead,* when they are belonging to the relay region of another node. Otherwise the node is marked as alive. After receiving all neighbor information node *u* determines its neighbor set. The neighbor set only includes *alive* members. The protocol then uses a function called *Flip-All-States-Down-Chain* to update the dead and alive node states.

In the second phase; the protocol uses *Bellman-Ford algorithm* on enclosures graph to compute the minimum energy reverse spanning tree, rooted at the master node. The energy cost metric is used to find the shortest path between any two nodes. Each node calculates the minimum energy to reach the master node. Once the minimum paths are established, data transmission starts in that path.

Though R&M protocol conserve energy and preserve network connectivity at worst case, but it increases message overhead in phase two, as it require global information to construct final topology.

Direction Based Approach

In this approach it is assumed that the nodes are able to determine the direction of signal received from the other nodes. There are many techniques used to estimates the transmitting direction of a node, some techniques are proposed and discussed in IEEE Antenna and Propagation community. The direction based approaches are based on less accurate information as compared to location based approaches. However these approaches are able to produce equally good topology control protocols. There are several techniques used to estimate the direction for neighbor nodes. Yao graph algorithm (Yao, 1980) is an earlier version of directional based technique. Here we are discussing two direction based protocols: CBTC (Li, *et al.*, 2005) and DRNG (Borbash& Jennings, 2002).

- CBTC

In cone based topology control (CBTC) protocol, each node increases its transmission power until it finds a neighbor in every cone of degree α. The protocol has two phases. In the first phase a node determines its minimum power needed to reach its neighbors in all directions.

After that node exchanges information with other nodes to identify energy efficient edges and removes the inefficient links from the final topology. The protocol uses two types of messages viz. *beacon* and *acknowledgement* (ACK). The beacon message is transmitted with certain power $p \leq Pmax$, where *Pmax* is the maximum power of the node. Beacon messages contains the sender identity and the transmitted power level at which the message is sent. Nodes which are reachable using power *p*, acknowledged the signal. A node transmits both beacon and ACK messages in same power. After getting the transmitting power levels of neighbor, node invokes $CheckGap(\rho, d(u))$ procedures. Where ρ is the required angular gap between neighbors and *d(u)* is the set of *u*'s neighbor directions. The procedure returns *true* if CBTC condition parameter is satisfied, otherwise node invokes another procedure *IncPower(p)* to increases power value to next level. This procedure receives the current power level *p* and returns the next power level *p1*. If the condition not satisfied, *IncPower(p)* increases the transmission power and repeats the procedure until the condition is met or the transmitted power reaches its maximum value. The algorithm has been proven to preserve the network connectivity when, $\alpha \leq (5\pi/6)$.

CBTC is probably the most popular fully distributed topology control protocol, which requires less accurate direction information to operate. It saves good amount of energy by removing energy inefficient links from the final topology, it also maintains network connectivity in worst case. However its message complexity is higher due to beacon-ACK message handshaking and mechanism to find neighbors based on beacon transmission at increasing transmission power. These issues are utmost importance for the new variants of CBTC protocols.

- DRNG

It is another topology control protocol based on directional information. In DRNG it is expected that the node has low degree, small transmitting power, the hop-diameter is close to max power communication graph, and topology supports the connectivity (bi-connectivity) property. Satisfying all these goals is a difficult task in protocol designing point of view. Borbash and Jennings, performed extensive simulation on different topology such as *MST, RNG* and *minR* and found that RNG (RNG is discussed earlier in this section) provides better results compare to other approaches. Distributed relative neighborhood graph (DRNG) is based on RNG. It computes the RNG in a fully distributed and localized manner. The node *u* in DRNG computes the RNG with small transmission power $p(u) \leq Pmax$. The computing RNG continue till the covered region is equal to 2л or the current transmission power *p(u)* reaches maximum power *Pmax*. The neighbor covered region is shown in Figure 11. The coverage region of node *X* shown in the Figure 11, which is the shaded area of cone width *byd.* The covered region of node *y* is the sum of the covered region of node *X* and *Z*.

The main difference between DRNG and CBTC is that, the former includes uncovered nodes in the set of selected neighbors and the message complexity is less than CBTC. The

Figure 11. Coverage area of node y is the union of coverage area of node Z and X (shown in shade)

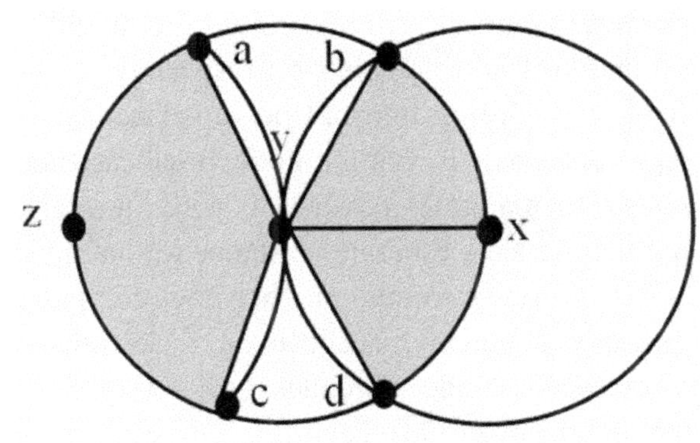

other nice features of DRNG are, it is fully distributed and it has maximal logical node degree equal to six.

Neighbor Based approach

The previous two approaches, (location and direction based,) requires extra information from their neighbors rather than their own presence. They based on very specific location information obtained from GPS or less accurate direction information based on *angel of arrival*. However these supporting technologies are expensive and may not support always. For examples the GPS enabled nodes may not get location information in indoor environments. Neighbor based approach in other way can overcome these problems. The protocols based on this approach are based on logical topology control approach, in which every node in the network is connected to the k-closest neighbor. Here we are discussing two neighbor based protocol called K-Neighbor (Blough, *et al.*, 2003)) and LFTC (Sheu, Tu, & Hsu, 2008)

- K-Neigh Protocol

The k-neighbor protocol is based on maintaining the number of neighbors of every node equal to or below a specific value *k*. It is assumed that initially all the nodes are transmitting with common maximum power *Pmax* and the wireless medium are symmetric. For a generic node *u*, *k* is taken as the input parameter represented as target number. *N(u)* and *KN(u)* are neighbor set and k-closest neighbor set respectively. The final broadcasting transmit power of node *u* is represented as *p(u)*. Initially nodes broadcast their *node ID* at maximum power. Any node on receiving the message estimates distance to that node and store these information. Then node computes its k-closest neighbor according to the estimated distance. The node *u* broadcast (u, *KN(u)*) with maximum power. On receiving this message other nodes store this information for preparing final topology. By exchanging the neighbor set each node keeps the symmetric neighbor and removes the asymmetric neighbor. From *KN(u)* transmit power *p(u)* is obtained which is the minimum power required to reach the farthest node in *KN(u)*.

K-Neigh is based on relatively low quality information and is light weight. Its message complexity is $O(n)$, and the total number of messages in the network is *2n* (as each node sends two messages only). As compared to CBTC it save 20% more energy and it perform better in terms of reducing node degree. However, its connectivity in worst case is a major issue of concern.

- LFTC

Location free topology control [LFTC] protocol, is based on neighbor information to determine the final topology. The algorithm has two phases: *link determination* and *interference announcement*.

Link determination phase: each node broadcast *hello* message with maximum power, *Pmax*. Each *hello* message contains the *sender-ID* and a specific data structure such as *direct communication cost, minimum communication cost*, and *link type*. Nodes in LFTC maintain a table known as *vicinity table* to keep these information. Table 4 shows the structure of vicinity table. The table is updated each time when a node receives a *hello* message from its neighbor. The *direct cost* field stores the energy consumption in the direct path to that particular neighbor. *Minimum cost* field stores the minimum energy cost to its neighbor through alternate path. Initially the minimum cost and direct cost are same, but on receiving any alternate path through other neighbor, the table is updated if the path is energy efficient as compared to direct path. *Link type* fields records *direct* or *indirect* depending on the minimum cost path. The field is marked as *d* for direct path and *i* for indirect path. Initially nodes derive their neighbor set which is obtained by transmitting

Table 4. Vicinity table of a node in LFTC

Neighbor-ID	Direct Cost	Minimum Cost	Link type

Table 5. Summary of topology control protocols

Protocol	Assumption	Approach
LMST	Homogeneous node	Location based
R & M	Master-node & Fmax	Location based
CBTC	Direction Information is known	Based on Angle of arrival (AoA)
DRNG	Network topology is a RNG	Graph based, Angle Based
K-Neigh	No prior information of location and direction is required	Distance based, Neighbor based, Probabilistic connectivity
LFTC	No prior information of location and direction is required	Distance based, Neighbor based

with maximum power. Then nodes update their vicinity table information by the exchange of *hello* message periodically among themselves. Finally a generic node *u* prepares its DCS(u) from the updated table information. In LFTC minimum transmission power to send the data packets to the farthest neighbor in the DCS (u) is calculated from the vicinity table.

The second phase of LFTC is *interference announcement phase.* This phase handles the hidden terminal problems at MAC layer to increase the network throughput. This phase determines the optimal power required to transmit control packets.

LFTC enjoy many nice features, the protocol has a smaller hop count and a low collision ratio. It is operated without location and directional information similar to XTC protocol (Wattenhofer & Zollinger, 2004). Each node in LFTC determines separate optimal powers for data and control packet. However it has higher message complexity like CBTC. In a network of *n* nodes its message complexity is *O(n2)* to construct DCS.

3.3.2. *Summarization*

We discuss topology control protocols for ad hoc network and classify them depending upon their mode of operation as: *location, direction* and *neighbor* based. Table 5 summarizes them based on their assumption and approach. These are few findings listed below:

- LMST handles the unidirectional links in the final topology either by converting the unidirectional link to bidirectional link or by deleting that completely. LMST requires *O(n2)* message exchange to perform bidirectional links in the final topology as each node sends *(n-1)* message to other node.
- R&M protocol has many good features. Its *Min-Energy-All-To-One* feature optimizes energy efficiency but it relies on radio propagation model to compute the enclosure graph.
- One of the major limitations of location based protocols (such as GPS based approach) incurs high delay in acquisition of location information which substantially decreases the network throughput.

- Direction based approaches does not requires more accurate information and they relies on less accurate information such as angel-of-arrival(AoA) but they produces very good topology control protocols.
- The CBTC is the most popular topology control protocol based on directional information. But the major limitation is increased message overheads, as higher numbers of messages exchange is required to construct the final topology.
- Node degree in DRNG is relatively low, and it preserves connectivity in worst case.
- K-neigh protocol is a light weight protocol with low message overheads.
- LFTC is based on vicinity table information to construct the final topology. It has good energy saving feature but it requires accurate distance estimation for constructing its DCS.

4. CONCLUSION AND FUTURE DIRECTIONS

In this chapter we discussed energy management in wireless ad hoc networks with considering the different issues and requirements. The different power conservation protocols are discussed to provide a complete overview of the energy managements in MANETs. It mainly focused on three power efficient techniques: *power management*, *power control*, and *topology control.* We discuss few protocols in each category to provide a basic idea about their operation. We summarize here highlighting important points

- Energy optimization can be done at all layer of the protocol stack but is focused mainly at MAC layer and routing layer.
- Power management techniques save energy by putting nodes in sleep state and their energy conservation is based on the time they spent in sleep mode. However, this saving comes at significant delay in the packet delivery.
- Power control techniques are used to derive transmission power at node level for transmitting data and control signals based on different criteria. Their performance depends on the appropriate selection of transmission power level; otherwise they will increase computational overhead.
- Topology control techniques reduces energy consumption of a network by reducing interference in network by adjusting network topology but they have to maintain the network connectivity otherwise their power saving will be no more applicable.
- IEEE 802.11 PS mode is one of the most popular energy saving protocol and has been appreciated by majority of studies; however it's variant for multi-hop network yet to prove its efficiencies.
- Cross layer design (CLD) approach is a potential area to be focused more not only for energy saving but to improve overall performance in the networks. It can also handle energy issue at different layer of protocol stack.
- Energy harvesting is a promising approach of energy saving which can harvest energy from the different sources such as solar power, electromagnetic waves, wind energy, etc. The characteristic of energy harvesters in wireless network differs from the conventional approach thus requires a redesign of network protocols and transceiver hardwires.

The aforementioned techniques and opportunities for power saving are yet to be complete, new methodologies and techniques continue and will continue to be developed to save as much energy as possible. With the new technologies and techniques we will be able to achieve much better results.

The approaches discussed in this chapter can provide benefits for particular environment and can be implemented at various layers in the protocol stack. They provide a solid basis for future developments; however are not definitive answers to the problem of optimal energy usage in wireless ad hoc networks.

REFERENCES

Abasgholi, B., Kazemi, R., Arezoomand, M., & Enayati, A. R. (2008). Neighbor detection power control mac protocol in mobile ad hoc networks. *Proceedings of 3rd International Symposium on Wireless Pervasive Computing*, Greece.

Blough, D., Leoncini, M., Resta, G., & Santi, P. (2003). The k-neighbors protocol for symmetric topology control in ad hoc networks. *Proceedings of 4th ACM International Symposium on Mobile Ad Hoc Networking & Computing,* Annapolis, MD, (pp. 141-152).

Borbash, S., & Jennings, E. (2002). Distributed topology control algorithm for multihop wireless networks. *Proceedings of IEEE International Joint Conference on Neural Networks*, Honolulu, HI, (pp. 355-360).

Chen, B., Jamieson, K., Balakrishnan, H., & Morris, R. (2002). Span: An energyefficient coordination algorithm for topology maintenance in ad hoc wireless networks. *ACM Wireless Networks Journal*, *8*(5), 481–494. doi:10.1023/A:1016542229220

Choi, H. S., Byun, H., & Lim, J. (2008). Adaptive power control MAC in wireless ad hoc networks. *IEICE Transaction on Communication . E (Norwalk, Conn.)*, *91–B*(1), 309–313.

Deng, J., & Hass, Z. J. (1998). Dual busy tone multiple access (DBTMA): A new medium access control for packet radio networks. *Proceedings of International Conference on Universal Personal Communication.*

Feeney, L. M. (2001). An energy-consumption model for performance analysis of routing protocols for mobile ad hoc networks. *Mobile Networks and Applications Journal, 6*(3).

Feeney, L. M. (2004). Energy efficient communication in ad hoc networks . In Basagni, S., Conti, M., Giordano, S., & Stojmenovic, I. (Eds.), *Mobile ad hoc networking* (pp. 301–327). IEEE Press and Wiley-Interscience. doi:10.1002/0471656895.ch11

Gupta, P., & Kumar, P. (2000). The capacity of wireless networks. *IEEE Transactions on Information Theory*, *46*(2), 388–404. doi:10.1109/18.825799

IEEE 802.11 Working Group. (1999). *Wireless LAN medium access control (MAC) and physical layer (PHY) specification.*

Jung, E. S., & Vaidya, N. H. (2002). A power control MAC protocol for ad hoc networks. *Proceedings of ACM International Conference on Mobile Computing and Networking (MOBICOM)*, (pp. 36-47).

Jung, E. S., & Vaidya, N. H. (2002). *An energy efficient MAC protocol for wireless LANs*. IEEE INFOCOM.

Kim, J., Lin, X., & Shroff, N. B. (2011). Optimal anycast technique for delay-sensitive energy-constrained asynchronous sensor networks. *IEEE/ACM Transactions on Networking, 19*(2), 484–497. doi:10.1109/TNET.2010.2072515

Labrador, M. A., & Wightman, P. M. (2009). *Topology control in wireless sensor networks.* Springer Science + Business Media B.V. ISBN 978-1-4020-9584-9

Li, L., Halpern, J. Y., Bahl, P., Wang, Y.-M., & Wattenhofer, R. (2005). A cone-based distributed topology-control algorithm for wireless multi-hop networks. *IEEE Transactions on Networking, 13*(1), 147–159. doi:10.1109/TNET.2004.842229

Li, N., Hou, J., & Sha, L. (2003). Design and analysis of an MST-based topology control algorithm. *Proceedings of IEEE Infocom 03*, San Francisco, CA, (pp. 1702-1712).

Lin, X. H., & Lau, V. K. N. (2003). Power control for IEEE 802.11 ad hoc networks: Issues and a new algorithm. *Proceedings of International Conference on Parallel Processing, ICPP'03*, Kaohsiung, Taiwan.

Rodoplu, V., & Meng, T. (1999). Minimum energy mobile wireless networks. *IEEE Journal on Selected Areas in Communications*, *17*(8), 1333–1344. doi:10.1109/49.779917

Santi, P. (2005). Topology control in wireless ad hoc and sensor networks. *ACM Computing Surveys*, *37*(2), 164–194. doi:10.1145/1089733.1089736

Sheu, J., Tu, S., & Hsu, C. (2008). Location-free topology control protocol in wireless ad hoc networks. *Elsevier . Computer Communications*, *31*(14), 3410–3419. doi:10.1016/j.comcom.2008.05.039

Singh, S., & Raghavendra, C. S. (1998). PAMAS Power aware multi-access protocol with signaling for ad hoc networks. *ACM Computer Communication Review*, *28*(3), 5–26. doi:10.1145/293927.293928

Sivalingam, K. M., Chen, J., & Agrawal, P. (2000). Design and analysis of low-power access protocols for wireless and mobile ATM networks. *Wireless Networks*, *6*, 73–87. doi:10.1023/A:1019152506607

Stemm, M., & Katz, R. H. (1997). Measuring and reducing energy consumption of network interface in handheld devices. *IEICE Transaction on Communications . E (Norwalk, Conn.)*, *80-B*(8), 1125–1131.

Tseng, Y., Hsu, C., & Hsieh, T. (2003). Power-saving protocols for IEEE 802.11-based multi-hop ad hoc networks. *Computer Networks . The International Journal of Computer and Telecommunications Networking*, *43*(3), 317–337.

Wattenhofer, R., & Zollinger, A. (2004). Xtc: A practical topology control algorithm for ad hoc networks. *Proceedings of Fourth International Workshop on Algorithms for Wireless, Mobile, Ad Hoc and Sensor Networks (WMAN)*, Santa Fe, NM.

Wu, S.-L., Tseng, Y.-C., & Sheu, J.-P. (2000). Intelligent medium access for mobile ad hoc networks with busy tones and power control. *IEEE Journal on Selected Areas in Communications*, *18*(9), 1647–1657. doi:10.1109/49.872953

Xu, Y., & Heidemann, J. & Estrin. D. (2000). *Adaptive energy-conserving routing for multi-hop ad hoc networks.* Technical Report 527, USC/ Information Sciences Institute.

Zhang, J., Fang, Z., & Bensaou, B. (2005). Adaptive power control for single channel ad hoc network. *Proceedings of IEEE ICC'05*, Seoul, (pp. 3156-3160).

Chapter 18
Modelling WSNs Using OMNeT++

Erwin Anggadjaja
Earth Observatory of Singapore, Nanyang Technological University, Singapore

Ian V. McLoughlin
Nanyang Technological University, Singapore

ABSTRACT

Wireless sensor networks (WSNs) have been used to observe and monitor many environments for specific purposes and in many ways over the past few years. A number of operational trade-offs are possible when planning a WSN, influencing coverage, bandwidth, redundancy, lifetime, expandability, and so on. However, for systems in potentially hazardous locations or those experiencing restricted access, system unreliability tends to be the greatest operational concern. In the process of creating reliable WSNs for hazardous locations, it is highly desirable to ensure both an accurate and a reliable system design prior to deployment, and with as little unnecessary trade-off as possible. Especially as sensing systems become larger and more complex, and potential failure modes increase, this becomes more difficult to achieve. In an attempt to answer the question of reliability assurance, the authors investigate WSNs in the context of accurate and fast modelling of such networks. A comprehensive comparison of three modelling tools (ns-2, OPNET, and OMNeT++) is explored in this chapter, concluding that OMNeT++ is worthy of study as an alternative to the other two more established tools. As an illustration of the use of OMNeT++, two modelling schemes are simulated and compared against the theory to determine both bit-level correctness, but also to demonstrate ease of modelling and analysis.

INTRODUCTION

Mobile ad hoc networks (MANET) are self-configuring networks of mobile devices intended for particular purposes related to communication, and usually associated with wireless multi-hop communication (Figure 1). Two fundamental research challenges exist related to MANETs: the aspect of self-configuration – which is essential since each device in a MANET is free moving autonomously, and the limited reach of wireless communication links. These ideas are shared between MANETs and wireless sensor networks (WSNs) in general.

DOI: 10.4018/978-1-4666-0321-9.ch018

Figure 1. Mobile ad hoc networks

To date, WSNs have been well used in many different application areas, including scientific experimentation, engineering validation, environmental monitoring, health-related monitoring, industrial sensing & control, and also numerous commercial applications. MANETs are starting to be applied across the same range of similar areas, experiencing similar issues, compounded by their mobile (and ad hoc placement) nature. When such systems operate in a real environment, system unreliability may become the principal concern. For such systems, we need to answer the following questions:

1. Can we simulate such system to some degree of accuracy that will ensure high reliability in the field (with some trade-offs accepted, or course)?
2. As sensing systems become larger and more complex, the "try it and hope it works well" approach becomes less likely to succeed; we need an assurance that a system will work, before it is deployed. Can we ensure this through modelling?

We propose the use of fast but accurate *scenario modelling* to answer these questions - and will describe the practical use of OMNeT++ in aiming to resolve answers to both of the above questions.

To easily and conveniently model WSNs deployments and their failure scenarios, a reliable simulation tool is required. The aim is obvious; to model and predict the system-level behaviour of the real world environment as precisely as possible. There are a number of network simulators related to sensor networks (as can be seen later). It is important, in order to effectively develop any protocols with the help of simulation, to know the benefits and weaknesses of the various simulators. This will aid us in choosing which simulators should be used in terms of flexibility, reliability, and budget.

We explore the features of the three most commonly used network simulator tools, which are ns-2, OPNET, and OMNeT++. Then we explore various related issues (i.e. availability of protocol and topology models, debugging & GUI environment, simulation comparisons, and open source & cost issues), to conclude that OMNeT++ is worthy of study. We restrict the discussion to the smaller, less expensive research tools, rather than the more professional, capable, and higher cost tools such as Qualnet. The endpoint is ideally a low cost, preferably open source, capable and fast tool with ease of use in planning WSN deployments for hazardous locations.

Within this chapter, we also simulate the concepts of simple wireless transmission, and built this step-by-step into a full WSN simulation using OMNeT++ as an illustration of the process. First, we will model a simple concept of lost and received packets with the condition of variable Bit Error Rate (BER). We then match this to analytical equations for BER, Packet Error Rate (PER), goodput and efficiency, and simulate various conditions. We the progress to a more advanced and realistic simulation, introducing the concept of Segmentation and Reassembly (SAR), defining a window size as a 'dimension' of a packet that is sent over the air, to explore various system trade offs. Next, we add an error control mechanism (inspired by TCP), with retries and reACKs schemes similar to the radio link protocol (RLP). Lastly we will consider overall system efficiency issues. Simulation results are compared against theory, and various trade-off issues explored briefly.

In summary, this chapter introduces the issues of modelling WSNs for hazardous locations, in particular through the use of the OMNeT++ simulation tool and provides readers with a useful reference for developing their own models using this tool. Section 2 will provide a brief WSN background, and state the case for modelling. Section 3 explores the candidate simulation tools, followed by narrowing the discussion to OMNeT++ in Section 4 and concluding in Section 5. Section 6 briefly notes a number of interesting opportunities for further research in this domain.

WSN BACKGROUND AND MODELLING

2.1 The Need for Wireless Sensor Networks

Wireless Sensor Networks (WSN) are intercommunicating sets of scattered digital radio devices hosting sensor nodes, and conveying sensor data. They are usually deployed across a geographically distributed area (which could range in extent between an office and a tectonic fault) and which cooperatively observe environmental situations for specific purposes, linked together by wireless networks. Usually these form an ad-hoc network, where each node performs multi-hop routing using some kind of in-built routing algorithm. The nodes 'inside' the network forward or disseminate data to a specific destination called the *sink-node* (depicted in Figure 2).

Each of sensor node is typically equipped with a transceiver capable of performing wireless com-

Figure 2. Wireless Sensor Networks

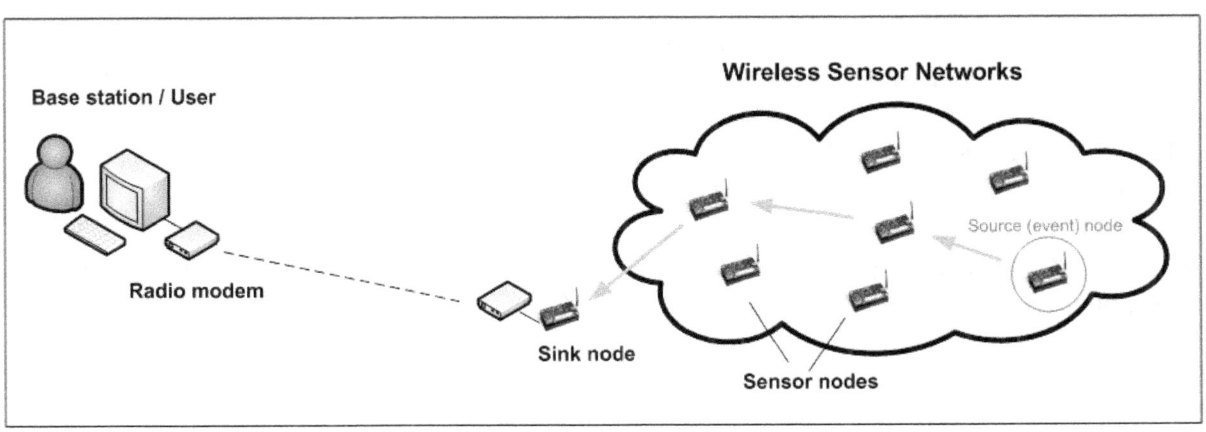

munications tasks. Its size might vary considerably; similarly the price may range from a few dollars upwards to hundreds or even thousands of dollars, depending upon the device complexity and capability. Both size and cost constraints impact the system in terms of energy, memory, computational speed and bandwidth (Romer and Mattern, 2004).

The original use of WSN was for military surveillance. However, WSN systems are now used in many different areas. For example, according to Akyildiz, Su, Sankarasubramaniam, and Cayirci (2002), these include environmental monitoring, health, and also commercial applications. Further afield uses also include industrial, civilian, machinery condition-based maintenance (Tiwari, Ballal, & Lewis, 2007), habitat condition monitoring (Szewczyk, Polastre, Mainwaring, Anderson, & Culler, 2004; Mainwaring, Polastre, Szewczyk, Culler, & Anderson, 2002), home automation, data logging (Das, Lewis, & Popa, 2009), and so on.

Wireless sensor networks also have been used specifically in many areas of scientific study, including environmental monitoring (Vardell-Oliver, 2005; Tolle et al., 2005), active volcano monitoring (Werner-Allen, Lorincz, Johnson, Lees, & Welsh, 2006), monitoring of debris flow (Lee, Cho, King, Fang, & Lee, 2009), and also structural monitoring (Paek, Chintalapudi, Caffrey, Govindan, & Masri, 2005; Xu et al., 2004; Pakzad et al., 2005).

Other areas of deployment for sensor networks are their use in harsh environmental conditions. For example, for monitoring weather conditions in rugged wild-land fire environments, FireWxNet (Hartung, han, Seielstad, & Holbrook, 2006) was developed. Similarly in a fire environment, Doolin and Sitar in 2005 deployed nodes network to measure humidity and temperature in post flame situations.

2.2 The Need for Modelling

The common approach in WSNs is by trying to deploy all the sensor nodes by placing them in the desired area (with perhaps some adjustment in location due to local topography), providing power, then expecting that they will work, with perhaps no more than minor adjustment. Following any critical activities, all the scattered sensor nodes should detect the event, collect the related data, and transfer it back to a base station via long-distance wireless radio link.

However, this approach has limitations. As usually, the environment is *dynamic*, meaning it is changing all the time, it may undergo accelerated change during an event. Some of the problems that may happen include:

a. Displacement of nodes.

Some activity or environmental conditions almost inevitably produce motion during, between, and after the events being sensed. These changes are likely to displace the nodes, with a further possibility of harming the nodes, causing the loss of collected data.

b. Node failure.

The dependability of this network on its nodes for both data collection and routing, also counts on the condition of the nodes, their sensors and antenna(s). For example, a power failure or overheating of nodes might contribute to both corrupted or missing data.

c. Unreliability of the wireless links.

It is well known that, although wireless networks have evolved significantly in capability over recent years, research shows that such links can still be quite unreliable (Zhang & Soong, 2005). The stability of the overall system results from many factors in wireless communication such as multipath fading and other forms of path-loss besides reflection, shadowing, diffraction, and scattering, which are statistically unstable (Rappaport, 1996). Not to mention some undesired obstructions, such as loss of LOS (line of sight) or interference caused by extreme weather conditions.

d. Deploying additional nodes.

Consider that when some of the above scenarios happen, deploying more nodes might be needed, however some extra node deployment may in fact cause the network to become unsustainable in the event of particular failure modes.

From the above explanation, much thought, and research is needed to make wireless sensing both more efficient and more effective. Therefore, we propose high-level *modelling* of such systems using discrete network simulator tools, in order to assist in answering a number of research questions:

1. Complexity of the system.

As sensing systems become larger and more complex, it becomes less and less possible to rely on a "try and see" or "try and hope it works well" approach. We need an assurance or correct operation in pre-deployment systems, which becomes both more important and more difficult to achieve as systems scale upwards in complexity. With modelling, this can be achieved with less effort. For example, a large scale deployment of nodes can be simulated much more easily in a simulator, than can be tested through a geographical test deployment (assuming, of course, that the simulator is accurate).

2. Data Reliability.

In addition, we can model the system to some degree of reliability, such:

a. to create a reliable simulation of how the data is carried or disseminated between all the sensor nodes.
b. to observe or develop the most suitable protocol for different scales and arrangements of wireless environmental monitoring systems.

We note that it is not possible to guarantee 100% reliability in reality, however, we can approach closer to this ideal with some trade-offs, usually cost, plus forethought (or fore-modelling).

With modelling or simulation, we are confident we can design high reliability and fault tolerant sensing systems, adapted to local conditions and supporting a wide range of sensors with different precisions to simulate a reliable, realistic, efficient and effective environment for wireless environment.

SIMULATION TOOLS OVERVIEW: WHY OMNET++ FOR WSNS

3.1 Motivation

To easily and conveniently model WSN deployments and failure scenarios, a reliable simulation tool is required. Typically, simulation tools test protocols being developed, or perhaps are used to weigh options against competing protocols. Further common uses are to develop communication or network protocols or architectures, when correctly modelled and specify the relevant operating behaviour of subcomponents, for example network nodes, channel links, etc. The aim is usually simple; it is to model and predict the behaviour of the real world system as precisely as possible.

We notice some disadvantages of using simulation tools:

a. Uniqueness and drawbacks of each simulator

Some simulators are little more than ongoing development projects and thus updates are expected, i.e. they are active research communities. Definitely a simplified or inaccurate modelling of library functions might therefore be a shortcoming of some co-called simulation tools. Moreover, not all simulators have relevant or well-known protocols included, and these would need to be developed before such tools can be used to model real-world scenarios.

b. Uncertain results

There also exists a danger when using simulation tools in that, the results could be uncertain or un-useful, and at times even critical detail may be missing. This is a particular risk with higher-level tools that may hide unrealistic models within a pretty GUI.

However, there are also a number of motives behind the use of simulation tool, such as:

a. Many researchers publish their papers, i.e. give results, based on experimental simulation. This type of investigation is both accepted and in some cases actually preferred in the research literature. This stems from the need for a reliable model to foresee the performance in a real world environment, especially in the field of wireless sensor networks (Varga & Hornig, 2008; Lessmann, Janacik, Lachev, & Orfanus, 2008; Zhang, Chen, Fan, Xu, & Sun, 2008) and particularly for complex high-level data transmission protocols. This approach is usually well suited to tackling scalability issues (large numbers of nodes) and especially for systems with costly sensors and difficulties in real-world deployment.
b. Practical implementation

When considering to test a large-scale network, deploying sensor nodes in the field can be a time consuming and difficult task. Even dangerous or extremely expensive in some locations.

c. Investment wise.

In terms of cost and time, simulation-based testing can help to suggest whether or not the time and monetary investments to build the system are wise, before undertaking a real-world deployment.

Therefore, we advocate the use of simulation and modelling tools as a useful and rapid approach to developing and testing new protocols and system arrangements for sensor networks.

3.2 Comparison of Several Simulation Tools

jadThere are a number of network simulators related to sensor networks, as suggested by Curren in 2008. All of them present both positive and negative features in different respects. Thus, it is important, in order to effectively develop any protocols with the help of simulation, to know the benefits and weakness of the various simulators, to aid us in choosing which simulators should be used in terms of flexibility, reliability, and budget-wise.

Several simulation frameworks have been widely used in networking and telecommunication research, including ns-2, GloMoSim, OPNET, SensorSim, J-Sim, OMNeT++, etc (Curren, 2008), with most of the available network simulation tools based on the paradigm of discrete event-based simulation (DES) (Fishman, 1978). Here, the simulated network is triggered when a message or a packet arrives at a particular node, then this successively processes more events in turn (queue operations, for example).

Some of the mainstream simulators that may be suitable for our current simulation needs are described as follow:

3.2.1. ns-2

Network Simulator (http://www.isi.edu/nsnam/ns), ns, is a simulation tool developed by Network Research Group at the Lawrence Berkeley

National Laboratory (LBNL) as a variant of the REAL network simulator in 1989 and has subsequently evolved over the past few years. It is a discrete event network simulator, mainly used for ad-hoc networking research as well as simulation of network protocols, both wired and wireless, especially under realistic channel conditions. Its simulation language is based on C++ and provides an interface through OTcl, an object-oriented dialect of Tcl (Tool Command Language).

At the present, it is popularly called ns-2 (as its current generation release), as all the functionality in version-1 has been combined into version-2, and that only ns-2 will be developed in the future. It also maintains full backward compatibility for almost all functionality to version-1, but provides more features for configuration and programming. It is very popular in the research area for its open source model and online documentation.

ns-2 is also one of the most popular network simulation tools for sensor networks (Downard, 2004; Naoumov & Gross, 2003). As an object-oriented discrete event simulator, ns-2 has a range of extensions of protocols already modelled, and allows users to create and write new protocols. This extensibility feature has possibly acted as the main draw for many developers to use, with its combination of ease and popularity and high number of different protocols built in.

However, according to Curren (2008), ns-2 posses some drawbacks which are:

a. ns-2 does not scale well for sensor networks

This is because of its sub-optimal object-oriented design methodology: each node stands as its own object, thus creates a large number of dependencies, which are a obstacle when applied to large environments. Moreover on this issue, Henderson, Roy, Floyd, and Riley (2006) and Xue et al. (2007) also report that a major deficiency of ns-2 is regards to memory usage and simulation run-time. This is a problem for the WSN research domain since some WSN systems have potentially thousands of nodes being involved.

b. Lack of customization

For example, packet formats, energy models, MAC protocols, are different for most sensors - there is a real lack of standardization in WSN hardware and software.

c. Lack of application model

This should not be a problem for most network environments, however in terms of sensor networks, there are many interactions between protocol levels: cross layer systems are common. Its successor, ns-3, which addresses many of these concerns, could be an alternative. However this is still in the early stages of development as reported by Weingrtner, Vom Lehn, and Wehrle (2009).

3.2.2. OPNET

As ns-2 has become established as one of the standard network simulation tools used among developers, other network simulators have gained more attention during recent years. Another high-level discrete event simulator, OPNET (Chang, 1999; Pan; Cai & Jia, 2005), has also emerged as an option to be considered.

OPNET was started in 1986 by MIL3 Inc. (known as OPNET Technologies Inc. now). It is a network level simulation tool which operating at a "packet-level". Originally built to simulate fixed networks, it is now available for wireless network hardware and protocols as well. It is also constructed using C and C++, and contains many network specific library functions.

Its hierarchical structure defines three domains which are:

a. Network/ topology level

This level defines an overall scope of the system, for example a demonstration of a global world network. Some of the purposes of this are to identify and define the objects in the system, such as configurations, interconnections, locations, etc.

b. Node level

This is where we define nodes such as routers, mobile devices, etc.

c. Process level

This domain summarizes the behaviour of the modules, such as data traffic, packet processing algorithms, protocols, source code, etc.

OPNET supports the use of modelling different sensor-specific hardware, such as physical-link transceivers and antennas. Its graphical interface can help developers quickly simulate various models in graphical form as well as provide an animation of the output (not only an eye-catching 'feature', but sometimes an important tool to aid understanding of such systems).

However, we found its drawbacks also to be in terms of object-oriented scalability, and its models always using a fixed topology, as stated by Xian, Shi, and Huang (2005), which tends to make it less flexible. It is perhaps not as popular as ns-2 and does not have a high number of pre-defined protocols. Additionally, another major issue is that OPNET is only available in commercial form, not open source at the present time.

3.2.3. OMNeT++

OMNeT++ (http://www.omnetpp.org) is the last simulation tool to be considered here. OMNeT++ is also built upon a component-based discrete event simulator. The modules are designed in a hierarchical style, similar to OPNET, where each module can contain or encapsulate several other modules.

After considering earlier reports by Xian et al. (2005) and others, introducing the features of OMNeT++ and comparing this with ns-2 and OPNET, we can summarise the following points:

a. Protocol models

OPNET has several prebuilt protocol models (i.e. Ethernet, TCP/IP, ATM), however it is expensive. Compared to ns-2, it is more centred around TCP/IP, SCSI, and FDDI. We found that OMNeT++ is more accommodating (flexible) and mature (Lessmann, Janacik, Lachev, & Orfanus, 2008) for protocol modification, besides with its other model/frameworks available such as INET (http://inet.omnetpp.org), MF (http://mobility-fw.sourceforge.net), MiXiM (Kopke et al., 2008), Castalia (Pham, Pediaditakis, & Voulis (2007), etc. Recent works by Zhang et al. (2009) and Maureira, Uribe, Dalle, Asahi, and Amaya (2009) also show that OMNeT++ has gained confidence as a simulator choice prior to implementing real models. It should be noted, however, that OMNeT++ is not yet approaching the level of popular use of either ns-2 or OPNET (both in terms of industrial applications, and published research).

b. Hierarchical / Network Topology models

ns-2 does not allow hierarchical models, which limits it for WSN use. OPNET allows hierarchical models, but only with a fixed topology. Since in WSN environments, networks tend to be organic (i.e. changing), especially for scenario modelling, thus the flexibility of different topologies and protocols is important, and OMNeT++ appears best suited to this. As noted by Xian et al. (2005), although OPNET allows hierarchical models with arbitrarily deep nesting, OMNeT++ has more flexibility by permitting customized topologies to be used.

c. Debugging and GUI environment

ns-2 debugging performance is considered relatively poor as reported by Xian et al. (2005) since it takes longer to run, and requires more memory to simulate. OMNeT++ offers three convenient tools for debugging which are: automatic animation, module output windows, and object inspectors. The first is mainly supported by most of simulators, however not the other two. OPNET possess a powerful command-line simulation debugger, but it does not boast a rich graphical runtime environment which makes OMNeT++ more convenient in modelling and rapidly visualizing WSN deployment (Weingrtner, Vom Lehn, & Wehrle, 2009).

d. Simulation comparison

Two different and important simulation studies have been carried out for comparing OMNeT++ and ns-2. First, in a work by Xian et al. (2005), a simulation of Directed Diffusion under simpleMAC and IEEE 802.11 MAC was performed. The parameters compared were delivery ratio, runtime, and memory usage. The concluding result showed an excellent performance indicated that OMNeT++ is much more scalable than ns-2 in terms of both execution time and memory overhead. Another simulation comparison by Albeseder and Fuegger (2005) also claimed that OMNeT++ outperforms ns-2, for moderate network sizes.

e. Open source and cost

OPNET provides the source code for protocol models only. Both OMNET++ and ns-2 are fully open-sourced. This is the last but not least consideration for choosing a WSN simulation tool. Furthermore, since OPNET is quite costly compared with the other two, we will not consider OPNET further.

To sum up this section, OMNeT++ was found to be a useful candidate for WSN simulation modelling software, based upon its features. Compared with ns-2, OMNeT++ exhibits better performance for certain bit- or packet-level network scenarios. Furthermore it also posseses significant advantages compared with the commercial OPNET software.

3.2.3.1. Modelling in OMNeT++

This section will briefly explain OMNeT++ modelling. As mentioned above, OMNeT++ (Objective Modular Network Testbed in C++) is an object-oriented modular discrete event simulator, mainly focusing on the following domains:

a. Modelling of wired and wireless communication networks.
b. Protocol and queuing network modelling.

Overall, it models and simulates any system where the discrete event approach is suitable, by conveniently mapped all the communications by exchanging messages. One of the fundamental ingredients of this infrastructure is a component architecture for simulation models. Components (modules) are programmed in C++, then assembled into larger components and models using a high-level language (*ned*).

Graphical, animating user interfaces are highly useful for demonstration and debugging purposes, and command-line user interfaces are best for batch execution. OMNeT++ supports both natively. Other available frameworks supported include INET (contains models for several Internet protocols like TCP, UDP, IP, etc), INETMANET (INET with mobile/ad-hoc protocols), Mobility Framework (contains a library of wireless sensor networks simulation models for OMNeT++), and the promising MiXiM project (Kopke et al., 2008).

The active modules are termed simple modules (written in C++, using the simulation class library). Simple modules can be grouped into compound modules (refer to Figure 3), and communicate between each other using so-called messages that contain data, via gates. Both gates, input and

Figure 3. OMNeT++ model structure

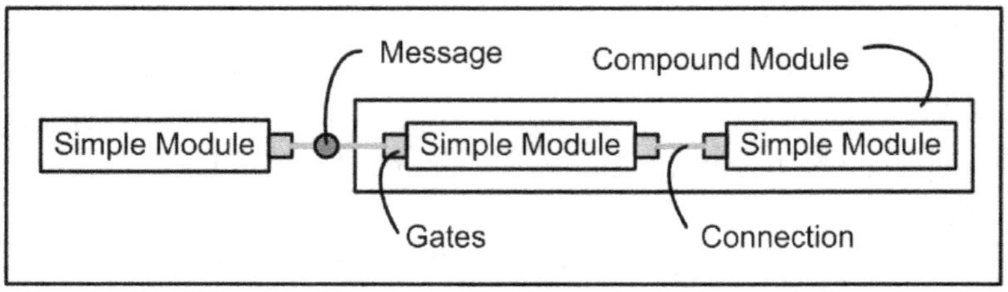

output, are linked with a connection (also called a link), to which we can assign parameters, such as bit error rate, data rate, propagation delay. Those parameters can take values as strings, boolean, character, and numeric constants.

Parameters can be assigned in either the *ned* file or the configuration file *omnetpp.ini*. Since it is based on C++, modules (and sub-modules) can pass parameters to each other and to those that inherit their properties.

USING OMNET++ TO MODEL WSNS: CONCEPT AND DISCUSSIONS

In this section, we present a study case of an OMNeT++ modelling simulator, with the condition of variable Bit Error Rate (BER). We derived analytical equations for BER, Packet Error Rate (PER), and efficiency, and compare these theoretical values against the simulation.

First, we model a simple concept of lost and received packets, Then, we progress with a more advanced and realistic simulation, introducing the concept of *Segmentation and Reassembly* (SAR), also defining a *window size* as a 'dimension' of a packet that is sent over the air. Next, we add an *error control mechanism* (inspired by TCP), with retries and reACKs schemes, and lastly consider overall *system efficiency* issues.

4.1. Reliable OMNeT++ Simulation of Lost and Received Packets

4.1.1. Concept

A simple concept of lost and received packets is simulated, primarily in order to gain confidence in using OMNeT++ as a network simulator. First, we derive analytical equations related to packet transmission over air, to be tested against the experiment to confirm that the OMNeT++ simulation is working correctly.

We start by assuming:

$$Np = Lp + Rp \qquad (4.1)$$

with Np is the total number of packets sent by sender, Rp is the number of received packets at receiver, and Lp is the lost packets over the air. We know the relation between BER (Bit Error Rate) and PER (Packet Error Rate) is:

$$PER = 1 - \left(1 - BER\right)^{N} \qquad (4.2)$$

with N being the uniform length of each packets in bits. Since:

$$Lp = PER \times Np \qquad (4.3)$$

then we can conclude that:

Figure 4. Simulation topology

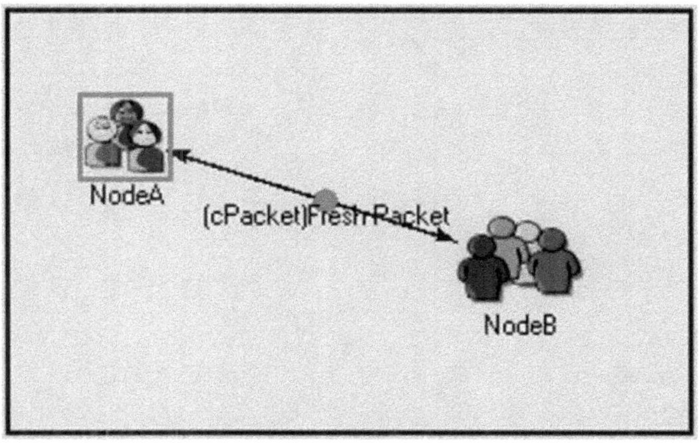

$$Np = \frac{Rp}{\left(1 - PER\right)} \quad (4.4)$$

and:

$$Lp = \frac{Rp}{\left(1 - PER\right)} - Rp \quad (4.5)$$

Later we test and prove the equations derived by theory and the results from the experiment of those equations as described below.

4.1.2. Experimental Scenario

We begin with a simulation which is established by defining two nodes, NodeA and NodeB which communicate through wireless transmission. NodeB will stop the simulation once it receives *n*-correct packets (equals to *Rp* value) regardless of the number of packets sent/resent by NodeA. The environment (depicted in Figure 4) is simulated under the condition of variable Bit Error Rate (BER), recording the link statistics as the simulation progress.

The simulation was run with the parameters stated as in Table 1. The results, confirming the theory of (4.4) & (4.5), against the OMNeT++ simulation are shown in Figure 5.

4.1.3. Discussion

We start by analysing the comparison of lost packets related to BER. Referring to (4.2), with a smaller packet length of $N = 1024$ bits, we know that PER will be lower compared with a larger packet length of N = 2048 bits. Thus small PER will result in the loss of fewer packets, as suggested in (4.3). In Figure 5 (left), a significant increase in the number of lost packets for bigger

Table 1. Simulation-1 parameters

Attribute	Value
Rp (packet)	125
BER	10^{-3}, 10^{-4}, 10^{-5}, 10^{-6}
N (bit)	1024, 2048

Figure 5. Lost packets (left) & sent packets (right), vs. BER, for N=1024 & 2048 bits

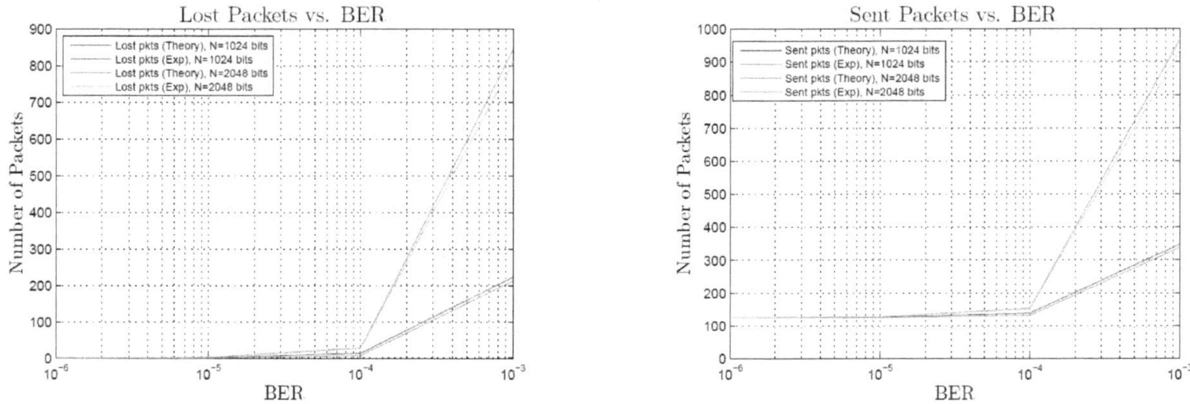

packet length (at constant BER value = 10^{-3}) can be observed.

In terms of BER, progressing from bad to good, the results show that most of the packets were lost (213 packets up to 811 packets) for large BER. Conversely, for good BER, the number of lost packet was very small (i.e. approximately zero). We also notice that just for BER value = 10^{-3}, as expected, bigger packets clearly showed a strong tendency to become lost. These observations are as anticipated.

From the angle of sent packets shown in Figure 5 (right), the outcome was similarly encouraging. For very bad BER, more packets were sent from NodeA (varied from 338 to 936 packets) to convey a fixed amount of data, since most of them were lost during the transmission. When it comes to a good BER value, the number of sent packets was equal to 125 (*Rp*), indicating that no packets were lost in the transmission. For both scenarios, a very encouraging conclusion could be observed that the discrepancies between theory and experiment are very minor, thus builds confidence in the reliability of OMNeT++.

4.2. OMNet++ Simulation: SAR with Error Control and System Efficiency

4.2.1. Segmentation and Reassembly (SAR)

Segmentation and Reassembly (SAR) is one of the concepts used to fragment and re-accumulate packets as it allows them to be safely transported across error-prone transmission links and networks. The process is straightforward, as the incoming packet to be transmitted is simply chopped up into smaller segments (according to the desired size), augmented with a header, and perhaps some error check or error code information.

At the other end these segments are then fitted back together to reform the original packet. Manageably sized packets are thus usually segmented in the source node and reassembled at the destination node. Furthermore, we also introduced a window size parameter, *w*, defined as the length of segments that we transmit over the air. The concept of SAR and window size can be seen in Figure 6.

4.2.2. Error Control Mechanism

To be more realistic in modelling, we add an error control mechanism, just as TCP does (Forouzan,

Figure 6. Segmented data

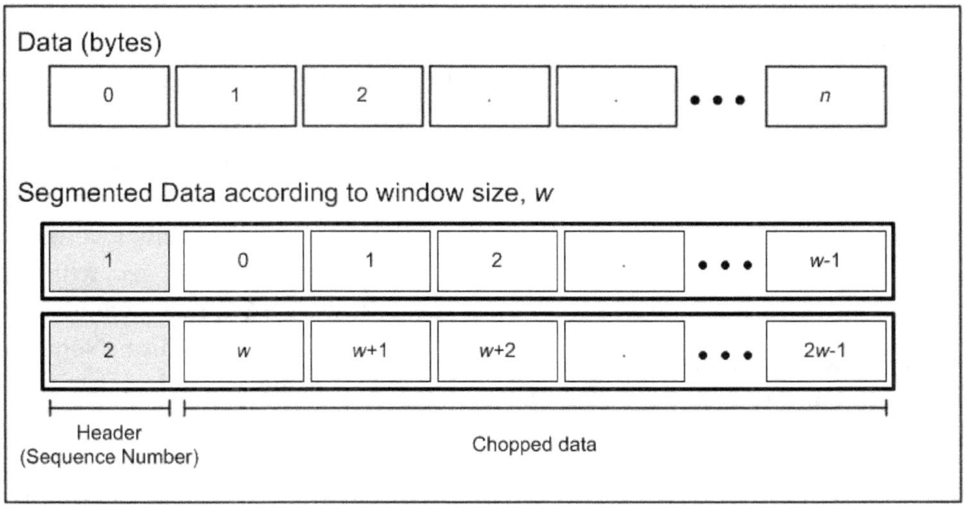

2003). In our work, we limited our implementation in terms of number of re-tries and reACKs (Figure 7). Both the retries and reACKs were used in conjunction with a timeout. This timeout counter will be set once a segment is sent. NodeA will send *retries* if it does not receive an appropriate ACK for the corresponding segments. NodeB will send *reACKs* if it gets the same segment from Node B (meaning its ACK is lost).

This mechanism was designed to ensure the segments were sent and received correctly, thus increasing the realism of the model and simulation. We also use set the number of per segment retries and reACKs parameters to 3 and 5, respectively. If both retries and reACKs exceed this value, then the segments are considered lost.

Figure 7. Error control mechanism

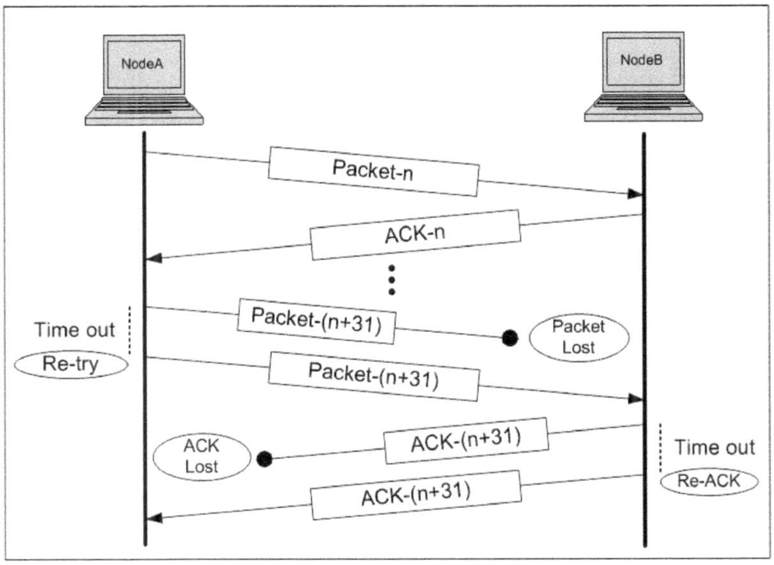

4.2.3. System Efficiency

In addition, we need to find out how many bits are received correctly over the total transmitted bits from the sender. We create another parameter, system efficiency:

$$\eta = \frac{Received\,bits}{Total\,sent\,bits} \tag{4.6}$$

This can be used to determine the maximum bits per second we can transmit over the air and to work out how many seconds we need in order to transmit a certain amount of data.

4.2.4. Experimental Scenario

The modelling starts with a simple concept of segmentation and reassembly (SAR). The same topology was used as previously, but with a slightly different scenario as follows:

a. NodeA defined the number of segments according to the length of the data, by breaking it into the appropriate window size (segmentation) then adding a sequence number before beginning to transmit to NodeB.
b. NodeB will be receiving all the segments until no more segments are left. If all the segments were correctly received, reassembly will occur and full data will be generated.
c. Besides SAR, now both nodes have an error control mechanism. NodeA must resend the segments (retries) that were lost in the transmission, while NodeB had to resend the acknowledgement (reACKs) when it received the same segment from NodeA (which means the previous ACK is lost). The maximum number of retries and re-ACK were set to 3 and 5 respectively (which can be changed easily in the simulation).

The simulation was carried on with the conditions stated in Table 2, while the flowchart of the simulation is depicted in Figure 8 for both NodeA and NodeB.

Figure 8. NodeA & NodeB flowchart

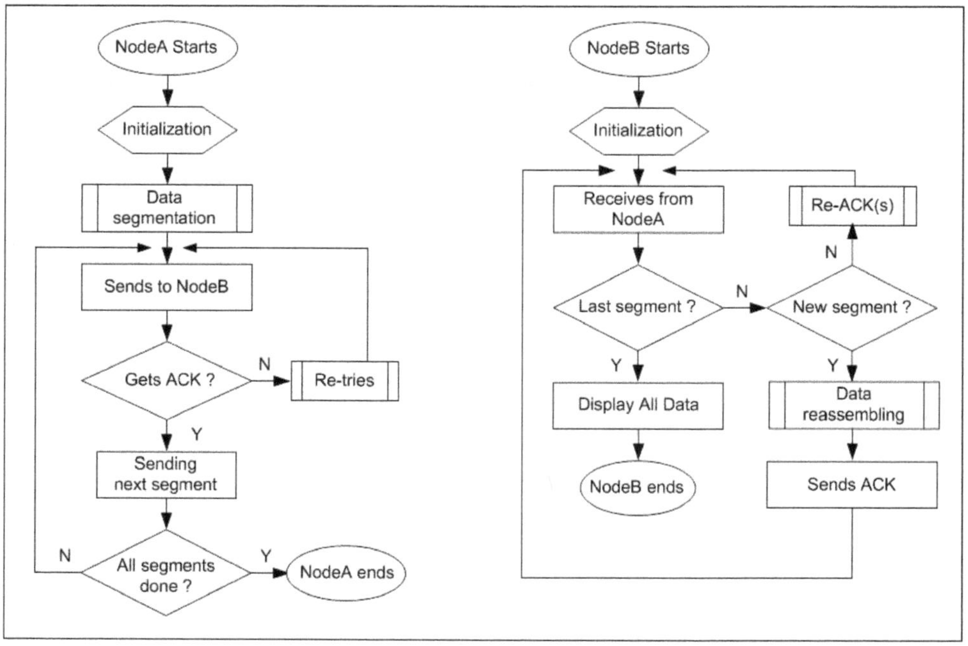

Table 2. Simulation-2 parameters

Attribute	Value
Window size, w	49, 196
Data length (bytes)	400, 1600, 4000, 8000, 12000, 20000
Retries and reACKs	3, 5
BER	10^{-3}, 10^{-4}, 10^{-5}

4.2.5. Results and Discussion

We plot the results in Figure 9 in terms of BER and PER for different data sets, while for number of retries & reACKs and window size were plotted in Figure 10.

Firstly, in the relation between BER, PER and window size, the results in Figure 9 showed that, as expected, low BER caused more data to be lost in transmission, thus showed large PER percentage. The percentage value itself increases gradually as we changed the window size from 49 to 196. In terms of this, a careful choice of window size has been suggested to help maximize system efficiency (Thakur, Sirisena, & Mcloughlin, 2005). A small window size implies a greater overhead in terms of segment header and ACK traffic.

Figure 9. PER vs. BER of various data sets, w=49 (left) and w=196 (right)

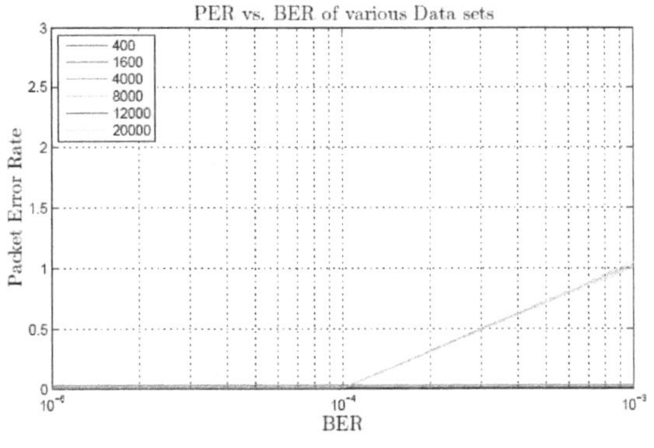

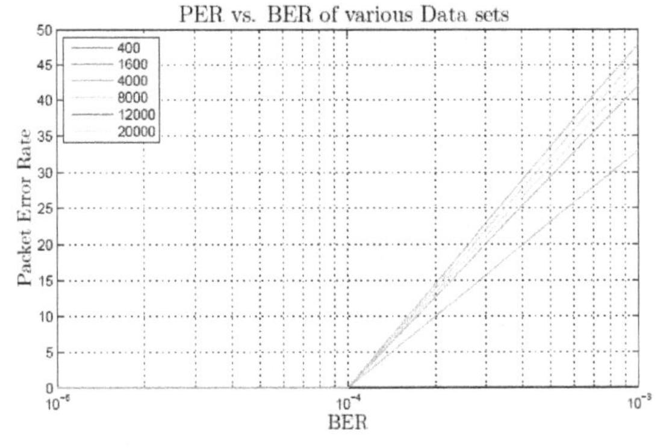

Figure 10. Total retries and reACKs vs. BER, w=49 (left) and w=196 (right)

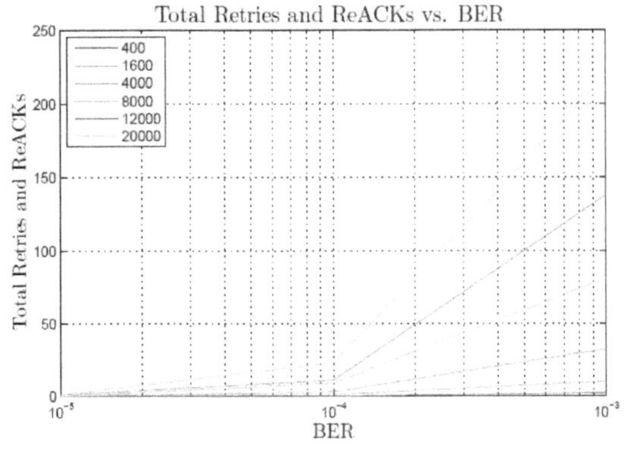

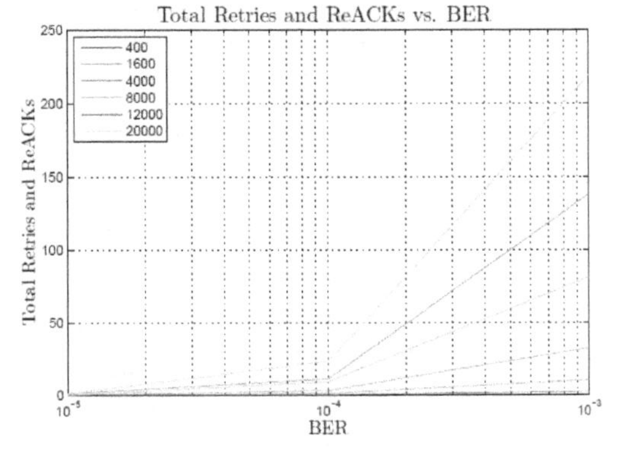

In terms of retries and reACKs, Figure 10 showed that more segments were obviously re-sent under bad BER conditions. It is interesting that for different window size graphs, the plots showed similar values, which can lead to the conclusion here that increasing window size does not automatically make the number of retries and reACKs reduced.

Lastly, we altered the parameters in our experiments to determine the relationship between numbers of retries or reACKs and:

1. PER of received segments,
2. total bits sent over the air, to analyze the system efficiency.

We modified simulation structure and operating parameters, changing the value of retries/reACK from 3 to 5 to identify the correlation. We narrowed our discussion to relatively poor BER value of 10^{-3}. The results are shown in Table 3.

When the maximum number of retries is increased from 3 to 5, we notice that PER reduced, meaning that all the segments had a greater probability of arriving at NodeB. We also notice that total bits sent over the air increased (this can be seen also from the increasing number of retries

Table 3. Comparison of packet error rate and total bits sent on 3 and 5 retries for BER = 10^{-3}

(a) With window size = 49					
Data Length (bytes)	**3 Retries**				
	PER (%)	**Bits received (bits)**	**NodeA Retries**	**NodeB ReACKs**	**Efficiency**
12000	1.22	94424	123	15	98.36%
50000	1.18	393696	499	68	98.42%
100000	1.37	785824	995	129	98.23%
Data Length (bytes)	**5 Retries**				
	PER (%)	**Bits received (bits)**	**NodeA Retries**	**NodeB ReACKs**	**Efficiency**
12000	0.00	95600	128	15	99.58%
50000	0.10	398008	516	68	99.50%
100000	0.10	796016	1037	129	99.50%
(b) With window size = 196					
Data Length (bytes)	3 Retries				
	PER (%)	Bits received (bits)	NodeA Retries	NodeB ReACKs	Efficiency
12000	41.94	55232	123	3	57.53%
50000	39.45	241632	490	9	60.41%
100000	38.75	487960	976	22	61.00%
Data Length (bytes)	5 Retries				
	PER (%)	Bits received (bits)	NodeA Retries	NodeB ReACKs	Efficiency
12000	27.42	69344	180	3	72.23%
50000	22.27	309048	671	12	77.26%
100000	23.48	608688	1324	26	76.09%

and reACKs). This makes sense since more retries and reACKs happened, more segments are sent over the air, and therefore the number of bits of air traffic increased. The simulation result showed similar tendencies when we increased the window size to 196 bytes.

The system efficiency is also analyzed by calculating the total bits received over the total data length. We found that the efficiency decreased when the window size increased for both retry value 3 and 5. It was due to the straightforward correlation that when part of any segment -determined by window size- is lost, so are all of the bits contained within it. So for bigger window sizes, the probability of a segment being lost is higher. In reality, even relatively weak error check coding (ECC) could improve this, but at the expense of greater node CPU speed requirement, greater RAM size, and increased transport latency.

However, we found something interesting in terms of the number of bits sent when we extended the window size from 49 to 196: the number of lost bits almost reached up to 42%; an extremely large overhead. From this we noted that there was a trade-off to be considered in terms of system efficiency and window size within any implemented system.

To sum up this section, we gained confidence in the OMNeT++ simulation model by being able to simulate the concepts of simple wireless transmission through developing a segmentation and reassembly simulation model incorporating error control mechanisms. This model is now to be used as the basis of further OMNeT++ WSN simulation.

CONCLUSION

This chapter has examined WSNs, and the need for accurate and fast simulation of networks. This was followed by an overview of several simulation tools, including a comprehensive comparison of what are arguably the three main comparative tools found in the research literature today: ns-2, OPNET and OMNeT++. Various related issues were explored in terms of availability of protocol models, topology models, debugging & GUI environment, simulation comparisons, and open source & cost issues. This study concluded that OMNeT++ may be worthy of further study.

As part of the discernment process, OMNeT++ was then simulated, and found to be a capable and flexible tool. The initial simulation was relatively straightforward; modelling the concept of lost and received packets by defining two nodes communicating simply through lossy wireless transmission. This took into account the condition of variable Bit Error Rate (BER), derived analytical equations for BER, Packet Error Rate (PER), and compared these against the simulation. A very encouraging conclusion was observed in that any discrepancies between theory and experiment were very minor, thus building confidence into OMNet++ reliability.

The initial simulation was then followed with a more realistic wireless scenario, introducing the concept of Segmentation and Reassembly (SAR), defining a window size as a 'dimension' of a packet that is sent over the air, and adding in an error control mechanism using retries and reACKs schemes. Finally, overall system efficiency issues were captured and explored. Again, this simulation was compared to theoretical and published work and the results discussed in terms of various trade-off issues.

In conclusion, OMNeT++, as far as has been demonstrated through these tests, and other published comparisons, is found to be a trustworthy, capable and relatively fast simulation tool. This chapter has demonstrated the tool for simple networks, and having confidence in these, is poised to address more complex systems.

5.1 Lessons Learned

We conclude this chapter by stating that, despite many competing system designs having been published in the literature, it is not difficult to conclude that there are still no absolute solutions for many of the issues plaguing WSNs systems, in particular the design of systems for use in hostile, remote or hazardous locations. Regarding the OMNeT++ simulator, we conclude that it operates as a reliable simulator for studying the concepts of wireless transmission, despite relatively little research literature adopting this simulation tool to date. We also find that it is relatively fast (the slowest simulations in this paper - TCP - completed in no more than a few tens of minutes). We believe it is possible to extend the system to more complex wireless topologies in which theoretical or closed-form solutions are not readily obtainable.

To be clear, significant ongoing research is required to make wireless sensing both efficient and effective. By proposing the modelling of such systems, we expect we can come up with something useful for the academic world - particularly the wireless modelling community, but more importantly, to work towards a reliable and accurate tool for designing networks for hazardous locations.

FUTURE RESEARCH

As mentioned in the beginning of this chapter, WSNs have been used in many areas of scientific study. To make the area of research more interesting, they have also been used for monitoring of active volcanos . Here, the challenges have been studied in a few key works, but with no absolute or scientifically-satisfying solutions: such systems are reported in the works of Aster et al. (2004), Werner-Allen, Johnson, Ruiz, Lees, and Welsh (2005), Werner-Allen et al. (2006), and Song, Huang, Xu, Shirazi, and LaHusen (2010).

The authors are motivated for testing and applying OMNeT++ to model such system with the purpose of simulating how data is disseminated between sensor nodes in a variety of dynamic and unreliable deployments. We therefore note some open research questions related to this proposition:

a. Is it possible to model, simulate, and apply an all-in-one protocol for WSN systems of different scales, or to devise a protocol that takes account of deployment scale, bandwidth and arrangement?
b. Can such a system then be applied to both simple and complex systems, and can it account for real life factors?
c. Ideally, WSNs which are key to scientific modelling of areas such as volcanos and active tectonic faults require an integrated perspective from both engineers and earth scientists: is this feasible?
d. In terms of data reliability, how can WSNs be designed to achieve as high a reliability as possible – with minimal cost trade-offs? Does OMNeT++ (or some other tool) possess the capabilities to model realistic scenarios such that it can be accurate in performing these trade-offs?

REFERENCES

Akyildiz, I. F., Su, W., Sankarasubramaniam, Y., & Cayirci, E. (2002). Wireless sensor networks: A survey. *Computer Networks: The International Journal of Computer and Telecommunications Networking, 38*(4), 393–422.

Albeseder, D., & Fuegger, M. (2005). *Small PC-network simulation - A comprehensive performance case study. Research Report*. TU Wien, Institut fur Technische Informatik.

Aster, R., McIntosh, W., Kyle, P., Esser, R., Bartel, B., & Dunbar, N. (2004). Real-time data received from Mount Erebus Volcano, Antartica. *Eos, Transactions, American Geophysical Union, 85*(10), 97–104. doi:10.1029/2004EO100001

Cai, J., & Jia, B. (2005). *Network simulation based on OPNET and application*. 1st International Workshop on Education Technology and Computer Science.

Cardell-Oliver, R. (2005). Rope: A reactive, opportunistic protocol for environment monitoring sensor network. *Proceedings of the 2nd IEEE Workshop on Embedded Networked Sensors.*

Chang, X. (1999). Network simulations with OPNET. *Proceedings of the 31st Conference on Winter Simulation.*

Curren, D. (2008). A survey of simulation in sensor networks. *Architecture (Washington, D.C.), 2008*, 867–872. Retrieved from http://citeseerx.ist.psu.edu/viewdoc/ download?doi=10.1.1.94.2626&rep=rep1&type=pdf

Das, A. N., Lewis, F. L., & Popa, D. O. (2009). Data-logging and supervisory control in wireless sensor networks. *International Journal of Sensor Networks*, *6*(1). doi:10.1504/IJSNET.2009.028022

Doolin, D., & Sitar, N. (2005). *Wireless sensors for wildfire monitoring*. SPIE Symposium on Smart Structures & Materials.

Downard, I. (2004). *Simulating sensor networks in ns-2*. NRL Formal Report 5522, April.

Fishman, G. S. (1978). *Principles of discrete event simulation*. New York, NY: John Wiley & Sons, Inc.

Forouzan, B. A. (2003). *TCP/IP protocol suite* (2nd ed.). McGraw-Hill.

Hartung, C., Han, R., Seielstad, C., & Holbrook, S. (2006). FireWxNet: A multi-tiered portable wireless system for monitoring weather conditions in wildland fire environments. *Proceedings of the 4th International Conference on Mobile Systems, Applications and Services.*

Henderson, T. R., Roy, S., Floyd, S., & Riley, G. F. (2006). ns-3 project goals. *Proceeding of the 2006 Workshop on ns-2: The IP Network Simulator.*

INET. (n.d.). *Framework*. Retrieved from http://inet.omnetpp.org/

Kopke, A., Swigulski, M., Wessel, K., Willkomm, D., Haneveld, P. T. K., Parker, T., et al. (2008). Simulating wireless and mobile networks in OMNeT++ - the MiXiM vision. *Proceedings of the 1st International Workshop on OMNeT++.*

Lee, H., Cho, C., King, C., Fang, Y., & Lee, B. (2009). *Design and implementation of non-autonomous mobile wireless sensor for debris flow monitoring*. IEEE 6thIInternational Conference on Mobile Adhoc and Sensor Systems.

Lessmann, J., Janacik, P., Lachev, L., & Orfanus, D. (2008). Comparative study of wireless network simulators. *Proceedings of the 7th International Conference of Networking.*

Mainwaring, A., Polastre, J., Szewczyk, R., Culler, D., & Anderson, J. (2002). *Wireless sensor networks for habitat monitoring.* ACM International Workshop on Wireless Sensor Networks and Applications.

Maureira, J., Uribe, P., Dalle, O., Asahi, T., & Amaya, J. (2009). Component based approach using OMNeT++ for train communication modelling. *9th Int. Conf. Intelligent Transport Systems Telecommunications*, (pp. 441-446).

Mobility Framework. (n.d.). Retrieved from http://mobility-fw.sourceforge.net/

Naoumov, V., & Gross, T. (2003). Simulation of large ad hoc networks. *Proceedings of the 6th ACM International Workshop on Modeling Analysis and Simulation of Wireless and Mobile Systems.*
OMNeT++. (n.d.). *Website*. Retrieved from http:/www.omnetpp.org/

Paek, J., Chintalapudi, K., Caffrey, J., Govindan, R., & Masri, S. (2005). A wireless sensor network for structural health monitoring: performance and experience. *Proceedings of the 2nd IEEE Workshop on Embedded Networked Sensors.*

Pakzad, S. N., Kim, S., Fenves, G. L., Glaser, S. D., Culler, D. E., & Demmel, J. W. (2005). Multi-purpose wireless accelerometers for civil infrastructure monitoring. *Proceedings of the 5th International Workshop on Structural Health Monitoring.*

Pan, Y. (2010). *Introduction to OPNET simulator*. Retrieved from http://www.scribd.com/doc/37951842/Introduction to OPNET Simulator

Pham, H. N., Pediaditakis, D., & Boulis, A. (2007). From simulation to real deployments in wsn and back. *Proceedings of the 2007 IEEE International Symposium on a World of Wireless, Mobile and Multimedia Networks.*

Rappaport, T. S. (1996). *Wireless communications: principles and practice*. Englewood Cliffs, NJ: Prentice-Hall.

Romer, K., & Mattern, F. (2004). The design space of wireless sensor networks. *IEEE Journals on Wireless Communications*, *11*(6), 54–61. doi:10.1109/MWC.2004.1368897

Song, W., Huang, R., Xu, M., Shirazi, B. A., & LaHusen, R. (2010). Design and deployment of sensor network for real-time high-fidelity volcano monitoring. *IEEE Transactions on Parallel and Distributed Systems*, *21*(11), 1658–1674. doi:10.1109/TPDS.2010.37

Szewczyk, R., Polastre, J., Mainwaring, A., Anderson, J., & Culler, D. (2004). An analysis of a large scale habitat monitoring application. *Proceedings of the 2nd International Conference on Embedded Networked Sensor Systems*, (pp. 214-226).

Thakur, M., Sirisena, H., & McLoughlin, I. V. (2005). *Link layer mitigation in rural UHF-MIMO linking systems*. TENCON 2005 IEEE Region 10.

The Network Simulator. (n.d.). *ns-2*. Retrieved from http://www.isi.edu/nsnam/ns

Tiwari, A., Ballal, P., & Lewis, F. L. (2007). Energy-efficient wireless sensor network design and implementation for condition-based maintenance. *ACM Transactions on Sensor Networks, 3*(1).

Tolle, G., Polastre, J., Szewczyk, R., Culler, D., Turner, N., Tu, K., et al. (2005). A macroscope in the redwoods. *Proceedings of the ACM Conference on Embedded Networked Sensor Systems.*

Varga, A., & Hornig, R. (2008). *An overview of the OMNeT++ simulation environment*. 1st International Conference of Simulation tools and Techniques for Communications, Networks and Systems.

Weingrtner, E., Vom Lehn, H., & Wehrle, K. (2009). A performance comparison of recent network simulators. *Proceedings of the 2009 IEEE International Conference on Communications.*

Werner-Allen, G., Johnson, J., Ruiz, M., Lees, J., & Welsh, M. (2005). Monitoring volcanic eruptions with a wireless sensor network. *Proceedings of the 2nd European Workshop on Wireless Sensor Networks*

Werner-Allen, G., Lorincz, K., Johnson, J., Lees, J., & Welsh, M. (2006). Fidelity and yield in a volcano monitoring sensor network. *Proceedings of the 7th Symposium on Operating Systems Design and Implementation.*

Werner-Allen, G., Lorincz, K., Ruiz, M., Marcillo, O., Johnson, J., Lees, J., & Welsh, M. (2006). Deploying a wireless sensor network on an active volcano. *IEEE Internet Computing, Special Issue on Data- Driven Applications in Sensor Networks, 10*(2).

Xian, X., Shi, W., & Huang, H. (2005). *Comparison of OMNET++ and other simulator for WSN simulation.* 3rd IEEE Conference of Industrial Electronics and Applications.

Xu, N., Rangwala, S., Chintalapudi, K., & Ganesan, D. Broad, A., Govindan, R., & Estrin, D. (2004). A wireless sensor network for structural monitoring. *Proceedings of the 2nd International Conference on Embedded Networked Sensor Systems.*

Xue, Y., Lee, H. S., Yang, M., Kumarawadu, P., Ghenniwa, H., & Shen, W. (2007). Performance evaluation of ns-2 simulator for wireless sensor networks. *Proceedings of the Canadian Conference on Electrical and Computer Engineering.*

Zhang, J., Chen, J., Fan, J., Xu, W., & Sun, Y. (2008). *OMNeT++ based simulation for topology control in wireless sensor network: A case study.* International Wireless Communications and Mobile Computing Conference.

Zhang, Y., & Soong, B. H. (2005). The effect of unreliable wireless channel on the call performance in mobile network. *IEEE Transactions on Wireless Communications*, *4*, 653. doi:10.1109/TWC.2004.842947

Zhang, Z., Pang, Z., Chen, J., Chen, Q., Tenhunen, H., Zheng, L., & Yan, X. (2009). *Two- layered wireless sensor networks for warehouses and supermarkets.* 3rd International Conference of Mobile Ubiquitous Computing, Systems, Services and Technologies.

ADDITIONAL READING

Al-Karaki, J. N., & Kamal, A. E. (2004). Routing techniques in wireless sensor networks: A survey. *IEEE Wireless Communications*, *11*(6), 6–28. doi:10.1109/MWC.2004.1368893

Chen, I., Speer, A., & Eltoweissy, M. (2010). Adaptive fault tolerant QoS control algorithms for maximizing system lifetime of query-based wireless sensor networks. *IEEE Transactions on Dependable and Secure Computing*, 99.

Chonggang, W., Sohraby, K., Bo, L., Daneshmand, M., & Yueming, H. (2006). A survey of transport protocols for wireless sensor networks. *IEEE Journals on Network*, *20*(3), 34–40. doi:10.1109/MNET.2006.1637930

Karl, H., & Willig, A. (2005). *Protocols and architecture for wireless sensor networks*. John Wiley & Sons. doi:10.1002/0470095121

Kawadia, V., & Kumar, P. R. (2005). A cautionary perspective on cross-layer design. *IEEE Journals on Wireless Communications*, *12*(1), 3–11. doi:10.1109/MWC.2005.1404568

Liu, J., & Singh, S. (2001). ATCP: TCP for mobile ad hoc networks. *IEEE Journal on Selected Areas in Communications*, *19*(7), 1300–1315. doi:10.1109/49.932698

Shakkottai, S., Rappaport, T. S., & Karlsson, P. C. (2003). Cross-layer design for wireless networks. *IEEE Communications Magazine*, *41*(18), 74–80. doi:10.1109/MCOM.2003.1235598

Sundaresan, K., Anantharaman, V., Hsieh, H., & Sivakumar, A. R. (2005). ATP: A reliable transport protocol for ad-hoc networks. *IEEE Transactions on Mobile Computing*, *4*(6), 588–603. doi:10.1109/TMC.2005.81

Xiang, C., Hongqiang, Z., Jianfeng, W., & Yuguang, F. (2004). TCP performance over mobile ad hoc networks. *IEEE Journals of Electrical and Computer Engineering, 29*(1).

Xylomenos, G., Polyzos, G. C., Mahonen, P., & Saaranen, M. (2001). TCP performance issues over wireless links. *IEEE Communications Magazine*, *39*(4), 52–58. doi:10.1109/35.917504

Zhang, Y., & Soong, B. H. (2005). The effect of unreliable wireless channel on the call performance in mobile network. *IEEE Transactions on Wireless Communications*, *4*, 653. doi:10.1109/TWC.2004.842947

KEY TERMS AND DEFINITIONS

BER (Bit Error Rate): A number of received bits that have errors at the receiver during a transmission.

Error Control: A mechanism for detecting corrupted, lost, out-of-order, as well as duplicated data packets.

PER (Packet Error Rate): An error percentage of a transmitted packet over the transmission link, defined as the number of incorrectly received packet over the number of transferred packets.

SAR (Segmentation and Reassembly): A process to fragment and reassemble packets into fixed length size prior to transfer over the transmission channel.

Wireless Sensor Networks: Networks consisting of dispersed sensor nodes, which considerately observe environmental situations for some particular purposes, and it is linked by wireless transmissions.

Chapter 19
Data Fusion in Wireless Sensor Networks:
Classification, Techniques, and Models

Lalit Agrawal
Indian Institute of Information Technology Allahabad, India

Alok Kumar
Indian Institute of Information Technology Allahabad, India

Jaya Nagori
Indian Institute of Information Technology Allahabad, India

Shirshu Varma
Indian Institute of Information Technology Allahabad, India

ABSTRACT

Many strategies have been devised over the years for improving performance of wireless sensor networks with special consideration to energy efficiency, accuracy of the sensed data, increasing network lifetime, providing QoS, et cetera. Such techniques include design of various MAC protocols, routing algorithms, aggregation techniques, and many more. Data fusion has been developed to be one of such techniques that has a span of benefits across all the above mentioned performance criteria. The basic concept of data fusion is to apply various mathematical concepts on the raw sensor data to achieve an output with better fidelity and accuracy. Various mathematical, signal processing, and probabilistic techniques can be implemented to fuse the data. This might be done for decreasing the volume of the data, improving the precision of the data, deriving an inference from the raw data that was not present earlier, and many other similar functions. The motive of this chapter is to introduce data fusion and why it is important in case of wireless sensor networks. Further, it classifies data fusion, explains the major techniques, models, as well as the implementation aspects of data fusion, and concludes with the challenges and limitations posed for data fusion to be useful.

DOI: 10.4018/978-1-4666-0321-9.ch019

1. INTRODUCTION

With the emergence of numerous interfaces as well as standards in the domain of wireless transmission and communication systems, the world has noticed a substantial amount of research towards developing high-end and compatible devices that are sought to be lighter and quicker in respect of seamless connectivity as well as user end requirement. While the initial development was restricted to infrastructure-based networks which were integration of wireless plus wired networks, the latest research have produced purely wireless networks without any fixed backbone called as ad-hoc networks in which wireless entities form a network among themselves on requirement. Various classes of adhoc networks have been profound namely Mobile Adhoc Network (MANET), Vehicular Adhoc Networks (VANET), etc. The underlying principle that differentiated these variations of networks are the entities involved in formation of the adhoc network. While VANET is formed between fixed devices mounted on high speed vehicles, MANETs are formed between low speed portable devices with low end computing needs. Interest in MANET arose due to its ability to provide distributed and instant wireless networking solution in situations where infrastructure based networks are unfeasible and expensive to be deployed. Moreover, they have greater flexibility towards congestion and robustness towards single-point failures.

A variation of MANET is Wireless Sensor Network (WSN) is a novel technique proposed for countering the numerous challenges of real-time, non-real time and various event-driven applications like habitat monitoring, surveillance, target tracking [Akyildiz et al. (2002)] with an ad-hoc network of cost effective and easily deployable micro-sensors. These micro-sensors made of MEMS technology are resource-constrained in the terms of power, processing ability as well as memory. The sensors are randomly spread into the field of operation and might remain unattended. Once deployed the sensors self-organize into a network without intervention of any other support. These specify unique characteristics of WSN, i.e. self-organization, random deployment and unattended operation. Moreover, WSNs are application-specific implying that the network and hence its various parameters vary with the application they are used for.

The resource-constrained nature of sensors and the unique characteristics of WSN imposed several difficulties in their practical realization, like battery consumption, scalability, sensing accuracy and coverage, to name a few. Rather than developing new protocols for these types of networks, researchers thought of modifying the existing network protocols to suit the demands of WSN, so that they can be implemented with ease and minimum of complexity. Accordingly, both hardware and software based approaches were devised towards forming a practically useful wireless sensor network.

The hardware-based approaches concentrate towards designing low power consuming, accurate and reliable computational modules of a sensor. These include design of low power consuming transceiver, decreasing size of the mote through VLSI (Very Large Scale Integration), adopting specific modulation techniques, choosing between various channel coding and error coding techniques for transmission over the unreliable wireless channel. The software-based approaches intend to modify the existing network protocols or developing newer algorithms to increase the efficiency of sensor network. Several such approaches are categorized under design of energy-efficient MAC protocols (Wang et al., 2007), developing various routing techniques (Al-Karaki & Kamal, 2004), clustering methods (Abbasi & Younis, 2007), in-network data aggregation methods Krishnamachari et al., 2002), and many-a-more. All these techniques use various optimization and protocol reduction techniques so as to remove the part of protocols that would have been an overhead for Wireless Sensor Network.

Data Fusion is such a technique where data from several sensors can be fused or integrated together to take a decision or draw an inference about a particular sensing task or any sensed event. In general terms, data fusion produces a new output data or feature or decision based on some input data that might be same or completely different to what the output actually is. This phenomenon is similar to the natural phenomena of sensing in human beings where they utilize various senses to infer the presence or absence of things, for example in case of food. Also known as Information Fusion, data fusion can also be defined as processing of the data by utilizing the correlation between the data sensed by various sensor nodes and producing a more accurate and reliable data or decision. In (Wald, 1999), data fusion is outlined as, "A formal framework that uses various means and tools for alliance of data originating from different sources so as to obtain information of greater quality; however the notion of 'greater quality' depending upon the application".

Data Fusion has been found of great value in many applications ranging from target tracking, military surveillance of battlefield, medical diagnosis and process monitoring task. A common data fusion scenario in WSN for target tracking application is shown in Figure 1. The common thing that has allowed data fusion to have an impact upon these varying range of applications is the presence of high amount of correlation among the data that can be exploited to provide an output of much higher value and accuracy. The need of discussion of data fusion for WSN lies in the fact that; multiple sensors are required to sense, collect, process and hence transmit a large amount of redundant data over to the sink. Hence, for every step of data acquisition, processing and communication, data fusion finds its application in: improving the quality of data, cutting down the amount of transmissible data or making a decision out of raw data. Various terminologies have been laid for task similar to data fusion in the literature. Multi-sensor integration (Luo et al., 2002), Multi-sensor Fusion, Data Aggregation (Cohen et al., 2001), etc., manipulate the data similar to what data fusion intends to, however with slightest of differences as shown in Figure 2.

1.1. Need of Data Fusion

Advent of advanced processing hardware and mathematical techniques, development of new advanced sensors have enabled data fusion to be carried out in real-time. However, the importance of data fusion need to be realized; to know which type of application would require data fusion, for what type of processing and when. The impor-

Figure 1. Basic data fusion for single target tracking in wireless sensor network

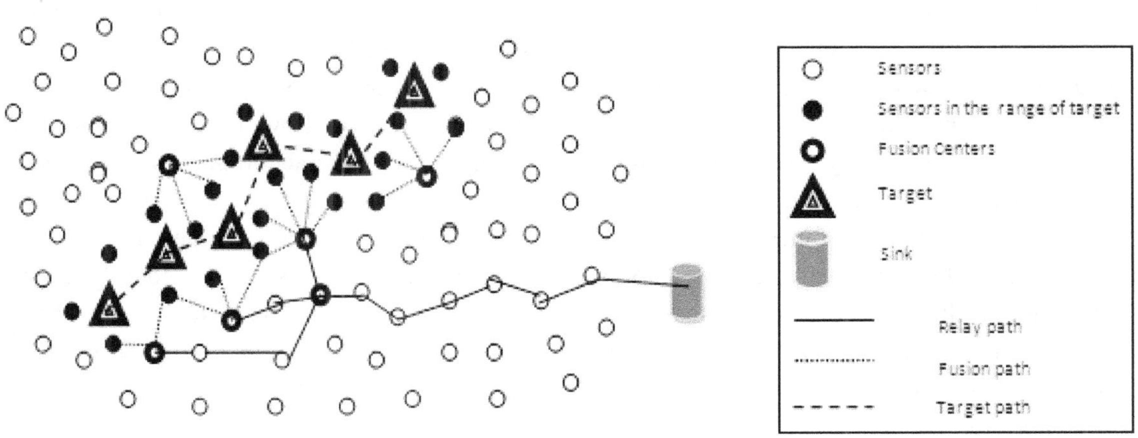

Figure 2. Relationship between various fusion terminologies

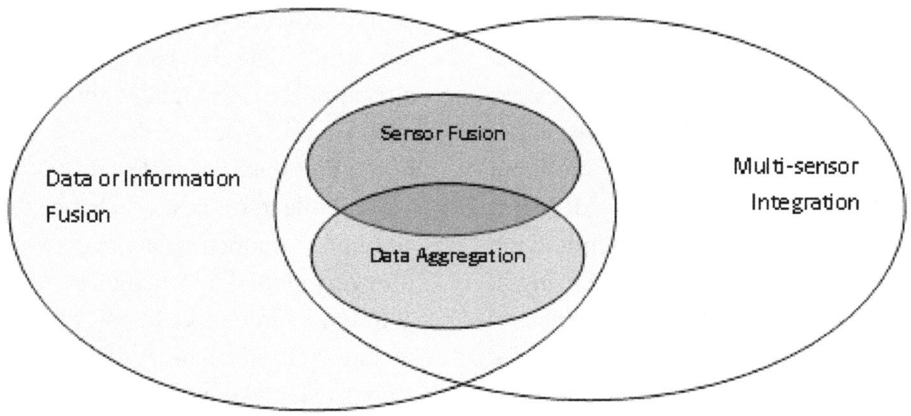

tance of data fusion in wireless sensor network is proven by the three enhancements it adheres to: complementarity, redundancy and co-operation (Durrent-Whyte, 1988).

Sensors are usually associated with imperfect measurements mostly due to environmental conditions that consist of electromagnetic radiation, electrical interference, temperature and pressure variations as well as manufacturing defects that results into imprecision in the sensed values, even when an ideal environment is assumed. Failure of sensor nodes might increase the inaccuracy of the data which might further results into infidelity of the data sensed or operation performed. Data Fusion enhances the sensor network to overcome failures that are most recurring phenomena in wireless environment.

Apart from imprecise sensor measurements and failures, there are other factors that restrict the performance of WSN. Spatial and temporal coverage are two of them. Spatial coverage implies the total physical area covered by the WSN, and thus defines the coverage of the network. Ideally, a sensor network is desired to have a wide spatial coverage in the area of interest, so that no events are missed. On the same grounds, temporal coverage defines the provisioning of coverage by sensor network over a certain period of time. In monitoring applications, sensor network needs to be alive for long periods, which is what temporal coverage refers to. Hence, spatial and temporal coverage are essential aspects to be taken care of, when the performance improvement is to be considered.

The three enhancements mentioned above allow multiple sensors to co-operate and complement each other to generate an accurate value by considering the measurements over a number of sensors, rather than a single sensor. Fusing data from various sensors allow an estimation of data which is highly accurate, thus eliminating the impact of imprecise readings from individual sensors. Redundancy allows multiple sensor nodes to perform the same task so that, in case one sensor fails, there is always a second one which is providing the system with the same data. Complementarity implies a situation where different sensors sense different aspect or part of the region, which are then later fused to form a coherent output.

Data fusion exhibits its importance in all these enhancements by allowing common sensor nodes to fuse the data in accordance to the demand of accuracy and energy efficiency, so that the final output obtained is always better than the raw data. While the nodes are redundantly deployed, co-operating and complementing each other, data fusion intervenes to aggregate the correlated data from this type of network and produce a decision of it. Taking an example, increase in redundancy implies increase in fault tolerance of the network while providing spatial coverage at the same time. Data Fusion is found to be effectively used in improving coverage of the wireless sensor network (Xing et al., 2009). Complementarity allows intelligent use of wireless nodes to perform a big task by dividing it into subtasks and allocating each subtask to individual sensor node. This would not have been possible, had a single sensor node been considered. Moreover, data fusion can reduce the volume of aggregated data to reduce the overhead of carrying large amount of data over wireless link (latency) could be significantly reduced.

2. BACKGROUD: CLASSIFICATION OF DATA FUSION

Data fusion being such a largely defined area, classifying it based on some single stringent criterion is impossible and hence the classification is made on the basis of different criteria. Here, we classify data fusion based on interaction among sources, levels of abstraction, and type of input and output data.

2.1. Based on Interaction among Source Sensors

Interaction among the sources whose data is fed to the fusion module or the relationship among the data inputted to data fusion technique can be used to classify data fusion into three categories namely complementary fusion, co-operative fusion and redundant fusion (Durrent-Whyte, 1988) and is shown in Figure 3.

i. **Complementary fusion:** The sources participating in data fusion contribute data that are not similar to any other among them. That is, each node is responsible for sensing different part of the environment, thus complementing each other to provide a full view of a complex environment. The data from these complementarily acting sensor sources are fused to form a complex structure that gives an integral view of the environment under observation. As shown in Figure 2, where two sensors S1 and S2 sense different parts of information and the fusion module integrates the results of both sensors to give a more broader result.

Figure 3. Types of data fusion on the basis of interaction among the sensors (Elmenreich, 2002)

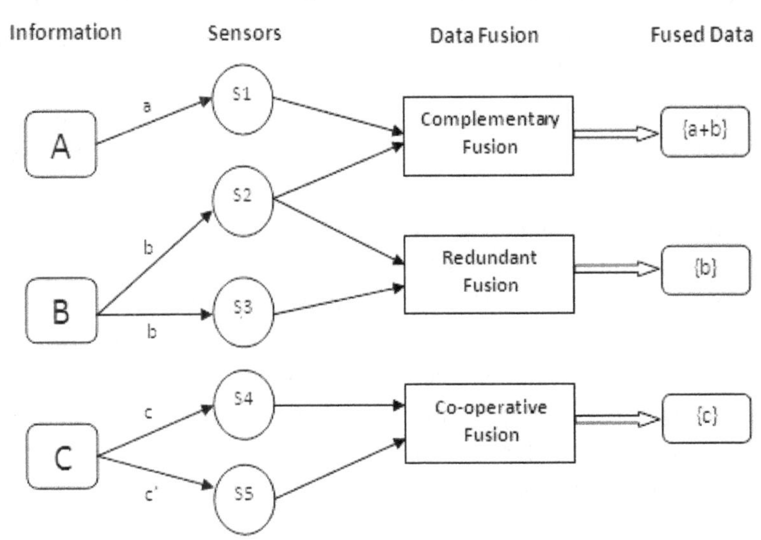

This type of fusion is important for an environment that is complex and large, with assumption that each sensor makes an accurate measurement of the part of environment it is assigned to. With sensors sensing different parts of environment, fusion combines the results from all of these sensors to get the full view of the entire observation area, thus removing the burden over single sensor to sense the entire environment. Zhao et al. (2002) used complementary fusion to integrate data from various sensors to build a feature map describing the entire sensor environment.

ii. **Redundant Fusion:** In this fusion technique (Figure 2), sensors S2 and S3 provide the same information to the fusion device which then integrates the data to produce a more accurate data. Redundancy is an important aspect of WSN that ensures reliability, fault tolerance and many other advantages. Hence multiple sensors are required to sense the same data so that the level of confidence on the readings from the sensors can be increased. Hence data fusion, on one hand, increases the accuracy of the data through fusion of redundant data; while on the other hand, eliminates redundancy from the sensor readings, so that least amount of data (though sufficient to represent the actual information) is transmitted. This minimizes the energy consumption implying data fusion as an important aspect for WSN. Filters are examples of such fusion modules, which will be explained later.
iii. **Co-operative Fusion:** In this type of fusion, different inputs coming from different sources (sensors) are combined into new data or information piece, completely different from the inputs but reflecting the application scenario in a more relevant manner than individual inputs. That in Figure 2 represents co-operative fusion, where the different data c and c' from sensors S4 and S5 are combined to form a new data (c), which represents the application scenario better than c and c'. An example to such type of application is calculation of location of a target based on angle and distance measurements from two different sensors. However, similar to complementary fusion, this fusion is heavily dependent on the assumption that each sensor makes its own readings perfectly, failure in which might result into inaccurate measurement computations.

2.2. Based on Abstraction Level of Raw Data

The data sensed by sensors are raw and can be used directly or abstracted to different levels in order to apply different processing techniques. For instance, raw data can be taken as signal to directly apply signal processing techniques on it. Arguing on same lines Luo et al. (2002) defined four levels of data abstraction: signal, pixel, feature and symbol, and accordingly classified data fusion into four categories, each operating at different levels of abstraction. As per the name suggests, signal level fusion deals with direct signal from sensors, pixel level fusion processes the image to improve the quality of the image. Feature level fusion works on integrating features or attributes of the signals or images generated by sources while the symbol level fusion deals with decision and hence is also termed as decision level fusion. However, such classification is unsuitable for wireless sensor network where signals as well as images are taken as raw data. Adding to it are many other sources of information like databases, user's input, etc. Finally, it also horizontally categorizes data fusion, not allowing a single technique to work at different levels of fusion concurrently. Hence Dasarathy et al. (1997) and Iyengar et al. (2001) proposed four categories of data fusion based on abstraction of raw data:

i. **Low-Level Fusion:** Nothing different to signal level fusion described above, raw data from multiple sensors are combined to generate more accurate data. Estimation of ambient noise and clarity of communication channel by using a moving average filter (Polastre et al., 2004), is an example of low-level fusion.

ii. **Medium-Level Fusion:** Also known as feature or attribute fusion, several features of sensed entity, like shape, area, etc. are combined to form a map that can be further used for other applications. Examples are energy maps (Zhao et al., 2002), or feature maps.
iii. **High-Level Fusion:** Known as symbol or decision fusion, several symbols or decisions from sensor sources are combined to take a confident global decision or a different symbol is carried out in high level fusion. Bayesian approach used for binary event detection and measurement of faults Krishnamachari et al., 2004) is an example of high-level fusion.
iv. **Multi-Level Fusion:** It combines the fusion at different abstraction levels. A possible example of such a fusion is combination of a measurement with a feature to imply a decision that can be taken. However, not much works specify this type of fusion method, Nakamura in (Sichitiu et al., 2004) demonstrated multilevel fusion through Dempster-Shafer inference to detect node failures on the basis of traffic decay features.

2.3. BASED ON ABSTRACTION LEVEL OF INPUT AND OUTPUT DATA

Dasarathy (1997) proposed another classification of data fusion, to avoid any possible ambiguities, based on type of abstracted data at input and output of the fusion technique, and hence identified five categories of data fusion:

i. *Data In–Data Out (DAI-DAO):* Similar to low-level fusion, both input and output data are raw signals or measurements, with the possibility of output being comparatively more accurate or reliable.
ii. *Data In–Feature Out (DAI-FEO):* Multilevel fusion, as the input data is raw data from sensors which is utilized to derive a feature about any entity or area under observation.
iii. *Feature In–Feature Out (FEI-FEO):* Medium-level fusion, where several features of the entity are combined to produce a better/new feature or feature set.
iv. *Feature In–Decision Out (FEI-DEO):* Multilevel fusion, in that features of an entity are inputted to fusion module which then makes a decision or symbolic representation out of it.
v. *Decision In–Decision Out (DEI-DEO):* Similar to high-level fusion where multiples decisions are fused to produce a single global decision as output.

3. METHODS OF DATA FUSION

Several techniques have been laid down for data fusion in wireless sensor networks. The general data fusion techniques use probabilistic theories, estimation techniques or signal processing algorithms, source coding algorithms, fuzzy logic, artificial neural network techniques, semantic fusion maps and many others. However, none of these techniques are suitable for all kind of applications in wireless sensor network. Each technique is different on its own, as they produce a different level of confidence in the output data obtained after fusion. While some work directly on integrating raw data into a complex data set, some intend to infer decisions based on data inputted to the fusion algorithm. Irrespective of technique used, all of the techniques work towards improving the quality of the sensed or raw data and consequently, performance of the network. Amongst a large pool of a variety of techniques, this section illustrates most popular techniques used so far for data fusion in wireless sensor networks and finally also outlines other techniques used for data fusion in an application-specific manner.

3.1 Inference-Based Techniques for Data Fusion

Inference is the term used for action of taking a conclusion or judgment through mathematical or logical reasoning by taking circumstantial evidence and prior conclusions in consideration, instead of direct observations. This philosophy suits to the need of data fusion perfectly. Inference based data fusion techniques are intended in making a decision based on prior results. The inferred value represents the desired output in most relevant manner, and might be completely different to the values it is calculated from. Though there are several complex inference methods for data fusion based on fuzzy logic, abductive reasoning, semantic information fusion and neural networks, here we explain how the conventional Bayesian Inference and Dempster-Shafer Inference techniques for data fusion work.

3.1.1. Bayesian Inference

It is a statistical inference based decision fusion technique and derives its name from the usage of Bayes' probability theorem for its calculations. As per Bayes' rule, Bayesian inference utilizes past evidences to calculate the likelihood of a hypothesis to be true. It does so by representing the uncertainties with conditional probabilities that take values from 0 (total disbelief) to 1 (absolute belief).

Bayesian inference relies on its interpretation of probability, which is different from other interpretations as it allows probabilities to be assigned to truth or falsehood of non-random events. It uses a scientific method to distinguish between hypotheses based on degree of confidence of a hypothesis, where the degree of confidence of a hypothesis is determines by the number of previously observed evidences with it. Higher the degree of confidence, more true a hypothesis is. This degree of confidence is represented as a numerical estimate based on observed evidences. However, this method heavily depends on the initial evidences that might not be true.

The calculation of conditional probabilities is done by Bayes' theorem, which can be mathematically given as:

$$P\left(\frac{H}{E}\right) = \frac{P\left(\frac{E}{H}\right).P(H)}{P(E)} \tag{1}$$

Where,

- $H \rightarrow$ A hypothesis, null or not null.
- $E \rightarrow$ The evidence observed.
- $P(H) \rightarrow$ The prior probability of hypothesis H, assumed before new evidences were observed
- $P(E/H) \rightarrow$ conditional probability of getting evidence E, provided hypothesis H is true
- $P(E) \rightarrow$ marginal probability of E, or can be taken as normalization constant.
- $P(H/E) \rightarrow$ posterior probability of hypothesis H being true provided evidence E
- P (E) is calculated as

$$P\left(E\right) = \sum P\left(\frac{E}{H_i} P(H_i)\right) \tag{2}$$

where, Hi is an element of a set of mutually exclusive hypotheses. The numerator in equation (1) evaluates how an hypothesis is impacted upon by an evidence.

Several works showcases the use of Bayesian Inference in general applications as well as wireless sensor networks. In (Coue et al., 2002), raw data from sensors on various vehicles were fused, using Bayesian Inference, to get more accurate and robust information required for high-level driving assistance. In a unique work, Bayesian inference was used for localization (Sichitiu et al., 2004) by estimating the most likely geographical coordinates of the node by using the information gained from the mobile beacon.

3.1.2. Dempster-Shafer Inference

Dempster-Shafer Inference (Wu et al., 2002) is a generalized version of Bayesian inference method and uses Dempster-Shafer Belief Accumulation Theory to perform decision fusion. It works with beliefs and mass functions in a manner similar to Bayes' rule with probabilities. It permits proposition support distribution to the union of propositions that contain it, rather than the proposition itself.

The basic idea of Dempster-Shafer reasoning system remains in "the frame of discernment Θ". It is outlined as the set of all possible mutually exclusive context facts of the same kind. To describe in more details, we consider Θ = {φ1, φ2, …., φn} where φ represent state that describes the system. So, Θ is an exhaustive as well as mutually exclusive set of all possible system states, as a system can be in exactly one state { φi } at any time. As the elements of Θ are used to discern the states of the system, hence the name follows frame of discernment. To further illustrate it with a real example, if we need to recognize a person in a room as user A or user B or someone else, the frame of discernment of the user would be:

$$\Theta = \{A, B, \{A, B\}, \{\text{somebody else}\}\} \quad (3)$$

Which means the user is 'A', 'B', 'either A or B', 'neither A nor B, somebody else'. Hence we can see the number of elements is corresponding to the size of power set {2Θ} with each element of the power set representing a hypothesis. Each hypothesis is assigned beliefs, by a sensor Si, based on "probability mass function" of sensor Si, denoted by mi. mi has values within [0,1] and satisfies,

$$m(\varnothing) = 0 \quad (4)$$

$$m(H) \geq 0, \ \forall \ H \in 2^{\Theta} \quad (5)$$

$$\sum_{H \in 2^{\Theta}} m(H) = 1 \quad (6)$$

Hence, a "confidence interval" is defined, by sensor Si to indicate the probability that "the detected person is A", and is given by

[Beliefi (A), Plausabilityi (A)] (7)

The overall belief for the hypothesis H, based on all evidences Ek supporting the hypothesis, is the lower boundary of confidence interval known as belief confidence:

$$\text{Belief}_i(A) = \sum_{E_k \subseteq A} m_i(E_k) \text{ for all } A \subseteq H \quad (8)$$

Where, the value of belief function ranges between [0,1] with Belief($\varnothing$) = 0, and Belief(Θ) = 1. Similarly, all the evidences that do not support the given hypothesis are accounted by "plausibility confidence", which is the upper edge of the confidence interval:

$$\text{Plausibility}_i(A) = 1 - \sum_{E_k \cap A = \varnothing} m_i(E_k) \quad \text{for all } A \subseteq H \quad (9)$$

Plausibility, in simple sense, means the greater the doubt in hypothesis H, the more plausible it is.

Hence, for each hypothesis, mi observations of Si sensors can be combined with mj observation of Sj sensors, using Dempster-Shafer theory, as per the following equations:

$$m_1 \blacklozenge\ m_2(\varnothing) = 0 \tag{10}$$

$$(m_i \blacklozenge\ m_j)(A) = \frac{\sum_{E_k \cap E_{k'}} m_i\left(E_k\right)\ m_j\left(E_{k'}\right)}{1 - \sum_{E_k \cap E_{k'}} m_i\left(E_k\right)\ m_j\left(E_{k'}\right)} \tag{11}$$

This combining function can be iteratively used for combining new observation to the already combined sensor's observations.

Dempster-Shafer theory has a greater reasoning capability, with its potentiality to assign uncertainty or ignorance to hypothesis, and is closer to human perception as compared to Bayesian Theory. Also its is more flexible than Bayesian theory due to three reasons

- It allows sensors to provide different levels of details.
- It allows fusion of data sensed by different types of sensors.
- There is no need of assigning prior probabilities to unknown hypothesis, as these are assigned when the information is fully available.

To illustrate with an example, if there are two sensors that can distinguish a male from female and a third sensor that can distinguish a student from a teacher, then Dempster-Shafer reasoning can be used to separate male/female students and male/female teachers. This would have been very difficult to compute with Bayesian Inference though Bayesian technique is found to be more accurate than Dempster-Shafer Inference (Bracio et al., 1997), (Buede, 1988). In (Nakamura et al., 2005), Dempster-Shafer inference is used to find routing failures, and initiating a route re-discovery when necessary for a Topology Rebuilding Algorithm (TRA).

3.2 Estimation-Based Techniques for Data Fusion

Estimation theory, derived from statistics and signal processing, uses the measured or observed data having a random component, to estimate the values parameters that describe a physical phenomenon and affect the distribution of empirical data. In estimation methods, an estimate of an unobservable parameter is calculated, based on a small random sample of measured values. The measured data is assumed to be random and having a probability distribution depending on the parameters of interest. Like inference, data fusion can also be implemented with the purpose of drawing an estimate of the desired parameters. For example, from a set of temperatures observed by sensors, estimation of temperature gradient can be made using various techniques. Here we explain some of commonly used estimation methods namely, Maximum Likelihood Estimation (MLE), Maximum A Posteriori (MAP), Moving Average Filter, Particle Filter, Kalman Filter and Least Squares method for data fusion.

3.2.1. Maximum Likelihood Estimation (MLE)

Maximum likelihood method estimates the parameters of interest, for a given sample of fixed data and probability model. The parameter values are selected so that they maximize the likelihood function, or in other sense generate a distribution which most likely would have outputted to the observed data. MLE is well-defined for normal or Gaussian distribution, and is found to be usable in the cases where estimated parameter is not a random variable (Brown et al., 1992).

The principle of MLE lies in forming a likelihood function which is to be maximized for finding the estimate of the parameter. Let us suppose a random set of independent and identically distributed (IID) observations of x (state of the system being estimated) of size 'n', given as x ={x1, x2, x3, ….., xn}, coming from a distribution d0(.). This distribution is assumed to obey a certain distribution called parametric model given by d0 = d (.|Φ0). Φ0 has an unknown value taken as true value of the parameter Φ0. Hence, the problem of MLE is to find an estimator $\hat{\Phi}$ that would be closest to the true value of Φ0.

For this, firstly a joint density function of all the observations is calculated for the IID sample 'x':

$$d\,(x1, x2, x3, \ldots.., xn \mid \Phi) = d\,(x1|\Phi) \cdot d\,(x2|\Phi) \cdot d\,(x3|\Phi) \cdots\cdots\cdots d\,(xn|\Phi) \quad (12)$$

Now, assuming observed vector 'x' be the fixed parameter and Φ be the function's variable, the function d would be called the likelihood function,

$$L.F.\,(\Phi \mid x1, x2, x3, \ldots.., xn) = d\,(x1, x2, x3, \ldots.., xn \mid \Phi) = \prod_{i=1}^{n} d(x_i|\Phi) \quad (13)$$

Generally, log of the likelihood function is formed, which is easier to work with, also named as log-likelihood or average log-likelihood

$$\ln \{\, L.F.\,(\Phi \mid x1, x2, x3, \ldots.., xn) \,\} = \sum_{i=1}^{n} \ln f(d(x_i \mid \Phi) \quad (14)$$

$$\hat{l} = \frac{1}{n} \ln L.F. \quad (15)$$

Finally MLE estimates Φ0 to a value Φ that maximizes $\hat{l}(\Phi \mid x)$, known as maximum likelihood estimator of Φ0, given as

$$\hat{\Phi}\ \text{mle} = \arg\max\ \hat{l}\ (\Phi \mid x1, x2, x3, \ldots.., xn) \quad (16)$$

This logarithm function does not affect the estimate. A distributed, localized and wireless link robust MLE was proposed in (Xiao et al., 2005), where a local unbiased estimate is computed by each node. This estimate follows the MLE value. It was further extended to support the observations/ measurements that are happening asynchronously in the network (Xiao et al., 2006). MLE has been mostly used for location discovery in WSN. Patwari et al. (2003) used MLE to estimate accurate distance for calculation of node's location.

3.2.2. Maximum A Posteriori (MAP) Estimation Method

Maximum A posteriori probability estimation method, based on Bayesian theory, utilizes the prior knowledge of distribution of the quantity to be estimated. It is a mode of posterior distribution and can be thought as a regularization of MLE. In general, MAP is used to find a point estimate of unknown quantity based on measured data set, when the parameter x to be estimated has a randomness with known probability density function.

For MAP estimation, let d be the distribution of sample observations 'x'. Then, d (x| Φ) represents the probability of x for the population parameter Φ. Then the likelihood function and MLE are given as

$$\Phi \rightarrow d\,(x|\,\Phi) \quad (17)$$

$$\hat{\Phi}\ ML(x) = \arg_{\Phi}\ \max f(x \mid \Phi) \quad (18)$$

Assuming a prior distribution f over Φ, thus taking Φ as a random variable (according to Bayesian theory), posterior distribution of Φ is given by Bayes' theorem as

$$\Phi \rightarrow d\,(x|\,\Phi) = \frac{d\left(x|\Phi\right)f(\Phi)}{\int_{\Phi' \in Æ} d\left(x|\Phi'\right)f\left(\Phi'\right)\,D\Phi'} \quad (19)$$

Where Æis domain of f. Finally Φ is estimated as the mode of posterior distribution of (19), as

$$\hat{\Phi}\ MAP(x) = \arg_{\Phi}\ \max \frac{d\left(x|\Phi\right)f(\Phi)}{\int_{\Phi' \in Æ} d\left(x|\Phi'\right)f\left(\Phi'\right)\,D\Phi'} = \arg_{\Phi}\ \max d\left(x|\Phi\right)f(\Phi) \quad (20)$$

It can be seen that the MAP estimate converges to ML estimate when the prior distribution f is uniform. While the distribution and pdf in case of MLE is unkown with x assumed fixed; in case of MAP, the pdf of x (a random variable) is known. MAP has been used by a collision resolution algorithm proposed in (Yuan & Kam, 2004) to regulate traffic among source nodes and the cluster head (fusion node). MAP is used to estimate number of nodes willing to transmit, so that the nodes can adjust their retransmission probability accordingly.

3.2.3. Kalman Filter

Kalman Filter, named after Rudolf E. Kalman (1996), is a linear and recursive filtering method used to estimate values closer to true values of measurements from their noisy measurements received over a communication channel. It also estimates uncertainty of the estimated value and calculates a weighted average of estimated and measured value. For doing so, it uses system's dynamics model, already known control inputs of the system, and measurements from the sensors and combines them to estimate the state of the system making it favorable for data fusion. A general Kalman filter is shown in Figure 4.

Where, x is the state vector of discrete-time controlled process governed by state-space model and d is the vector of measurements. Kalman filter estimates x and d as,

$$x(k+1) = \theta(k)\,x(k) + \mathbf{S}(k)\,c(k) + u(k) \quad (21)$$

$$d(k) = \mathbf{M}(k)\,x(k) + v(k) \quad (22)$$

Here, θ(k) is the state transition matrix, c(k) is the input vector representing the location of sensor nodes, **S**(k) is the input transition matrix, **M**(k) is the measurement matrix, u(k) and v(k) are random

Figure 4. Block diagram of Kalman filter

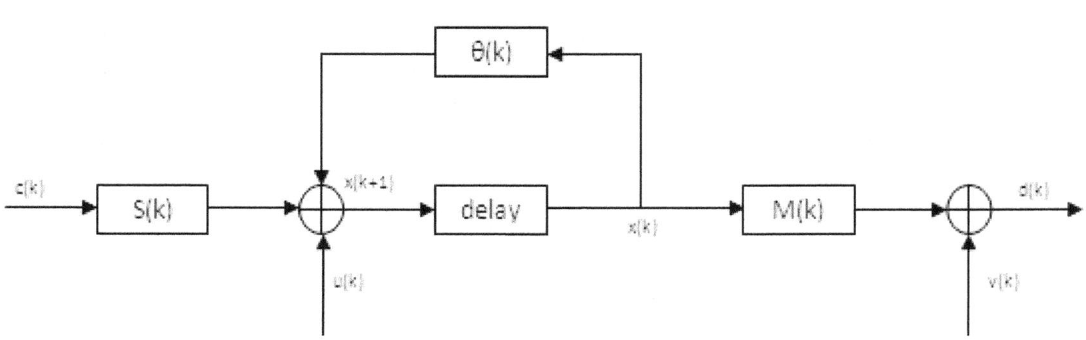

variable vectors following zero-mean Gaussian distribution with covariance matrices **I**(k) and **J**(k).

With the knowledge of system parameters and measurement vector d, estimate of x(k) and x(k+1) can be given, respectively, by the following two equations,

$$\hat{x}(k) = \hat{x}(k \mid k\text{-}1) + \mathbf{KG}(k)\,[\,d(k) - \mathbf{M}(k)\,\hat{x}(k|k\text{-}1)\,] \quad (23)$$

$$\hat{x}(k+1 \mid k) = \theta(k)\,\hat{x}(t|t) + \mathbf{S}(k)\,c(k) \quad (24)$$

Where, **KG** is the gain of Kalman filter, given as

$$\mathbf{KG}(k) = Pr(k \mid k\text{-}1)\,\mathbf{MT}(k)\,[\,\mathbf{M}(k)\,P(k|k\text{-}1)\,\mathbf{MT}(k) + \mathbf{J}(k)\,]\text{-}1 \quad (25)$$

Where, P(k|k-1) is prediction covariance matrix given as,

$$P(k+1|k) = \theta(k)\,Pr(k)\,\theta T(k) + \mathbf{I}(k) \quad (26)$$

$$P(k) = P(k|k\text{-}1) - \mathbf{KG}(k)\,\mathbf{M}(k)\,P(k|k\text{-}1) \quad (27)$$

Kalman filter is found to be ideal for linear systems with Gaussian noise model, as it can be recursively utilized to generate statistically optimal estimates. However, for non-linear systems with non-linear measurement models, variances of KF like Extended Kalman Filter (Welch & Bishop, 2001), Unscented Kalman Filter (Julier & Uhlmann, 1997), and others can be used.

Kalman Filter has found many applications in WSN; location refinement and distance estimation (Savvides et al., 2003), tracking different sources (Li et al., 2006), prediction of frame size for MAC protocol (Ci et al., 2004), etc. However there are a few challenges that affect the performance of Kalman filters; loss of critical observations over wireless link and clock synchronization among sensor nodes (Ganeriwal et al., 2003).

3.2.4. Moving Average Filter

The moving average filter, developed by Steven W. Smith [Yang et al. (2005)], is used to minimize noise level from temporal data series as well as acting as a low pass filter, thus smoothening out short-term fluctuations and highlighting long-term trends or cycles. However, for robustness and preserving effect, kalman filter is preferred. Mathematically, it is similar to convolution in signal processing, thus due to its simplicity is used for time domain encoded signals. Consequently, it is unable to separate frequencies from each others, making it highly unsuitable for frequency domain encoded signals.

For an input signal x = (x(1), x(2), …….,), moving average filter find the arithmetic mean to produce the output signal points. The true signal $y = (\hat{y}(1), \hat{y}(2),)$ is estimated as,

$$\hat{y}(k) = \frac{1}{W}\sum_{i=0}^{W-1} x(k-i) \text{ for all } k \geq W \quad (28)$$

Where, W is the filter's window size, number of inputs that are to be fused and also the number of steps chosen for detecting the change in signal level. Low value of W gives sharper step edges while large value yields into less noisy signal. In case of step signal having random white noise, the filter is able to reduce noise variance by a factor of M, thus emphasizing the need to choosing M appropriately, according to application needs. In WSN, moving average filter has been used to predict the data traffic for routing-failure detection (Bracio et al., 1997), reduce errors for tracking applications by filtering target locations (Yang et al., 2005), etc. A variation of moving average filters used in WSN is exponentially weighted moving average (EWMA) filter. This filter assigns different, exponentially decreasing, weights to different data points.

3.2.5. Particle Filters

Particle filters are alternatives to Kalman filters for environment with non-Gaussian noise, low sampling rate and cheap processing power (Nordlund et al., 2002). Also popularly known as sequential Monte Carlo methods (SMC) (Crisan & Duocet, 2002), these filters are used to estimate Bayesian models containing latent variables connected in a Markov chain. The basic goal of particle filters is to form a posterior pdf for a large sample of random values known as particles. These particles propagate over time; some of them get discarded at different time steps by the resampling. With the sequential combination of sampling and resampling process, a higher posterior probability based regions becomes more important. Each particle/sample of the same state variable x is assigned a weight that implies the quality of the particle. Hence, the estimated result is a weighted sum of all particles.

Basic particle filtering algorithm consists of two phases: prediction phase, where the particles are modified as per the existing model while intentionally adding random noise so that effect of actual noise can be simulated; and update phase, where latest sensory information available is used to re-evaluate the weight of each particle. Those with lesser weights are rejected.

Applications of particle filters for data fusion in WSN have been multi-target tracking (Crisan & Duocet, 2002), location discovery (Gunnarsson & Gustafsson, 2003), multi-user parameter tracking in communication systems (Guo et al., 2005), and many others.

3.2.6. Least Squares Method

Least squares method is a simple mathematical curve fitting technique, which works on deriving a function that best fits the set of observations or measurements. Optimization is carried out by minimizing summing the squares of error between points of the function and actual input measurements. While different such optimization metrics like Huber loss function and root mean square are present, here we explain the simplest of all, the ordinary least squares method (Brown et al., 1992).

This method is much similar to Maximum likelihood in the sense that the parameter to be estimated is considered fixed with no prior known probability. For a sensor model 'm', noise sequence 'n' and set of observations $1 \leq i \leq k$, state of the system is given by

$$s(i) = m(i,x) + n(i) \quad (29)$$

Least squares method hence tried to find a value of x that results into the least sum of squares of error between actual and forecasted observations.

$$\hat{x}(k) = \arg\min_{x} \sum_{i=1}^{k} [s(i) - m(i,x)]^2 \quad (30)$$

With the noise n(i) having series of outputs which are iid random variables with symmetric zero-mean probability density function, least squares and maximum likelihood function turn out to be equivalent estimation methods. This method has been used for node localization (Liu et al., 2006) while considering noisy measurements. Santini & Romar (2006) uses Least squares method to reduce the traffic being transmitted from source to sink, by using the method at both ends; a value is transmitted to the sink only when the difference between actual value and the predicted value is bigger than the threshold error value.

3.3 Other Useful Data Fusion Techniques

3.3.1. Semantic Information Fusion

It is a highly resource efficient inference-based fusion technique, introduced by Friedlander and

Phoha (2002) initially for target classification. The basic idea of semantic fusion is to process the raw data, extracting semantic information from the data, and then using this information for all the future processes (collection, storage and processing). As can be understood, this semantic information would be of much lesser volume than actual data. This makes semantic information fusion technique as highly efficient technique in terms of resource utilization.

In a more technical concept of semantic information fusion: a formal language is used to represent the aggregation sensor data which is then compared with languages of known behaviors stored in a knowledge database, as it is assumed that known behaviors of similar formal languages have similar semantics. In WSN, this formal language can be employed to infer the behavior of application or to produce back data that are statistically comparable to the actual raw measurement data. However, all this requires maintenance of a complex database of known behaviors. Semantic Information fusion is different to the conventional pattern-matching algorithms which compare known behaviors and feature maps of observed entity by measuring distances between them.

It basically consists of two phases: first is knowledge base construction where all the semantic information is extracted from the relevant knowledge abstractions and second phase is the pattern matching phase where these attributes are fused to generate a semantic interpretation of raw sensor data. While the knowledge construction phase is usually done off-line, the pattern matching phase is carried out online. Friedlander (2005) used semantic information fusion to infer robot behavior based on their trajectories.

3.3.2 Data Aggregation

As defined in (Cohan et al., (2001), data aggregation is "collection of raw measurements from sensory sources, while reducing the volume of the refined data using a flexible and programmable composition, and delivering the refined data to users within specified time limit". Having a concept pretty much similar to data fusion, some of aggregation methods can also be categorized under the data fusion techniques.

Aggregation techniques are general mathematical functions used by query languages for retrieving summarized data from databases. It was originally developed in WSN to solve the problem of implosion (duplication of data due to routing technique) and overlap (dissemination of same data from two different nodes) (Kulik et al., 2002). Several data aggregation techniques are proposed: summarization (Madden et al., 2002), suppression (Intanagonwiwat et al., 2000), packaging (He et al., 2004), average, etc. As the name suggests, in summarization, data from multiple sensors are summed up without disturbing the integrity of the information held by the data. In suppression technique, redundant data is suppressing through elimination of duplicates. Packaging reduces volume of data by grouping multiple observations into single packet before transmitting it, so that the MAC overhead is reduced. But as the correlation among the data is not utilized, packaging is not considered as a data fusion technique.

Significant research work has been done on data aggregation techniques in WSN; in (Krishnamachari et al., 2002), it is shown that forming an optimal aggregation tree is a NP-complete problem, whereas (Intanagonwiwat et al., 2002) evaluated impact of data aggregation in dense sensor networks, tradeoff between energy consumption and accuracy for a summarization based data aggregation technique has been evaluated in (Boulis et al., 2003).

3.3.3. Miscellaneous Techniques

Apart from all of these frequently used techniques, it is also important to outline some application specific approaches to data fusion. Fuzzy Logic is one inference techniques of data fusion that utilizes approximate reasoning to derive conclusions from imprecise measurements. It does so by fuzzyfying the input with a membership function, thus producing fuzzy outputs which are further defuzzyfied as per certain rules. It has been used for updating location of the node, in (Shu & Liang, 2005), though derivation of an updating rule, by fuzzyfying number of neighbors of each sensor as well as average distance among them. The feature of Neural Networks to learn from samples of input/output pairs in a systematic manner has made them another choice for data fusion, especially in applications like automatic target recognition using multiple complementary sensors (Luo & Kay, 1992), fusing audio-visual data for audio-visual speech recognition (Lewis & Powers, 2002).

Feature Maps based methods have also been proposed as methods for data fusion, especially for applications where usage of raw data is not feasible, e.g. guidance, resource management, etc. Feature maps are computed from joint effort of estimation and inference based methods (explained above), and are formed of features which are nothing but representations of environment. On these lines, Occupancy Grid is one of feature maps based technique which represents the environment in a multidimensional map containing of representations of spaces that are either free or occupied by an object. The space under consideration is split into square or cubic cells with the occupancy of each cell indicated by a probability. Though initially being used for deriving internal structure of static environments on the basis of ultrasonic data (Elfes, 1987), occupancy grids have also found their applications in localization (Wongngamnit & Angluin, 2001), navigation (Pagac et al., 1998), robot perception (Hoover & Olsen, 2000), etc.

Network Scan (Zhao et al., 2002) is another type of feature map based method, which represents geographical distribution of activities or resources of WSN using resource/activity map. In (Zhao et al., 2002), network scan (named eScan) extracts information of remaining energy in the network. It does so by forming an aggregation tree of communicating nodes where each sensor calculates its local eScan. Nodes transmit their eScan to the sink on seeing significant drop in the energy level since last receipt of any data packet. On receiving topologically identical eScans having similar energy levels, the same are aggregated. With energy level assigned a color level, nodes with full charge can be distinguished from dead ones by viewing at the so obtained 2D image map. This can allow network designed or identify regions with low energy levels and hence deploy sensors accordingly. However, this method assumes ideal MAC layer, which is unlikely.

Marzullo (1990) introduced the concept of Reliable Abstract Sensor for data fusion and defined three types of sensors: concrete sensor that senses the actual physical state of environment variable, abstract sensor that represents the measurements of concrete sensor through an interval of values, and reliable abstract sensor that gives the interval containing the real value of actual physical state of variable. In same work, he introduced Fault-Tolerant Averaging to derive a reliable abstract sensor through the fusion of n abstract sensors, and demonstrated that few imprecise actual sensors do not affect the result. Another variation of abstract sensor based data fusion technique, named Fault-Tolerant Interval Function, was introduced in (Schmid & Schossmaier, 2001) that considered width of the intervals going to be fused. They also demonstrated that slight variations in these intervals sharply affect the result of fault-tolerant averaging function.

Compression Based Approaches can also be used for data fusion. With these techniques, the high correlation among the measurements, provided by the group of neighboring sensors, are processed to generate codes smaller than individual observations. Along these facts, Xiong et al. used distributed source coding to compress the correlated data provided by multiple physically separated sensors that are not communicating with each other (Xiong et al., 2004).

With such a rich variety of techniques at disposal, it is significant for the designer to opt for the technique that gives best performance for the given application and environmental scenario. Proper selection of data fusion technique requires a careful inspection of various parameters desired by the application; QoS, latency, bandwidth, energy consumption, accuracy, etc.

4. DATA FUSION MODELS AND ARCHITECTURES

This section describes the various architectures and models that provide guidelines for design of data fusion systems. They provide rules about specifications, proposal and use of data fusion in WSN. These models have evolved and been classified on various bases; based on abstraction level of output of data fusion (called information-based models), based on activities to be carried out by a data fusion system (called activity-based models), based on roles of fusion and relationships among them (called role-based models).

4.1. Information-Based Models

Information-based models have been categorized on the basis of levels of abstraction of data (information) used in the data fusion process and can be used to specify what methods can be integrated in the data fusion system for any given application. The most popular information-based data fusion models are Joint Directors of Laboratories (JDL) Model (Kessler et al., 1992) and Dasarathy Model (Dasarathy, 1997), which is explained in this sub-section.

4.1.1 JDL Model

JDL Model, standing for Join Directors of Laboratories, was proposed jointly by U.S. JDL and U.S. Department of Defense (Kessler et al., 1992). Being developed for military purpose, the terminology used is defense-oriented. A block diagram of the JDL model is illustrated in Figure 5. The model consists of five processing levels, a database and a data bus for interconnection of all the components. These components are described below.

Figure 5. The JDL data fusion model; figure adopted from (Singh et al., 2006)

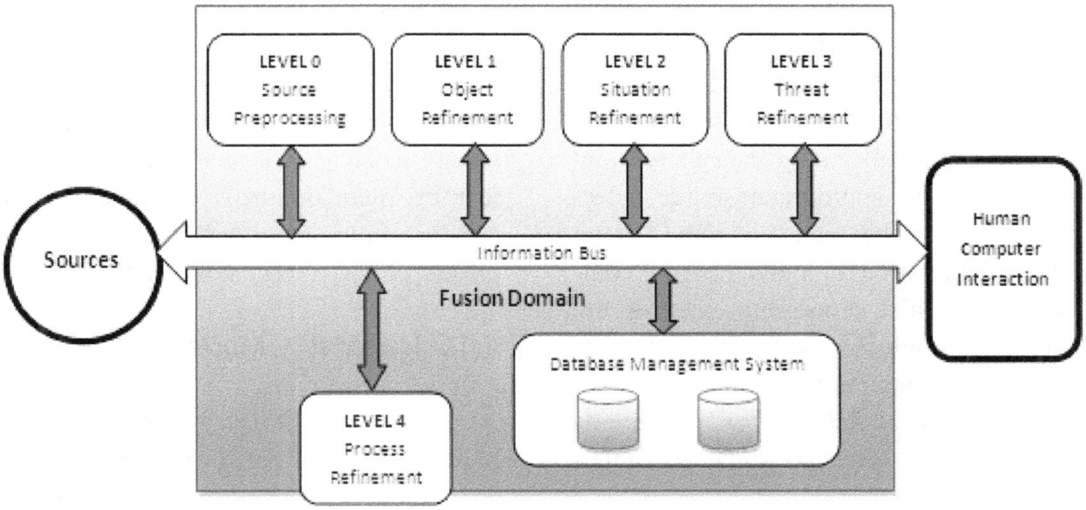

- **Sources:** These are the sensors or other devices (human input, database, assumed information, etc.) that provide input data for fusion.
- **DBMS:** Database management system, it stores and maintains all the data of the data fusion system, whether it is input or output. It plays a critical role in overall fusion environment by handling huge amount of different data sets and also restricts sensors from accessing critical resources, as per some guidelines. For this purpose, data-centric storage has been a recent technique where data is stored in a set of nodes base by a name (Ratnasamy et al., 2003). This permits users to fetch the required information easily either by directly querying the network or the node containing required type of data, instead of each individual node.
- **Human Computer Interaction (HCI):** This component provides an interface to the human or user for interacting with the system. User can input commands or queries and get the results via display screens, graphics, audio-video output, alarms, etc. Usually these interfaces are provided by query-based interfaces as in TinyDB (Madden et al., 2005).
- *Level 0 (Source preprocessing or Process Alignment):* This level tries to solve the issue of source selection in WSN through intelligent resource usage, while still maintaining the quality of data fusion technique. It does so by allocating data to relevant processes and appropriate source selection. Such a technique is given in (Zhao et al., (2002), where for a given cost of communication and processing, sources are chosen through dynamic optimization of relevant information in the data.
- *Level 1 (Object Refinement):* It converts the data from raw state (images, angles, acoustic data, etc.) into consistent state (target location, etc.)
- *Level 2 (Situation Refinement):* It uses advance knowledge and environmental data to identify a situation and describe the relationship between objects and observed events in contextual manner. An example of it is (Chen et al., 2006), where the acoustic signals from birds and prior knowledge of bird sound patterns, are processed to describe the inter-relationship between the acoustic signals and that pattern, in order to infer the class of birds.
- *Level 3 (Threat Refinement):* A computationally complex task, threat refinement tries to predict any future threats or vulnerabilities based on evaluation of current situation. An example of it a target tracking applications, where the future location of the target is forecasted based on its current movement so that evaluation of it being a threat or not can be done.
- *Level 4 (Process Refinement):* This process actually is outside the data fusion process but plays a significant role in it. It allocates the sources to various processes based on specified goals and current system performance. For this a system should be continuously monitored, management and exposure information should be aggregated while providing QoS.

JDL helps designer to identify major solutions for each part of the data fusion system so that can be installed into the system. However, it does not identify interaction between processing element and how future iterations will be done from the present and past results of data fusion.

4.1.2 Dasarathy Model

Also known as DFD (Data-Feature-Decision) Model (Dasarathy, 1997), it is another information-based model with fine granularity with elements

Figure 6. The Dasarathy data fusion model

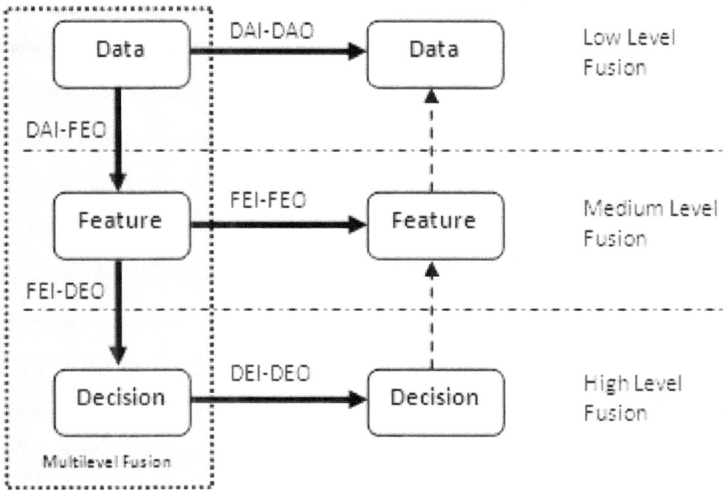

of data fusion being specified on the basis of the respective input and outputs. As shown in Figure 6, raw data is taken as input to generate a decision as an output. DAI-DAO, DAI-FEO, FEI-FEO and DEI-DEO are the different components of several fusion stages. Accordingly, the types of fusion can be specified and categorized based on their input and output data. With a feedback based system, decision blocks can be used to tune up the decision ability. This enhances system performance significantly.

A limitation of DFD model is the complexity in providing complex fusion blocks through its basic functional blocks. To explain this with an example; in a feature fusion where a feature is to be fused with two signals to obtain a refined feature, a DAI-FEO block is required first to fuse the signals that produce a feature output which is then combined with given feature using a second FEI-FEO block. However, its fine-grained approach to fusion has been widely used in WSN for ambient noise estimation using DAI-DAO (Polastre et al., 2004), building feature maps using FEI-FEO (Singh et al., 2006), event detection using DEI-FEO (Luo et al., 2006), and failure detection using FEI-FEO (Bracio et al., 1997).

Both of these models do not define the execution sequence of data fusion process nor do they specify the distributed nature of network in WSN. While the JDL model is more defense-oriented, the DFD model is application-independent and also allows users to explicitly define the abstraction level of input and output data. Another aspect of difference in both models is; while JDL provides a system-oriented perception of data fusion useful for designing data fusion systems, DFD gives an input-output-oriented based perception of data fusion which can be used for knowing relationship between fusion tasks and the manipulated data.

4.2 Activity-Based Models

These models are specified on the basis of activities that take place in a data fusion system and also mention proper execution sequence of these activities. Three models have been popular so far; Boyd Control Loop (Boyd, 1987), Intelligence Cycle (Shulsky & Schmitt, 2002) and Omnibus Model (Bedworth & O'Brien, 1999). Here we explain all the three models in a brief manner.

Figure 7. The OODA or Boyd-control loop

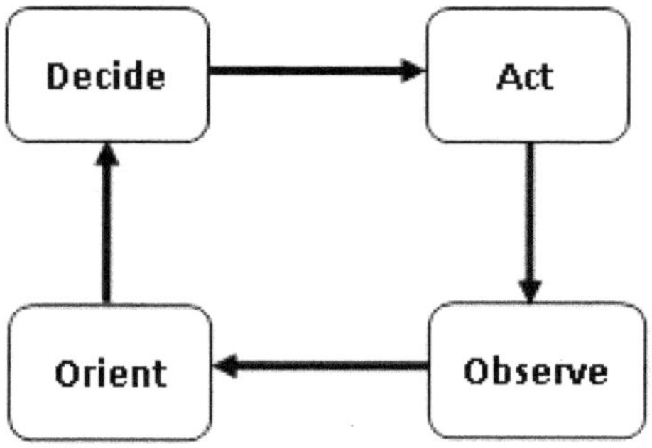

Figure 8. The intelligent cycle

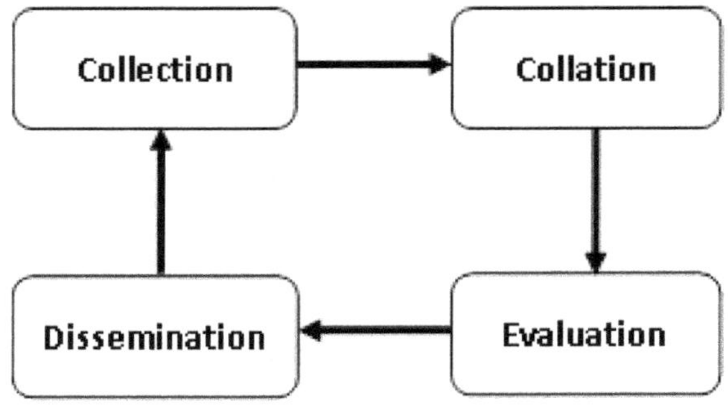

4.2.1. Boyd Control Loop

This Boyd Control loop, illustrated in Figure 7, is also known as OODA Loop (Boyd, 1987) for its four stages: Observe Orient, Decide and Act. It is similar to the conventional decision-support mechanism used for military purposes. The main activities of data fusion that form the various stages of OODA loop are,

- Observe, in which information is aggregation from all the available sources, i.e. sensors.
- Orient, where the aggregated information from observe stage is fused to form an interpretation of present scenario.
- Decide, where the interpretation is utilized to decide on future action plan.
- Act, where the plan is actually executed.

Taking example of SCAR routing algorithm in (Mascolo & Musolesi, 2006), sensor node aggregates information about the mobility and resources of the neighbor sensors (Observe), uses Kalman filter to estimate future value and updating current estimations (Orient), elect the best neighbor based on these estimations (Decide) and finally forwards the packet to that neighbor node (Act). Comparing it with JDL Model: Observe corresponds to level 0, Orient to level 1, 2 and 3, Decides matches with level 4, while no level corresponds to Act process. While Boyd Control Loop allows modeling of various main activities related to fusion in simple way, it fails to represent specific activities related to the same.

4.2.2. Intelligence Cycle

Along the similar lines to OODA loop explained above, Intelligence cycle (Shulsky & Schmitt, 2002) too is a four-staged cycle, specified by UK Intelligence Community as the process of transforming raw information from sensors into finished intelligence that can be used for making decision and taking actions. The four steps of Intelligence Cycle, as shown in Figure 8, are:

- Collection: raw sensory information is gathered from the environment.
- Collation: Collected data is analyzed, compared and then correlated while discarding irrelevant and imprecise information at the same time.
- Evaluation: The output data of collation stage is fused and analyzed.
- Dissemination: The results of evaluation are provided to users for making decision and taking actions based on detection situation.

Usually tracking applications are best exploiters of Intelligence Cycle. Again corresponding it with JDL Model; Collection step matches level 0, Collation to level 1, evaluation matches levels 2 and 3, while dissemination matches level 4. The difference between OODA model and Intelligence Cycle is the absence of Decide and Act phases in Intelligence Cycle, where they are assumed to be done by Evaluation and Dissemination steps. However, similar to OODA loop, this model too is unable to represent specific fusion related activities.

4.2.3. Omnibus Model

This model is similar to OODA loop and Intelligence Cycle in the sense that it too follows a cyclic sequence of fusion relation activities. However, here the specific tasks related to each stage is explicitly stated as shown in Figure 9. This model can be used to produce an overall perception of the data fusion system and then utilizing it to specify the subtasks, provides a fine-grained understanding of data fusion system and eases out the design process through a repetitive use.

While this model was actually developed for information gathered by sensor nodes, few modifications have made it customized for data fusion. Various strategies can be used for different specific data fusion tasks at each level depending upon the application being used for. Example applications are surveillance, target tracking, etc.

4.3. Role-Based Models

These models changes the way of modeling and design of data fusion systems. Here the fusion roles and relationships between them are utilized for designing fusion systems. Object-Oriented Model (Kokar et al., 2000) and Frankel-Bedworth Architecture (Frankel & Bedworth, 2000) are two types of Role-based models, which are explained below.

4.3.1. Object-Oriented Model

As shown in Figure 10, object-oriented model also follows cyclic architecture, though unlike in case of information-based or activity-based models, this model outlines a set of roles and relationships among them. Four roles described are:

Figure 9. Omnibus model

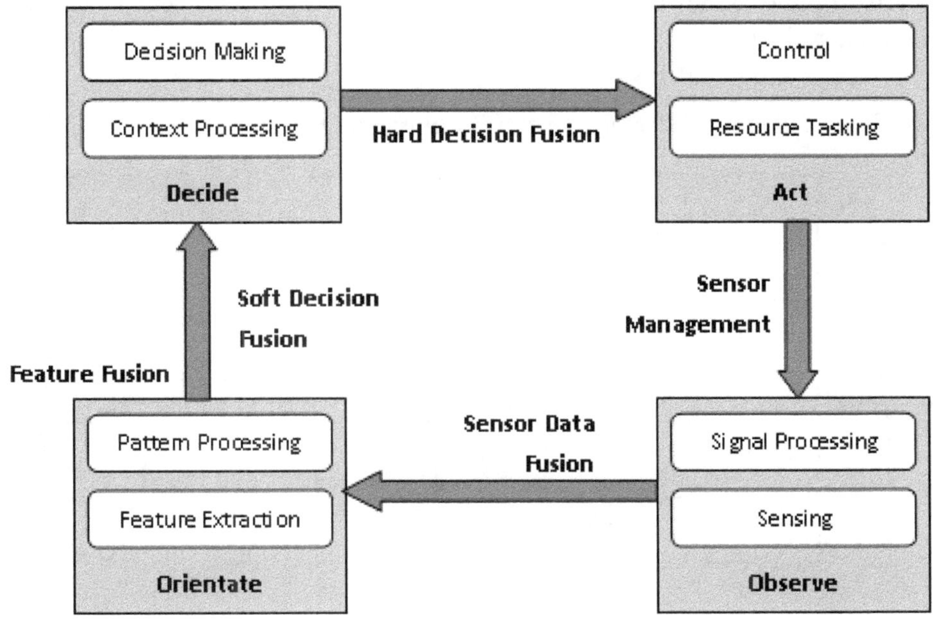

Figure 10. The object-oriented data fusion model; figure taken from (Kokar et al., 2000)

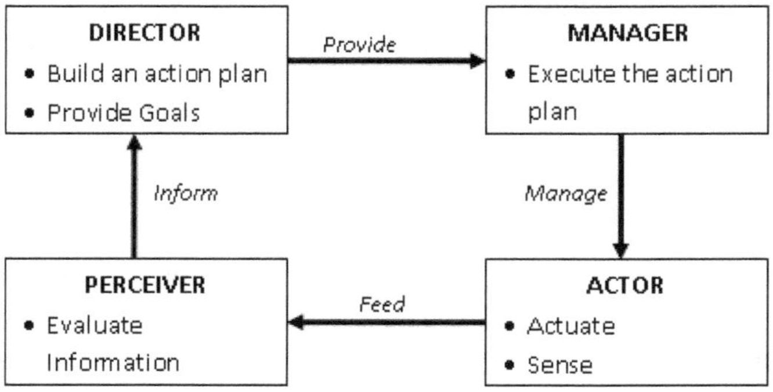

- Actor: It interacts with users, collects the information and executes the actions on the environment.
- Perceiver: It measures the collected information to provide a contextualized analysis for the director.
- Director: It builds an action plan, based on analysis provided by perceiver that can achieve system goals.
- Manager: As the name implies, it manages the actors in execution of plans built by the director.

However, the actual system implementation of this model through object-oriented programming languages is difficult due to indistinct nature of human and computer objects.

4.3.2. Frankel-Bedworth Architecture

Frankel (1999) proposed a human fusion consisting of two processes: global and local. While the global process uses the feedback returned by local process to update its goals and timetables, the local process on the other hand uses these goals and timetables to manage the execution of current activities.

This architecture when fed to a machine fusion architecture that divides control and estimation, goal-achieving and goal-setting behaviors, the resulting model is known as Frankel-Bedworth architecture shown in Figure 11. Both global and local processes are self-regulatory having different objectives and hence different roles.

Figure 11. The Frankel-Bedworth data fusion architecture; figure taken from (Frankel & Bedworth, 2000)

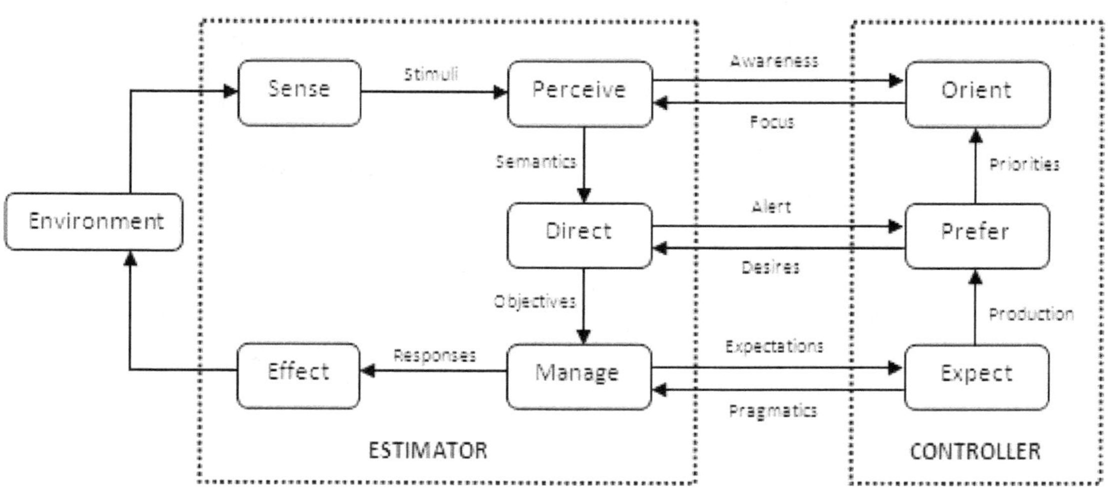

Local process has the role of an Estimator, which performs following tasks:

- Sense: Gathering of raw information by sources.
- Perceive: Stimuli generated by sensors are relevantly dealt (focus) and the Controller is made aware of the stimuli being used (awareness).
- Direct: On the basis of perception (semantics), a feedback (alert) is sent to the Controller. Difference between desired goals and current situation is calculated and the new goals and objectives (desires) are forwarded to the estimator.
- Manage: Estimator provides appropriate response when the controller is activated to specify practical aspects (pragmatics). Controller gets activated on the basis of objectives. Estimator sends this response as a feedback to controller by reporting anticipations about the response (sensitivity).
- Effect: The decisions (responses) are executed and the control loop is terminated through sensing of environmental changes.

Global control process plays the roles of Controller which controls and manages the role of Estimator. The basic tasks of Global Control Process are:

- Orient: Configures the importance of sensed stimuli.
- Prefer: Those with greatest relevance to goal-achieving behavior are given priority. This priority also explains the local goals (desires).
- Expect: The Estimator pragmatics are determined by making predictions and filtering of intentional objectives.

This architecture varies from previous models by providing global process which monitors the performance and manages all the goals of the local processes that are responsible for implementation of actual fusion related tasks. In the real scenario, the global processes will be the queries disseminated by human beings who will send the commands to direct the local estimation process which itself is implemented within the network.

5. PERFORMANCE EVALUATION OF DATA FUSION SYSTEMS

Here, we see the implementation aspects of data fusion systems by observing some of the parameters that can be considered while evaluating the performance and effectiveness of the data fusion system. These parameters are important as they decide how different data fusion algorithms will behave under different scenario, so that the best algorithm for chosen application can be implemented. While evaluating data fusion processes, one must realize, whether the data fusion process was integrated within the system or explicitly added to the system as an enhancement. As we have already seen, the major applications of data fusion are target tracking and localization; we explain some of the performance parameters related to these major applications of data fusion:

Parameters providing measures for evaluating system performance (Hall & Llinas, 2001),

- Detection probability: Probability of target detection or estimating locations as a function of range, SNR, etc.
- False alarm rate: Rate of false identification of noisy or spurious signals as valid targets or locations.
- Location estimate accuracy: Accuracy with which the position of an entity is determined.
- Identification probability: Probability of identifying the target or location correctly.
- Identification range: The range between sensors and the target within which, the

probability of accurate identification of target is larger than a certain threshold.

- Time from transmission to detect: Difference in the time when the signal was emitted by a target and the target was detected by the fusion system.
- Target classification accuracy: Accuracy with which the sensing system can identify the target and classify it into some pre-defined class or group.

These parameters determine how ably the fusion system can transform the signal from sensors or target to the location or attributes of the target.

Some of the parameters for measuring systems effectiveness (Hall & Llinas, 2001) (ability of fusion system to assist in completion of the operation) are:

- Target nomination rate: The rate with which fusion system can identify the targets and assign them to other functional systems for actions.
- Timeliness of information: Ability of the system to provide the result within desired time to the command center so that decisions can be taken within time.
- Warning time: Time to provide user with a warning about future danger or events.
- Countermeasure immunity: Ability of fusion system to countermeasure any future degradation by the targets.

Some of the metrics that can be computed for the above mentioned measures are:

Percentage fit error (PFE) in position: It is the ratio of the norm (a MATLAB function) of the difference between actual and estimated position coordinates. For accurate estimations, PFE will be zero and will increase as difference between the estimated value and true value increases. Ideally, algorithm giving least PFE is chosen for location based applications. Mathematically, PFE is calculated as

$$\mathrm{PFE} = \frac{\mathrm{norm}(x_a - x_e)}{\mathrm{norm}(x_a)} \times 100 \tag{31}$$

where, x_a is the actual position and x_e is the estimated position. Norm is an operator from MATLAB function to find Euclidean length of the vector. Similarly PFE can be calculated for positions, velocities, accelerations, and other parameters of interest. In case, true values is not known PFE can be calculated between measured value and estimated value. This parameter is useful in target tracking and any other locations finding applications.

Root mean square error in position: Similar to PFE, the only difference is calculation of root mean square error of the difference rather than norm as in case of PFE. Again, a lower value is preferable for choosing between algorithms. Mathematically it is given as

$$\mathrm{RMSPE} = \sqrt{\frac{1}{N}\sum_{k=1}^{N} \frac{(x_a(k) - x_e(k))^2 + (y_a(k) - y_e(k))^2 + (z_a(k) - z_e(k))^2}{3}} \tag{32}$$

where, N is number of samples. Similarly, this parameter can also be evaluated for acceleration, velocity and other parameters of interest.

Root sum square error (RSSE) in position: It is a discrete time or index dependent variable, which produced a sequence of results that can be plotted to predict the results. These values will be zero if the estimated value is same as the actual value and increase with the increasing deviation of estimated values from actual values. For comparison of data fusion techniques, one producing lowest time history of the RSSE is preferred. It is given as,

Figure 12. Performance parameters for different fusion levels

Level 1 or Low level fusion	Level 2 and 3 or Medium/High Level Fusion
• Accuracy • Repeatability/consistency • Robustness • Computational complexity • Energy Consumption	• Correctness in reasoning • Quality or relevance of decisions/advice/recommendations • Intelligent behavior • Adaptability in reasoning (robustness)

$$\mathrm{RSSPE}(k) = \sqrt{(x_a(k) - x_e(k))^2 + (y_a(k) - y_e(k))^2 + (z_a(k) - z_e(k))^2} \tag{33}$$

where, $k = 1,2,3, \ldots N$. Similar processes can be carried out for velocity and acceleration as well.

As per the different levels of data abstraction based data fusion discussed earlier, we can further categorize various parameters for different levels.

6. FUTURE CONSIDERATIONS

After learning the various aspects of data fusion through its importance, techniques, models and performance evaluation parameters, still there are unanswered questions that a system designer should consider before using a certain fusion algorithm in the system. Hall and Garga (1999) discussed some of the limitations of data fusion systems:

- No data fusion technique, used for whatever amount of time, can replace a single and accurate sensor that can measure the desired physical entity.
- The errors in upstream processing cannot be made up by downstream processing. That is, errors or lack of preprocessing of data at individual sensor nodes cannot be corrected by a data fusion process.
- For the case where sensor performance cannot be accurately modeled, i.e. wrong information about the performance of sensor is used; data fusion might lead into corrupted results.
- Various assumptions made for implementation of data fusion techniques, none of these techniques give optimal results in all conditions. For example, while MLE requires Gaussian distribution with unknown probability, MAP works with a known probability case.
- Data Fusion techniques are application specific, as same technique cannot be used for all applications. In target tracking applications kalman filtering techniques are preferrred, while localization applications prefer MLE estimators. Hence, a user is always concerned with knowing the demands of the application to choose the specific technique.
- In case of techniques like neural networks, there is never enough training data to support the fusion algorithms.
- Though there are several performance measurement metrics, the value of data fusion in the overall operation cannot be quantified entirely. This is a desired aspect in military applications where effectiveness of mission has to be measured.

- Complexity of various fusion algorithms makes them difficult for general user to understand them and choose between them for the required operation.
- Unknowledgeable use of data fusion algorithms might increase network traffic as well as the energy consumption.

These limitations pose some challenges in design of a data fusion system which should be considered by the designer before proceeding to implement the data fusion system.

- Which data fusion technique would be best suitable for the desired application? What are the needs and attributes of the application that should be considered while selecting the data fusion technique? This question can be understood with an example of a target tracking application where accuracy is primary need, so that fusion techniques that provide better accuracy has to be implemented.
- What parameters (energy consumption, accuracy, reliability, robustness, complexity, etc.) a data fusion system should be evaluated at, before using it in the system? Decision of parameters is an essential aspect, as one might end up wasting all the time and resources to find that the technique does not provide an evaluation of desired parameter.
- Will the addition of data fusion degrade or enhance performance of the operational system? For this, the performance of the system has to be assessed in both cases: with and without data fusion employed in the system.
- With several models into perspective, which model should be followed to design the system? The features of model and their limitations have to be considered before proceeding to design a fully fledged information system.

7. CONCLUSION

Data fusion is a complex mathematical process of distilling useful information from raw data, coming from different sensors sensing the physical phenomena, to produce a meaningful result (a signal, symbol, image or decision). In such a process, the aim would be to provide enhancements to the system so that the overall performance of the information provisioning system can be bettered. In WSN domain, it has found wide variety of applications ranging from target tracking to localization, MAC overhead reduction, source failure recovery, and many more. Moreover, it has been found to easily fit into the distributed nature of the WSN, while providing the same accuracy compared to the centralized one. Data fusion supports complementarity, redundancy and co-operative operation of sensors and hence hides imprecise measurements from the sensors, failure of sensors while also providing spatial and temporal coverage.

Data fusion algorithms are classified on various bases; on the basis of interaction among sensors (complementary, co-operative, redundant), level of abstraction of raw data (low, medium, high and multilevel), abstraction of input-output data (DAI-DAO, DAI-FEO, FEI-FEO, DEI-DEO, FEI-DEO). Various techniques of data fusion based on inference, estimation, compression, aggregation, abstract sensors are described. Of these, Bayesian inference, Dempster-Shafer inference, Maximum Likelihood, Maximum A Posteriori, Kalman Filtering are the most common techniques. With each technique, a work related to usage of the technique in WSN has been mentioned. It is essential to consider the benefits and limitations of using the technique, the type of application it is suited to, and the complexity with which it can be implemented. Choice of data fusion technique is also application dependent as the technique must fulfill the demands of the applications while also remaining within the constraints imposed by the application.

The design aspects of data fusion system are explained through various models; information-based models, activity-based models and role-based model. While JDL and Dasarathy Model are information-based models and also popular ones, Boyd-Control or OODA loop, Intelligent Cycle, etc. are some of activity-based models. Choice of models depends on implementation complexity desired by the designer and way one would like to represent the data fusion system in the future for any further modifications.

Finally, various parameters of measuring system performance and effectiveness, and metrics related to them are explained. Overall these intend to allow a user to compare various algorithms before making a choice on using particular data fusion algorithm in the information fusion system. Limitation and Challenges in data fusion informs data fusion system designer to take extra care while making a choice on technique.

Data fusion is an ever-growing field of research due to its application in wide variety of areas, and having techniques from a lot of different domains make it easier to people having understanding of specific areas to choose a data fusion technique on the basis of his understanding. For example a guy working on robotics can easily understand fuzzy logic and hence can develop fuzzy logic based data fusion system. Hence, data fusion is an important part of information processing.

REFERENCES

Abbasi, A., & Younis, M. (2007). A survey on clustering algorithms for wireless sensor networks. *Computer Communications*, *30*, 2826–2841. doi:10.1016/j.comcom.2007.05.024

Akyildiz, I. F., Su, W., Sankarasubramaniam, Y., & Cayirci, E. (2002, March). Wireless sensor networks: A survey. *Computer Networks*, *38*, 393–422. doi:10.1016/S1389-1286(01)00302-4

Al-Karaki, J. N., & Kamal, A. E. (2004, December). Routing techniques in wireless sensor networks: A survey. *IEEE Wireless Communications*, *11*(6), 6–28. doi:10.1109/MWC.2004.1368893

Bedworth, M. D., & O'Brien, J. C. (1999). The omnibus model: A new model for data fusion? In *Proceedings of the 2nd International Conference on Information Fusion (FUSION)*, ISIF, Sunnyvale, (pp. 437–444).

Boulis, A., Ganeriwal, S., & Srivastava, M. B. (2003, September). Aggregation in sensor networks: An energy-accuracy trade-off. *Ad Hoc Networks*, *1*, 317–331. doi:10.1016/S1570-8705(03)00009-X

Boyd, J. R. (1987). *A discourse on winning and losing. Unpublished set of briefing slides available at Air University Library*. Alabama: Maxwell AFB.

Bracio, B. R., Horn, W., & Moller, D. P. F. (1997). Sensor fusion in biomedical systems. In *Proceedings of the 19th Annual International Conference of the IEEE Engineering in Medicine and Biology Society*, Vol. 3, (pp. 1387–1390). Chicago, IL: IEEE.

Brown, C., Durrant-Whyte, H., Leonard, J., Rao, B., & Steer, B. (1992). Distributed data fusion using Kalman □ltering: A robotics application . In Abidi, M. A. (Ed.), *Data fusion in robotics and machine intelligence* (pp. 267–309).

Buede, D. M. (1988). Shafer-Dempster and Bayesian reasoning: A response to Shafer-Dempster reasoning with applications to multisensor target identi□cation systems. *IEEE Transactions on Systems, Man, and Cybernetics*, *18*, 1009–1011. doi:10.1109/21.23099

Chen, C., Ali, A. M., & Wang, H. (2006). Design and testing of robust acoustic arrays for localization and enhancement of several bird sources. In *Proceedings of the 5th International Conference on Information Processing in Sensor Networks (IPSN)*, (pp. 268–275).

Ci, S., Sharif, H., & Nuli, K. A. (2004). UKF-based link adaptation scheme to enhance energy efficiency in wireless sensor networks. In *Proceedings of the 15th IEEE International Symposium on Personal, Indoor and Mobile Radio Communications (PIMRC'04),* Vol. 4, (pp. 2483–2488). IEEE.

Cohen, N. H., Purakayastha, A., Turek, J., Wong, L., & Yeh, D. (January 2001). *Challenges in □exible aggregation of pervasive data.* IBM Research Report RC 21942 (98646), NY.

Coue, C., Fraichard, T., Bessiere, P., & Mazer, E. (2002). Multi-sensor data fusion using Bayesian programming: An automotive application. In *IEEE/RSJ International Conference Intelligent Robots and Systems,* Vol. 1, Switzerland, (pp. 141–146).

Crisan, D., & Doucet, A. (2002, Mar). A survey of convergence results on particle □ltering methods for practitioners. *IEEE Transactions on Signal Processing, 50,* 736–746. doi:10.1109/78.984773

Dasarathy, B. V. (1997, Jan). Sensor fusion potential exploitation-innovative architectures and illustrative applications. *Proceedings of the IEEE, 85,* 24–38. doi:10.1109/5.554206

Durrant-Whyte, H. F. (1988, December). Sensor models and multisensor integration. *The International Journal of Robotics Research, 7,* 97–113. doi:10.1177/027836498800700608

Elfes, A. (1987, June). Sonar-based real-world mapping and navigation. *IEEE Journal on Robotics and Automation, 3,* 249–265. doi:10.1109/JRA.1987.1087096

Elmenreich, W. (2002). *Sensor fusion in time-triggered systems.* Ph.D. thesis, Institut fur Technische Informatik, Vienna University of Technology, Vienna, Austria.

Frankel, C. B. (1999). *Such order from confusion sprung: Adaptive competence and affect regulation.* Ph.D. thesis, Pacific Graduate School of Psychology, Palo Alto, CA.

Frankel, C. B., & Bedworth, M. D. (2000). Control, estimation and abstraction in fusion architectures: Lessons from human information processing. In *Proceedings of the 3rd International Conference on Information Fusion (Fusion),* Vol. 1. IEEE.

Friedlander, D. S. (2005). Semantic information extraction . In Iyengar, S. S., & Brooks, R. R. (Eds.), *Distributed sensor networks* (pp. 409–417). Boca Raton, FL: CRC Press.

Friedlander, D. S., & Phoha, S. (2002). Semantic information fusion for coordinated signal processing in mobile sensor networks. *International Journal of High Performance Computing Applications, 16,* 235–241. doi:10.1177/10943420020160030401

Ganeriwal, S., Kumar, R., & Srivastava, M. B. (2003). Timing-sync protocol for sensor networks. In *Proceedings of the 1st International Conference on Embedded Networks Sensor Systems (SenSys'03),* (pp. 138–149).

Gunnarsson, F., & Gustafsson, F. (2003). Positioning using time-difference of arrival measurements. In Proceedings of the *IEEE International Conference on Acoustics, Speech, and Signal Processing. Vol. 6,* IEEE, (pp. 553–556).

Guo, D., Wang, X., & Chen, R. (2005, Apr.). New sequential Monte Carlo methods for nonlinear dynamic systems. *Statistics and Computing, 15,* 135–147. doi:10.1007/s11222-005-6846-5

Hall, D. L., & Garga, A. K. (1999). New perspectives on level four processing in data fusion systems. In *Proceedings of SPIE AeroSense'99 Conference on Digitization of the Battle□eld IV.*

Hall, D. L., & Llinas, J. (1997, January). An introduction to multi-sensor data fusion. *Proceedings of the IEEE*, *85*, 6–23. doi:10.1109/5.554205

Hall, D. L., & Llinas, J. (2001). *Handbook of multisensor data fusion*.

He, T., Blum, B. M., Stankovic, J. A., & Abdelzaher, T. (2004). AIDA: Adaptive application independent data aggregation in wireless sensor network. *ACM Transactions in Embedded Computer Systems*, *3*, 426–457. doi:10.1145/993396.993406

Hoover, A., & Olsen, B. D. (2000). Sensor network perception for mobile robotics. In *Proceedings of the IEEE International Conference on Robotics and Automation*, Vol. 1, (pp. 342–347).

Intanagonwiwat, C., Estrin, D., Govindan, R., & Heidemann, J. (2002). Impact of network density on data aggregation in wireless sensor networks. In *Proceedings of the 22nd IEEE International Conference on Distributed Computing Systems (ICDCS)*, IEEE, Vienna, Austria, (pp. 457–458).

Intanagonwiwat, C., Govindan, R., & Estrin, D. (2000). Directed diffusion: A scalable and robust communication paradigm for sensor networks. In *Proceedings of the 6th Annual International Conference on Mobile Computing and Networking (MobiCom)*, (pp. 56–67). Boston, MA: ACM Press.

Iyengar, S. S., Chakrabarty, K., & Qi, H. (2001, September). Introduction to special issue on Distributed sensor networks for real-time systems with adaptive con□guration. *Journal of the Franklin Institute*, *338*, 651–653. doi:10.1016/S0016-0032(01)00027-8

Julier, S. J., & Uhlmann, J. K. (1997). New extension of the Kalman □lter to nonlinear systems . In *Signal Processing, Sensor Fusion, and Target Recognition VI* (*Vol. 3068*, pp. 182–193). San Diego: SPIE.

Kalman, R. E. (1960). A new approach to linear □ltering and prediction problems. *Transactions of ASME Journal of Basic Engineering*, *82*, 35–45. doi:10.1115/1.3662552

Kessler, J., et al. (January 1992). *Functional description of the data fusion process*. Tech. rep., Naval Air Development Center, Warminster, PA, USA.

Kokar, M. M., Bedworth, M. D., & Frankel, C. B. (2000). A reference model for data fusion systems . In *Sensor Fusion: Architectures, Algorithms and Applications IV* (pp. 191–202). Orlando, FL: SPIE. doi:10.1117/12.381632

Krishnamachari, B., Estrin, D., & Wicker, S. (2002). The impact of data aggregation in wireless sensor networks. *Proceedings of the 22nd International Conference on Distributed Computing Systems*.

Krishnamachari, B., & Iyengar, S. (2004, Mar). Distributed Bayesian algorithms for fault-tolerant event region detection in wireless sensor networks. *IEEE Transactions on Computers*, *53*, 241–250. doi:10.1109/TC.2004.1261832

Kulik, J., Heinzelman, W., & Balakrishnan, H. (2002, March). Negotiation-based protocols for disseminating information in wireless sensor networks. *Wireless Networks*, *8*, 169–185. doi:10.1023/A:1013715909417

Lewis, T. W., & Powers, D. M. W. (2002). Audio-visual speech recognition using red exclusion and neural networks. In *Proceedings of the 25th Australasian Conference on Computer Science. Australian Computer Society*, Melbourne, Victoria, Australia, (pp. 149–156).

Li, T., Ekpenyong, A., & Huang, Y. (2006, Oct). Source localization and tracking using distributed asynchronous sensor. *IEEE Transactions on Signal Processing*, *54*, 3991–4003. doi:10.1109/TSP.2006.880213

Liu, J., Zhang, Y., & Zhao, F. (2006). Robust distributed node localization with error management. In *Proceedings of the 7th ACM International Symposium on Mobile Ad Hoc Networking and Computing (MobiHoc'06)*, Florence, Italy, (pp. 250–261).

Luo, R. C., & Kay, M. G. (1992). Data fusion and sensor integration: State-of-the-art 1990s . In Abidi, M. A., & Gonzalez, R. C. (Eds.), *Data fusion in robotics and machine intelligence* (pp. 7–135). San Diego, CA: Academic Press, Inc.

Luo, R. C., Yih, C.-C., & Su, K. L. (2002, April). Multisensor fusion and integration: Approaches, applications, and future research directions. *IEEE Sensors Journal, 2*, 107–119. doi:10.1109/JSEN.2002.1000251

Luo, X., Dong, M., & Huang, Y. (2006, Jan). On distributed fault-tolerant detection in wireless sensor networks. *IEEE Transactions on Computers*, 55.

Madden, S. R., Franklin, M. J., Hellerstein, J. M., & Hong, W. (2002). TAG: A tiny aggregation service for ad-hoc sensor networks. *ACM SIGOPS Operation Systems Review, 36*, 131–146. doi:10.1145/844128.844142

Madden, S. R., Franklin, M. J., Hellerstein, J. M., & Hong, W. (2005, March). TinyDB: An acqusitional query processing system for sensor networks. *ACM Transactions on Database Systems, 30*, 122–173. doi:10.1145/1061318.1061322

Marzullo, K. (1990, November). Tolerating failures of continuous-valued sensors. *ACM Transactions on Computer Systems, 8*, 284–304. doi:10.1145/128733.128735

Mascolo, C., & Musolesi, M. (2006). SCAR: Context-aware adaptive routing in delay tolerant mobile sensor networks. In *Proceeding of the International Conference on Communications and Mobile Computing (IWCMC)*, ACM, Vancouver, Canada, (pp. 533–538).

Nakamura, E. F., Nakamura, F. G., Figueiredo, C. M., & Loureiro, A. A. (2005, November). Using information fusion to assist data dissemination in wireless sensor networks. *Telecommunication Systems, 30*, 237–254. doi:10.1007/s11235-005-4327-y

Nordlund, P.-J., Gunnarsson, F., & Gustafsson, F. (2002). Particle □lters for positioning in wireless networks. In *Proceedings of the XI European Signal Processing Conference (EURSIPCO'02). Vol. II*, TeSA, Toulouse, France, (pp. 311–314).

Pagac, D., Nebot, E. M., & Durrant-Whyte, H. (1998, August). An evidential approach to map-building for autonomous vehicles. *IEEE Transactions on Robotics and Automation, 14*, 623–629. doi:10.1109/70.704234

Patwari, N., Hero, A. O., Perkins, M., Correal, N. S., & O'Dea, R. J. (2003, Aug). Relative location estimation in wireless sensor networks. *IEEE Transactions on Signal Processing, 51*, 2137–2148. doi:10.1109/TSP.2003.814469

Polastre, J., Hill, J., & Culler, D. (2004). Versatile low power media access for wireless sensor networks. In *SenSys'04*, (pp. 95–107).

Ratnasamy, S., Karp, B., Shenker, S., Estrin, D., Govindan, R., Yin, L., & Yu, F. (2003, Aug). Data-centric storage in sensornets with GHT, a geographic hash table. [MONET]. *Mobile Networks and Applications, 8*, 427–442. doi:10.1023/A:1024591915518

Santini, S., & Romer, K. (2006). An adaptive strategy for quality-based data reduction in wireless sensor networks. In *Proceedings of the 3rd International Conference on Networked Sensing Systems (INSS)*, TRF, Chicago, IL, (pp. 29–36).

Savvides, A., Han, C., & Strivastava, M. B. (2003, August). The n-hop multilateration primitive for node localization. *Mobile Networks and Applications, 8*, 443–451. doi:10.1023/A:1024544032357

Schmid, U., & Schossmaier, K. (2001, April). How to reconcile fault-tolerant interval intersection with the Lipschitz condition. *Distributed Computing, 14*, 101–111. doi:10.1007/PL00008927

Shu, H., & Liang, Q. (2005). Fuzzy optimization for distributed sensor deployment. In *IEEE Wireless Communications and Networking Conference (WCNC), Vol. 3*, IEEE, New Orleans, LA, (pp. 1903–1908).

Shulsky, A. N., & Schmitt, G. J. (2002). *Silent warfare: Understanding the world of intelligence* (3rd ed.). New York, NY: Brasseys, Inc.

Sichitiu, M. L., & Ramadurai, V. (2004). Localization of wireless sensor networks with a mobile beacon. In *Proceedings of the 1st IEEE International Conference on Mobile Ad Hoc and Sensor Systems (MASS 2004)*, IEEE, Fort Lauderdale, FL, (pp. 174–183).

Singh, A., Nowak, R., & Ramanathan, P. (2006). Active learning for adaptive mobile sensing networks. In *Proceedings of the 5th International Conference on Information Processing in Sensor Networks (IPSN)*, (pp. 60–68).

Wald, L. (1999, May). Some terms of reference in data fusion. *IEEE Transactions on Geoscience and Remote Sensing, 13*, 1190–1193. doi:10.1109/36.763269

Wang, J. C. P., Abolhasan, M., Safaei, F., & Franklin, D. (2007). A survey on control separation techniques in multi-radio multi-channel MAC protocols. In *the 7th IEEE International Symposium on Comm. and Information Technologies* (IEEE ISCIT), Sydney, Australia.

Welch, G., & Bishop, G. (2001). *An introduction to the Kalman □lter*. In SIGGRAPH 2001 Course Notes. *ACM*, Los Angeles, CA. Course 8.

Wongngamnit, C., & Angluin, D. (2001, Mar). Robot localization in a grid. *Information Processing Letters, 77*, 261–267. doi:10.1016/S0020-0190(00)00154-X

Wu, H. D., Siegel, M., Stiefelhagen, R., & Yang, J. (May 2002). *Sensor fusion using Dempster–Shafer theory.* In IEEE Instrumentation and Measurement Technology Conference, Anchorage, AK, USA.

Xiao, L., Boyd, S., & Lall, S. (2005) A scheme for robust distributed sensor fusion based on average consensus. In *Proceedings of the 4th International Symposium on Information Processing in Sensor Networks (IPSN'05)*, (pp. 63–70).

Xiao, L., Boyd, S., & Lall, S. (2006). A space-time diffusion scheme for peer-to-peer least-squares estimation. In *Proceedings of the 5th International Conference on Information Processing in Sensor Networks (IPSN'06)*, (pp. 168–176).

Xing, G., Tan, R., Liu, B., Wang, J., Jia, X., & Yi, C. W. (2009). *Data fusion improves the coverage of wireless sensor networks.* In MobiCom.

Xiong, Z., Liveris, A. D., & Cheng, S. (2004, September). Distributed source coding for sensor networks. *IEEE Signal Processing Magazine, 21*, 80–94. doi:10.1109/MSP.2004.1328091

Yang, C.-L., Bagchi, S., & Chappell, W. J. (2005). Location tracking with directional antennas in wireless sensor networks. In *IEEE MTT-S International Microwave Symposium Digest.* Long Beach, CA: IEEE.

Yu, B., Sycara, K., Giampapa, J. A., & Owens, S. R. (2004). Uncertain information fusion for force aggregation and classi□cation in airborne sensor networks. In *AAAI-04 Workshop on Sensor Networks*. AAAI Press.

Yuan, Y., & Kam, M. (2004, August). Distributed decision fusion with a random-access channel for sensor network applications. *IEEE Transactions on Instrumentation and Measurement, 53*, 1339–1344. doi:10.1109/TIM.2004.830598

Zhao, J., Govindan, R., & Estrin, D. (2002). Residual energy scans for monitoring wireless sensor networks. In *Proceedings of the IEEE Wireless Communications and Networking Conference (WCNC'02)*, Vol. 1, IEEE, Orlando, FL, (pp. 356–362).

KEY TERMS AND DEFINITIONS

A Posteriori: A knowledge dependent on experience.

A Priori: A knowledge independent of experience.

Data Aggregation: It is the method of reducing the volume of data to be transmitted by processing all the incoming data using some mathematical technique to produce a value that represents all the input readings with desired accuracy.

Data Fusion: Integration or combination of data from various sources to produce an inference or a mathematical conclusion, that is more decisive and higher degree of trust associated with it or a result that was not present in the original data.

Fusion Model: An overview of various processes, stages or independent modules involved in a fusion process. It facilitates design of a fusion system, through independent selection of optimal techniques for each module or process.

Fusion: Integration or combination of multiples pieces of information into one.

IID variables: Independent and identically distributed variables are the random variables, independent of each other, yet having same probability distribution.

Inference: Technique of drawing conclusion from given propositions (observations), one which is not directly indicated in the propositions itself.

Information: Any piece or combination of raw data that yields into changing the state of knowledge while being useful at the same time.

Localization: The process of finding the location of any object under observation.

Mutually Exclusive Events: The events those are impossible to occur at the same time.

Probability Density Function: Calculated for a continuous random variable, it indicated the likelihood of a random variable to occur within some interval of time.

Probability Distribution: Determines the probability of how value of a random variable is distributed. They are of various forms: namely Gaussian distribution, Poisson distribution, Binomial distribution, etc., with each specifying the probability of presence of the value of a random variable at some time or within some interval of time.

Sampling: The process of picking up discrete set of values from a continuously changing continuous signal.

Target: Any object that is to be monitored and is different from the instruments employed to monitor it.

Target Tracking: A common application scenario under military applications, where the path followed by target is to be estimated, based upon the detected or estimated locations of the target at various instants of time.

Wireless Sensor Network: A self-organized network of large number of sensors, deployed over an area, for some specific application.

Chapter 20
Agility Improvement in Cognitive Radio under Bluetooth Paradigm Using Ant Colony Metaphor

Ashraf Darwish
Helwan University, Egypt

ABSTRACT

Cognitive radio is a paradigm for wireless communication in which either network or wireless node itself changes particular transmission or reception parameters to execute its tasks efficiently. This parameter alteration is based on observations of several factors from external and internal cognitive radio environment, such as radio frequency spectrum, user behavior, and network state. The chapter presents the relationship between cognitive radio and ant colony. The bio-inspired cognitive radio algorithm is presented.

1. INTRODUCTION AND BACKGROUND

Cognitive radio is an emerging wireless technology that allows dynamic system reconfiguration, promising dramatic improvement in spectrum utilization. The basic theme of sharing unlicensed spectrum drives cognitive radio and ultra wideband technologies. By its inherent ability and conceptualization, cognitive radio is capable of sensing and learning from the environment. This feature finally authenticates its service. Since the cognitive (unlicensed) users are utilizing the license band, they must detect the presence of licensed (primary) users in a very short span of time and must vacate the band of primary user. Hence, it has become phenomenal to explore that how the cognitive radio (unlicensed) sense the presence of the (licensed) primary user?

This rudimentary communication initiative and inquisitiveness has prominently encouraged the community to apply the concept more into reality. Therefore, a new trend and plethora of applications were evolved encompassing satisfied spectrum utilization more authentic and adaptive way. Although, initially the concept was incepted from software defined radio.

DOI: 10.4018/978-1-4666-0321-9.ch020

While the cognitive radio community has had significant success popularizing the concept of cognitive radio and developing prototypes, applications, and critical components, the community has had a surprisingly difficult time agreeing upon exactly what is and is not a cognitive radio beyond. Perhaps echoing the sentiment of former Supreme Court Justice Potter Stewart, many members of the cognitive radio community believe that "they know it when they see it," even if a precise definition is ineffable. Some have succeeded in formulating a definition of cognitive radio but have found their definitions at significant variance with others' definitions.

The rest of the chapter is organized as follows. In the section 2 the background of cognitive radio is introduced. In section 3 the components and the functions of cognitive radio protocol was presented. Section 4 presents the Agility Management in cognitive radio for Bluetooth.In section 5 the problem of ant colony optimization for cognitive radio is presented. Section 6 presents some open research challenges in bio-inspired cognitive radios and section 7 concludes the chapter.

2. BACKGROUND

Tautologically, a cognitive radio could be defined as "A radio that is cognitive," or paraphrasing Descartes, "*Cogitat, ergo est*cognitive radio."[1] In the absence of a Turing test for radios, applying this definition is nontrivial and implies a level of functionality that many researchers consider excessive. Indeed, while many researchers and public officials agree that upgrading a software radio's control processes will add significant value to software radio, there is currently some disagreement over how much "cognition"is needed which results in disagreement over the precise definition of a cognitive radio. The following section provides some of the more prominently offered definitions of cognitive radio.

In the 1999 research paper that first coined the term "cognitive radio", Joseph Mitola III defines a cognitive radio as (J. Mitola, 1999)**:** "*A radio that employs model based reasoning to achieve a specified level of competence in radio-related domains.*"

However, in his recent popularly cited chapter that surveyed the state of cognitive radio, Simon Haykin defines a cognitive radio as (S. Haykin, 2005)**:** "*An intelligent wireless communication system that is aware of its surrounding environment (i.e., outside world), and uses the methodology of understanding-by-building to learn from the environment and adapt its internal states to statistical variations in the incoming RF stimuli by making corresponding changes in certain operating parameters (e.g., transmit-power, carrier frequency, and modulation strategy) in real-time, with two primary objectives in mind:*

- Highly reliable communications whenever and wherever needed;
- Efficient utilization of the radio spectrum.

Coming from a background where regulations focus on the operation of transmitters, the

FCC has defined a cognitive radio as a radio that can change its transmitter parameters based on interaction with the environment in which it operates.

Meanwhile, the other primary spectrum regulatory body in the US, the NTIA adopted the following definition of cognitive radio that focuses on some of the applications of cognitive radio. While aiding the FCC in its efforts to define cognitive radio, IEEE USA offered (IEEE USA Position, 2003) the definition including the term o "intelligent detection". The broader IEEE tasked the IEEE 1900.1 group to define cognitive radio which has the working definition of a type of radio that can sense and autonomously reason about its environment and adapt accordingly. This radio could employ knowledge representation, automated reasoning and machine learning mechanisms in establishing, conducting, or terminating communication or networking functions with other radios. Cognitive radios can be trained to dynamically and autonomously adjust its operating parameters.

Accumulating all these technical paradigms of definitions, we identify that the following capability- focused attributes of cognitive radio:

- Awareness of its environment and its own capabilities
- Goal driven autonomous operation
- Understanding or learning how its actions impact its goal
- Recalling and correlating past actions, environments, and performance.

The differences in the definitions for cognitive radio can be largely attributed to differences in the expectations of the functionality that a cognitive radio will exhibit. In his dissertation (J. Mitola, 2000), Joseph Mitola III considers the nine levels of increasing cognitive radio functionality.

A reference for how a cognitive radio could achieve these levels of functionality, (J. Mitola, 2000) introduces the cognition cycle, shown in Figure 1, as a "top- level control loop for cognitive radio." In the cognition cycle, a radio receives information about its operating environment (*Outside world*) through direct observation or through signaling. This information is then evaluated (*Orient*) to determine its importance. Based on this valuation, the radio determines its alternatives (*Plan*) and chooses an alternative (*Decide*) in a way that presumably would improve the valuation. Assuming a waveform change was deemed necessary, the radio then implements the alternative (*Act*) by adjusting its resources and performing the appropriate signaling. These changes are then reflected in the interference profile presented by the cognitive radio in the *Outside world*. As part of this process, the radio uses these observations and decisions to improve the operation of the radio (*Learn*), perhaps by creating new modeling states, generating new alternatives, or creating new valuations.

The role of intelligence and significantly artificial intelligence and its associated tools redefine furnish and further advance the cognitive radio scenario in recent years. These techniques broadly influenced the efficient and intelligent spectrum management leading cognitive radio network research a niche. Implementation of cognitive radio networks will prove that the single greatest cause of interference will be the presence of other cognitive radios. For the factors associated with the CR networks, mainly the imperfect information each radio will have and the potential "moves" it could make through frequency adaptations, the application of game theory is very logical (J. Neel, R. Menon, A. MacKenzie, J. Reed, 2005). Defining a CR as a complex state machine, it possesses the six basic stages of cognition: Orient, Plan, Decide, Act, Learn, and Observe. Through interactions with the "Outside World," the radio builds its basis of knowledge (which has been extended to ultimately predicting how the User may utilize his/her CR in routine applications). These cognitive states can be easily mapped into a game model, where each state is a player and the possible actions are the "action set." Different game models must be applied to the states/agents, as Learn requires a more complex model than Act. The potential and super modular game models are both relevant to the CR application. Three physical layer algorithms are given as proof of concept to the two aforementioned models, for Frequency Selection Adaptive Interference Avoidance, OFDM Channel Filing, and Distributed Power Control.

Developing from the work of Joe Mitola and others, a main source of CR application lies in spectrum pooling (also spectrum holes) where a primary user's licensed spectrum is dormant. A spectrum pooling architecture based on Orthogonal Frequency Division Multiplexing has been developed at the University of Karlsruhe, and is detailed in the text. DARPA has also developed architecture for policy-based CRs (R. Lowdermilk and F. Harris, 1996).

The current governmental regulation of the frequency spectrum forces a scarcity of spectrum, as a result of treating the spectrum like land. A more appropriate analogy of spectrum management is one of ocean shipping lanes, considering a world where boats could only travel at a given longitude. This would make no sense in a situation where a certain lane is utilized only twice a day, and today's technology can be made to be intelligent enough to "hop lanes" according to the amount of, or lack of, traffic that may be present. To facilitate the realization of this technology, Werbach recommends the following policy changes: Develop rules to foster unlicensed user cooperation, allocate more unlicensed user spectrum, eliminate restrictions on underlay technology, and promote research in unlicensed wireless technology. Wi-Fi has already proved the commercial and technological success of open spectrum implementation, becoming a huge industry. The two main approaches now to achieving open spectrum are unlicensed spectrum device-specific frequency bands (like Wi-Fi) and the underlay of transmissions in licensed bands (enter cognitive radio) (M. J. Marcus, 2005).

There are couples of significant contribution in intelligent computing deployed in cognitive radio in yesteryears. These works implicate the relevance of artificial intelligence in this context.

Successful cognitive radios must be aware, learn, and take action for any situation that might arise. Applications range from voice communications under low power conditions, communications in high interference zones, to more complex and critical, hostile military networks of interoperating vehicles and soldiers with many different network needs. A radio must respond to any of these scenarios and adapt its many different parameters that define the radio's waveform and protocols. These radios do not only require learning; instead, they need highly sophisticated learning and decision making capabilities.

Machine learning has been well documented with both criticisms (D.E. Goldberg, 1989) as well as successes (T. W. Rondeau, C. J. Rieser, T. M. Gallagher, C. W. Bostian, 2004). Successful applications of artificial intelligence (AI) tend to occur specifically in narrowly defined, well-bounded applications. While the radio application is still bounded, the technical demands for intelligence in a radio exceed those normally associated with successful applications of classic artificial intelligence techniques such as expert systems or neural networks.

Signal modulation classifier design using artificial neural networks has already been implemented for cognitive radio. Practically, Modulation classification consists of signal feature extraction and pattern recognition. Different algorithms are useful depending on which features are extracted. Modulation-related signal features are either waveform-level features extracted at digital IF or symbol-level extracted at baseband. Mostly multi-layer perceptron networks (MLPNs) that trade network complexity and computational cost for flexibility and robustness in classification (G. P. Zhang, 2000).

The novelty of this series of new intelligent methods lies in the learning and adaptation processes. These processes are augmented by a feedback system that compares the radio's actions to its performance to help direct future adaptations. The conventional cognitive engine works with evolutionary techniques that provide powerful methods of analyzing, predicting, and discovering new solutions and then using the feedback information to adjust itself intelligently for more accurate analysis and predictions that lead to improved solutions. The system uses a tiered approach to learning and optimization that mixes aspects of different AI techniques together to make a more robust computational intelligence. The machine learning techniques exploit the major benefits of evolution through random searching. The techniques also make skilled use of prior knowledge to direct the adaptation towards areas most likely to contain solutions. The general system topology is a two-layered evolutionary algorithm: the bottom layer discovers new solutions through trial, and the top layer monitors the behavior of the bottom layer as well as the feedback information from the environment. In the case of the cognitive radio, the solutions are new radio parameters. The feedback information consists of performance measurements such as bit error rate, frame error rate, throughput, etc. The feedback information is then compared to the performance predicted by the first layer algorithm. As the observed performance deviates from the predictions, the top layer genetic algorithm adjusts the bottom layer to compensate for areas with large errors. Such directed adaptation has been studied and previously used in evolutionary algorithm work through techniques like chromosome tagging and templates or in knowledge-based evolutionary techniques (L. Kallel, B. Naudts, and A. Rogers, 2001). The present proposal incorporates truly bio-inspired technique in the form of Ant colony optimization to be used in searching and learning phenomena. Prior to introduce bio-inspired motivation of cognitive radio, it would be worthy to discuss about the un-licensed spectrum management on which the cognitive radio is being positioned.

2.1 Unlicensed Spectrum Management in Mobile Network

It is commonly believed that there is a crisis of spectrum availability at frequencies that can be economically used for wireless communications. This misconception has arisen from the intense competition for use of spectra at frequencies below 3 GHz. At higher frequencies and thus makes it extremely difficult with respect to time place and direction. This corresponds to the unlicensed UNII spectra, which has only minimal constraints from the regulatory standpoint. What is clearly needed is an extension of the unlicensed usage to other spectral bands, while accommodating the present users who have legacy rights and also to insure that future requirements can be met. An approach, which can meet these goals, is to develop a radio that is able to sense the spectral environment over a wide available band and use the spectrum only if communication does not interfere with licensed user. These un-licensed low priorities Secondary Users (SU) would thus be using Cognitive Radio (CR) techniques, to ensure non-interfering coexistence with higher priority users and thus reduce concerns of a general allocation to unlicensed use.

The access to unlicensed spectrum is open but its utilization is strictly regulated. An unlimited number of users are sharing the same unlicensed spectrum. Spectrum usage is allowed to all devices that satisfy certain technical rules or standards in order to mitigate potential interference. Examples for these technical rules are the limitation of transmission power or advanced coexistence capabilities. The usage rights at unlicensed spectrum are flexible and no concrete methods to access spectrum are specified. Besides the Industrial, Scientific and Medical (ISM) bands at 900 MHz and 2.4 GHz, the FCC opened for unlicensed operation in 1990 during PCS rule making 20 MHz at 1.9 GHz for Unlicensed PCS (UPCS). Additionally, FCC reserved in 1997 300 MHz and in 2003 255 MHz at 5 GHz for unlicensed operation. This frequency band at 5 GHz is referred to as Unlicensed National Information Infrastructure (U-NII) band. During 2005, operators in the telecommunications industry announced an increasing number of fixed-mobile convergence (FMC) initiatives, many of which are expected to materialize in 2006. Most of these initiatives are focused around UMA (Unlicensed Mobile Access), the first standardized technology enabling seamless handover between cellular and WLAN. Clearly, in one way or another, FMC is a key agenda item for many operators. Fixed operators are looking at FMC as a defensive move to protect their customer base against churn and traffic losses due to fixed-to-mobile substitution. Mobile operators are considering FMC to address the indoor voice market (i.e. accelerating the rate of fixed-to-mobile substitution) as well as to enhance the performance of mobile services indoors. Furthermore, fixed and mobile operators are looking at FMC as a means to fend off the growing threat from new Voice-over-IP (VoIP) operators. As the first standardized technology enabling cellular/WLAN convergence, UMA is receiving significant interest from all types of operators. Consequently, most indicators point too many deployments of UMA beginning in 2006. Keeping these commercial standards in the context, the relevance of UMA and GSM Network evaluation could be realized and one of the major concerns in this aspect is For mobile operators, adding UMA access to existing GSM networks is not a major undertaking. Since UMA constitutes a part of the radio access network, it is introduced by simply adding a UMA Network Controller (UNC) to connect the WLAN access network with the GSM core network using standard A and Gb interfaces. Therefore, UMA allows operators to use broadband and WLAN together with the GSM/GPRS core network to provide alternative access for mobile voice and data services.

Fundamentally, the inception of cognitive radio assists the mobile operator's extendible liberty as it refers to the regime where the end-to-end communications system is aware of its propagation environment and adapts to it and it is continuous dynamic process. This continuity keeps on adding certain features on the evolving wireless technologies and wireless devices have proliferated over past decade dramatically increasing the demand for electromagnetic spectrum. Because of the current approach to spectrum access, spectrum supply has not kept up with spectrum demand leading to the appearance of scarcity in the electromagnetic spectrum. The opportunistic use of spectrum presents significant challenges to the technical and regulatory communities. From a technical perspective, the devices must be able to autonomously resolve conflicts over spectrum access and when operating opportunistically should be able to avoid interfering with incumbent signals. While avoiding introducing interference to a signal that is continuously present, it is more difficult with regularly structured, but intermittent signals such as a TDMA signal, and impossible to guarantee with a random access signal. To some extent this implies that the control processes of cognitive radios will need to be able to operate over multiple time scales as is proposed in recent work (N. Devroye, P. Mitran, and V. Tarokh, 2006). Since there is a technical problem, there is also a regulatory problem. Current spectrum licensees are generally only amenable to opportunistic spectrum access when they can be assured that their signals will not be degraded. Figuring out how to achieve the 10-fold gain in capacity while limiting the impact of existing services is currently a topic of much debate in the regulatory community. Hopefully, the role of different optimization algorithms helps to quantify the problem in more amicable way.

3. COGNITIVE RADIO PROTOCOL: COMPONENTS AND FUNCTIONS

A cognitive radio develops waveforms and chooses protocols in real-time using artificial intelligence. These actions require three components:

- *Perception:* Sensors that collect data on both external factors (channel conditions, other radios, regulations, user needs) and internal factors (waveform capabilities, computational power available, battery life).
- *Conception:* An intelligent core that learns and understands how to combine knowledge from the sensing mechanism to instruct the adaptation mechanism.
- *Execution:* An optimization and adaptation mechanism that alters the radio's behavior.

According to the work in the AAF project will fork into two tracks named *Cognitive System* and

Cognitive Radio System. First track *aims at results/ products on a short time notice* and focuses on existing standards IEEE 802.11 and IEEE 802.16. Research will be made only on layers above link layer and work will concentrate on utilization of existing cognitive capabilities of aforementioned standards (however it is questionable whether such capabilities already exist since IEEE 802.11k group responsible for radio resource measurements in WLAN is still working on a draft version of its document). Second track focuses on long term scale and just like in track one utilizes both IEEE standards. However it will try to alter many parameters in physical and link layer such that full radio cognition is met. Below we will distinguish those two tracks while describing issues of routing protocols adaptation to cognitive radio ad hoc network. There are certain specific requirements to be incorporated in cognitive radio protocol.

- *Differentiate routes depending on channel type due to changing propagation characteristics of various radio links:* This leads to better QoS when compared to algorithm taking into account number of hops only.
- *Increase nodes connectivity due to wider set of available radio links and available longer transmission distances:* Any cognitive radio node is capable of transmitting with broad set of frequencies, i.e. UNII and USM band or UNII, USM and TV band (R. W. Brodersen, A. Wolisz, D. Cabric, S. M. Mishra, D. Willkomm, 2004). By utilizing simple measure that the higher the frequency the shorter the transmission distance, routing algorithm may decide which radio link should be used for specific hop. It has very important implications to emergency network since high frequency signals have bigger problems with thick objects penetration. It is why routing has to utilize the channel information and send high priority packets on highly resilient channels (lower frequency channels).
- *Detect faster link failures.*
- *Perform more efficient multicast due to increased connectivity.*

Some of the multi carrier MAC protocols and outline their important features relevant to cognitive radio network could be mentioned.

- While using 802.11 MAC scheme RTS/ CTS control packets in separated control channel packets will encounter less number of collisions (DATA and ACK packets will only be transmitted in group frequency bands (R. W. Brodersen, A. Wolisz, D. Cabric, S. M. Mishra, D. Willkomm, 2004)). Moreover we will encounter smaller probability of hidden and exposed terminal phenomenon. What's also important is that all group bands are assigned for data traffic transmission so we will encounter no throughput decrease due to transmission of signaling traffic.
- *Dimensioning of control channel.* The gain from introducing more traffic channels to CRN is not linearly proportional (S. L. Wu, C.-Y. Lin, Y.-C. Tseng and J.-P. Sheu, 2000) due to the fact that common control channel becomes "saturated" with signaling packets. We might construct dynamic common control channel which will have different bandwidth pending on the current number of users in all bands in CRN. Such behavior will decrease overall SINR (Signal to Interference Ratio) in traffic channels (assuming that UWB channel resides in the same part of the spectrum as most of the traffic channels in CRN).
- *Time of traffic channel switching in CR node must be negligible.* Channel switching cannot introduce additional latency to transmission
- *Each CR node can only transmit or receive at the same time and can either send/receive data or send/receive signaling.* This

assumption comes from the fact that usually not every node can have spectrum sensing capability (in other words equipped in two antennas – one for transmission/receive and one for spectrum sensing/signaling

- *Either UWB or dedicated ISM band.* Using a UWB as a common control channel has this advantage that we can simultaneously provide nodes with the information about position. Such information might be used by all of the protocols (especially by routing and MAC).
- *Use of localization capabilities for cognitive radio.* We should utilize information about position of the node in the MAC protocol. One of the proposed schemes: if nodes know the exact locations of transmitting stations of the licensed bands (this information will be stored in node's internal memory and periodically updated by means of software download) it can always compare its position with stored coordinates and utilize this information as one of the criterion in frequency selection process.
- *Multiple criterions for channel switching.* The one and only approach for design of physical and link layer is through cross layer design because cognitive radio MAC (CR MAC) protocol must utilize information from physical layer to assign resources for wireless nodes. Assigning policies must be based on various requirements and data for making decision must be provided only by physical layer. Some of those policies are: usage diversity, spatial reuse, common frequency use, minimize use of licensed bandwidth. Table 1 presents most recent Multi Channel MAC protocols proposed in the literature.
- *Assigning particular bands groups to particular group of users.* We define a band group as a set of frequencies which might be utilized by only some users due to the construction of RF front end.
- *Even when saturated common control channel must inform about licensed users occurring in the area.* Every packet about primary user of the used licensed band must be transported immediately to each cognitive node in the receiving vicinity of the primary node (node that has always the right to access the channel as the owner). When we assume that information about occupied channels is transported through CSMA/CA protocol of the IEEE 802.11 we should construct a signaling scheme

Table 1. Cognitive radio parameters modeled in pheromone value

Pheromone ID	Description
1	Carrier Frequency
2	Bandwidth
3.	Symbol Rate
4.	Modulation
5.	Fo Forward Error Correction
6.	Paylaod/Frame length
7.	Automatic Repeat Request
8.	DynamicRange
9.	Equalization
10.	Encryption
11.	Antenna Configuration
12.	Voice
13.	Noise cancellation
14.	TimeDivision Duplex

in which packets with information about available licensed users must be always served first no matter how many other packets are waiting in the queue.

In a more recent work the architecture of CogNet which is a framework for research into architectural tradeoffs and protocol design approaches for cognitive radio networks at both local network and the global inter network levels (Dipankar Raychaudhuri, Narayan B. Mandayam, Joseph B. Evans, Benjamin J. Ewy, SriniSeshan, and Peter Steenkiste, 2006). Similarly there are substantial efforts to implementimproved spectrum sharing over the past decade through manifold avenues like property rights regimes (G. Faulhaber and D. Farber, 2003), spectrum clearinghouse (Y. Benkler. 2002), unlicensed bands with simple spectrum etiquette (D. Satapathy and J. M. Peha, 1998), open access etc. They are called spectrum utilization protocol.

The "agile wideband radio" scheme (KiranChallapali, Stefan Mangold and ZhunZhong, 2004) is the most prevalent concept for cognitive radio in which transmitters scan the channel and autonomously choose their frequency band and modulation waveform to meet interference minimization criteria without any protocol-level coordination with neighboring radio nodes.

The distinctions between unlicensed spectrum regimes, open access and cognitive radio approaches are relatively subtle as they are all based on the concept of technology neutral bands to be used by a variety of services using radio transceivers that meet certain criteria. For example, cognitive radio may be viewed as a special case of open access or unlicensed regimes in which radio transceivers are required to meet a relatively high standard of interference avoidance via physical and/or network layer adaptation.

The IEEE 802 community is currently developing two standards the directly relate to cognitive radio – 802.22 and 802.11h. Additionally, 802.11k is developing techniques for incorporating radio resource management information into WLAN operation – in effect incorporating knowledge about the environment and the radios. There are three applications typically discussed for coexistence with initial trial deployments of cognitive radios: television, microwave point-to-point links, and land mobile radio. Each of these applications has been shown to dramatically underutilize spectrum on average. However, only television signals have the advantage of incumbent signals that are easy to detect (as opposed to a microwave point-to-point links) and not involved in life-critical applications (as would be the case for many land mobile radio systems).

Unlike 802.22, 802.11h is not formulated as a cognitive radio standard. However, the World Wireless Research Forum has noted that a key portion of the 802.11h protocol – dynamic frequency selection – has been termed a "cognitive function". To investigate why an 802.11h WLAN might be considered a cognitive radio, consider that the 802.11h protocol requires that a WLAN be capable of the following tasks, which closely model the cognitive radio phenomena.

- *Observation* 802.11h requires WLANs to estimate channel characteristics such as path loss and link margin and 5.4.4.2 further requires the radios estimate channel characteristics such as path loss and link margin.
- *Orientation*: Based on these observations, the WLAN has to determine if it is operating in the presence of a radar installation, in a bad channel, in band with satellites, or in the presence of other WLANs.
- *Decision*: Based on the situation that the WLAN is encountering, the WLAN has to decide to change its frequency of operation (*Dynamic Frequency Selection*), adjust the transmit power (*Transmit Power Control*), or both.
- *Action*: The WLAN has to then implement this decision.

These events are analogous with Cognitive radio activities In the context of BlueTooth, the orientation of protocol needs to be redefined. Bluetooth is a specification for the wireless communication of voice and data using a short-range radio. It is defined by a number of protocols residing in the physical and data link layers in the OSI model. Bluetooth uses an ad-hoc, piconet structure with a single master and up to seven slaves.

4. AGILITY MANAGEMENT IN COGNITIVE RADIO FOR BLUETOOTH

Agility is an interesting proposition in cognitive radio and cooperation network. In its basic form agility is expressed as the reduction of detection time by allowing the cognitive radios operating in the same band to cooperate. The concept could contemplate further with multiple user case and show that the inherent asymmetry of the network can lead to faster detection with cooperation (Ghurumuruhan Ganesan and Ye Li, 2005). There are many impressive results obtained from cooperation schemes (J. N. Laneman and D. N. C. Tse, 2004) (A. Sendonaris, E. Erkip and B. Aazhang, 2003) strongly the agility has been expressed as:

$$\mu nl\ \ c(2) = Tn\ /\ Tc$$

where,

- T_n is the total time taken by both users to vacate the band (considering the multiple users)
- T_c isthe total average detection time of the cooperation scheme
- μn represents agility gain

Cognitive radio introduces a new level of sophistication to wireless communications technology. Basic cognitive radios operate autonomously and depend on highly sensitive receivers and device learning to know when and how spectrum can be accessed. An alternative approach, the UWAN, employs an intelligent network to distribute more information to the cognitive devices so that better decisions can be made (William Krenik and AnujBatra, 2005).

Converging, this technology a device to another with a universal short-range radio link. Beyond untethering devices by replacing cables, Bluetooth provides a universal bridge to existing data networks, a peripheral interface, and a mechanism to form small private ad-hoc groupings of connected devices away from fixed network infrastructures. Bluetooth has a number of distinctive features compared to existing wireless LANs. These are:

- Support for both data and voice traffic.
- Frequency hopping to avoid interference.
- A master driven Time Division Duplex (TDD) system at the Medium Access Control (MAC) layer to support full duplex transmission.
- Segmentation and Reassembly (SAR) to handle large data packets.
- Support for link level Automatic Repeat Request (ARQ) and Forward Error Correction (FEC) schemes.

Bluetooth technology allows for the formation of network units called *piconets*and a connected set of piconets, called *scatternets*without the involvement of a central infrastructure, in an ad-hoc fashion. This is made possible using distributed algorithms for inquiry and paging. There are many features of Bluetooth that may significantly impact the performance of data traffic. Fragmentation of large data packets by performing SAR, which allows them to be transmitted in small base band packets, may increase their end-to-end delay. Master driven scheduling at the MAC level will affect throughput and queuing delay. The success rate of data transmissions will be affected by the presence of FEC and ARQ mechanisms at the link level. The bandwidth available for data traffic will be reduced in the presence of voice connections. The effect of these issues needs to be better understood in order to enhance the performance of asynchronous data traffic over Bluetooth. When multiple data transfers share the wireless link, as in a Bluetooth piconet, MAC scheduling algorithms are needed to achieve fair sharing of bandwidth, high link utilization and low queue occupancy. It is observed from several works (M. Kalia, D. Bansal and R. Shorey, 1999) (Das, A. Ghose, A. Razdan, H. Saran and R. Shorey, 2001) that that Round-Robin scheduling is unable to meet these requirements and propose three new scheduling algorithms which meet these criteria adequately. In his phenomenal work Mitolla pioneered plenty of new conceptual level communication corollaries and thus his primary inclusion of the term self *organized agents* etc could be viewed in the tune of bio-inspired agents in the next section.

5. ANT COLONY OPTIMIZATION: A STEP TOWARDS BIO-INSPIRED COMPUTING

In the early 1990s, ant colony optimization (ACO) (M. Dorigo, V. Maniezzo, A. Colorni, 1991) was introduced by M. Dorigo and colleagues as a novel nature-inspired metaheuristic for the solution of hard combinatorial optimization (CO) problems. CO belongs to the class of metaheuristics (I. Rechenberg, 1973), which are approximate algorithms used to obtain good enough solutions to hard CO problems in a reasonable amount of computation time. Other examples of metaheuristics are tabu search, (F. Glover, 1989), simulated annealing (S. Kirkpatrick, C.D. Gelatt, M.P. Vecchi, 1983) (V. C. ern, 1985) and evolutionary computation (J.H. Holland, 1975) (I. Rechenberg, 1973) (L.J. Fogel, A.J. Owens, M.J. Walsh, 1966). The inspiring source of ACO is the foraging behavior of real ants. When searching for food, ants initially explore the area surrounding their nest in a random manner. As soon as an ant finds a food source, it evaluates the quantity and the quality of the food and carries some of it back to the nest. During the return trip, the ant deposits a chemical pheromone trail on the ground. The quantity of pheromone deposited, which may depend on the quantity and quality of the food, will guide other ants to the food source. As it has been shown in (J.L. Deneubourg, S. Aron, S. Goss, J.-M. Pasteels, 1990), indirect communication between the ants via pheromone trails enables them to find shortest paths between their nest and food sources. This characteristic of real ant colonies is exploited in artificial ant colonies in order to solve CO problems.

According to Papadimitriou and Steiglitz (C.H. Papadimitriou, K. Steiglitz, 1982), a CO problem $P = (S, f)$ is an optimization problem in which, given a finite set of solutions S (also called *search space*) and an objective function $f: S \mapsto \square^+$ that assigns a positive cost value to each of the solutions, the goal is either to find a solution of minimum cost value, 3 or—as in the case of approximate solution techniques—a good enough solution in a reasonable amount of time. ACO algorithms belong to the class of metaheuristics and therefore follow the latter goal. The central component of an ACO algorithm is a parameterized probabilistic model, which is called the *pheromone model*. The pheromone model consists of a vector of model parameters T called *pheromone trail parameters*. The pheromone trail parameters $\tau i \in \tau$, which are usually associated to components of solutions, have values τi, called *pheromone values*. The pheromone model is used to probabilistically generate solutions to the problem under consideration by assembling them from a finite set of solution components. At runtime, ACO algorithms update the pheromone values using previously generated solutions. The update aims to concentrate the search in regions of the search space containing high quality solutions. In particular, the reinforcement of solution components depending on the solution quality is an important ingredient of ACO algorithms. It implicitly assumes that good solutions consist of good solution components. Broadly, the ACO approach attempts to solve an optimization problem by repeating the following two steps:

- Candidate solutions are constructed using a pheromone model, that is, a parameterized probability distribution over the solution space;
- The candidate solutions are used to modify the pheromone values in a way that is deemed to bias future sampling toward high quality solutions. For detail algorithm and results one could refer (M. Dorigo,V. Maniezzo, A. Colorni, 1991).

5.1 Cognitive Radio and Ant Colony: A Metaphoric Relation

Since its inception cognitive radio has been functionally compared with ant colony and similar types of algorithms (C. Rieser, 2004). The term cognitive becomes too crucial as ant agents are capable ofevolving their own paths through the spectrum and intervening nodes to ferry voice, data, video, and multimedia packets through the ether. These radio-ants might move packets from the smaller power-starved radios through multimode vehicular radios and on to conventional cell sites. If the etiquette is too strict, very little additional benefit will come from spectrum pooling because the control overhead will be too high to be workable. If, on the other hand, the etiquette is too liberal, there will be much interference and universally poor quality of service. Such complex adaptive systems operate best "at the edge of chaos" (Kaufman, S. 1995). These types of complex adaptive situations may lead to NP hard problems, where optimization on bandwidth, spectrum and other parameters seems difficult to be achieved. Thus, in some sense an ant-colony of cognitive radios left to evolve spectrum use among them might be the most efficient way to achieve high value from limited radio spectrum in a reasonable time.

Symbolically on the same context the present chapter envisages the fundamental analogy with ant colony and Bluetooth paradigm. Further a novel ant colony model has been proposed to detect the inherent asymmetry in the network of two cognitive users and thereby could enhance the agility of the radio significantly. As the cognitive radio is self organized hence many structures built by social insects are the outcome of a process of self organization. A major governing factor is being stigmergy (Theraulaz, G., Bonabeau, E., 1999). The distribution of the pheromone represents the memory of the recent history of the swarm (his social cognitive map), and in a sense it contains information which the individual ants are unable to hold or transmit (Ramos, V., Fernandes, C., Rosa, A.C., 2005). There is no direct communication between the organisms but a type of indirect communication through the *pheromonal* field. The real metaphor of ant colony with cognitive radio could be this indirect communication deployed in this chapter. Ideally, the algorithmic framework has been motivated by the fact that the main requirement of cognitive radio architecture is to detect the presence of the primary or licensed users as quickly as possible. For that reason, the cognitive users should continuously monitor the spectrum for the primary (licensed) users. Keeping these anomalies in front, the proposed model would like to present a brief mathematical stint to support the bio-inspired cognitive radio concept in terms of agility inspection. The genesis of this novel idea was through the agent technology and followed by the ant colony system. Communicating and deciding about the spectrum utilization through ant colony system is modeled as transition probability. As described by *Chialvo*and *Millonas*, (Chialvo, D.R., Millonas, M.M., 1995) (Millonas, M.M., 1994) the state of an individual ant can be expressed by its position *r*, and orientation q. Since the response at a given time is assumed to be independent of the previous history of the individual, it is sufficient to specify a transition probability from one place and orientation (r,q) to the next (r^*,q*) an instant later. In previous works by Millonas, (Millonas, M.M. 1992) transition rules were derived and generalized from noisy response functions, which in turn were found to reproduce a number of experimental results with real ants. The response function can effectively be translated into a two parameter transition rule between the cells by use of a pheromone weighting function expressed as follows:

$$W(\sigma) = (1 + \frac{\sigma}{1 + \gamma\sigma})^{\beta} \tag{1}$$

where,

- $w(\sigma)$ is the pheromone weighing function
- σ stands for pheromone density
- β is the physiological inverse-noise parameter or gain.

The model assumes a situation where cognitive radio users $U1$ and $U2$ operating in a fixed TDMA mode for sending data to a common receiver and a primary user starts using the band soon after. Then the two cognitive users need to vacate the band as soon as possible to make way for the primary user. However, the detection time becomes significant if one of if one of the users is in the boundary of decidability of the primary user. The signal received from the primary user is so weak that the cognitive user $U1$ takes a lot of time to sense its presence. We assume that the position of ant in the channel is r and its orientation is being q in a slotted transmission system where $U1$ and $U2$ transmit in successive slots according to the AF protocol (IEEE USA Position, 2003) in the same frequency band. Accordingly in time slot $T1$, $U1$ transmits and in $T2$, $U2$ relays the information of the previous slot. Unknown to both these users, a primary user has started using the band. This primary user must be detected as quickly as possible. At time slot $T1$, $U1$ transmits its data. Hence the signal received by $U2$ in the time-slot $T1$ is given by,

$$Y = w(\sigma)\theta + aw((\Delta\theta) - \beta \tag{2}$$

where,

- $w(\sigma)$ is the pheromone gain between the primary user and user Ui
- $w(\Delta\theta)$ is the change in direction at each time step,
- a denotes the message sent from $U1$ to the common receiver
- θ indicates the presence of the primary user. Hence $\theta = 1$ implies the presence of the primary user and $\theta = 0$ implies its absence.
- β is the inverse-noise parameter

As an additional condition, each individual leaves a constant amount η of pheromone at the cell in which it is located at every time step t. This pheromone decays at each time step at a rate k. Then, the normalized transition probabilities on the lattice to go from cell k to cell i are given by

$$P_{ik} = \frac{W(\sigma_i)w(\Delta i)}{\sum_{j/k} W(\sigma j)w(\Delta j)} \tag{3}$$

Where the notation j/k indicates the sum over all the surrounding cells j which are in the local neighborhood of k. Δ_i measures the magnitude of the difference in orientation for the previous direction at time t-1.

After investigating the pheromone gradient across channel, the model demonstrates the cooperation with the help of two ant agents (e.g. transmitter ant and detector ant). The algorithm and model proposed here is different compared to conventional ant colony. As the cooperation is crucial for the improvement of agility of cognitive radio network, therefore a primary level comparison of pheromone deposition for transmitter ant and detector ant could be simulated. The following parameters are considered while cooperation in the cognitive radio takes place:

All these parameters are mapped into synthetic pheromone and the value of pheromone is measured on the basis of the proposed algorithm mentioned in the next section.

5.2 Proposed Bio-Inspired Cognitive Radio Algorithm

The following out line of the proposed bio-inspired cognitive radio is presented.

1. $t \leftarrow 0$, r, q, $\theta \leftarrow 0$/**Observe and Orient* */
   - /* *Values of parameters used in experiments* */
   - $k = 0.015$, h = 0.07, b=3.5, _=0.2,
   - p = 1.9, *tmax*= 500, 600, 1000 or 1150 steps.
2. Set the initial density $\tau\, ij(t) = C$ /* an arbitrary large value, e.g. 10)*/
3. Sample two individuals randomly for each unit k, then choose the best one in the unit and set it as $ant^*_{k,0}$ (k=1,2,. . ., m)

Update $\tau\, ij(t)$ according to enhancement of pheromone at cognitive site r:

$$P_r = P_r + [\eta + p(\Delta[r] / \Delta \max) \tag{4}$$

Sample $T\text{-}ant^*_{k,t+1}$ for k=1,2,. . ., m according to $D\text{-}antk,t(: ant^*_{k,t})$ and $\tau_{ij}(t+1)$ /* *T-Ant → Transmitter Ant*

D- Ant → Detector Ant */

6. Compare $T\,ant_{k,t+1}$ and $D\text{-}ant_{k,t,}$ set the best one as $ant^*_{k,t+1}$ for i=1,2,. . ., m
7. $t \leftarrow t+1$, modify all parameters as per the channel model
8. Reconfigure Channel /**all users experience Rayleigh fading that is independent from user to user**/
9. If the termination criteria are met, terminate the algorithm. Otherwise, go to 4

Traditional AI research has focused software implementations of cognition at the application (APP) layer. Current software radio approaches to cognition have been notably layer three or application (APP) centric due to the AI legacy. Unfortunately, assuming the presence of a workstation class application computational host can result in the acceptance of levels of complexity not accepted in PHY and MAC layer cognitive functions. These layers may be limited in power consumption, size, or digital architecture and processing complexity. This research assumes that the cognitive functions operating to control an agile radio may be resource or time limited. Such functions should be designed not search for "the solution" but instead "a solution" that meets the balance of users requirement. Depending upon the set benchmark the next subsection has made comparison of result compared with standard *Ant Colony System (ACS)* with *Bio-Inspired Cognitive Radio algorithm (BCR)*.

6. OPEN RESEARCH CHALLENGES IN BIO-INSPIRED COGNITIVE RADIOS

Cognitive radio initiates a revolution regarding the spectrum allocation considerations by putting forward a new concept called opportunistic spectrum usage, which involves the soft usage of the current licensed and unlicensed available

Table 2. Results of BCRBestavg is average best solution over 25 runs and Error indicates average excess (%) of Bestavg from optimal in 25 runs

BCR BCR with γ =0.4			Non BCR γ =1			Ant Colony System (Gambardella, L.M., Dorigo, M., 1996)	
Best $_{avg}$	**SD**	**Error in %**	**Best** $_{avg}$	**SD**	**Error in %**	**Best**	Error
426.6	0.5	0.065	4.27	0.6	0.31	428.1	0.48
21280.0	0.0	01.10.000	213224	48.44	0.27	21420.0	0.65
159594.4	35.6	1.123	158985	10.07	1.14	16054.0	1.74
14465.5	34.9	0.30	14509.76	47.32	0.62	14565.5	0.99
38736.1	77.3	0.16	39105.86	169.5	1.12	39099.1	1.10
36303.2	120.3	0.20	36743.1	261.1	1.30	36857.0	1.73
2827.1 8	8.73	2.64	2820.76	14.8	2.38	38.736.1	1.81

spectrum. This concept proposes that licensed bands can be utilized by secondary users at times when they are not being used by their owners, leading to the most efficient exploitation of the entire spectrum. In this opportunistic way of spectrum usage, it has to be guaranteed by the unlicensed systems that their operation does not affect the primary users. Many works, white bands are directly taken as the spectrum opportunities.

- However, there are spectral, temporal and spatial requirements that a white band has to satisfy in order to be useful and to be considered an opportunity. These opportunities endorse several research initiatives keeping the following points as key factors:
- Opportunity is not an instantaneous white space in spectrum. It is necessary to monitor a white space continuously over a time frame (in the order of seconds) and ensure that it does not display an erratic behavior, i.e. for a reasonably long time the noise temperature in that band resides below a certain threshold, and the band remains as a white space.
- It may not be reliable to consider a white band an opportunity if it is detected by only one single cognitive radio device. The reasons include that the device has a limited sensing range, as well as that it may be experiencing shadowing. Optimally, the spectrum has to be sensed by a number of cognitive nodes over a region that goes well beyond the range of a single cognitive device. A band can be considered a candidate for being a spectrum opportunity only if it is detected as white by many
- Cognitive nodes at different locations that exchange the spectrum sensing information with each other.
- If a frequency band to be utilized is too narrow, it may be hard for the cognitive radio to generate a temporally limited pulse shape that fits into that band. Therefore, a white band has to be wider than a certain bandwidth for being targeted for opportunistic spectrum usage.

7. CONCLUSION

The bio-inspired cognitive radio algorithm is presented in this chapter. the application of bi-inspired computation in cognitive radio give very good results.

The chapter present information about the cognitive radio and the ant colony and try to introduce a relation between two fields. Future research will extend the application of ant colony in cognitive radio.

REFERENCES

Benkler, Y. (2002). Some economics of wireless communications. *Harvard Journal of Law & Technology*, 16.

Brodersen, R. W., Wolisz, A., Cabric, D., Mishra, S. M., & Willkomm, D. (2004). *CORVUS: A cognitive radio approach for usage of virtual unlicensed spectrum.* University of California, Berkley. Retrieved from http://bwrc.eecs.berkeley.edu/Research/MCMA/CR_White_chapter_final1.pdf

Challapali, K., Mangold, S., & Zhong, Z. (2004). *Spectrum agile radio: Detecting spectrum opportunities*. 6th Annual International Symposium on Advanced Radio Technologies, Boulder, Colorado, 2 - 4 March.

Chialvo, D. R., & Millonas, M. M. (1995). How swarms build cognitive maps. *Autonomous Agents*, *144*, 439–450.

Das, A., Ghose, A., Razdan, H. S., & Shorey, R. (2001). *Enhancing performance of asynchronous data traffic over the Bluetooth wireless ad-hoc network.* In IEEE INFOCOM'2001.

Deneubourg, J. L., Aron, S., Goss, S., & Pasteels, J. M. (1990). The self-organizing exploratory pattern of the Argentine ant. *Journal of Insect Behavior*, *3*, 159–168. doi:10.1007/BF01417909

Devroye, N., Mitran, P., & Tarokh, V. (2006). Achievable rates in cognitive radio channels. *IEEE Transactions on Information Theory*, 52.

Dorigo, M., Maniezzo, V., & Colorni, A. (1991). *Positive feedback as a search strategy.* Tech. Report 91-016, Dipartimento di Elettronica, Politecnico di Milano, Italy.

Ern, V. C. (1985). A thermodynamically approach to the travelling salesman problem: An efficient simulation algorithm. *Journal of Optimization Theory and Applications*, *45*, 41–51. doi:10.1007/BF00940812

Faulhaber, G., & Farber, D. (2003). Spectrum management: Property rights, markets and the commons. In *Proceedings of the Telecommunications Policy Research Conference*, Alexandria, VA.

Fogel, L. J., Owens, A. J., & Walsh, M. J. (1966). *Artificial intelligence through simulated evolution*. New York, NY: Wiley.

Gambardella, L. M., & Dorigo, M. (1996). Solving symmetric and asymmetric TSPS by ant colonies. *Proceedings of the IEEE International Conference on Evolutionary Computing* (ICEC'96), (pp. 622–627).

Ganesan, G., & Li, Y. (2005). *Agility improvement through cooperative diversity in cognitive radio*. IEEE GLOBECOM Proceedings.

Glover, F. (1989). Tabu search—Part I. *Computers*, *1*(3), 190–206.

Goldberg, D. E. (1989). *Genetic algorithms in search, optimization, and machine learning*. Addison-Wesley Pub Co.

Guntsch, M., & Middendorf, M. (2001). Pheromone modification strategies for ant algorithms applied to dynamic TSP. In *Applications of Evolutionary Computing: Proceedings of Evo Workshops 2001, Lecture Notes in Computer Science, Vol. 2037*, (pp. 213–222).

Haykin, S. (2005). Cognitive radio: Brain-empowered wireless communications. *IEEE Journal on Selected Areas in Communications*, *23*(2), 9–16. doi:10.1109/JSAC.2004.839380

Holland, J. H. (1975). *Adaption in natural and artificial systems*. Ann Arbor, MI: The University of Michigan Press.

IEEE USA Position. (2003). *Improving spectrum usage through cognitive radio technology.* Retrieved from http://www.ieeeusa.org/policy/positions/cognitiveradio.asp

Kalia, M., Bansal, D., & Shorey, R. (1999). MAC scheduling and SAR policies for Bluetooth: A master driven TDD pico-cellular wireless system. In *IEEE International Workshop on Mobile Multimedia Communications (MoMuC)*, (pp. 384–388).

Kallel, L., Naudts, B., & Rogers, A. (Eds.). (2001). *Theoretical aspects of evolutionary computing.* New York, NY: Springer.

Kaufman, S. (1995). *At home in the universe.* Wiley.

Kirkpatrick, S., Gelatt, C. D., & Vecchi, M. P. (1983). Optimization by simulated annealing. *Science, 220*(4598), 671–680. doi:10.1126/science.220.4598.671

Krenik, W., & Batra, A. (2005). *Cognitive radio techniques for wide area networks.* DAC 2005, Anaheim, California, USA.

Laneman, J. N., & Tse, D. N. C. (2004). Cooperative diversity in wireless networks: Efficient protocols and outage behaviour. *IEEE Transactions on Information Theory, 50,* 3062–3080. doi:10.1109/TIT.2004.838089

Lowdermilk, R., & Harris, F. (1996). Interference mitigation in orthogonal frequency division multiplexing (OFDM). In *Proceedings of IEEE International Conference on Universal Personal Communications,* Vol. 2, Cambridge, MA, (pp. 623–627).

Marcus, M. J. (2005). Unlicensed cognitive sharing of the TV spectrum: The controversy at the Federal Communications Commission. *IEEE Communications Magazine*, 51.

Millonas, M. M. (1992). A connectionist-type model of self-organized foraging and emergent behavior in ant swarms. *Journal of Theoretical Biology, 159,* 529. doi:10.1016/S0022-5193(05)80697-6

Millonas, M. M. (1994). Swarms, phase transitions and collective intelligence . In Langton, C. G. (Ed.), *Artificial Life III, Santa Fe Institute, Studies in the Sciences of Complexity* (*Vol. XVII*, pp. 417–445). Reading, MA: Addison- Wesley.

Mitola, J. (1999). Cognitive radio for flexible multimedia communications. *1999 IEEE International Workshop on Mobile Multimedia Communications,* (pp. 3 –10)

Mitola, J. (2000). *Cognitive radio: An integrated agent architecture for software defined radio.* Ph.D. dissertation, Computer Communication System Laboratory, Department of Teleinformatics, Royal Institute of Technology (KTH), Stockholm, Sweden, May 2000.

Neel, J., Menon, R., MacKenzie, A., & Reed, J. (2005). *Using game theory to aid the design of physical layer cognitive radio algorithms.* Conference on Economics, Technology and Policy of Unlicensed Spectrum, May 16-17, Lansing, Michigan.

Papadimitriou, C. H., & Steiglitz, K. (1982). *Combinatorial optimization—Algorithms and complexity.* New York, NY: Dover Publications Inc.

Ramos, V., Fernandes, C., & Rosa, A. C. (2005). Social cognitive maps, swarm collective perception and distributed search on dynamic landscapes. *Brains, Minds & Media – Journal of New Media in Neural and Cognitive Science.*

Raychaudhuri, D., Mandayam, N. B., Evans, J. B., Ewy, B. J., Seshan, S., & Steenkiste, S. (2006). *CogNet – An architecture for experimental cognitive radio networks within the future Internet.* MobiArch'06 - First International Workshop on Mobility in the Evolving Internet Architecture.

Rechenberg, I. (1973). *Evolutionsstrategie: Optimierung technischer Systeme nach Prinzipien der biologischen Evolution.* Frommann-Holzboog.

Rieser, C. (2004). *Biologically inspired cognitive radio engine model utilizing distributed genetic algorithms for secure and robust wireless communications and networking.* PhD Dissertation, Virginia Tech, August.

Rondeau, T. W., Rieser, C. J., Gallagher, T. M., & Bostian, C. W. (2004). Online modeling of wireless channels with hidden Markov models and channel impulse responses for cognitive radio. *IEEE Proceedings of IMS 2004.*

Satapathy, D., & Peha, J. M. (1998). Etiquette modification for unlicensed spectrum: Approach and impact. In *Proceedings of the IEEE Vehicular Technology Conference*, (pp. 272–276).

Sendonaris, A., Erkip, E., & Aazhang, B. (2003). User cooperation in diversity - Part I: System description. *IEEE Transactions on Communications, 51*, 1927–1938. doi:10.1109/TCOMM.2003.818096

Theraulaz, G., & Bonabeau, E. (1999). A brief history of stigmergy. *Artificial Life, 5*(2), 97–116. doi:10.1162/106454699568700

Wu, S.-L., Lin, C.-Y., Tseng, Y.-C., & Sheu, J.-P. (2000). *A new multi-channel MAC protocol with on-demand channel assignment for multi-hop mobile ad hoc networks.* International Symposium on Parallel Architectures. Algorithms and Networks.

Zhang, G. P. (2000). Neural networks for classification: A survey. *IEEE Transactions on Systems, Man, and Cybernetics, 30*(4), 451–462. doi:10.1109/5326.897072

ENDNOTE

1 It thinks, therefore it's a cognitive radio.

Chapter 21
Potential Area of Research in MANET

Sudha Singh
Bengal College of Engineering and Technology, India

D. K. Singh
Birsa Institute of Technology Sindri, India.

Mr. Sudipta Mondal
Bengal College of Engineering and Technology, India

ABSTRACT

The growth of laptop and 802.11/Wi-Fi wireless networking have made Mobile Ad hoc NETworks (MANETs) a popular research topic since the mid to late 1990s. Many academic papers evaluate protocols and abilities assuming varying degrees of mobility within a bounded space, usually with all nodes within a few hops of each other and usually with nodes sending data at a constant rate. Different protocols are then evaluated based on the packet drop rate, the overhead introduced by the routing protocols, and other measures. Research in MANET is like playing in a vast playground with only few known rules and large number of unknown rules applied to its different areas. Some areas are still unknown and some areas are at its early stage. Some of the research challenges that MANET present are dynamic topologies, battery lifetime, disconnected operations, security, et cetera. In general there are three types of MANET: vehicular ad hoc networks, intelligent vehicular ad hoc network, and Internet based mobile ad hoc networks. Each of these three types has its own research areas.

I. INTRODUCTION

The growth of laptop and 802.11/Wi-Fi wireless networking have made mobile Ad hoc NETworks(MANETs) a popular research topic since mid to late 1990s. Many academic papers evaluate protocols and abilities assuming varying degrees of mobility within a bounded space, usually with all nodes within a few hops of each other and usually with nodes sending data at a constant rate. Different protocols are then evaluated based on the packet drop rate, the overhead introduced by the routing protocols and other measures. Over the last few years MANET has attracted a

DOI: 10.4018/978-1-4666-0321-9.ch021

lot of research interest. This has led to creation of a working group at the IETF that is focusing on mobile ad hoc networking called MANET(2002). The purpose of MANET working group is to standardize IP routing protocols functionality suitable for wireless routing application within both static and dynamic topologies with increased dynamics due to node motion and other factors.

Research in MANET is like playing in a vast playground with only few known rules and large number of unknown rules applied to its different areas. Some areas are still unknown and some areas are at its early stage. Some of the research challenges that MANET present are dynamic topologies, battery lifetime, disconnected operations, security etc. In general there are three types of MANET-vehicular ad hoc networks, intelligent vehicular ad hoc network and internet based mobile ad hoc networks. Each of these three types has its own research areas. This can be better defined by the following in Box 1.

This chapter will be fully devoted to find and demonstrate various potential research areas of MANET. This chapter is divided into the following seven sections. Section II discusses research topics related to routing in MANET. Section III will have research topics related to quality of service in MANET. Section IV discusses the research topics related to energy management and battery power in MANET. Section V discusses research topics related to security in MANET. Section VI will discuss new research areas in MANET. Section VII will have table of open problems in MANET.

II. ROUTING IN MANET

A Mobile Ad Hoc Network (MANET) is a collection of wireless mobile nodes that out of necessity form a network required to facilitate communication. Because the nodes have limited energy storage, computing power and communication ability, designing an energy efficient protocol is a challenge(Roche, Westphall, Pack and Mecklenburg, 2002).

The focus of most research is on Routing Protocols. Protocols are divided into three categories:

Box 1.

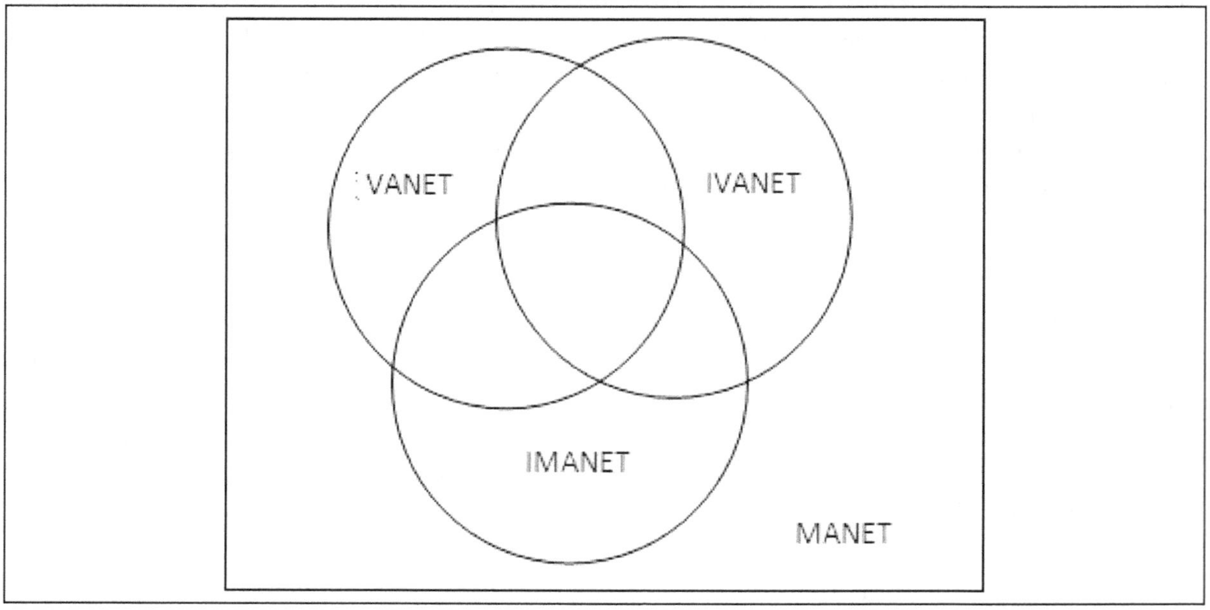

Proactive, Reactive and Hybrid. In Proactive protocols each mobile node maintains a routing table where control packets are broadcast periodically across the whole network. The proactive approach is to establish the route before communication occurs. This approach uses up bandwidth unnecessarily, where control packets are often sent in order to maintain the network map, where the majority of the routes are unused (Weniger and Zitterbart, 2004). In Reactive routing protocol, the routes are discovered only when needed. The broadcast overhead is reduced because control packets are only broadcast when there is data to transmit. This type protocol requires route discovery and route maintenance. If the network is highly mobile, the maintenance phase is used to keep the route available (Abusalah, Khokhar and Guizani,2008). Hybrid routing protocols attempt to use the advantages of proactive and reactive approaches. The hybrid protocols separate the network into two layers. The inner layer is comprised of the easy to maintain routes, where a routing table is easily maintained. The outer layer is comprised of the reactive type network nodes, where the table must be updated by a maintenance phase (Zhao, Aggarwal, Frost and Bai, 2011). Routing protocols also fall into two other categories. If the location information is not used, the routing is topology based. If the position information is used, the routing is position based. There are also two methods of transmitting data in position based protocols: greedy forwarding and directional flooding. GPSR (Greedy Perimeter Stateless Routing) is an example of greedy forwarding where the next hop is the closest node in distance to the destination. LAR (Location Aided Routing) is an example of a directional flooding protocol. The data is flooded in the direction of the destination node (Weniger and Zitterbart, 2004; Vidhyasanker, Manoj and Murthy 2006). In position-based routing, the mobile nodes calculate their position. There are two ways to do this. The nodes may use a GPS system. The second way is to use a directional antenna approach based on directional arrival and signal strength. We will focus on GPS systems due to their prevalence and increased usage (Weniger and Zitterbart,2004). Position based routing protocols offer the following advantages. Due to recent developments, they provide better scalability and performance by using geographical position to improve routing decisions & efficiency. Many of the current protocols have a low packet overhead meaning they use small packets and a reduced number of signaling packets (Mikki,2009). There are several advantages of GPS technology. GPS technology is owned and maintained by the US Government. It is a global satellite system that provides reliable location information in all types of weather. It is by design, the best option for tracking and identifying nodes. It is also a free service to any device with a receiver. The technology first became available for mobile phones in November of 2004. There are currently thirty satellites available and a plan for increasing the number to thirty-six (Vidhyasanker, Manoj and Murthy, 2006; Mikki,2009). Examples of position based protocols are (Zhao, Aggarwal, Frost and Bai, 2011): MFR, LAR,GRID,TERMINODES,LABAR, AODPR,PNR(Mikki,2009) etc.

In MANET, the data packet may fail to be delivered for various reasons such as nodes movement, packet collision and bad channel condition. In that case, source node must perform a re-route discovery process. Re-route discovery of traditional on-demand routing protocols involves large control message overhead and high latency. Therefore re-route discovery reduces efficiency of the networks. In particular, routing performance degenerates in dense environments because packet collisions happen frequently (Jeon, Lee and Kim,2011).

Another definition of mobile ad hoc network (MANET) is " It is an adaptive, self-configurable, self- organizing, infrastructure-less multi-hop wireless network with unpredictable dynamic topologies" (Qabajeh, Kiah and Qabajeh, 2009). By adaptive, self-configurable and self-organizing, we mean that an ad hoc network can be formed, merged together or partitioned into separated networks on the fly depending on the networking needs. This means that a formed network can be deformed on the fly without the need for any system administration. By infrastructure-less, we mean that an ad hoc network can be promptly deployed without relying on any existing infrastructure such as base stations for wireless cellular networks. By multi-hop wireless, we mean that in an ad hoc network the routes between end users may consists of multi-hop wireless links. In addition, each node in a mobile ad hoc network is capable of moving independently and forwarding packets to other nodes. Before proper routing can be possible, each node needs to be configured to a unique address. In the conventional network, such as Internet, the network address of a device can be preconfigured statically or assigned through the dynamic protocol, such as Dynamic Host Configuration Protocol (DHCP). The major list of the objectives of an optimal mobile ad hoc network address auto configuration protocol is as follows.

Dynamic Address Configuration

Nodes should be able to dynamically obtain IP addresses without manual or static configuration.

Uniqueness

Nodes should obtain unique addresses for correct routing and communication.

Robustness

The addressing protocol should adapt to the dynamics of the network, including partitions and merges.

Scalability

The protocol should avoid significant performance degradation as the size of the network increases.

The main task of an address auto configuration protocol is to manage the resource address space. It must be able to select, allocate, and assign a unique network address to an auto configured node. When a node leaves the network, the corresponding address must eventually be deallocated to prevent exhaustion of the address space. A major challenge is network partitioning and merging. An auto configuration protocol can at most guarantee unique addresses within a single network partition. If two partitions merge, address conflicts may occur. The resolution of a conflict requires at least one node to acquire a new address.

The auto configuration protocols proposed for mobile ad hoc networks can be classified into three categories (Yagan and Tham, 2005) stateful approaches, stateless approaches and hybrid approaches. A stateful approach assumes the existence of a central entity to keep an address allocation table for whole network and assigns unique address for unconfigured node. However, when subnet merging occurs, it is still a great challenge to synchronize multiple central entities. Furthermore, most of these protocols rely on either periodic or reliable flooding, which usually consumes a considerable amount of bandwidth. *Stateless approaches* do not need a central entity to maintain an address allocation table. Instead, each node selects an address by itself and verifies its uniqueness with the so-called *duplicate address detection* (DAD) procedure. If duplication is detected, at least one of the nodes with duplicate addresses must change its address. *Hybrid approaches* combine both stateful and stateless approaches by maintaining an allocation table and performing a DAD, which can increase the robustness, but may result in higher complexity and higher protocol overhead.

In mobile ad hoc networks, pre-configuration is not always possible and has some drawbacks. So, an autoconfiguration protocol is required to provide dynamic allocation of node's address. Most current ad hoc routing architectures use flat addressing which uses fixed length for addressing each node. The overhead in control packets of source routing is reduced highly with variable addressing scheme. The length of this routing address is variable, but remains same for all the nodes in the network at one point of time. Our scheme allocates addresses in such a way that nearby nodes have same prefixes, thereby helping the routing protocols to make use of the location information of nodes from their addresses. The main challenge of address auto configuration in mobile ad hoc networks is the provision of both address uniqueness and efficiency in terms of communication overhead and configuration latency (Yang, Luo, Ye, Lu and Zhang, 2004).

III. QUALITY OF SERVICE IN MANET

QoS provisioning in wireless networks is a difficult problem, particularly due to the time varying and unreliable channel medium. Supporting QoS in an ad hoc network is even more difficult mainly due to the absence of a centralized infrastructure. Mobile nodes rely on multi-hop communica- tion. In addition, these nodes have limited resources, such as battery, transmission capacity and buffer memory, which pose additional technical challenges for QoS support (Govindaswamy, Blackstone and Balasekaran,2011).

There are three major issues in QoS provisioning for ad hoc networks. First. since the topology of the network is highly dynamic, flow relying on a pre-established path and resource reservation along the path will suffer traffic interruptions due to frequent path changes. A better option is to have reactive QoS treatment coupled with the flow instead of the proactive QoS approach proposed in RSVP. Second, QoS parameters like throughput or latency are time-variant. Throughput can dramatically change in a short time due to radio interference or movement of hosts. The QoS mechanism needs to be adaptive, even when confronted with rapidly changing environments, in order to provide relatively stable QoS guarantees. Third, a MANET has less communication bandwidth. smaller processing and power capacity than a fixed network. Given the limited resources, the QoS in mechanism and signaling should not exhaust these resources.

A mobile ad-hoc network is an independent system of mobile nodes connected by wireless links forming a short, live, on-the-fly network even when access to the Internet is unavailable. Nodes in MANETs generally operate on low power battery devices (Cheng, Huang and Du,2006). These nodes can function both as hosts and as routers. As a host, nodes function as a source and destination in the network and as a router, nodes act as intermediate bridges between the source and the destination giving store-and- forward services to all the neighboring nodes in the network. Generally, QoS for a network is measured in terms of guaranteed amount of data which a network transfers from one place to another during a certain time (Zhang, Deering and Estrin,1993). There are several service models in wired networks. The two QoS models are the Integrated Services (IntServ) (Crawley, Nair, Rajagopalan and Sandick, 1998) and the Differentiated Service (DiffServ) models (Luo, Chen, Huang and Chen,2010). Both of these models require accurate link state such as available bandwidth, packet loss rate, delay, and topology information. The time-varying low-capacity resources of the network make maintaining the accurate routing information very difficult. The IntServ model provides QoS on a flow basis. It means IntServ architecture allows sources to communicate their QoS requirements to routers and destinations on the data path by means of a signaling protocol. The DiffServ model overcomes the difficulty in implementing and deploying IntServ model and Resource Reservation Protocol (RSVP) (Braden,Clark and Shenker,1994) in the Internet. The RSVP is used for reserving the resources along the route. In DiffServ model, flows are aggregated into a limited number of service classes. This solves the scalability problem in the IntServ model, but it does not guarantee services on per-hop basis. This problem makes DiffServ model difficult to use in the Internet, and will be a weakness for MANETs. Quality of Service providing a set of service requirements to the flows while routing them through the network (Swarnapriyaa, Vinodhini,Anthoniraj and Anand,2011). The widespread use of wireless technologies has increased QoS for multimedia applications in wireless networks and traditional internet QoS protocols like RSVP (Crawley, Nair, Rajagopalan and Sandick, 1998) can not be used for wireless environment due to the error-prone nature of wireless links and the high mobility of mobile devices in MANETs. Therefore, providing QoS in MANETs is more challenging than in fixed and wireless networks. A number of research have been conducted on required QoS in internet and traditional wireless networks, but current results are not appropriate for MANETs and still quality of service for MANETs is an open problem. Suitable QoS for delivery of real time communications such as audio/ video creates a number of different technical challenges (Blake, Black, Carlson, Davies, Wang and Weiss 1998).

IV. BATTERY POWER AND ENERGY MANAGEMENT IN MANET

Battery power is a limited resource, and it is expected that battery technology is not likely to progress as fast as computing and communication technologies do. Hence, how to lengthen the lifetime of batteries is an important issue, especially for MANET, which is all supported by batteries. Solutions addressing the power saving issue in MANETs can generally be categorized as follows.

Transmission Power Control

In wireless communication, transmission power has strong impact on bit error rate, trans- mission rate, and inter-radio interference. These are typically contradicting factors. Power control is adopted to reduce interference and improve throughput on the MAC layer. Isseues are: How to determine transmission power of each mobile host so as to determine the best network topology?, How to increase network throughput by power adjustment for packet radio networks?

Power-Aware Routing

Power-aware routing protocols have been proposed based on various power cost functions, when a mobile host's battery level is below a certain threshold, it will not forward packets for other hosts.

Low-Power Mode

More and more wireless devices can sup- port low-power sleep modes. IEEE 802.11 has a power- saving mode in which a radio only needs to be awake periodically. HyperLAN allows a mobile host in power-saving mode to define its own active period. An active host may save powers by turning off its equalizer according to the transmission bit rate. Bluetooth also has low-power modes, but is based on a master-slave architecture, so time synchronization is trivial. A lot of works have focused on multi-hop MANETs on issues such as power-aware routing, topology control, and transmission power control (as classified above), but how to design PS mode is left as an open problem.

There are two major issues that need to be addressed in the power mode management problem in a multi-hop MANET.

Wakeup Prediction

Since a host entering the PS mode will reduce its radio activity, other hosts who intend to send packets to the PS host need to know when the host will turn its radio on so as to correctly deliver packets to it at the right time.

Neighbor Discovery

Because hosts' transmission/reception activities are reduced under the PS mode, a host may take longer time, or may be even unable, to detect the arrival and departure of other hosts in its radio covered range. Thus, hosts may become less sensitive to neighborhood change. Neighbor discovery is essential for route discovery in a MANET. A host may incorrectly report that another host is unreachable if the route to this host has to go through some PS hosts that are not detectable by their neighbors on the path.

V. SECURITY IN MANET

Unlike Wired networks, MANET pose a number of challenges to security solutions due to their unpredictable topology, wireless shared medium, heterogeneous resources and stringent resource constraint etc. Security is not a single layer issue but a multilayered issue over here. MANET requires a multifence security solution that provides complete security spanning over the entire protocol stack.

The mobile nodes in MANETs are highly susceptible to malicious damage. Security issues are important in MANETs to prevent potential attacks, threats and system vulnerabilities. The fundamental aspects of computer security (Zhao, Aggarwal, Frost and Bai, 2011) –confidentiality, integrity, authentication and non-repudiation are valid when the protection of routing in MANET network is discussed. Confidentiality ensures the survivability of network services in case of denial of service attacks. Integrity guarantees that a message being transferred is never corrupted. A message could be corrupted because of beginning failure because of malicious attacks. Authentication guarantees every piece of routing information, which can always be confirmed to be valid and to have originated from the correct sender. On-repudiation means that the node cannot deny having sent or handled certain piece of routing information in the past. Authentication mechanisms allow partial non-repudiation.

Finally, it should be noted that authentication gets often confused with authorization issues that involve more or less the control of policy and are typically implemented only in the application layer and not in the network layer (Vidhyasanker, Manoj and Murthy, 2006) . Attack model is given in security chapter. Layer wise security issues are given below (Govindaswamy, Blackstone and Balasekaran, 2011):

Research on security of MANETs remains active, in spite of years of exploration, in both academia and industry. It is partially due to the fact that no mature solution is widely accepted and the growing availability of small, personalized mobile devices with peer to peer communication capability through wireless channels (Blazevic, Boudec and Giordano,2005). General security requirements for MANETs include (Ashtiani, Alirezaee, Hosseini and Hamid,2009): *Data Confidentiality* that keeps data secret to outsiders, *Data Integrity* that prevents data from being altered, *Data Freshness* that keeps data in the correct order and up-to-date, *Data Availability* that

Table 2.

Application Layer	Detecting and preventing viruses, worms, malicious codes, and application abuses
Transport layer	Authenticating and securing end-to-end communications through data encryption
Network layer	Protecting the ad hoc routing and forwarding protocols
Link layer	Protecting the wireless MAC protocol and providing link-layer security support
Physical layer	Preventing signal jamming denial-of-service attacks

ensures data to be available on request, *Data & Identity Authentication* that verifies that the data or request came from a specific, valid sender, and *Non-repudiation* that ensures a node cannot deny sending a message. Security mechanisms that are widely used and proven to be effective in wired networks are not always applicable to MANETs. Attacks that can be effectively detected and prevented in wired networks have been big security challenges in MANETs. Examples include, but are not limited to, identity/address spoofing, message tampering and forgery, message replay, etc. Compared to wired networks, the combination of the following characteristics of MANETs make it especially difficult to achieve security requirements:

- Lack of a network infrastructure and online administration.
- Network topology and node membership dynamics.
- The potential insider attacks.
- Computing and communicational capacity constrained
- Resources.
- Wireless link vulnerabilities.

Security proposals in early research are typically attack- oriented. They often first identify several security threats and then enhance the existing protocol or propose a new protocol to thwart them. Such solutions are designed explicitly against limited attack models. They work well in the presence of designated attacks but may collapse under combined or unanticipated attacks (Karimi and Pan2009).

Cryptography is then used to provide a general design framework. Cryptography techniques used in MANETs can be classified into two categories, namely, Symmetric Key based and Asymmetric Key based. In symmetric key based schemes, if an attacker compromises the symmetric key of a group of users, then all encrypted messages for that group will be exposed. Asymmetric key based schemes can provide more functionalities than symmetric ones, e.g., key distribution is much easier, authentication and non-repudiation are available, compromise of a private key of a user does not reveal messages encrypted for other users in the group. However, they are generally computationally expensive. Traditional asymmetric cryptography widely and effectively used in the Internet relies on a Public Key Infrastructure (PKI). The success of PKI depends on the availability and security of a Certificate Authority (CA), a central control point that everyone trusts. In general MANETs, applying PKIs by maintaining a central control point is clearly not always feasible. Another obstacle that impedes PKI's employment in MANETs is the heavy overhead of transmission and storage of public key certificates (PKCs). Identity-based cryptography (IBC) is a special form of public key cryptography. It is an approach to eliminate the requirement of a CA and PKCs. Since 2001, IBC has attracted more and more attention from security researchers. Some properties of IBC make it especially suitable for MANETs. Fang et al. summarize the advantages of IBC to MANETs (Fang, Zhu and Zhang, 2009 ; Zhang, Liu, Lou and Fang, 2006):

- Easier to deploy without any infrastructure requirement. This saves certificate distribution, while bringing "free" pair wise keys without any interaction between nodes.
- Its resource requirements, regarding process power, storage space, communication bandwidth, are much lower.
- The public key of IBC is self-proving and can carry much useful information.

VI. RECENT RESEARCH AREAS IN MANET

Recent research areas in MANET include the following:

1. **Routing:** Routing is an essential protocol in this field, because changes in network topology occur frequently. An efficient routing protocol is required to cope with highly fluid network conditions.
2. **Multicasting/ Broadcasting:** Multicast service supports users communicating with other members in a multicast group. Broadcast service supports users communicating with all members on a network.
3. **Location Service:** Location information uses the Global Positioning System (GPS) or the network-based geo-location technique to obtain the physical position of a destination.
4. **Clustering:** Clustering is a method to partition the hosts into several clusters and provide a convenient framework for resource management, routing and virtual circuit support.
5. **Mobility Management:** In the ad-hoc network environment, mobile hosts can move unrestricted from place to place. Mobility management handles the storage, maintenance and retrieval of the mobile node position information.
6. **TCP/ UDP and Multiple access:** TCP and UDP are the standard protocols used in the Internet. Data applications running over MANETs, such as http and real audio, need transport layer protocols like TCP and UDP to send packets over the links.
7. **IP Addressing:** One of the most important issues is the set of IP addresses that are assigned to the ad-hoc network. IP addressing and address auto-configuration have attracted much attention in MANETs.
8. **Radio Interface and Bandwidth Management:** Mobile nodes rely on the radio interface or antenna to transmit packets. Packet forwarding or receiving via radio interface or antenna techniques in MANETs are useful investigations. Bandwidth management in MANETs is a typical characterization. Because the bandwidth is usually limited, effectively managing and using it is a very important issue.
9. **Power Management:** A power management approach would help reducing power consumption and hence prolonging the battery life of mobile nodes. Because most devices operate on batteries, power management becomes an important issue.
10. **Security and Fault Tolerance:** The mobile nodes in MANETs are highly susceptible to malicious damage. Security issues are important in MANETs to prevent potential attacks, threats and system vulnerabilities. Fault Tolerance involves detecting and correcting faults when network failures occur.
11. **QoS/ Multimedia and Standards/ Products:** Quality of Service (QoS) and Multimedia require high bandwidth, low delay, and high reliability. The standards and products issues that allow the development of small scale is emerging for this field. For instance, Bluetooth is a low-cost technology for short-range communications techniques.

VII. OPEN PROBLEMS IN MANET

CONCLUSION

1. Research in MANET is like playing in a vast playground with only few known rules and large number of unknown rules applied to its different areas. Some areas are still unknown and some areas are at its early stage.
2. Some of the research challenges that MANET present are dynamic topologies, battery lifetime, disconnected operations, security etc.
3. Recent research areas of MANETs are routing, multicasting, clustering, mobility management, TCP & UDP(to send packets over the links), IP addressing, multiple access, Quality of service, Multimedia, bandwidth

Table 3.

Sl. No.	Problems	Details
1	• Practical considerations • Technical issues	o Consumer demand or lack thereof o Standardization o Government regulations
2	• Traditionally, not much interaction between physical layer and upper layers • Many physical layer mechanisms not beneficial without help from upper layers Example: Adaptive modulation	Adaptive Modulation: Channel conditions are time-varying • Choose modulation scheme as a function of channel conditions • If physical layer chooses the modulation scheme transparent to MAC o MAC cannot know the time duration required for the transfer • Must involve MAC protocol in deciding the modulation scheme o Some 802.11-compliant implementations use a sender-based scheme for this purpose o Receiver-based schemes can perform better.
3	• Sender-Based "Autorate Fallback" MAC Protocol	• Sender decreases rate after N consecutive ACKS are not received • Sender increases rate after Y consecutive ACKS are received D C B A 1Mbps 2Mbps DATA 2Mbps.
4	• Performance of Sender-Based "Autorate Fallback" Expected ARF CCK (11Mbps) CCK (5.5Mbps) QPSK (2Mbps) BPSK (1Mbps).	• Sender sends RTS containing its best rate estimate • Receiver chooses best rate for the conditions and sends it in the CTS • Sender transmits DATA packet at new rate • Information in data packet header implicitly updates nodes that heard old rate Receiver-Based Autorate MAC Protocol 1Mbps 2Mbps D C B A 2 1 CTS (1) RTS (2).
5	Physical Layer • Several other physical layer capabilities call for changes to upper layers of protocol stack Example: Power control	Power Control: • Transmit power determines o "Range" of a transmission o Interference caused at other nodes Benefits of Power Control • Transmit a packet with least transmit power necessary to deliver to the receiver o Save energy: Important benefit to battery-powered hosts o Reduce interference o Can allow greater spatial reuse • Power control introduces asymmetry • D transmits to C at low power, but B uses high transmit power to transmit to A • B may not about D-to-C transmission, but can interfere with it • Proposals for medium access control and routing with power control exist • Do not solve the problem satisfactorily • Ideal solution will o Reduce energy consumption o Maximize spatial reuse
6	Directional / Smart Antennas	• Various capabilities o Sectored antennas (fixed beam positions) o Beam steering o Tracking a transmitter • MAC and routing protocols for ad hoc networks using such antennas • How to take into account antenna capabilities? ▪ Network may be heterogeneous

continued on following page

Table 3. Continued

Sl. No.	Problems	Details
7	• Physical Layer • Are ad hoc networks benefiting from the progress made at physical layer? • Other interesting areas o Efficient coding schemes o Various diversity techniques. • Physical Layer: Simulation Models	• Insufficient accuracy in commonly used physical layer models • Physical link state is not binary as often assumed • Reliable packet reception does not depend just on distance o Transmit power o Interference level o Fading • Need to use realistic models o Modulation scheme o Coding
8	Interesting Link Layer Issues • Medium access control • Retransmission mechanisms • Transmission scheduling o Which pending packet should a node attempt to transmit? • Adaptive parameter selection o Frame size o Retransmission limit	QoS in Medium Access Control • Many proposals for achieving fairness • Fair scheduling schemes attempt to provide equitable sharing of channel • Unpredictable nature of transmission errors makes it difficult to make hard guarantees • Need to develop a probabilistic framework • Easier in a centralized protocol (such as 802.11 point coordination function), than in a distributed protocol • Distributed MAC appears more suitable for ad hoc networks, however • Perhaps a hybrid protocol will be best o How to design such a protocol?
9	Transmission Scheduling • When multiple packets pending transmission, which packet to transmit next? • Choice should depend on o Receiver status (blocked by some other transmission?) o Congestion at receivers o Noise level at receivers o Tolerable delay for pending packets • Need interaction between upper layers and MAC	MAC for Multiple Channels • How to split bandwidth into channels? • How to use the multiple channels? • Dedicated channel for control?
10	Network Layer • Reactive versus Proactive Routing • Reactive protocols o Maintain routes between nodes that need to communicate • Proactive protocols o Maintain routes between all node-pairs • Lot of activity on routing protocol design	Routing • Reactive and proactive protocols are quite • Well-understood • Designing reactive protocols: "Solved" problem • Designing proactive protocols: "Solved" problem o At least, when using common assumptions about the network o Interesting problems exist when other issues are considered (such as QoS or physical layer properties)
11	Reactive versus Proactive • Choice of protocol depends on o Mobility characteristics of the nodes o Traffic characteristics • How to design adaptive protocols? • Existing proposals use a straightforward combination of reactive and proactive o Proactive within "radius" K o Reactive outside K o Choose K somehow • Need a more flexible way to manage protocol behavior • Assign proactive/reactive tag to each route (A,B)? • How to determine when proactive behavior is better than reactive?	Address Assignment • How to assign addresses to nodes in an ad hoc network? • Static assignment o Easier to guarantee unique address • Dynamic assignment o How to guarantee unique addresses when partitions merge? • Do we need to guarantee unique addresses?

continued on following page

Table 3. Continued

Sl. No.	Problems	Details
12	Transport Protocols • TCP • TCP performance degrades in presence of route failures • TCP cannot distinguish between packet losses due to route change and due to congestion • Reduces congestion window in response • Unnecessary degradation in throughput • Several solutions have been proposed to fix this • These techniques somehow inform TCP sender that the packet losses are due to route failure • TCP does not decrease congestion window in response	• New route may differ significantly from old route • Proposals for TCP-over-ad-hoc tend to use old timeout and congestion window after a route change • Does not seem like a good idea • How to choose appropriate timeout and congestion window after detecting a route change?
13	Distributed Algorithms • Rich body of work on distributed algorithms in traditional distributed environments o Shared memory o Message ordering o Clock synchronization o Leader election • Existing algorithms can usually be used on ad hoc networks without affecting correctness • Performance on ad hoc networks may not be good • Existing algorithm treat link repairs/failures as random events • With mobility, link failure/repairs are correlated with host movement • How to design distributed algorithm exploiting the correlation between mobility and link failure/repair ?	• Traditionally, complexity is measured as a function of problem "size" o Number of nodes o Number of failures • How to analyze algorithm complexity as a function of mobility ? • What measure of mobility is amenable to such an analysis ? o Need to capture the correlation without making the measure too complex
14	Security Issues • What's New? • Wireless medium easy to snoop on • With ad hoc networking, hard to guarantee connectivity • Easier for intruders to insert themselves into network • Authentication • How to authenticate a node? • May not have access to a certification authority • Resource Depletion Attack • Intruders may send data with the objective of congesting a network or depleting batteries	Routing Attacks o Intruders may mis-route the data not delivering it to the destination at all or delaying it significantly • How to detect such attacks? • How to tolerate such attacks? • Traffic Analysis • Despite encryption, an eavesdropper can identify traffic patterns • Traffic patterns can divulge information about the operation mode • Traffic analysis can be prevented by presenting "constant" traffic pattern • Insert dummy traffic • How to make this cheaper?
15	Other Issues • Incentives for Ad Hoc Routing • Why forward packets for some other nodes? • Need some incentive mechanism • Policies to determine reward for performing each operation.	Applications • New applications for ad hoc networks? • Hybrid Environments • Use infrastructure when convenient • Use ad hoc connectivity when necessary or superior
16	• Plenty of interesting research problems • Research community disproportionately obsessed with routing protocols	• Interesting problems elsewhere at the two ends of the protocol stack • How to design algorithms and applications ? • How to exploit physical layer techniques ? • Increase interaction between physical layer and upper layers. L

management, radio service, power management, security, standards etc.

4. In general there are three types of MANET-vehicular ad hoc networks, intelligent vehicular ad hoc network and internet based mobile ad hoc networks.
5. A Mobile Ad Hoc Network (MANET) is a collection of wireless mobile nodes that out of necessity form a network required to facilitate communication. Because the nodes have limited energy storage, computing power and communication ability, designing an energy efficient protocol is a challenge.
6. The focus of most research is on Routing Protocols. Protocols are divided into three categories: Proactive, Reactive and Hybrid.
7. The major objectives of an optimal mobile ad hoc network address auto configuration protocol are Dynamic Address Configuration,-Uniqueness,Robustness and Scalability.
8. QoS provisioning in wireless networks is a difficult problem, particularly due to the time varying and unreliable channel medium. Supporting QoS in an ad hoc network is even more difficult mainly due to the absence of a centralized infrastructure.
9. Mobile nodes rely on multi-hop communication. In addition, these nodes have limited resources, such as battery, transmission capacity and buffer memory, which pose additional technical challenges for QoS support.
10. General security requirements for MANETs include: *Data Confidentiality* that keeps data secret to outsiders, *Data Integrity* that prevents data from being altered, *Data Freshness* that keeps data in the correct order and up-to-date, *Data Availability* that ensures data to be available on request, *Data & Identity Authentication* that verifies that the data or request came from a specific, valid sender, and *Non-repudiation* that ensures a node cannot deny sending a message. Security mechanisms that are widely used and proven to be effective in wired networks are not always applicable to MANETs.
11. Security proposals in early research are typically attack- oriented. They often first identify several security threats and then enhance the existing protocol or propose a new protocol to thwart them. Such solutions are designed explicitly against limited attack models. They work well in the presence of designated attacks but may collapse under combined or unanticipated attacks.
12. Research on security of MANETs remains active, in spite of years of exploration, in both academia and industry. It is partially due to the fact that no mature solution is widely accepted and the growing availability of small, personalized mobile devices with peer to peer communication capability through wireless channels.

REFERENCES

Abusalah, L., Khokhar, A., & Guizani, M. (2008). A survey of secure mobile ad hoc routing protocols. *IEEE Communication Surveys & Tutorials*, *10*(4), 78–93.

Ashtiani, H., Alirezaee, S., Hosseini, S. M. M., & Hamid, K. (2009). PNR - New position based routing algorithm for mobile ad hoc networks. *Proceedings of the World Congress on Engineering 2009*, Vol. I, WCE 2009, London, UK.

Blake, S., Black, D., Carlson, M., Davies, E., Wang, Z., & Weiss, W. (1998). *An architecture for differentiated services*. IETF RFC 2475.

Blazevic, L., Boudec, J. Y. L., & Giordano, S. (2005). A location-based routing method for mobile ad hoc networks. *IEEE Transactions on Mobile Computing*, *4*(2). doi:10.1109/TMC.2005.16

Braden R., Clark, D., & Shenker, S. (1994). *Integrated services in the Internet architecture: An overview*. IETF RFC 1633.

Cheng, X., Huang, X., & Du, D. Z. (2006). *Ad hoc wireless networking* (pp. 319–364). Kluwer Academic Publishers.

Crawley, E., Nair, R., Rajagopalan, B., & Sandick, H. (1998, August). *A framework for QoS based routing in the Internet.* RFC 2386.

Fang, Y., Zhu, X., & Zhang, Y. (2009). Securing resource-constrained wireless ad hoc networks. *Wireless Communications*, *16*(2), 24–29. doi:10.1109/MWC.2009.4907556

Govindaswamy, V., Blackstone, W. L., & Balasekaran, G. (2011). Survey of recent position based routing mobile ad-hoc network protocols. *Proceedings of* 2011 UKSim 13th International Conference on Modelling and Simulation, (pp. 467-471).

IETF. (2000). *MANET terms*. Retrieved from http://tools.ietf.org/html/ draft-ietf-manet-term-00

Jeon, J., Lee, K., & Kim, C. (2011). Fast route recovery scheme for mobile ad hoc networks. Proceedings of *ICOIN*, (pp. 419-423).

Karimi, M., & Pan, D. (2009). Challenges for quality of service (QoS) in mobile ad-hoc networks (MANETs). *Proceedings of IEEE 10th Annual Wireless and Microwave Technology Conference,* WAMICON '09, (pp. 1-5).

Luo, Q., Chen, G., Huang, C., & Chen, G. (2010). *Study of MANET routing evaluation model based on best-first.* The Telecommunication Engineering Institute, Air Force Engineering University, Xi' an, China. 978-1-4244-5849-3/10

Mikki, M. A. (2009). Energy efficient location aided routing protocol for wireless MANETs. *International Journal of Computer Science and Information Security*, *4*(1-2).

Qabajeh, L. K., Kiah, L. M., & Qabajeh, M. M. (2009). A qualitative comparison of position-based routing protocols for ad-hoc networks. *International Journal of Computer Science and Network Security*, *9*(2).

Roche, A., Westphall, C. B., Pack, S., & Mecklenburg, G. V. (2002). Quality of service for ad-hoc wireless network. *Proceedings of the 22nd International Conference of the Chilean,* (pp. 100-105).

Slideshare. (n.d.). *Open problems in mobile ad hoc networking*. Retrieved from http://www.slideshare.net/networkingcentral/ open-problems-in-mobile-ad-hoc-networking

Swarnapriyaa, U. G., Vinodhini, V., Anthoniraj, S., & Anand, R. (2011). Auto configuration in mobile ad hoc networks. *Proceedings of the National Conference on Innovations in Emerging Technology*, (pp. 61-66).

Vidhyasanker, V., Manoj, B. S., & Murthy, C. S. R. (2006). Slot allocation schemes for delay sensitive traffic support in asynchronous wireless mesh networks. *The International Journal of Computer and Telecommunications Networking*, *50*(15), 2595–2613.

Weniger, K., & Zitterbart, M. (2004). Address auto-configuration in mobile ad hoc networks: Current approaches and future directions. *IEEE Network*, *18*(4), 6–11. doi:10.1109/MNET.2004.1316754

Wikipedia. (n.d.). *Mobile ad hoc network.* Retrieved from http://en.wikipedia.org/wiki/ Mobile_ad_hoc_network

Yagan, D., & Tham, C. K. (2005). *Adaptive QoS provisioning in wireless ad hoc networks: A semi-MDP approach. WCNC 2005* (pp. 2238–2244). IEEE Communications Society.

Yang, H., Luo, H., Ye, F., Lu, S., & Zhang, S. (2004). Security in mobile ad hoc networks: Challenges and solutions. *IEEE Wireless Communications*, *11*(1), 38–47. doi:10.1109/MWC.2004.1269716

Zhang, L., Deering, S., & Estrin, D. (1993). RSVP: A new resource reservation protocol. *IEEE Network*, *7*(5), 8–18. doi:10.1109/65.238150

Zhang, Y., Liu, W., Lou, W., & Fang, Y. (2006). Securing mobile ad hoc networks with certificateless public keys. *IEEE Transactions on Dependable and Secure Computing*, *3*(4), 386–399. doi:10.1109/TDSC.2006.58

Zhao, S., Aggarwal, A., Frost, R., & Bai, X. (2011). A survey of applications of identity-based cryptography in mobile ad-hoc network. *IEEE Communication Surveys and Tutorials*, *99*, 1–21.

ADDITIONAL READING

Aggelou, G. (2005). *Mobile ad hoc network.* McGraw-Hill Professional Publishing.

Aggelou, G. (2005). *Mobile ad hoc networks from wireless LAN to 4G networks*. McGraw Hill Companies.

Azzedine, B. (2005). *Handbooks of algorithm for wireless networking and mobile computing*, 1st edition. Chapman and Hall/CRC Computer and Information Science Series, Chapman and Hall/CRC.

Azzedine, B. (2008). *Algorithms and protocols for wireless mobile ad hoc network.* Wiley Series on Parallel and Distributed Computing-Wiley-IEEE Press.

Basagni, S., Conti, M., Giordano, S., & Stojmenović, I. (2004). *Mobile ad hoc networking*. IEEE Press, Wiley-Interscience Publication.

Boban, M., & Vinhoza, T. T. V. (2011). *Modeling and simulation of vehicular networks: Towards realistic and efficient models*. Intech Publication.

Bulent, T., & Wendi, H. (2006). *Mobile ad hoc networks: Energy efficient real time data communications* (1st ed.). Springer.

Caballero-Gil, P. (2011). *Security issues in vehicular ad hoc networks*. Intech Publication.

De Rango, F., & Socievole, A. (2011). *Meta-heuristics techniques and swarm intelligence in mobile ad hoc networks*. Intech Publication.

Della Rosa, F., Leppäkoski, H., Ghalib, A., Ghazanfari, L., Garcia, O., Frattasi, S., & Nurmi, Y. (2011). *Ad hoc networks for cooperative mobile positioning*. Intech Publication.

Forouzan, B. A. (2008). *Data communication and networking* (4th ed.). Tata McGraw Hill.

Galaviz Mosqueda, G. A., Santos, R. A., & González, L. A. V. (2011). *Survey on multi-hop vehicular ad hoc networks under IEEE 802.16 technology*. Intech Publication.

Gonzalez-Cañete, F. J., & Casilari, E. (2011). *Impact of the mobility model on a cooperative caching scheme for mobile ad hoc networks*. Intech Publication.

Hekmat, R. (2006). *Ad hoc networks: Fundamental properties and network topologies*. Springer.

Hutchison, D., Mariani, J. A., & Shepherd, W. D. (1985). Lecture Notes in Computer Science: *Vol. 184. Local area networks: An advanced course*. Springer.

Ilyas, M. (2003). *The handbook of ad hoc wireless networks*. New York, NY: CRC Press.

Ivan, S. (2002). *Handbook of wireless network and mobile computing* (1st ed.). Wiley-Interscience Publication.

Mishra, S., Issac, W., & Misra, S. C. (2010). *Guide to wireless ad hoc networks*. Springer Computer Communications and Network Series.

Rawashdeh, Z. Y., & Mahmud, S. M. (2011). *Communications in vehicular ad hoc networks*. Intech Publication.

Sharieh, S. (2011). *Ad-hoc networks as an enabler of brain spectroscopy*. Intech Publication.

Toh, C.-K. (2007). *Ad hoc mobile wireless networks: Protocols and systems*. Prentice Hall.

Wang, X. (2011). *Mobile ad hoc network: Applications*. Intech Publication.

KEY TERMS AND DEFINITIONS

Asymmetric Link: A link with transmission characteristics which are different depending upon the relative position or design characteristics of the transmitter and the receiver of data on the link. For instance, the range of one transmitter may be much higher than the range of another transmitter on the same medium.

Bandwidth: The total capacity of a link to carry information.

Bandwidth Utilization: The actual amount of information delivered over a link,expressed as a percent of the available bandwidth on that link.

Base station: A centralized node coordinating the channel access of a population of mobile nodes within its transmission range.

Beacon: A control message issued by a node (especially, a base station) informing all the other nodes in its neighborhood of the continuing presence of the node, possibly along with additional status information.

Channel: A subdivision of the physical medium allowing possibly shared independent uses of the medium. Channels may be made available by subdividing the medium into distinct time slots, or distinct spectral bands, or decorrelated coding sequences.

Channel Access Protocol: A protocol for mediating access to, and possibly allocation of, the various channels available within the physical communications medium. Nodes participating in the channel access protocol can communicate only when they have uncontested access to the medium, so that there will be no interference.

Cluster: A group of nodes located within close physical proximity, typically all within range of one another, which can be grouped together for the purpose of limiting the production and propogation of routing information.

Control Message: Information passed between two or more network nodes for maintaining protocol state which is not associated to any specific application.

Convergence: The process of approaching a state of equilibrium in which all nodes in the network agree on a consistent collection of state about the topology of the network, and in which no further control messages are needed to establish the consistency of the network topology.

Convergence Time: The time which is required for a network to reach convergence after an event (typically, the movement of a mobile node) which changes the network topology.

Distance Vector: A style of routing protocol in which, for each desired destination, a node maintains information about the distance to that destination, and a vector (next hop) towards that destination.

Fairness: A property of channel access protocols whereby a medium is made fairly equal to all eligible nodes on the link. Fairness does not strictly imply equality, especially in cases where nodes are given link access according to unequal priority or classification.

Flooding: It is the process of delivering data or control messages to every node within the ad hoc network.

Forwarding Node: It is a node within an ad hoc network which performs the function of forwarding datagram from one of its neighbors to another.

Goodput: The total bandwidth used, less the volume of control messages and protocol overhead from the data packets.

Hidden-terminal Problem: The problem whereby a transmitting node can fail in its attempt to transmit data because of destructive interference which is only detectable at the receiving node, not the transmitting node.

Laydown: The relative physical location of the nodes within the ad hoc network.

Link: A physical medium which can sustain data communications between multiple network nodes.

Link State: A style of routing protocol in which every node within the network is expected to maintain information about every link within the network topology.

Link-level Acknowledgement: A protocol strategy, typically employed over wireless media, requiring neighbors to acknowledge receipt of packets (typically unicast only) from the transmitter. Such strategies aim to avoid packet loss or delay resulting from lack of, or unwanted characteristics of, higher level protocols.

Local Broadcast: The delivery of data to every node on a link (i.e., within range of the transmitter).

Loop-free: A property of routing protocols whereby the path taken by a data packet from source to destination never transits the same intermediate node twice before arrival at the destination.

MAC-layer Address: An address (sometimes called the link address) associated with the link interface of a node on a physical link.

Mobility Factor: The relative frequency of node movement, compared to the convergence time of the routing protocols used in the ad hoc network.

Neighborhood: All the nodes which can receive data on the same link from one node whenever it transmits data.

Next Hop: A neighbor which has been designated to forward packets along the way to a particular destination.

Pathloss: A reduction in signal strength caused by traversing the physical medium constituting the link.

Pathloss Matrix: A matrix of coefficients describing the pathloss between any two nodes in an ad hoc network. When the links are asymmetric, the matrix is also asymmetric.

Payload: The actual data within a packet, not including network protocol headers which were not inserted by an application.

Scalability: Wide applicability of a protocol to large as well as small populations of nodes participating in the protocol.

Scenario: The tuple <laydown, pathloss matrix, mobility factor, traffic> characterizing a class of ad hoc networks.

Signal Strength: The detectable power of the signal carrying the data bits, as seen by the receiver of the signal.

Source Route: A route to a source (i.e., a path with indications of intermediate forwarding nodes) made available to a receiver by the source of the data arriving at the receiver.

Spatial Re-use: Simultaneous use of channels with identical or close physical characteristics, but located spatially far enough apart to avoid interference (i.e., co-channel interference)

System-wide Broadcast: Same as flooding, but used in contrast to local broadcast.

Throughput: The amount of data from a source to a destination processed by the protocol for which throughput is to be measured for instance, IP, TCP, or the MAC protocol.

Chapter 22
SimuMANET:
An Open-Source Based Remote Tool in the Lab

Ana Vazquez Alejos
University of Vigo, Spain

Paula Gómez Pérez
University of Vigo, Spain

Manuel Garcia Sanchez
University of Vigo, Spain

Muhammad Dawood
New Mexico State University, USA

ABSTRACT

Simulation software in MANET research is vital. Such a tool provides a versatile mechanism to understand all the involved aspects of these particular systems, from the radio interface to the last communication layer. In this chapter, the authors present the SimuMANET project, a tool for both simulation and field tests purpose. It allows the deployment of wireless reconfigurable ad-hoc networks and MANETs, assisted by a real-time graphical user interface (GUI) for network traffic monitoring and management of radio electric features of the links established between the active network nodes. Due to a set of functionalities, such as GUI, network topology visualization, traffic and motion pattern configuration, and real-time network status analysis, the simulator introduced here becomes a valid tool for both research and education targets. Two scenarios with different types of motion and traffic are simulated using the SimuMANET tool, and the results are shown and commented to illustrate some capabilities of this software.

DOI: 10.4018/978-1-4666-0321-9.ch022

INTRODUCTION

The SimuMANET tool was developed as the central part of two Master Thesis (MT) realized with two main goals:

1. Firstly, an educational objective leaded to the realization of MTs by means of implementing collaborative strategies between students. The project was divided in tasks, mapping the different program modules.

The individual tasks were assigned to two students of such way that the realization of the modules was interleaved: the module *n+1* to be implemented by the student A depends on the module *n* to be implemented by the student B. By this way we achieved a work division that required team work and, at the same time, it allowed the development of the professional responsibility sense in the student.

2. The technical objective of the project consisted on the development of a tool for both simulator and field tests. It allows the deployment of wireless reconfigurable ad-hoc networks and MANETs, assisted by a real-time graphical interface for network traffic monitoring and management of radio electric features of the links established between the active network nodes. The obstacle presence has also been considered to provide realistic and accurate propagation conditions.

The project start point consisted of a collection of software tools in form of scripts or pieces of non-executable programs, freely distributed and provided by the *US Navy Research Lab* (NRL). This available open source code was oriented to Linux operating systems, in this case *Fedora 5*.

These tools provided a set of individual source code that should be analyzed, corrected and updated to be able to start it up. Moreover, it was also necessary to perform changes in the source, according to the project needs to achieve module interoperability. One main script would act as dispatcher of the updated modules, and other many were exclusively developed to reach the required specifications. Finally, another open source code was used for designing the graphical interface, based on the *Qt library*.

The purpose of this chapter is to introduce developed tool. For that, firstly we discuss the requirements of a MANET to be simulated. Later, we describe the background software used in the SimuMANET tool. After that, we detail two implemented case studies. Conclusions close this chapter with an overall evaluation.

BACKGROUND

A Mobile Ad hoc Network can be considered as an autonomous distributed system that consists of a set of identical mobile nodes that move arbitrarily and use wireless links to communicate with other nodes that reside within its transmission range.

Variable link conditions are intrinsic characteristics in most mobile ad hoc networks. Rerouting among mobile nodes causes network topology and traffic load conditions to change dynamically. Given the nature of a MANET, it is difficult to support real-time applications with appropriate quality of service (QoS). In some cases, it may be impossible to guarantee strict QoS requirements. However, at the same time, QoS is a truly important feature in MANETs since it can improve performance and enable critical information to flow even under difficult conditions.

There exist several approaches to operate a MANET, each one providing different advantages. Regarding QoS and MANETs, the lack of centralized entities and the constraints introduced by the own nature of MANETs, such as changes in network topology, power consumption and noise interferences, emphasizes the need to achieve optimized results in all the aspects involved in MANET management. In the next points, some important issues regarding MANETs and QoS are presented:

Figure 1. Example of a multilayered power MANET

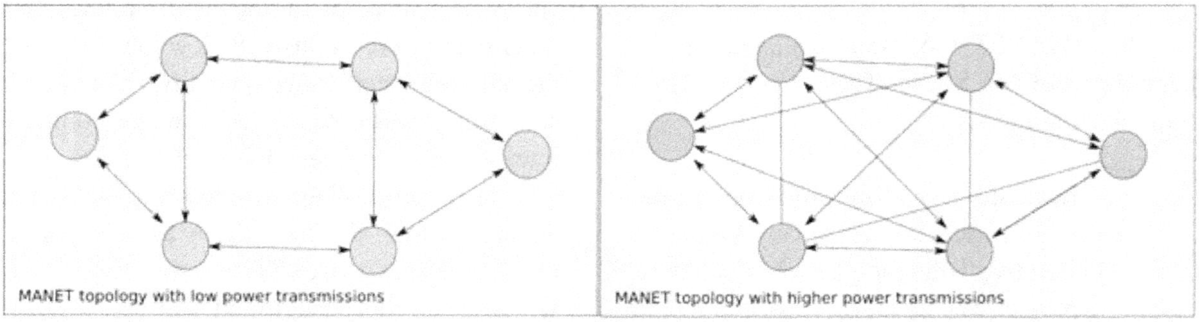

1. Find a route between two communication end-points is a major problem in MANETs. The problem worsens due to the node mobility. Many different approaches to handle the variable topology problem have been proposed in recent years, however none routing algorithm has been resulted suitable for all situations. There exist several well known routing algorithms such as the ant algorithms, Destination-Sequenced Distance Vector, Ad-Hoc On-Demand Distance Vector Protocol or Dynamic Source Routing.
2. Design of architecture valid to allow network topology prediction. This is an important issue since nodes may move in unpredictable ways. This task can be carried out in several ways, from a fully centralized approach to a fully decentralized one. Centralized approaches produce better results because they receive information from all the nodes. However, centralized solutions have some disadvantages that cannot be neglected, such as larger susceptibility to network congestion and vulnerability in case of node and link failures. On the other hand, decentralized approaches avoid these problems; however they hardly achieve optimal QoS performance due to the limited quantity of information available in the nodes.

Regardless of the adopted approach, to get a reliable QoS, the anticipation to future network states is a must and should be conducted in parallel with other tasks:

1. A multilayered power schema where larger transmission power is used for the traffic with the highest requirements. A common problem in MANETs is device power consumption, thus decreasing power transmission rates allows larger battery lifetimes. As a result, there will not only be one logical network topology but several, each one corresponding to a power layer. This is known as multilayer power topology networks. An illustration can be seen in Figure 1.

2. Introduction of network coding routing techniques to improve network reliability and optimize resource utilization. Network coding approaches can asymptotically achieve optimal capacity in multi-source multicast networks. Network coding has demonstrated to outperform traditional routing strategies when problems such as reliability or link bandwidth arise. Thus, network coding appears to be one option to consider when solving specific QoS problems such as bandwidth allocation or packet loss.
3. An important optimization that has been recently proposed is virtual MIMO. This technique is very useful in the area of node cooperation since that, given a cluster of

nodes, it is possible to transmit or receive data cooperatively decreasing the per-node power consumption.

4. Resource reservation. Paths are calculated using partial topology, and because of congestion and mobility, the validity of the path must be checked to admit a new flow in the source node. This is usually referred to as Call Admission Control (CAC). A CAC can be driven in the source or in a distributed mode for each hop in the path. A node can proceed to (soft and later hard) resource reservation on the links during the CAC process. This ensures that two flows initiated simultaneously will not contend for the same resource.
5. Radio channel propagation estimation determines maximum range of one link. The radio link budget estimation allows determining if the path presents a bidirectional or unidirectional behavior. Better link quality results in shortest path determination, as well as a low packet loss and rejection rate. The advantages of shorter multiple hops are less required power, greater range, and better link quality (QoS).
6. GPS – Assisted Routing Protocols. Recent advances in global positioning systems (GPS) have introduced a new form of scalable MANET routing, categorized as geographic-position-assisted routing. GPS technology utilizes a node's precise location and universal timing to assist in directional routing and network synchronization, respectively. In a mobile environment where nodes are traveling at unknown speeds, a special care must be taken with the accuracy and validity of a node's reported position. In this type of scenario, the GPS system must support frequent position updates or incorporate an algorithm that predicts a node's position based on previous mobility data. The protocol to implement for GPS-assisted routing assumes that nodes know their positions. In general GPS-assisted routing schemes allow for effective and scalable routing, but at the cost of overhead required to continually update position coordinates of all the nodes in the network.
7. QoS requirements are not only related to the bandwidth parameters. They are also related to also to delay and power consumption parameters.

The above considerations summarize the points to take into account to implement a MANET simulator or even a real system. Many of them represent important novelties in the field of MANETs, and were not included in our project.

In the following sections we describe the software modules selected and/or developed to achieve a simulator tool provided of mechanisms to analyze the involved issues when dealing to get a reliable QoS-MANET.

SIMUMANET: OPEN SOURCE SOFTWARE TOOLS

The simulation tool should comprise the following functionalities:

- Graphical interface
- Automatic dispatcher script
- Network topology geographical representation
- Basic radio propagation model
- Node set-up (motion, traffic,...)
- Network traffic status analysis

In the following sections we describe the open source software packages selected to achieve a simulator tool with the above listed functionalities.

Mobile Network Emulator (MNE)

The MNE scripts are a set of bash scripts for Linux that are used to automate IPv4 and IPv6 Mobile

Ad-hoc Networking (MANET) laboratory tests using NRL's Mobile Network Emulator (*NRL, Chao 2003*). These scripts are designed to emulate a MANET using one Linux machine (laptop or desktop) for each node in your network.

Each LINUX machine in the network should ideally have two interfaces or channels:

- *Control Channel*: one hard-wired interface with static addresses for backchannel traffic due to the simulation software: commands, MNE backchannel control info, ...
- *Test Channel*: one wired or wireless interface for test traffic which simulates the wireless network and is used by the nodes for data interchange: MGEN data, routing protocol packets, …

Although not required, the use of the control channel is highly recommended as it could otherwise reach saturation on the network interface and spoils the simulation. In this project we have used both channels assigning the wireless interface for the test channel and the wired interface for the backchannel.

The MNE scripts can also consider one machine in the network as a special node designated "XCom." XCom would act as a master control node for the MNE scripts, becoming the machine used to start tests, gather results, among others. In Figure 2 we illustrate the network structure considered in MNE.

MNE operates in distributed mode, so that each node controls its own virtual or real position.

In order to simulate the link loss, each user can block traffic packets coming from the remote nodes. The packet blockage is carried out by using the Linux IPTABLES to filter by MAC addresses.

The simulator also includes software to process the GPS positions written in a shared memory block which can be accessed by other MGEN modules. The GPS positions can be contained in *gpsLogger* files from real tests for "real" motion model.

MNE can be used also as a generator of movement pattern of one node that can be communicated to the other network members via the

Figure 2. MNE schematic

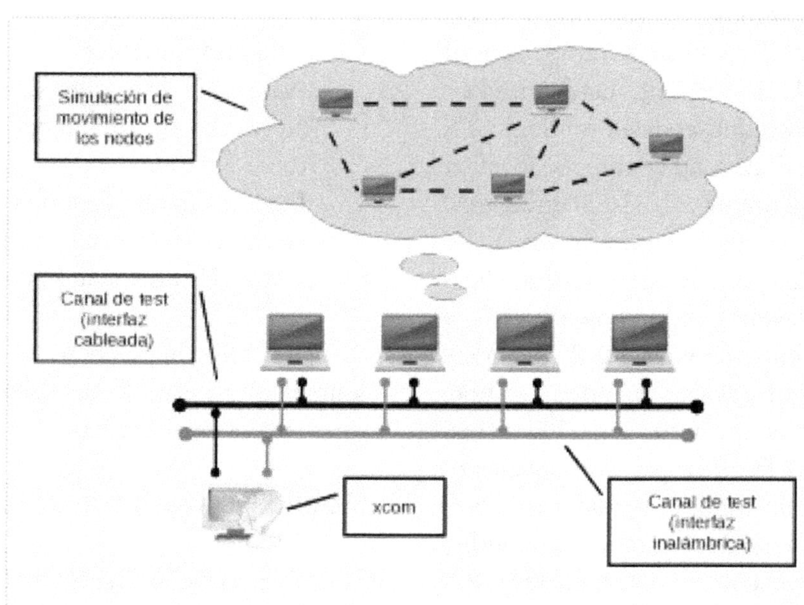

Table 1. MNE configuration parameters

Option	Description	Option	Description
-i	control interface	-lmargin	limit link margin (dBm)
-A	multicast address and port	-agthresh	air-ground threshold (m)
-m	movement model	-sens	sensitivity threshold (dBm)
-N	node unique identifier	-f	movement file
-M	MAC address	-S	start point for waypoints
-4	IPv4	-s	waypoints speed
-o	obstacles file	-R	stop time for waypoints
-freq	network operating frequency	-staticpos	coordinates for static node
-syslos	system loss (dB)	-upperleftpos	upper left corner for NS2

control channel using a multicast IP address. Thus, each node can compare its position with others, in order to decide which nodes are too far apart to maintain the link and so blocking their MAC address to simulate the communication loss.

Other significant feature of MNE is that it provides support for IPv4 and IPv6, multicast traffic, 3D motion, asymmetric communication links (nodes with different transmission parameters), and additionally to the movement patterns (static, linear, circular, waypoints) it also accepts NS2 scripts.

However, this tool shows some limitations. The only traffic filter available is the one provided by Linux IPTABLES which restricts its use on other platforms. Another important limitation is that the blockage is exclusively made according to IP packets and never selectively (blocking all or none).

The regular use of MNE is running the scripts from the command line, although it can also be used to facilitate the implementation and integration with other programs needed for the simulation. In the SimuMANET project we have used this second approach consisted of the individual scripts provided by NRL which have been integrated in a software tool as detailed in the following sections. In this case, additionally to the wireless network interface, both test and control channels must be used. MNE supports an extensive set of input arguments, which were used as listed in Table 1.

MNE is open source and free distribution software developed by the Naval Research Laboratory PROTEAN and the version used in the SimuMANET project is MNE-3.0.

MNE-Scripts

This software package consists of a set of individual scripts to automate the network simulation with MNE. They allow deploying different kinds of tests on the simulation. They are flexible and allow the user easily adapting them to develop new functionalities and run options.

As aforementioned, if MNE is intended to be used by the set of *mne-scripts*, additionally to the wireless network interface for the test channel, a wired interface must be used for the control channel. Otherwise, the single graphical interface provided by MNE must be executed from the command line.

The mne-scripts are designed choosing the node Xcom to centralize the control information. In addition, Xcom starts and stops the simulation, copies the configuration to other nodes and provides the result files. This special node only connects to other nodes using SSH, so it must be configured to not require password.

Both the installation and implementation of *mne-scripts* require root privileges. Most of the programs used in the development of this project have also this requirement, not representing this fact a constraint to select these scripts as the most appropriate mechanism to use MNE.

According to their functionality, the *mne-scripts* can be classified in four groups:

1. *Start-Scripts*: to start and stop the simulation
2. *Commands*: heterogeneous set of commands to allow checking the network connectivity status, establishing multicast addresses, file copy on remote nodes, and others
3. *Setup*: Personal automatic generation of configuration files
4. *Post-Analysis*: to create and show different plots with the outcomes of the simulation

Many of these scripts have been modified in order to correct or improve their functionalities.

Motion Generator: BonnMotion

BonnMotion is free distributed software developed by the *Institute of Computer Science IV* of the University of Bonn. It is distributed under GNU-GPL. The version used for the SimuMANET project was the BonnMotion-1.3a.

BonnMotion is a tool for the generation and analysis of mobile scenarios, even that in this project it has been used only as a motion generator later modified as to obtain files under NS2 format.

In SimuMANET only three of the five possible motion patterns have been incorporated, as following described:

1. *Waypoints*: random points are generated within a simulation area. The input parameters for this movement model are: maximum speed; minimum speed; maximum stop time; spatial dimension (1D, 2D or 3D).
2. *Manhattan Grid*: in this model, the node movement on a grid is simulated. The configurable parameters are: number of blocks along x and y to build the grid; minimum and average speed; probability and standard deviation of speed variation; probability of stop and the maximum stop time; shift/turn probability and update rate.
3. *Gauss-Markov*: the generated points are the result of the variation undergone by the velocity speed. The configurable parameters are: maximum speed; update rate, in meters; deviation of the velocity vector, in module and angle. It is also possible to follow an uniform distribution instead of the Gaussian distribution inherent to the Gauss-Markov model.

There are many other parameters shared by all the motion patterns or models: number of nodes, simulation duration, initial time to ignore, width and length of the simulation area, initial seed, and circular movement enforcement.

It is needed to convert the output files into NS2 format in order to use BonnMotion with MNE. It is important to remark that the numeration of nodes in MNE starts in 0 (identifier for XCom node), whilst in BonnMotion this identifier is for the node 1. To solve this conflict, in SimuMANET, the XCom has been taken as the node with the last identifier.

Traffic Generator: MGEN

For the traffic generation we have counted only on the program Multi-Generator (MGEN), open source tool developed by the PROTocol Engineering Advanced Networking (PROTEAN) group of the NRL. The version here used was MGEN-4.2b6.

MGEN provides the system with the capacity of analyzing and measuring the features of an IP network. The only supported protocol is UDP, even that future versions will likely implement also TCP.

This software generates real time network traffic patterns which allow simulating different cases with diverse traffic network loads. In the reception node, tracing files with the received packets are created, so that it is possible to analyze the load of the whole network as well as to generate plots with the most meaningful metrics.

MGEN combines the line command run with the use of individual scripts which describe the traffic pattern for each node. Anyway, with the purpose of automating the simulation launch, the *start-scripts* from MNE integrate the MGEN invocation in the input parameters configured via SimuMANET.

Even that MGEN presents multiple commands for running, only the commands indicated in the Table 2 have been used.

The MGEN scripts consist of text files containing commands and scheduled events that describe the generated traffic pattern. Each line of the file corresponds to an event and follows the following format:

< Time of the event > < Type of event > <Parameter> [<Options>]

where each items indicates:

1. *Time of the event*: sets the moment the event occurs. It is considered
2. relative time since the start of execution.
3. *Type of event*: Three groups of events are considered: transmission, reception and global commands. In the SimuMANET scripts have been used only for transmission and reception events, since that global commands needed for the simulation are passed as input parameters in the call from the *start-scripts MGEN*. There are three transmission events: ON, OFF MOD; and, at the reception are available the events LISTEN and IGNORE.

Each event type has its own parameters and options. As indicated, there are three patterns of traffic for transmission events. For each one, it must specify the type of traffic, the transmission frequency and size of each package. If more complex traffic patterns are required, the three different patterns offered by MGEN can be combined that flows using traveling to the same node.

Each traffic pattern is following describes, indicating as well the syntax:

1. *PERIODIC* [<rate> <size>]: Generates messages of a fixed size to a specified periodic rate, in messages per second. The maximum size for UDP packets is 8192 bytes.
2. *POISSON* [<average rate> <size>]: Generated messages have a fixed size specified and submitted to random time intervals which average given in the field <average rate>. The size of the messages in any case cannot be greater than the maximum supported by UDP.
3. *BURST* [REGULAR/RANDOM <average-Interval> <pattern> [<params>] FIXED/

Table 2. MGEN configuration parameters

Option	Description
-ipv4	forces MGEN to open sockets to operate only with IPv4
-input	makes MGEN to check the script file at the startup and to schedule each event specified in that file
-interface	specifies the interface used to transmit and receive traffic
-flush	clean up the output buffer after writing each line in the tracing file
-ttl	sets the maximum hop count

EXPONENTIAL <averageDuration>]: this model can generate bursts of traffic based on the specified pattern. This pattern may be selected from any of the three ones that offers MGEN, i.e., PERIODIC, POISSON or BURST, although the latter option has not been implemented in SimuMANET. According to the chosen pattern, the appropriate parameters should be completed. The burst will occur at intervals that can be configured as regular (REGULAR) or random (RANDOM) always within an average interval given by <averageInterval>. The duration of the bursts can be fixed (FIXED) of <averageDuration> seconds, or vary randomly following an exponential distribution (EXPONENTIAL) with an average duration <averageDuration>.

Routing Protocol: NRLOLSR

Although MNE supports multiple routing protocols, it was selected the NRLOLSR-7.8.1. It is open source and free distribution software developed by NRL.

NRLOLSR is an adaptation of Optimized Link State Routing (OLSR) routing protocol defined according to the specifications given in RFC 3626. In its design has dominated the need to reduce redundant information in the network. This design optimizes the use of bandwidth, turning it into largely suitable for MANET.

The basic functionality of NRLOLSR is to detect one-hop neighbor nodes and to choose among a group of Multipoint Relay (MPR) nodes, which will broadcast control messages which the topological information of the emitter node. These messages communicate the link status of each node, which may have four values:

1. *Symmetric*: it means that the link has been verified as bidirectional.
2. *Asymmetric*: the node can only listen to the one-hop neighbor; however the link has not been confirmed as bidirectional.
3. *MPR*: indicates that the node has been selected as MPR. This status indicates that the link is bidirectional.
4. *Lost*: there is no link.

The route between nodes is calculated according to their link status. This control information is updated periodically via the messages broadcasted by the MPR nodes without the intervention of any central authority, i.e. operating in a complete distributed mode.

Some major improvements implemented by NRL on the original OLSR are: support for IPv6 and multi-platform, interval for HELLO messages, update interval of link status, configurable trace level... NRLOLSR runs from the command line, although in this project it is released through the *start-scripts* of MNE. NRLOLSR supports many input arguments and in Table 3 we detail the parameters that have been used.

Table 3. NRLOLSR configuration parameters

Option	Description
-ipv4	IPv4 addresses will be used
-i	interface protocol used for the broadcast of control messages
-qos	QoS for OLSR packets
-d	tracing level
-l	path for trace file
-al	*All link options.* Sets TC_REDUNDANCY[1] equal to 2
-flooding	sets options for forwarding messages

[1] level of information in control messages; if equals 2, lists all nodes with symmetrical links

Visualization: CMap-1.0a10 and Sdt-1.1a2

With the purpose of illustrating the network status, in SimuMANET we have used two tools developed by the NRL group PROTEAN: CMap-1.0a10 and Sdt-1.1a2. In both cases, these tools consist of open source and free distributed software.

CMap takes information from the trace files of MNE, MGEN and NRLOLSR and poured into a file format compatible with SDT (Scripted Display Tool), which is the tool that deals with the graphical display. CMap supports two operating modes: *real*, under real-time conditions, or *play*, based on stored simulations.

In addition to obtaining the status information of the network nodes, CMap sets some properties such as the images to use in the visualization, coordinates of the simulation area, and addresses of the nodes. For this purpose, two configuration files are used:

1. *cmap.cfg*: to set the node name, its IP address and MAC, the icon that represents it on the map and its size in pixels.
2. *sdtsetup.cfg*: to choose the images of the map, to determine if there are physical obstacles in the radio link, and to set the coordinates of the upper-left and bottom right of the area to be displayed.

Among all of the possible commands available in CMAP, SimuMANET has made use of those described in Table 4. In order to automate its use in SimuMANET, CMAP runs through the *cmapAuto* script obtained from the *mne-scripts*. This script has been modified to allow as an input parameter the path to the trace files generated by MNE, MGEN and NRLOLSR. This modification facilitates the visualization of the saved simulations by using the stored results for a given scenario.

The second tool used for visualization is SDT (Scripted Display Tool) which graphically shows the network status. It is developed in C++ and makes use of the *wxWidgets* library, so it is only available for Linux platforms. By means of a window, SDT plots the network status.

Figure 3 illustrates how a simulation scenario is visualized in SDT, in this case from the point of view of node01. The visualization window shows the link status between the network nodes. We distinguish between symmetrical and asymmetrical links for the color that represents the link. The color for each case is user-configurable.

On the other hand, the color of the name tags indicating whether the node from being monitored network traffic received by other users. Green indicates that message exchange of between the two, red and yellow no traffic represents the moment of uncertainty when you start to lose the link between the nodes. In the example of Figure 3, we see that the only node in red is the actual node01 because it does not receive traffic from itself.

For more detailed information about any user participating in the simulation, a double click on the icon that represents it will bring up a window. This window displays the node name, status, their

Table 4. CMAP configuration parameters

Option	Description
-realtime	real time operating mode
-speed	visualization speed
-olsr	path for NRLOLSR trace file
-mne	path for MNE trace file
-mgen	path for MGEN trace file

Figure 3. Example of network status visualization using the SDT tool

IP addresses for each interface, the MAC address corresponding to the test interface and the two attached networks, i.e. the MNE test and control networks.

We must clarify that the visualization is sometimes not entirely accurate, as they SDT refresh rate of maybe not enough to track the rapid changes undergone by a link or a node. Another limitation to consider is that the obstacle representation is merely indicative. Obstacles in MNE are defined as a line between GPS coordinates (longitudes and latitudes initial and final) and, on the display window, they are drawn at the midpoint of that line.

Post-Analysis: TCPDUMP, TRPR, and Gnuplot

The analysis functionality has been implemented by combining three free distributed tools: TCPDUMP, TRPR and GNUPLOT.

TCPDUMP is a tool that captures and displays the user traffic flowing through the network. It was developed in the *Network Research Group Lawrence Berkeley Laboratory* and distributed under a free software license. It was originally developed to run on Linux systems, although that a version for Windows (WinDump) is available. In Linux, root privileges must be ensured prior to run. TCPDUMP supports a variety of commands that enables to set filters to select packets according to a particular attribute: source address, destination port, protocol, etc. This project has exclusively used the Xcom node to capture all packets received at the test interface.

TRPR (Trace Plot Real-Time) is a tool developed by NRL. The version used in this project is trpR-2.0b2.

TRPR analyzes the output files from other programs such as TCPDUMP or MGEN and generates a suitable output to be plotted.

This program has been used through the MNE post-analysis scripts. These scripts are designed to create files later used by GNUPLOT to generate graphs with metrics of interest. These files are

Table 5. TRPR configuration parameters

Option	Description
-drec	TRPR expects to process a *drec* trace file instead of *tcpdump*
-input	specifies the full path of the input file
-output	specifies the full path of the file used to save the TRPR output
-auto	calculated values differ by flows
-flow	total values are calculated (without differentiating by flows)
-latency	estimate the latency
-loss	estimate the losses

the result of running TRPR with the appropriate options for each case. The Table 5 shows the commands of TRPR used in the SimuMANET implementation.

GNUPLOT is a multiplatform software tool valid to generate graphs from functions and data. It is distributed under a free software license that allows copying and modifying its code. However, any change or modification of the original code can be distributed only in patches making it incompatible with the General Public License (GPL). The SimuMANET project has used GNUPLOT to generate graphics from the output files of TRPR.

Qt4

For the development of SimuMANET we have made use of Qt 4.4.2 integrated into the open source toolset *Eclipse* by including the appropriate Qt plug-in.

The Qt tool consists of libraries developed by the TrollTech group. It is distributed under free or commercial license. The SimuMANET project has used the free license version. The Qt plug-in for Eclipse provides a graphical editor which facilitates the design of the dialogues that form the application.

The Qt libraries, have been developed in C++ and its main objective is to build graphical user interfaces. They also include support for 2D drawings, threads, databases, and others.

One of the most interesting features of Qt is that it adds to standard C++ the concepts of 'signal' and 'slot', which implement a functionality similar to *callback* functions. A 'signal' is the indication emitted by an object when an event occurs. The 'signals' are associated with 'slots' consisted of functions that are executed when the 'signal' associated with a 'slot' is captured.

The dialogs were built using *QWidgets* calls, consisted of graphical components equivalent to the 'controls' in other programming environments.

SIMUMANET: TOOL DEVELOPMENT

In order to map the aforementioned functionalities, the SimuMANET project accomplished the implementation of six main modules:

- Networking setting module
- Motion module
- Traffic module
- Visualization
- Post-analysis module
- Start-up module

The description offered in the following subsections tries to illustrate the main features of each module and its relation to the NRL software. The set of these modules is the SimuMANET tool grouped under just one star-up interface.

Network Setting Module

The network configuration module integrated into the application SimuMANET provides a graphical user interface for generating the network configuration files "nodelist," which lists all the nodes participating in the ad hoc mobile network, as well as for setting some of the parameters required to launch various NRL programs herein used for simulating the MANET (MNE, NRLOLSR and MGEN).

In the "Network" menu of the SimuMANET main screen, it is possible to access the dialogs to configure the network and NRL software parameters respectively clicking on "Nodelist setting" and "NRL SW setting."

Network Setting (Nodelist)

Selecting "Nodelist setting" enables the "Network creation and configuration (nodelist)" menu, shown in Figure 4. The MNE software for MANET simulation reproduces a MANET from the information contained in the nodelist files. A nodelist file defines the network to be simulated, listing all the nodes and the related information necessary for configuring MNE scripts, IP addresses, MAC, and others.

Within this dialog, "Network creation and configuration (nodelist)," it is possible to create, edit, copy, delete and save nodelist files.

When a nodelist file is selected, the list with the names of the nodes and the information associated to them appears in a table. Clicking on each node it is possible to create, modify and delete nodes. Later, SimuMANET can create, copy, edit, delete and save the nodelist file, selecting "Files Nodelist."

Node Configuration Options

A nodelist file configures the MANET. It lists all the nodes with their IP addresses from test and control, MAC address of test and test and control interfaces.

The first node of the list corresponds to the node "Xcom." This is the node that launches the network simulation, and also the node in charge of centralizing and collecting all the information of the MANET once the simulation is done. The other nodes are called node*nn*, with *nn* consecutive integers, starting with node01. SimuMANET can create new nodes, edit and delete nodes, using the editor seen in Figure 5.

Create New Nodes

To create a new node must click on the table nodes and then in the menu button "New." SimuMANET displays a node editor screen where all the fields can be inserted except the node name. This field is automatically filled in by the application. The fields shown in this editor are:

1. Node name: not user-editable parameter.
2. Test IP Address: IPv4 address of the interface to be analyzed, i.e. the IP address associated with the WiFi interface.
3. Control IP Address: IPv4 address of the network used by MNE for control. We recommend the use of a not congested reliable network.
4. Test MAC address: MAC address blocked by MNE to simulate missed connections. MAC address associated with the test interface.
5. Interface test: name of the WiFi interface.
6. Control Interface: name of the interface of the control network.

NRL Software Settings

As aforementioned, SimuMANET is a graphical application that allows the simulation and configuration of MANET networks, using programs developed by the NRL. The execution of the NRL software necessary to simulate a MANET, is accomplished using scripts. These scripts take some

Figure 4. "Network" menu of SimuMANET and "Network creation and configuration (nodelist)" dialog and new node editor

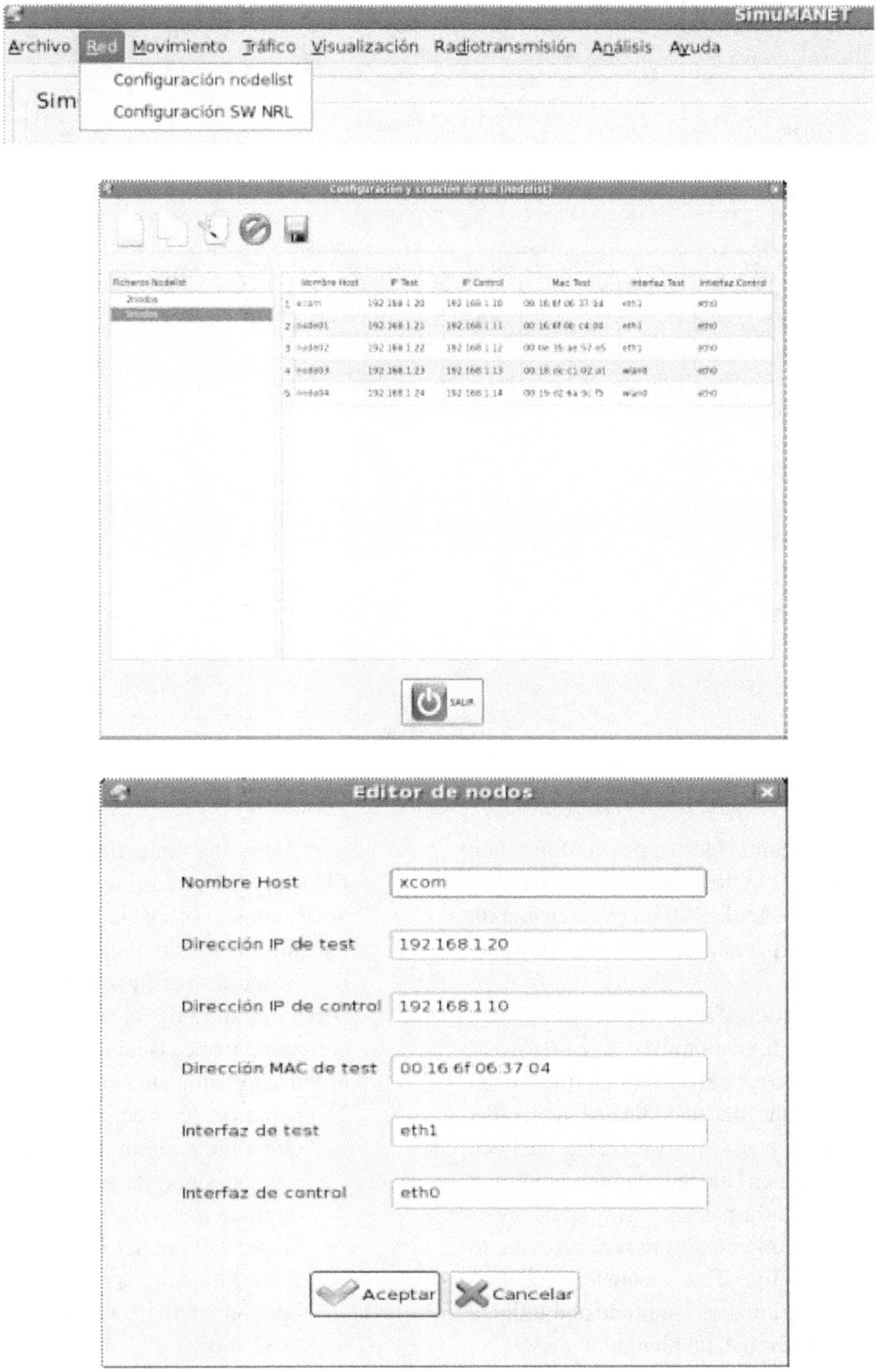

Figure 5. "MNE SW setting" dialog

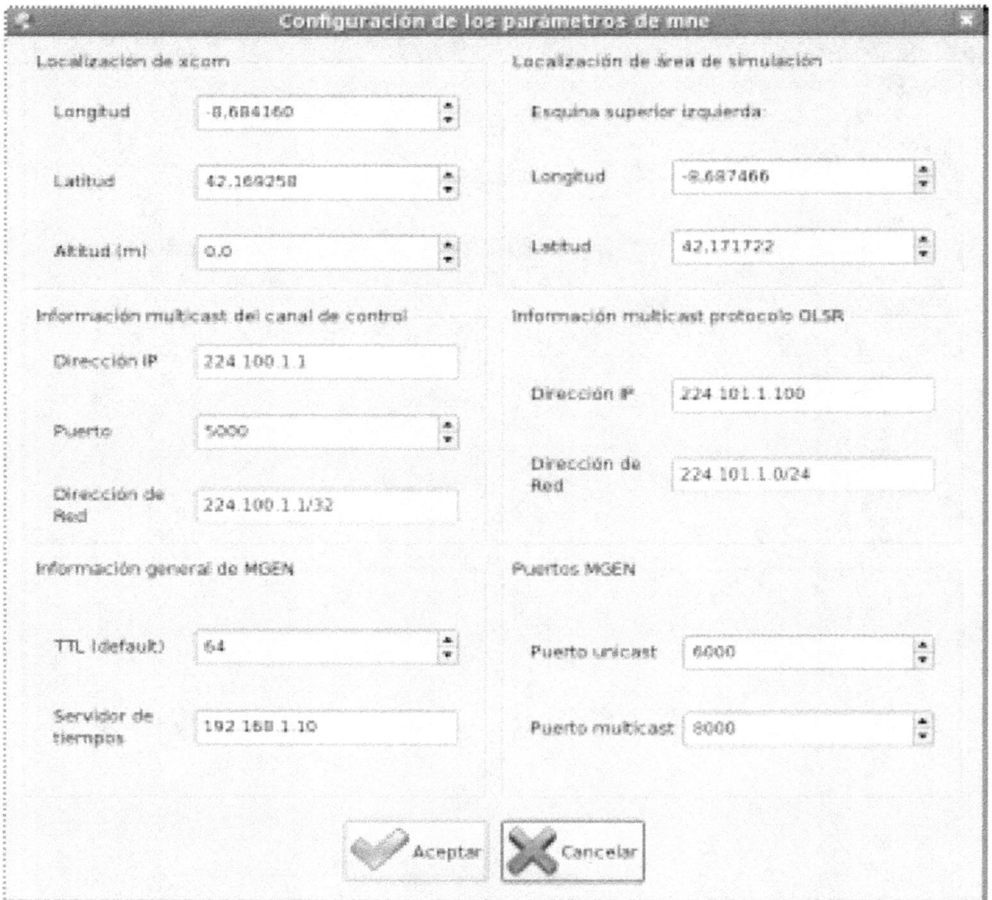

of the execution commands from a configuration file called *mne.cfg* in the "setup" directory.

The "NME SW setting" dialog mne enables to view and edit the parameters of the *mne.cfg* file:

1. Default location of the node "Xcom." When launching a simulation all nodes have associated a motion type and a motion file. A node will follow the motion indicated in this file. There is the possibility of forcing the node "Xcom," which launches the simulation and centralizes information, to stay static at one point while other nodes move according to the motion file. The parameters required for the location are: longitude and latitude in decimal format, and height in meters.
2. Location of the simulation area. In a motion file with NS2 format, the points defining the node path are not absolute but relative to the location of the simulation area. In a NS2 file, the location of the area simulation is not specified, but only its size. Then, to make correct network visualization, the location of the simulation area must be specified.
The parameter that defines the location of the simulation area is the upper left corner, since its size is known. The latitude and longitude of the corner upper left are set in decimal format. This parameter is only relevant if the motion type selected for simulation is NS2 format.

3. Multicast information of the channel control for the MNE tool. The MNE software runs distributed on every nodes of the network. The MNE nodes use a channel control to send messages about their position, MAC address, parameters radio, and others. This message is performed by multicast traffic on the control channel. In *mne.cfg* file it is specified the multicast address, port and network used by the nodes for this purpose.
4. Multicast information of the routing protocol OLSR. The routing protocol uses multicast to disseminate control information, network topology information, to find out neighbor nodes, etc. The network and address used by NRLOLSR are parameterized in the *mne.cfg* file.
5. Default TTL value. The time to live 'Time To Live' (TTL) indicates how many nodes can spend a bundle before being discarded by the network. This value is a posteriori editable for each traffic flow through the traffic generation file.
6. IP address of time server that will be used to synchronize the network. It is important that the nodes have their clocks synchronized to reliably analyze the network latency.
7. Receive ports taken by default for the reception of traffic generated by MGEN. This parameter has no relevance since that in the files of MGEN traffic generated by SimuMANET the receive ports should be specified.

After modifying the data, clicking on the OK button will update the changes in the file *mne.cfg*.

Motion Module

One of the main features of a MANET is the mobility of its users. To analyze a scenario that uses this type of network, it must the movement of its nodes must be simulated and SimuMANET allows setting different motion patterns for this purpose.

In a previous section we detailed the basis open source software tool considered in the SimuMANET project, the BonnMotion. Each motion model of the BonnMotion tool has been implemented through a dialog that creates the configuration files used by MNE to know the node positions in each moment of the simulation. The access to these dialogs is performed through the main screen menu, on the item 'Motion', as shown in Figure 6.

Traffic Module

Sending and receiving traffic in MANET simulations is done using the MGEN software developed by the NRL. This software uses script files that define the traffic flows for each node. At present, MGEN only sends and receives UDP traffic.

The SimuMANET traffic module provides a graphical user interface to generate, modify and delete script files which define traffic patterns, by accessing the "Traffic" menu in the main application screen.

Figure 6. Motion module in SimuMANET

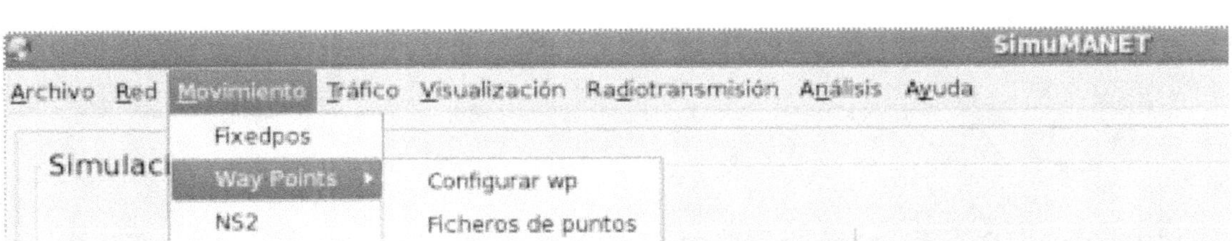

There must be a traffic generator script file on each node to perform a simulation with the traffic generation software MGEN. All traffic generator files must have the same name and extension **.mgen* in the "mgen_scripts" directory.

In order to allow each node having its own traffic pattern, to distinguish the network flows, and more, SimuMANET creates script files with the extension **. mgen.<node_name>*. The developed application stores in the system, within folder that defines the traffic type, the traffic scripts files of every node included in a given nodelist. Hereinafter, we define as "traffic type" the folder where the nodelist traffic scripts files are stored. The folder is stored on disk as *typeTraffic.nodelist*. In the network configuration process of the SimuMANET start-up module, the traffic generation script for the selected traffic type will be copied to each node.

Traffic Dialog

Traffic types are associated with the nodelist, which provides the number of nodes, node names, and IP addresses to copy in every node with the appropriate file name. This dialog lists the existing nodelist, the traffic types and the associated scripts files, as can be seen in Figure 7.

Once you select a nodelist, clicking on the control "Traffic types" enables menu buttons to copy, delete, save and edit traffic generation scripts for each node. Clicking on the "New" button, SimuMANET displays a dialog to enter a name for the type of traffic.

This creates a type of traffic and an empty script file for each node. The folder "Traffic type" and its files are not stored in the system until you press the save button. This feature of "unsaved" is marked with a red square to the left of the type of traffic.

To store a traffic type, all the related files must be edited. The "Traffic" dialog represents the empty or unedited nodes with icons different from the edited files or not empty, as shown in Figure 8.

Traffic Generation Script File

This dialog enables to individually edit traffic generation scripts files for each node in the network. Edition is the only action allowed on generating traffic scripts. There are two types of events: reception events and events related to the transmission flows. This dialog allows viewing and editing the reception events and events of transmission flows. For each flow, this dialog

Figure 7. "Traffic" menu

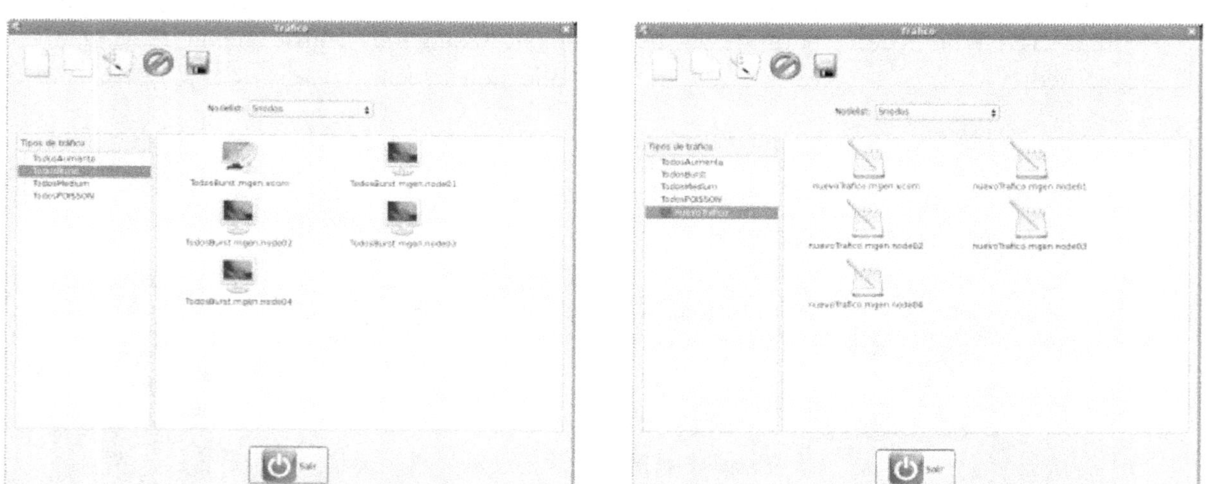

Figure 8. Transmission flows and flow editor

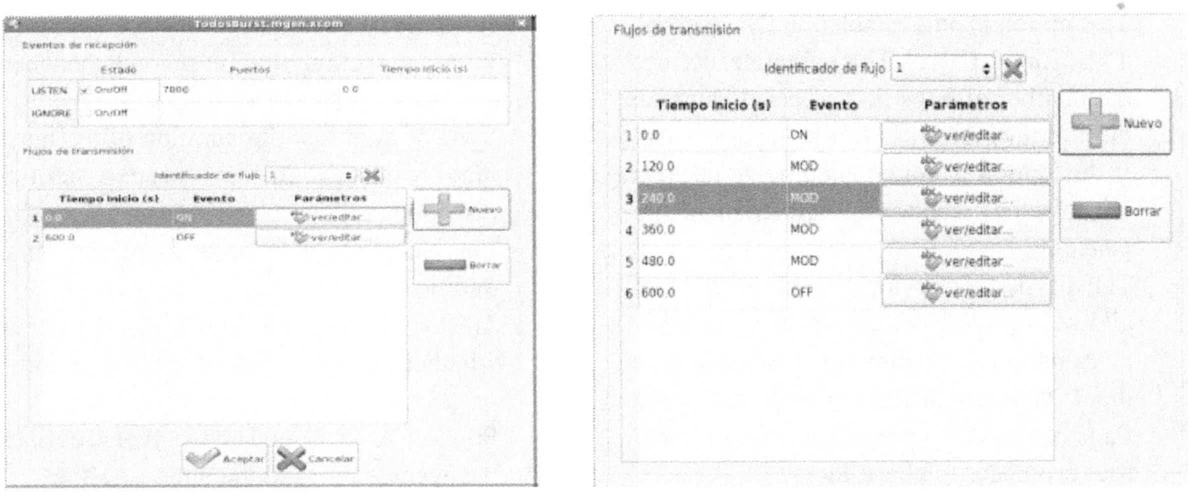

lists the event types and the time of simulation when they occur.

Reception Events

There are two reception events, IGNORE and LISTEN, whose parameters are the ports and instants when the event starts, in both cases.

1. LISTEN, opens at the specified instant a listening UDP socket on each input port. To enable the LISTEN event, the checkbox should be selected.
2. IGNORE, from the time specified in the time field, the traffic sent to the input ports is ignored. IGNORE checkbox should be selected to introduce this type of event in the traffic generation script.

Start times are specified in seconds and decimal format separated by commas. The ports are separated by a comma or a hyphen, if a range is defined.

Transmission Flows

Transmission events are related to traffic flows. A flow is exclusively identified on the network by their unique identifier, allowing analyzing the results individually for each flow once the simulation is finished.

We call 'flow' to a set of events with the same identifier.

In the option "transmission flows," the transmission flows defined in the related script can be created, modified and displayed. There are three types of transmission events to view, add, delete or modify: ON (onset of flow), MOD (modification of a parameter of a flow) and OFF (order flow).

General Options of Events

The options of the events define different parameters which can be mandatory, optional or not applicable:

1. DST (Destination): configures the IP address and port of destination node that is the node sending the traffic flow. The drop-down control "destination node" lists all nodes defined in the nodelist with their respective test IP addresses.
2. SRC (Source): optionally, the traffic transmitter node can be forced to do it on a specific port. To do this, select the SRC parameter and enter the port. If this option is enabled

the traffic is sent through any available port. This option is only enabled in ON events.

3. TTL (Time To Live): This parameter defines the number of hops that a packet can take through the network before being discarded. If this option is not enabled, UDP packets will take the default value defined in the file *mne.cfg*.
4. TOS (Type-Of-Service): IP byte header that allows a packet prioritization over other marking some of their flags. You must enter the TOS value in hexadecimal and some flags of the TOS byte require "super user" root privileges to check them.

Traffic Patterns

The traffic pattern defines the type of traffic, the number of UDP messages, message size, and so on. The parameter "traffic pattern" is a required field ON event. SimuMANET can generate three different types of traffic:

1. PERIODIC: this option generates a periodic traffic pattern, which specifies the number of UDP messages per second that are sent and the size in bytes of messages.
2. POISSON: this option generates a traffic pattern with a Poisson distribution. The "rate" indicates the average frequency of the Poisson distribution for sending messages. The message size in bytes must also be specified.
3. BURSTS: this option generates burst traffic. This type of traffic is generated by intervals of continuous traffic delivery, time which we call burst, and intervals of silence. An example of this type of traffic pattern is generated by VoIP applications. The bursts start at random instants and their duration is also random.

Different fields to be completed for this type of traffic are:

a. Burst type: defines whether the burst is scheduled or random.
b. Interval: average time in seconds between two burst traffic deliveries.
c. Type of duration: the duration of the burst may follow a uniform or exponential distribution.
d. Length: average in seconds of the burst duration.
e. Traffic pattern, rate and size: it defines whether the traffic pattern will be periodic or
f. Poisson, as explained above. It also defines the average rate and the omessage size.

Characterization of Events

An event is mainly characterized by the type and the moment when it occurs (Figure 8-11). There are three different types of events with three very different functions:

1. ON: event marking the start of the traffic flow delivery. Their mandatory fields are therefore the destination (DST), the node that receives the traffic, and the traffic pattern to send.
2. MOD: event to adjust a parameter of the flow. From that moment, sending
3. traffic is done by modifying the parameter. This event can change all parameters (TSS, TTL, etc.) except the source port (SRC). MOD event can modify just one parameter. If it is necessary to modify more than one, MOD would create another event to be launched at the same time modifying the other parameters.
4. OFF: event for sending traffic flow. This event has no options, but the time when it occurs.

The Table 6 summarizes which fields are mandatory, not optional and configurable for each of the events.

Figure 9. Editor for transmission events and reception events

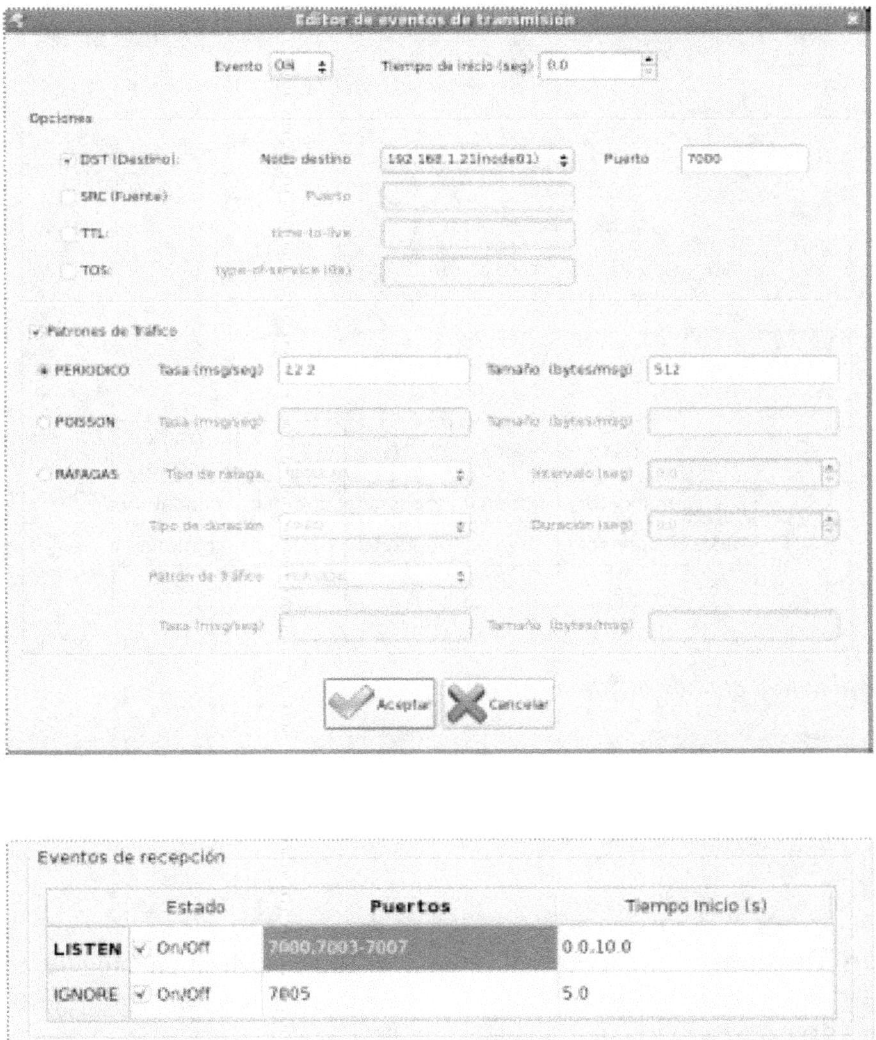

Figure 10. General options event and TOS byte

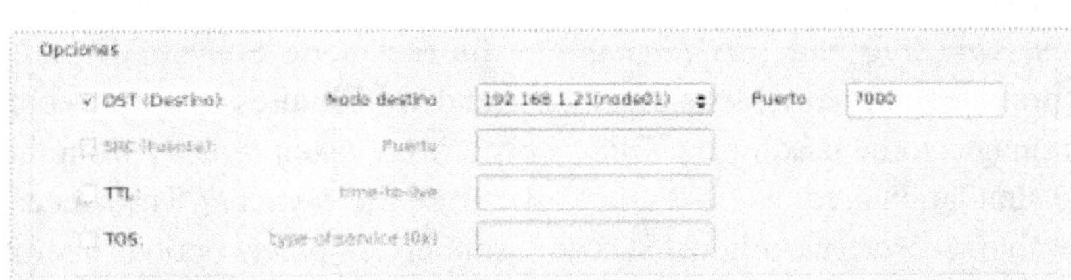


```
   0     1     2     3     4     5     6     7
+-----+-----+-----+-----+-----+-----+-----+-----+
|   PRECEDENCE    |          TOS          | MBZ |
+-----+-----+-----+-----+-----+-----+-----+-----+
```

Figure 11. Traffic generation editor and Table 6

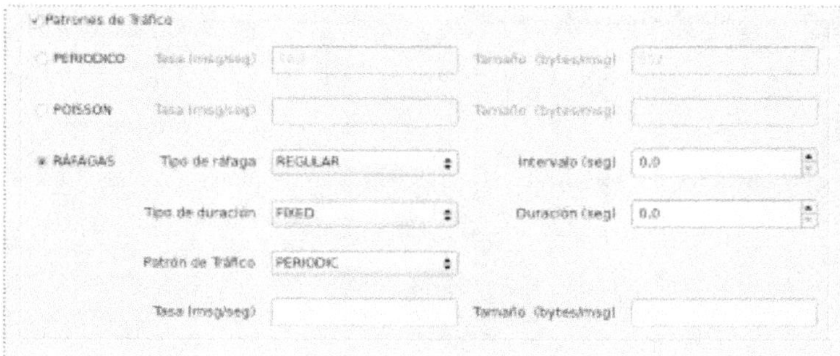

	DST	SRC	TTL	TOS	Traffic pattern
ON	mandatory	optional	optional	optional	mandatory
MOD	optional	---	optional	optional	optional
OFF	---	---	---	---	---

Figure 12. Visualization module in SimuMANET

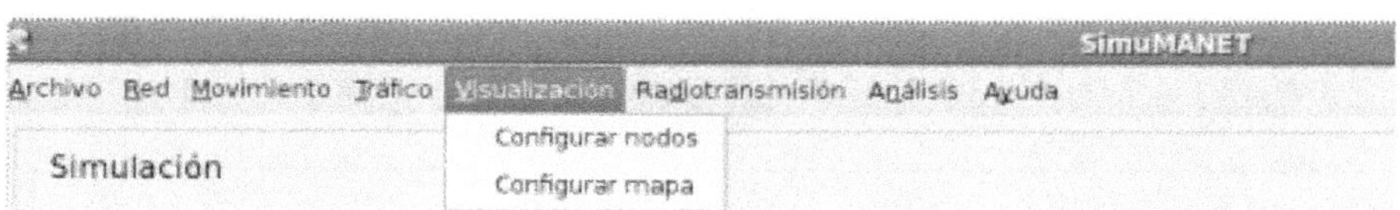

Visualization Module

We have used the SDT and CMAP tools to graphically monitor the simulation evolution, enabling to control CMAP and SDT with *mne-scripts*. Both SDT and CMAP programs require certain configuration files (*cmap.cfg* and *sdtsetup.cfg* respectively) that provide some necessary parameters, such as the images to be used or the GPS coordinates of the simulation area.

It was indispensable to correct the original SDT code which did not properly read the configuration file *sdtsetup.cfg* causing a runtime error and that the visualization window stayed empty. The *genSdtCfg* and *genCmapCfg* scripts were also modified to admit as an input argument the path where the generated configuration file is stored.

The visualization module consists of two graphical dialogs valid to create the configuration files, as shown in Figure 12. The 'Node visualization setup' dialog creates the file *sdtsetup.cfg* and the 'Map setup' the file *cmap.cfg*.

The file *cmap.cfg* sets the visualization feature for each node. Some of these features, such as IP and MAC addresses are not configurable because they are taken directly from the nodelist related to the file *cmap.cfg*. The user can just modify the merely graphical options on this file, such as the icon representing each node in the display layout and its size. The *cmap.cfg* file is located in the path: */home/mne/tools/cmaphome/*. The syntax for *cmap.cfg* is:

Figure 13. 'Node setup' and 'Map setup' dialogs in the SimuMANET visualization module

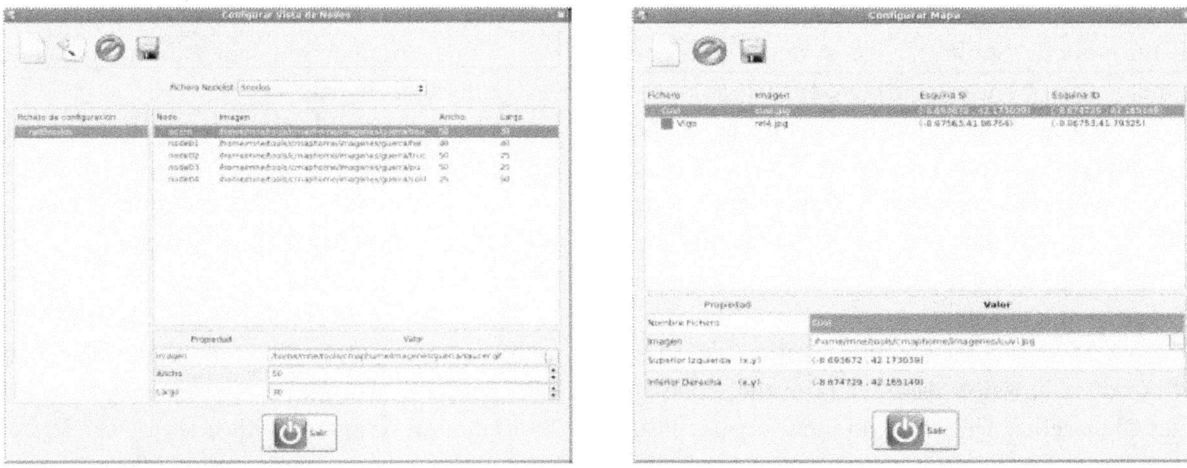

IP node name dirTest dirControl dirMacTest MAC IMAGE IMGSIZE pathImage width length

The other file required for a correct visualization is *sdtsetup.cfg*. The 'Map Setup' dialog enables to configure it. This dialog chooses the image of the map where the nodes are represented, as well as the GPS coordinates that define the simulation area. It can be linked to any nodelist, as the map representation is independent of the network configuration. The simulation software searches the file *sdtsetup.cfg* in the path: */home/mne/tools/cmaphome/*. The syntax of this file is:

imagen_mapa bgimage

bgbounds longitude_cornerSI, latitude_cornerSI, longitude_cornerID, latitude_cornerID

where the symbols 'SI' and 'DI' mean upper left and lower right corners, respectively. The longitudes and latitudes are entered as decimal numbers in English notation (with decimal point) and separated by commas.

Radio Transmission Module

The MNE software for MANET generation enables through the radio and obstacles files, to parameterize the radio values of the system and each of the nodes, as well as the location of obstacles in the simulation area.

SimuMANET implemented in the radio module, two dialogs that enable the creation and configuration of radio and obstacles files. These dialogues are accessed from the menu on the main screen by clicking on "Settings" and "Obstacles," respectively, as shown in Figure 14.

Figure 14. Radio transmission module in SimuMANET

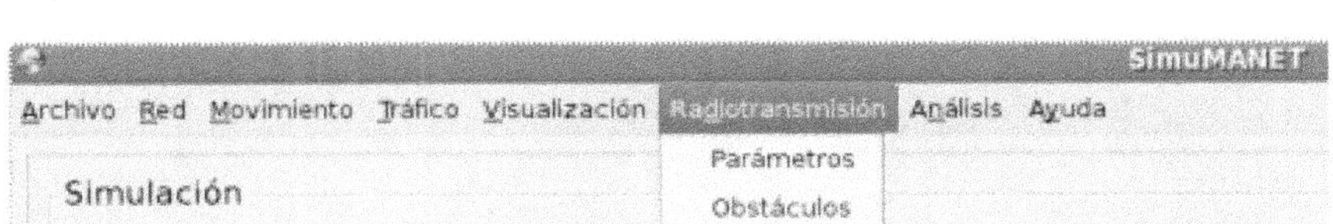

Radio Transmission Parameters

In the dialog "Settings," the system radio parameters can be user-configured, such as operating frequency, loss, sensitivity threshold, the link margin and the ground/air threshold. The user-configurable radio parameters of each node participating in the MANET are: transmission power, node antenna height, transmission and reception antenna gain.

SimuMANET relates the creation of files of radiotransmission parameters to the existing files of nodelist. Therefore, to view, create, edit, or delete radio files, a nodelist should be firstly selected, which defines the MANET network whose radio parameters can be set up.

The default values for the system and nodes radio parameters, used by the MNE software, can be seen in Figure 15.

System Parameters

SimuMANET user-configurable parameters are:

- System working frequency in MHz. Default value is 2.4GHz, as in WiFi systems.
- System losses, in dB.
- Sensitivity threshold or minimum receiving power level, in dBm.
- Connection margin or sensitivity threshold tolerance, in dB.
- Ground/Air threshold sets the height in meters from which it is considered that a node is in the air. Link range of a node in the air is greater than on ground.

The system connection margin ensures minimum values of QoS. The modulation used for WiFi networks is Orthogonal Frequency Division Multiplexing (OFDM) with a set of carriers at different frequencies transferring information by a BPSK, QPSK, 16QAM or 64QAM modulation. Since the parameter "connection margin" guarantees a minimum signal level, then a maximum bit error probability is guaranteed.

Node Parameters

The antenna parameters of each node can be individually set. The default values for these parameters defined by MNE can be seen in Figure15: antenna height, in meters; antenna transmission power, in mW; antenna gain, in dBi (if this value is nonzero, MNE equals the transmission and

Figure 15. 'Radio parameters setup' and 'Obstacles setup' menus in the SimuMANET radiotransmission module.

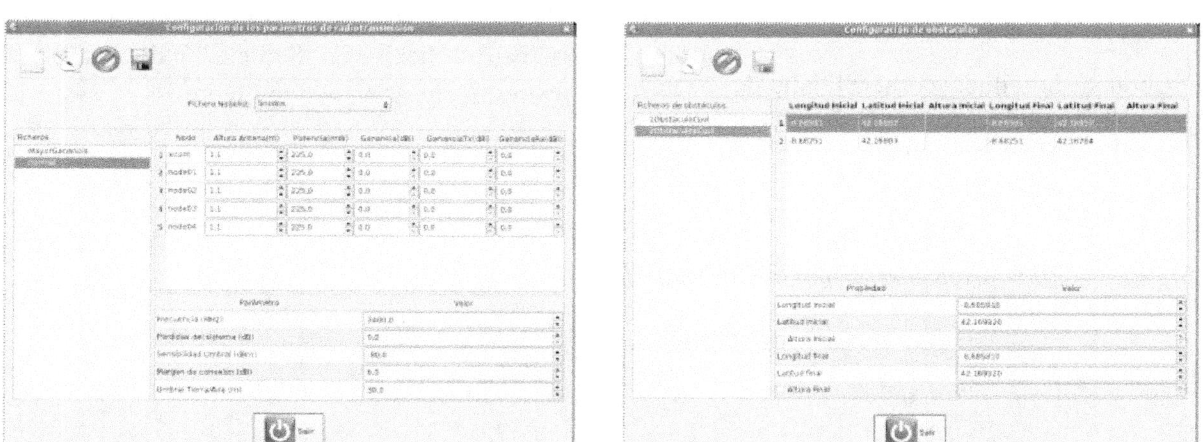

receive gains to this value); transmission gain, in dBi; receiver gain, in dBi.

Obstacles

To simulate the presence of obstacles in the node motion area, the MNE obstacles files must be defined. In a simulation, all nodes know the location of obstacles, enabling to block a link between two nodes when there is no line-of-sight (LoS) between them.

An obstacle file defines the obstacles by specifying the initial and final location, i.e. how long an obstacle extends. This definition forces to think that an obstacle can be represented by a straight line. Therefore, for SimuMANET, one obstacle is a straight line. If a link crosses this line, then it will be blocked.

In addition to an obstacle initial and final location, the height must be also considered. By default and if not specified otherwise, the height of obstacles is endless. In the dialog "Obstacles," it is enabled editing the obstacles parameters that may change their values. By default, the obstacles are not high (infinite). This setting can be modified to set a height, both initial and final. In this menu, it is also possible to create new obstacles or delete an existing one.

Post-Analysis Module

The implemented post-analysis module, Figure 16, allows studying the scenario outlined in the simulation through some of the most significant metrics. For this purpose, the dialogue 'View Results' has been implemented to generate graphs which analyze the network performance. It results also helpful to visualize the simulation again, especially if the work of analysis is performed some time after the test. The dialog 'Saved simulations' enables to recreate the scene observed from the point of view of any node in the network. The access to both dialogs is carried out, as in previous cases, through the menu available on the main screen of SimuMANET.

SimuMANET uses the post-analysis scripts of MNE. These scripts are concerned with creating files of points to be represented by GNUPLOT from the traces given by MNE, MGEN and NRLOLSR.

The mne-script *cmapAuto* was used to implement this functionality. This script integrates the two visualization programs developed by NRL: CMAP and SDT. The script was changed to automate the display function SimuMANET retrospectively. It has been necessary to include files sdtsetup.cfg and cmap.cfg the information stored for each scenario.

Displaying Results

Once simulated a scenario, it is useful to graphically observe some achieved results. The dialog 'Simulation results' generates graphs for the following metrics:

1 *Latency*: delay caused by the holdup in the propagation and transmission of messages. Although there are many factors affecting the delay, in a MANET the positions of the nodes and the network load will determine the delay. It is necessary that the nodes are

Figure 16. Post-analysis module in SimuMANET

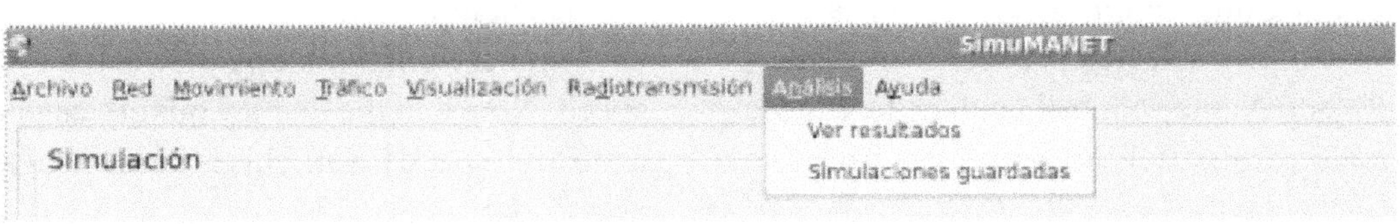

Figure 17. SimuMANET post-analysis module: 'Simulation results' and example of outcome plot.

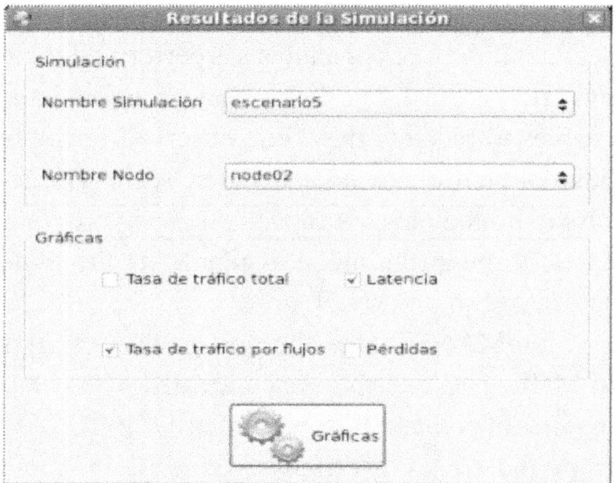

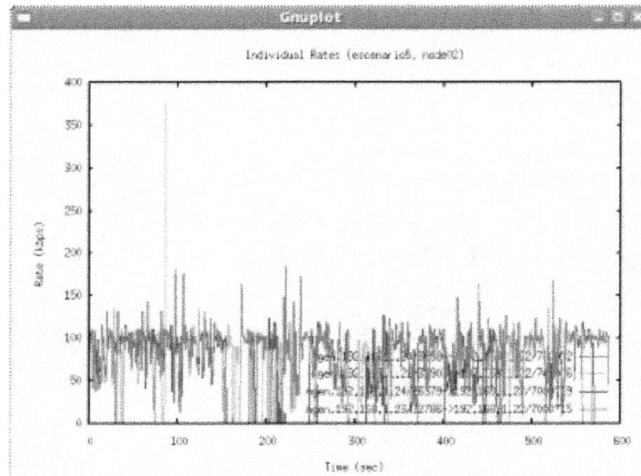

synchronized to obtain an accurate latency value; otherwise, the differences between the clocks of the network users can distort the measurement of the delays and even reach negative values.

2 *Total traffic rate and flow*: measures the bandwidth used at each instant of simulation. It can be obtained the values for each flow or for the whole network. This measure is useful to know the level of network congestion and the effective bandwidth. Obviously, by increasing the number of nodes, the probability of achieving an overload situation also increases.

3 *Losses*: reflects the portion of packet lost when the link between nodes breaks. In this case, different values can be obtained by flows. It is important to know the losses for the system design, since it is a clear indicator of whether the proposed model would be valid in a real application.

Since SimuMANET saves the results of each simulation, the analysis work can be performed at any time. To do this, simply select the scenario to be analyzed. Once selected the scenario of interest, the 'Node Name¡ control enables to choose whether the results will be presented from the point of view of the entire network or the node indicated in the selection control. The user also decides which metrics are displayed. In Figure 17 we show a graphic example of traffic flow rate, which represents the traffic rate of each stream received at node02.

Saved Simulation Results

At the time of study and interpret the results of a simulation, it can be very useful to see again the movement of the nodes, the link status at every moment or the position of the obstacles. SimuMANET implements this functionality in the dialog 'View Saved Simulations."

The application enables to select the scenario to visualize by the control 'Simulation name'. The simulation data that are no longer interested in being kept can be deleted.

Due to the features of a MANET, each user observes a different network topology. Therefore, it is interesting to have the possibility to choose the point of view from which to recreate the simulated scenario. For this reason, it is available a control 'Node name'.

Another configurable parameter is the playback speed. Note that the display quality can be low if a too high speed is set. This is because of a limita-

Figure 18. SimuMANET post-analysis module: 'Saved simulation' and playback example

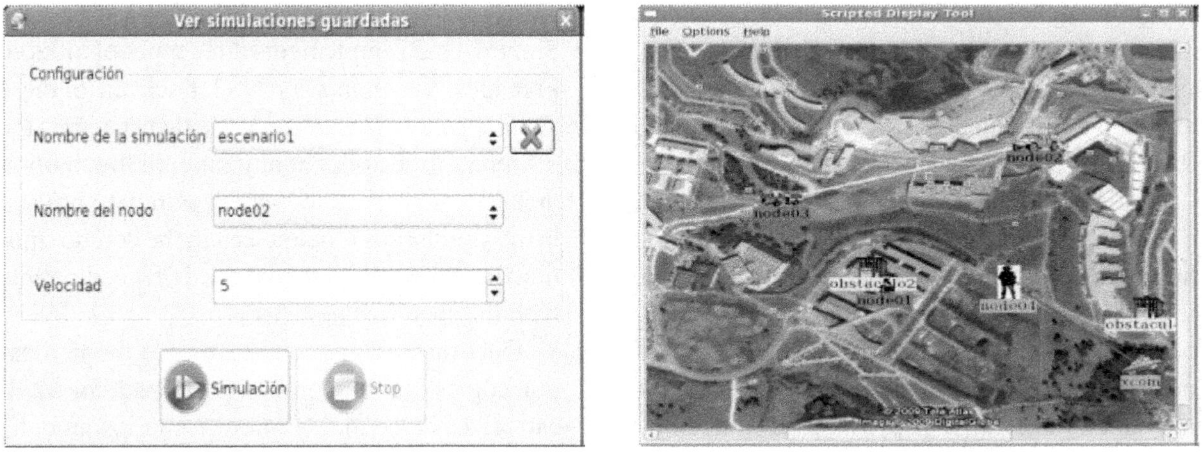

tion of CMAP and SDT regarding the refreshing time which is sometimes not enough to accurately represent the evolution of the simulation. Figure 18 shows an example of visualization of a saved scenario.

Simulation Start-up Module

Once you have created the setup files to configure the network, the movement and traffic of the nodes, visualization, and others, it is possible to start the simulation. The start-up module shown in Figure 19 allows the access to the remaining modules.

SimuMANET can store different settings for diverse scenarios and networks, so that before launching the simulation it is necessary to choose the patterns and parameters of interest in each case.

The main screen of the start-up module is divided in three areas: simulation, visualization

Figure 19. Main screen of the simulation start-up module in SimuMANET

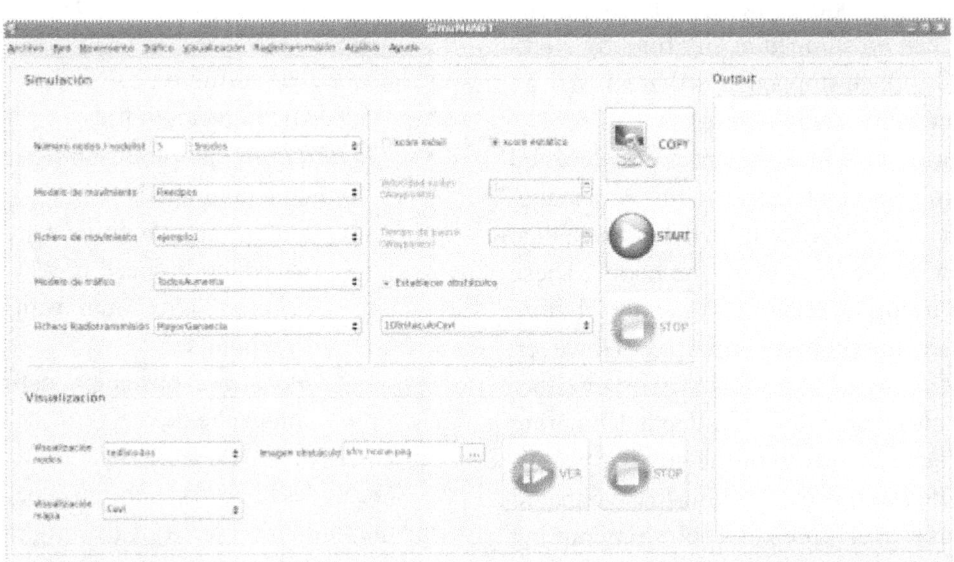

and simulation output. In this main screen it is possible set the specific scenario to be emulated. This screen also provides controls that let copy the configuration to the remaining nodes, start and stop the simulation, as well as start and stop the visualization.

Initially, all the controls except the 'Number of nodes' textbox are disabled. After establishing how many users will participate in the simulation, the other controls are enabled and loaded with appropriate existing configurations to the number of nodes indicated. The user can then establish the movement and traffic patterns, as well as other optional parameters for the simulation launch: Xcom mobility (static or mobile), maximum speed and stop time for Waypoints, and consideration of obstacles in the simulation scenario.

Once set the desired configuration, it is copied to other nodes using the button COPY. After checking that the information was properly distributed by the network, the simulation can be started by pressing START.

As aformentioned, SimuMANET allows graphically monitoring the network status in real time. For this purpose, it is required to select the files that set the visualization of the nodes, the map and the obstacles, if they are any defined.

SEE and STOP buttons at the bottom of the screen, starts and stops the visualization. The simulation can be stopped at any time by clicking on STOP. Once stopped, a SAVE RESULTS dialog is shown to save the results.

In the original NRL software it was detected a critical problem in the shared memory management used by both MNE and MGEN to exchange information on the position of the nodes. These tools were trying to access that memory before it was created and properly initialized. However, this did not cause the programs to abort or either to alert on any error. After an in-depth debugging process to state the cause of the error, it was finally corrected simply by checking that the shared memory was properly prepared before attempting any access.

Motion Pattern Configuration

SimuMANET implements three types of motion: Fixedpos, Waypoints and NS2. Each one of these generic patterns generates a file that specifies the position of the nodes at any time. In the 'motion model' control it can be set the pattern to be used in the simulation. Once selected, the combo 'motion file' shows the pattern related files compatible with the specified nodelist.

Once these two parameters are set, the application copies the 'motion file' into a path for MNE can read: /home/mne/motion_scenarios/model/

Start/Stop the Simulation

The simulation can be started once the configuration is set. With the button START, all the setting parameters are read and the MNE script *startAll* runs with the appropriate commands, as shown in Figure 20.

startAll modMov modTraf -file fichMov -radio fichRadiotx <optional commands>

where:

- *modMov*: movement type (Fixedpos, WP o NS2)
- *modTraf*: traffic type
- *fichMov*: movement file
- *fichRadiotx*: radiotransmission file
- <optional commands>:
    - *mobilexcom*: mobile/static Xcom
    - *speed*: node speed in Waypoints
    - *rotation*: maximum waiting time in Waypoints
    - *blockage*: obstacles defined in the 'blockage file'.

The nodes have to synchronize before launching the simulation software to mitigate the effects of latency. For the tests conducted in this project,

Figure 20. Motion simulation parameters load from the start-up module in SimuMANET

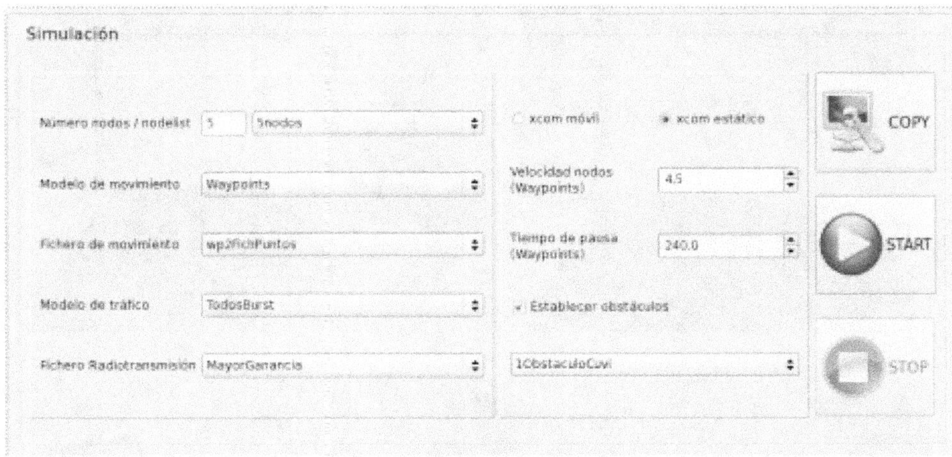

it has been used a local time server (Xcom) with an IP address configured in the file *mne.cfg*.

Saving Results

With the action SAVE dialog, the application gets the outcome data from the simulation. It makes use of the *mne-getDataAll* script that establishes a SSH connection with the remaining nodes to obtain a copy of the trace files generated by MNE, MGEN and NRLOLSR. Information obtained is saved into the path /home/mne/results/simulation_name/.

Within this directory, a folder is created with the name of each node to store their respective trace file. Additionally, the name chosen to save the simulation identifies the simulated scenario in the dialogues 'View results' and 'Saved Simulations' in the post-analysis module.

Visualization Configuration

The visualization software requires specific configuration files. With the control 'nodes visualization' it is chosen the file *cmap.cfg* containing information on nodes. Similarly, in 'map visualization' the file *sdtsetup.cfg* is selected which provides a map depicting the simulation area and the GPS coordinates for upper left and lower right corners.

After selecting the files, they are copied to /*home/mne/tools/cmaphome/* and it results possible to start the visualization by pressing VIEW. This action launches the script *cmapAuto* with the option '*real*', which in turn runs the programs *Sdt* and *CMap*. The '*real*' option indicates that the visualization is done in real time (Figure 20).

Figure 21. Visualization parameters load from the start-up module in SimuMANET

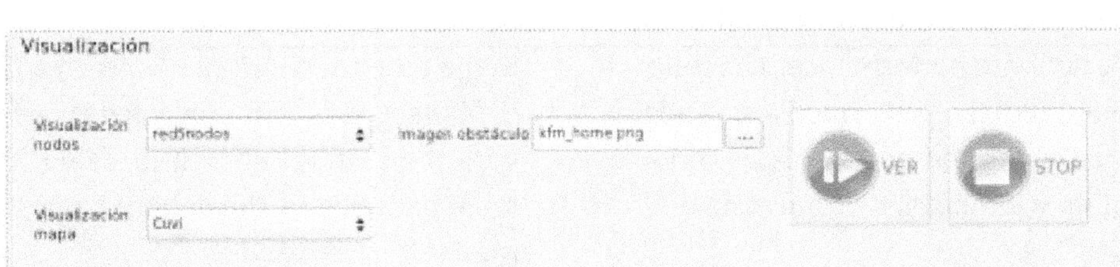

Figure 22. Laboratory test

SIMUMANET: CASE STUDIES

In this section we show the results of the simulations performed in order to check the validity of the tool developed by simulating two different scenarios to discuss results due to motion models, traffic patterns and transmission parameters. The tests were conducted in a five-node network. The picture in Figure 22 shows a basic setup using laptops.

The proposed scenarios emphasize the motion patterns implemented in SimuMANET. The same traffic type, burst, was combined with different motion patterns, waypoints and NS2. These combinations tried to estate the effect of the node mobility in the network performance. The following subsections detail the performed tests and the results obtained using the post-analysis module to generate plots with metrics of interest.

Scenario 1

This scenario presents a situation in which nodes move according to a two-dimensional pattern of waypoints. The routes set in the pattern files correspond to schools and faculties located at the campus of the University of Vigo. The nodes move at a constant speed of 3m/s and stop a maximum time of 240 seconds when they reach the destination point. In this case, obstacles are not included in the simulation area.

Traffic being exchanged by node-users is of type BURST. We have chosen this model because it characterizes much of the traffic present in a real network. Each node transmits four flows destined to the other network nodes, with a time duration if 600 seconds and consisted of periodic bursts with the features shown in Table 7. In terms of radio parameters, the default values set by SimuMANET were used.

Table 7. Scenario #1 traffic pattern and radio interface parameters

Traffic parameter	Value	Radio parameter	Value
burst interval	random, 10s average	**antenna power**	225mW
burst duration	exponential, 5s average	**antenna height**	1.1m
traffic type	periodic	**Gain (tx, rx)**	0dBi
traffic rate	10 messages per second	**Frequency**	2.4GHz
message size	10 bytes	**System losses**	0dB
		Threshold sensitivity	-80dBm
		Link connection margin	6dB
		Air/ground threshold	30m

Results for the Network

Analysis can be performed and plotted from the point of view of the network, or for each node. Following are the results for the network:

- **Total traffic rate:** Figure 23 shows the contribution of the set of nodes to the total traffic rate present in the network. There is a considerable variation, due to both the burst nature of the traffic and the mobility of the nodes. The greatest achieved value was 1.1 Mbps.
- **Traffic rate per flow:** Figure 23 shows the individual contribution of each flow in the network. In total there are twenty flows, four per node, so that the plot can result somewhat confusing. The flow of a single node, given the uniformity of the network, can provide clearer information of the network status.
- **Network latency:** The latency or delay between received messages is due to both the level of network load and the time variable position of the users, as well as the number of hops that were necessary so that each message reaches its destination. Negative latencies are due to the margin of error in the node synchronization. Network latency for scenario 1 is shown in Figure 24.
- **Losses per flow:** In this case, since there are no obstacles and that a situation of congestion is not reached at any time, the losses are only due to the failure of the link between nodes because of the separation distance and/or radio-propagation impair-

Figure 23. Scenario #1: Network total traffic rate and traffic rate per flow

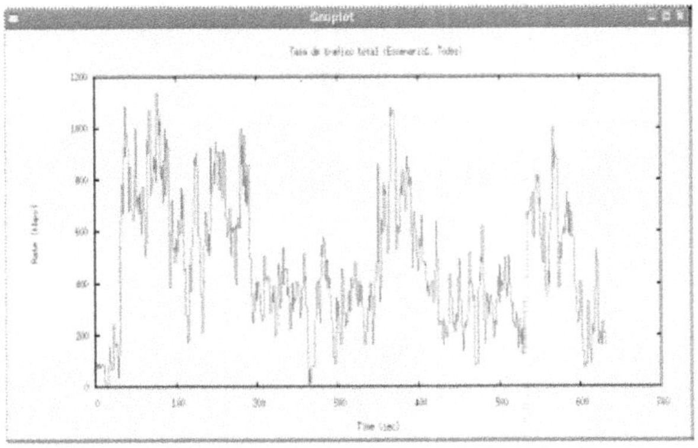

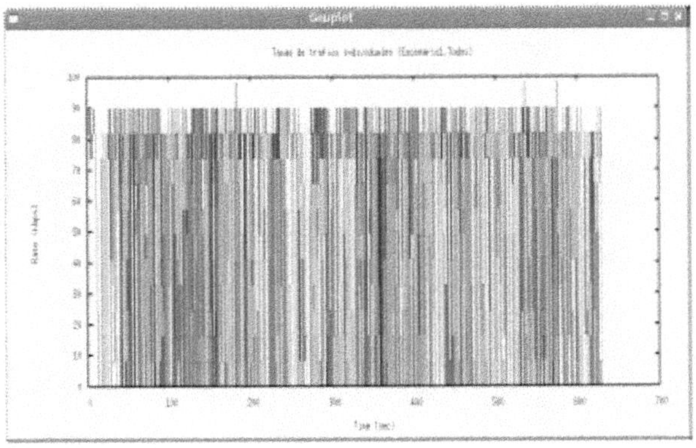

Figure 24. Scenario #1: Network latency and losses per flow

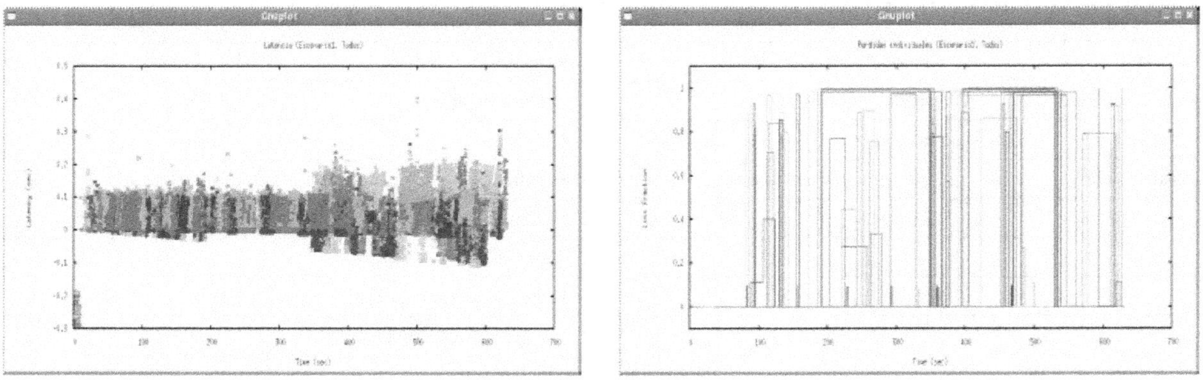

ments. Network losses per flow are shown in Figure 24.

Results for Node01

Below are the results for node01. Since the traffic generated is the same for all node-users and the route tracked is very similar in all cases, the graphs presented are representative of the state network that monitors any network node.

- **Total traffic rate:** in Figure 25 we observe in a clearer way the burst traffic received by node01.
- **Traffic rate per flow:** There is an average traffic rate about 82kbps. In Figure 25, the time interval with no detected traffic from any flow corresponds to the instants when node01 is in a location too far to have a direct link with other users.
- **Latency:** Figure 26 also presents an interval of time when no traffic is received from the other users. The negative latency due to the messages received from node04 is due to the error margin inherent to the synchronization process. Although the difference between the clocks of the two users is the same throughout the simulation, the increased delay in receiving a message is because the path followed by each node makes longer the distance between them.
- **Losses per flow:** The losses experienced in the node01 are an indication of how the MANET connectivity is affected by the node mobility. The plots in Figure 26 show

Figure 25. Scenario #1: Node01 total traffic rate and traffic rate per flow

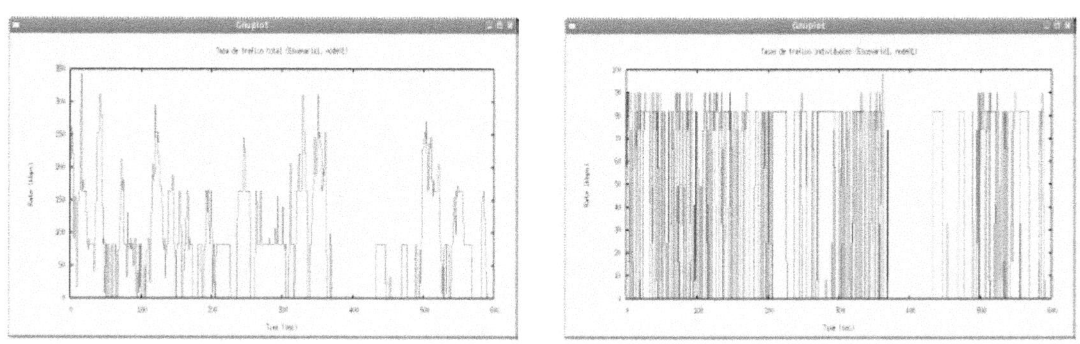

Figure 26. Scenario #1: Node01 latency and losses per flow

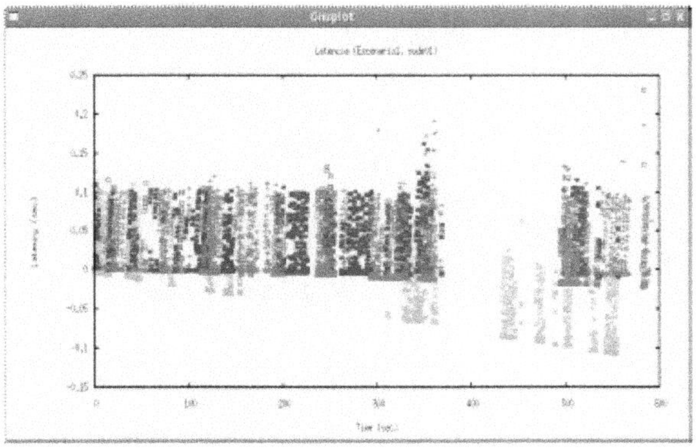

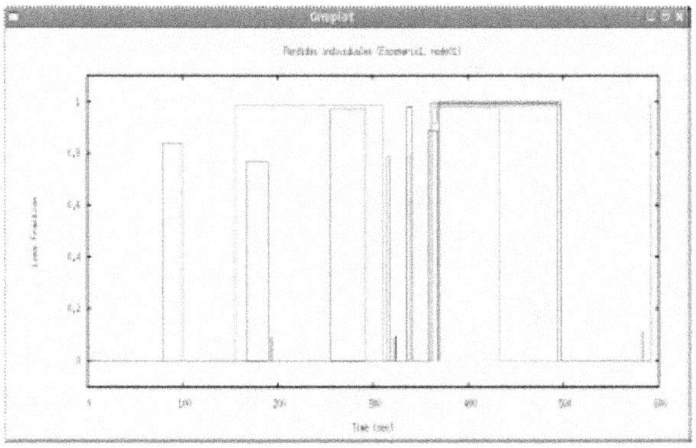

the losses that occur for each flow received on node01.

Scenario 2

This scenario was newly located on the campus of Vigo and the traffic pattern is the same as detailed in the previous section. The radio interface has been fitted with an antenna gain of 3dB keeping the values of scenario 1 for the other parameters. An obstacle was located at the following GPS coordinates (latitude, longitude): (-8.68546, 42.16850) for initial point, (-8.6854, 42.16990) for final point. In this scenario, we have chosen a Manhattan Grid motion pattern. The parameters used to generate the file NS2 are presented in Table 8.

While in Scenario 1, Waypoints motion pattern enables a complete control over the coordinates followed by the nodes, in the case of scenario 2, with a Manhattan Grid model, the motion points are spread randomly over the entire grid set for the simulation area. This can lead to a greater variability in the routes taken by the users that may lead to larger connectivity problems. This issue was tried to be mitigated through the antenna gains.

Since the motion modeling is the main difference with respect to Scenario 1, the results of this simulation allow comparing the validity for both motion patterns, Waypoints and Manhattan Grid.

Results for Network

Below are the results for the network:

Table 8. Scenario #2 motion pattern parameters

Traffic parameter	Value	Radio parameter	Value
simulation time	2000s	**average speed**	5.5m/s
time to ignore	1800s	**speed standard deviation**	0.5m/s
width – simulation area	650m	**probability of speed variation**	0.5
length – simulation area	450m	**probability of turn**	0.6
horizontal blocks number	13	**probability of stop**	0.4
vertical blocks number	9	**maximum stop time**	100s
minimum speed	4.0m/s		

Figure 27. Scenario #2: Network total traffic rate and latency

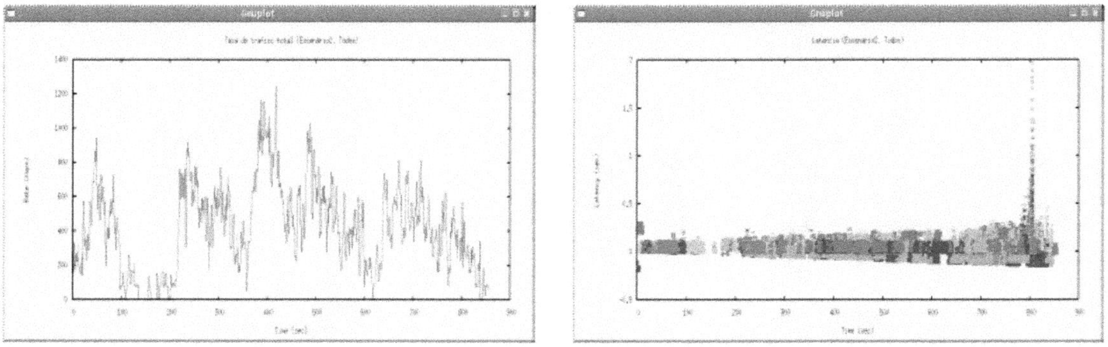

- **Total traffic rate:** while in Scenario 1, in no case a traffic rate of 1.1 Mbps was exceeded, in this case, the peak rate is slightly higher. However, in this scenario the average is less than the average obtained in the first test, as per Figure 27.

- **Latency:** as shown in Figure 28 and as in the previous case, there are negative values due to the error margin of timing. However, in this scenario the delay reaches a maximum value of 2 seconds, whilst for the scenario 1, in no case was exceeded a value of 0.4 seconds.
- **Losses per flow:** given the disparity in the movements and the presence of obstacles, the losses for each node are higher than in the previous scenario.

Results for Xcom

Below are the results for the XCom node:

- **Total traffic rate:** The interval with the traffic absence is due both to both the link loss by the distance from other nodes and the obstacle presence. No significant differences are detected for both the maximum and the average values with respect to the results for node01 in Scenario 1, as shown in Figure 27.
- **Latency:** a homogeneous delay is detected except for the 2-second peak that corre-

Figure 28. Scenario #2: Xcom total traffic rate and latency

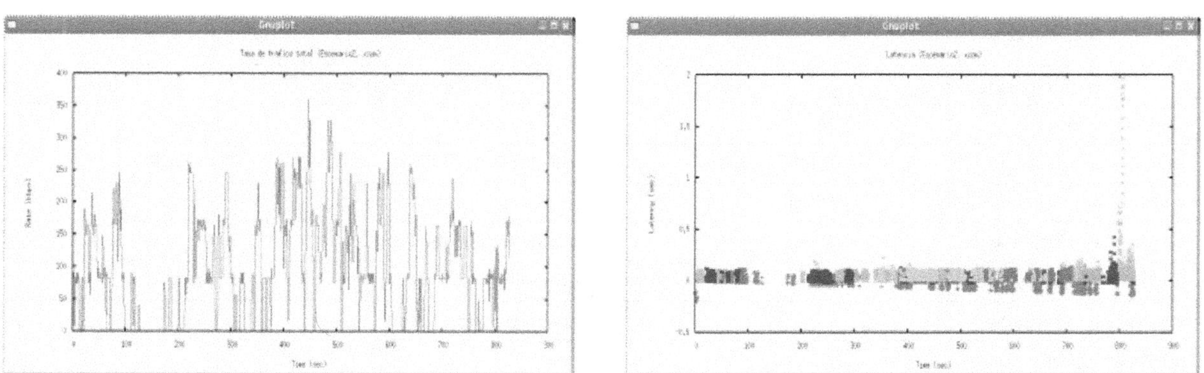

sponds to a maximum rate of one traffic flow (node03). Again, negative values appear due to the error margin that exists in the synchronization process carried out at the simulation start, as per Figure 28.

- **Losses per flow:** with respect to the results obtained in the previous simulation, largest losses are observed in all flows despite of considering an antenna gain of 3dB in transmission and reception.

CONCLUSION

The proposed tool allows analyzing the delay, traffic flows and packet losses for both the full network and single nodes. This functionality enables to determine the causes constraining the QoS accomplishment for a particular network topology, scenario, routing algorithm, and so on.

Despite of its doubtless research value, the tool presents a significant educational dimension. Then, from the educational point of view, the main advantages derived from this experience can be summed-up as follows:

- Multidisciplinary educational features involving part of the student's academic curriculum: mathematics, physics, computer, electricity …
- Enabling students' access to experimental work.
- Curricular reward for involved students through academic publications.
- Introducing students to the methodology of open-source technology, with arising importance in the professional technological fields.
- Enabling the experience to be used in class as a teaching material with large and probed quality.
- Scalability: possibility of extension by incorporating or defining new experiments (new traffic and motion models, routing protocols, radio interfaces...).
- New teaching methodology possibilities based on problem solving, collaborative or project learning.

At present the experience returned to the laboratory in form of practices and it continues in evolution. We will try to incorporate novelties that allow the evolution of the tool, such as differentiating elements for each course that uses this tool.

From the research point of view, it is intended to incorporate in the radio propagation model, the influence of impairments such as vegetation. This kind of networks can find a practical application in environments such as wildfires or some other similar, where a large communication network is needed to ensure information delivery from source nodes to reception ones, regardless of path and traffic conditions.

Real mobile nodes can be incorporated in our tool by using nodes equipped with GPSs and creating the *GPSlogger* files. These nodes retransmit their position to the other nodes, so that it makes possible to estimate the optimal route at any time for a confident delivery of information packets.

The SimuMANET code is be freely available through the website of the Radio Systems research group of the University of Vigo (http://www.sistemasradio.com).

REFERENCES

Chao, W., Macker, J. P., & Weston, J. W. (2003). *NRL mobile network emulator*. (RL/FR/5523--03-10,054).

Naval Research Laboratory. (n.d.). *Products*. Retrieved from http://cs.itd.nrl.navy.mil/products/

ADDITIONAL READING

Ariza, A., Casilari, E., & Sandoval, F. (2000). Strategies for updating link states in QoS routers. *Electronics Letters*, *36*, 1749–1750. doi:10.1049/el:20001154

Ariza, A., Casilari, E., & Sandoval, F. (2001). QoS routing with adaptive updating of link states. *Electronics Letters*, *37*, 604–606. doi:10.1049/el:20010373

Badis, H. (2006). *CEQMM: A complete and efficient quality of service model for MANETs.* 3rd ACM International Workshop on Performance Evaluation of Wireless Ad Hoc PE-WASUN, Spain, October 2006.

Badis, H., & Al Agha, K. (2003). *Distributed algorithm for multiple-metric link state QoS routing problem.* IEEE MWCN2003, October 2003.

Badis, H., & Al Agha, K. (2005). QOLSR: QoS routing for ad hoc wireless networks using OLSR. *European Transactions on Telecommunications*, *15*(4), 427–442. doi:10.1002/ett.1067

Badis, H., Munaretto, A., Al Agha, K., & Pujolle, G. (2004). *Optimal path selection in a link state QoS routing protocol.* IEEE VTC2004-spring, May 2004.

Blake, S., Black, D., Carlson, M., Davies, E., Wang, Z., & Weiss, W. (1998). *An architecture for differentiated services.* IETF RFC 2475.

Braden, R., Clark, D., & Shenker, S. (1994). Integrated services in the internet architecture. *RFC*, *1633*, 1994.

Braden, R., Zhang, L., Berson, S., Herzog, S., & Jamin, S. (1997). *Resource reservation protocol (RSVP) version 1 functional specification.* IETF RFC 2205, September 1997.

Bradner, S. (1997). *Key words for use in RFCs to indicate requirement levels.* IETF RFC 2119, March 1997.

Caro, G. D., Ducatelle, F., & Gambardella, L. M. (2004). AntHocNet: An ant-based hybrid routing algorithm for mobile ad hoc networks. *Proceedings of the 8th International Conference on Parallel Problem Solving from Nature (PPSN VIII),* (pp. 461–470).

Clausen, T., & Jacquet, P. (2003). *Optimized link state routing protocol.* IETF RFC 3626, October 2003.

Gulder, S., & Déziel, M. (2003). *Quality of service mechanism for MANET using Linux.* INSC Symosium, NATO C3 Agency, The Hague, November 4-6, 2003.

Gunes, M., & Spaniol, O. (2002). Routing algorithms for mobile multi-hop ad hoc networks. *International Workshop NGNT,* (pp. 20-24).

Ho, T., Médard, M., Koetter, R., Karger, D., Effros, M., Shi, J., & Leong, B. (2003). *Toward a random operation of networks*. Retrieved from http://web.mit.edu/trace/www/

Lee, S., & Campbell, A. (1988). *INSIGNIA: In-band signaling support for QoS in mobile ad hoc networks.* MoMuC'98, Berlin, Germany, October 1998.

Trzec, K., & Huljenic, D. (2002). Intelligent agents for QoS management. *Proceedings of the First international Joint Conference on Autonomous Agents and Multiagent Systems: Part 3,* Bologna, Italy, July 15 - 19, 2002.

Xiao, H., Seah, W. K. G., Lo, A., & Chua, K. C. (2000). A flexible quality of service model for mobile ad-hoc networks. Proceedings of IEEE VTC Fall, (pp. 445-449). May 2000.

Xiao, H., Seah, W. K. G., Lo, A., & Chua, K. C. (2001). On service prioritization in mobile ad-hoc networks. *Proceedings of the IEEE International Conference on Communications 2001* (ICC'01), (pp. 1900-1904).

Yamazaki, T., & Matsuda, J. (1999). On QoS mapping in adaptive QoS management for distributed multimedia applications. In *Proceedings of ITCCSCC'99*.

Zhao, M., Zhu, H., Li, V. O. K., & Ma, Z. (2005). A stability-based link state updating mechanism for QoS routing. *Proceedings of, ICC*, 33–37.

KEY TERMS AND DEFINITIONS

Latency: measure of time delay experienced by two events in a system. In a network, the latency measures the time used to send information between a source and a destination.

Link: virtual or physical path established between two ends.

Motion Generator: software that reproduces the movement of mobile elements. This movement can be random or fit a known or predictable pattern.

Node: each component of a network system able to generate or well retransmit data traffic.

Open Source Code: software whose source code is published and made available to the general public not subject to copyright, license or domain issues.

Rate: velocity of data traffic exchanged, sent or received on a network interface.

Traffic: data interchange of messages packed according to a definition, protocol or standard.

XCom: central node in a MANET used to observe the system from a general scope.

Compilation of References

Aad, I., Hubaux, J. P., & Knightly, E. W. (2004). Denial of service resilience in ad hoc networks. *Proceedings of ACM MobiCom 2004*, Philadelphia, PA, September 2004, (pp. 202-215).

Abasgholi, B., Kazemi, R., Arezoomand, M., & Enayati, A. R. (2008). Neighbor detection power control mac protocol in mobile ad hoc networks. *Proceedings of 3rd International Symposium on Wireless Pervasive Computing*, Greece.

Abbasi, A., & Younis, M. (2007). A survey on clustering algorithms for wireless sensor networks. *Computer Communications*, *30*, 2826–2841. doi:10.1016/j.comcom.2007.05.024

Abdul-Rahman, A., & Hailes, S. (1998). A distributed trust model. *NSPW '97: Proceedings of the 1997 Workshop on New Security Paradigms,* (pp. 48-60). New York, NY: ACM.

Abusalah, L., Khokhar, A., & Guizani, M. (2008). A survey of secure mobile ad hoc routing protocols. *IEEE Communication Surveys & Tutorials*, *10*(4), 78–93.

Agarwal, S., Ahuja, A., Singh, J. P., & Shorey, R. (2000). Route-lifetime assessment based routing protocol for mobile ad hoc networks. In *Proceedings of the International Conference on Communications* (pp. 1697-1701). New Orleans, LA: IEEE.

Agarwal, A., Ahuja, A., Singh, J., & Shorey, R. (2000). *Route-lifetime assessment based routing (RABR) protocol for mobile ad-hoc networks* (pp. 1697–1701). IEEE ICC.

Aggarwal, D., Dubey, C. K., & Mehta, S. K. (2006). Algorithms on graphs with small dominating targets. *ISAAC 2006 . LNCS*, *4288*, 141–152.

Agha, A. K., & Omidyar, G. C. (2003). *Mobile and wireless communications networks* (2nd ed.). Singapore: World Scientific.

Agi, I., & Gong, L. (February 1996). An empirical study of secure MPEG video transmission. *Proceedings of the Internet Society Symposium on Network and Distributed System Security*, San Diego, CA, (pp. 137-144).

Ahmed, S., & Alam, M. S. (2006). Performance evaluation of important ad hoc network protocols. *EURASIP Journal on Wireless Communications and Networking*, *2*, 42–53.

Akbari Torkestani, J., & Meybodi, M. R. (2010). Mobility-based multicast routing algorithm in wireless mobile ad hoc networks: A learning automata approach. *Journal of Computer Communications*, *33*(6), 721–735. doi:10.1016/j.comcom.2009.11.019

Akbari Torkestani, J., & Meybodi, M. R. (2010b). Weighted Steiner connected dominating set and its application to multicast routing in wireless MANETs. *Wireless Personal Communications*, *60*(2).

Akbari Torkestani, J., & Meybodi, M. R. (2011). A link stability-based multicast routing protocol for wireless mobile ad hoc networks. *Journal of Network and Computer Applications*, *34*(4), 1429–1440. doi:10.1016/j.jnca.2011.03.026

Akyildiz, A. F., Su, W., Sankarasubramaniam, Y., & Cayirci, E. (2002). Wireless sensor networks: A survey. *Computer Networks*, *38*, 393–422. doi:10.1016/S1389-1286(01)00302-4

Alattar, A. M., Al-Regib, G. I., & Al-Semari, S. A. (1999). Improved selective encryption techniques for secure transmission of MPEG video bit-streams. *Proceedings of the 1999 International Conference on Image Processing (ICIP '99)*, Vol. 4, (pp. 256-260). Kobe, Japan, October 24-28.

Albeseder, D., & Fuegger, M. (2005). *Small PC-network simulation - A comprehensive performance case study. Research Report*. TU Wien, Institut fur Technische Informatik.

Al-Karaki, J. N., & Kamal, A. E. (2004, December). Routing techniques in wireless sensor networks: A survey. *IEEE Wireless Communications*, *11*(6), 6–28. doi:10.1109/MWC.2004.1368893

Alzoubi, K. M., Li, X. Y., Wang, Y., Wan, P. J., & Frieder, O. (2003). Geometric spanners for wireless ad hoc network. *IEEE Transactions on Parallel and Distributed Systems*, *14*(4), 408–421. doi:10.1109/TPDS.2003.1195412

Alzoubi, K. M., Wan, P. J., & Frieder, O. (2002). Distributed heuristics for connected dominating sets in wireless ad hoc networks. *Journal of Communications and Networks*, *4*(1), 22–29.

Alzoubi, K. M., Wan, P. J., & Frieder, O. (2003). Maximal independent set, weakly connected dominating set, and induced spanners for mobile ad hoc networks. *International Journal of Foundations of Computer Science*, *14*(2), 287–303. doi:10.1142/S012905410300173X

Ammar, H., & El-Rewini, H. (2004). *A multicast protocol for mobile ad hoc networks using location information*. Boston, MA: Boston University.

An, B., & Papavassiliou, S. (2003). MHMR: Mobility-based hybrid multicast routing protocol in mobile ad hoc wireless networks. *Wireless Communication and Mobile Computing*, *3*, 255–270. doi:10.1002/wcm.115

Anderson, I. (1990). *Combinatorial designs: Construction methods*. Ellis Horwood Limited.

Anna, H. (2003). *Mobile telecommunication protocol for data networks* (1st ed.). John Wiley & Sons Ltd.

Argyroudis, P. G., & Mahony, D. O. (2005). Secure routing for mobile ad hoc networks. *IEEE Communications Surveys & Tutorials*, *7*(3), 2–21. doi:10.1109/COMST.2005.1610547

Ashtiani, H., Alirezaee, S., Hosseini, S. M. M., & Hamid, K. (2009). PNR - New position based routing algorithm for mobile ad hoc networks. *Proceedings of the World Congress on Engineering 2009*, Vol. I, WCE 2009, London, UK.

Aster, R., McIntosh, W., Kyle, P., Esser, R., Bartel, B., & Dunbar, N. (2004). Real-time data received from Mount Erebus Volcano, Antartica. *Eos, Transactions, American Geophysical Union*, *85*(10), 97–104. doi:10.1029/2004EO100001

Athanasios, B., et al. (2006). A mobility sensitive approach for efficient routing in ad hoc mobile networks. In *Proceedings of the 9th ACM International Symposium on Modeling Analysis and Simulation of Wireless and Mobile Systems*. Terromolinos, Spain: ACM Press.

Atzori, L., & Raccis, A. (2004). Network Capacity assignment for multicast services using genetic algorithms. *IEEE Communications Letters*, *8*(6), 403–405. doi:10.1109/LCOMM.2004.831328

Badarneh, O. S., & Kadoch, M. (2009). Multicast routing protocols in mobile ad hoc networks: A comparative survey and taxonomy. *EURASIP Journal on Wireless Communications and Networking*, Article ID 764047, 42 pages.

Bae, S., Lee, S., Su, W., & Gerla, M. (2002). The design, implementation, and performance evaluation of the on-demand multicast routing protocol in multi-hop wireless networks. *IEEE Network*, *4*(1), 70–77.

Bai, F., Sadagopan, N., & Helmy, A. (2003). IMPORTANT: A framework to systematically analyze the impact of mobility on performance of routing protocols for Ad hoc networks. *IEEE International Conference on Computer Communications*, (pp. 825-835).

Bai, F., & Helmy, A. (2004). A survey of mobility modelling and analysis in wireless ad hoc networks . In *Wireless ad hoc and sensor network*. Kluwer Academic Publishers.

Bai, F., Sadagopan, N., & Helmy, A. (2003). The IMPORTANT framework for analyzing the impact of mobility on performance of routing protocols for adhoc networks. *Elsevier Ad Hoc Networks*, *1*, 383–403. doi:10.1016/S1570-8705(03)00040-4

Ballardie, T., Francis, P., & Crowcroft, J. (1993). Core based trees (CBT): An architecture for scalable inter-domain multicast routing. In *Proceedings of ACM SIGCOMM '93*, (pp. 85-95).

Banerjee, S. (2008). Detection/removal of cooperative black and gray hole attack in mobile ad-hoc networks. *Proceedings of the World Congress on Engineering and Computer Science 2008 WCECS 2008,* October 22 - 24, 2008, San Francisco, USA.

Bani Yassein, M., Ould Khaoua, M., Mackenzie, L. M., Papanastasiou, S., & Jamal, A. (2006). *Improving route discovery in on-demand routing protocols using local topology information in MANETs. PM2HW2N'06, October 2, 2006, Torremolinos*. Malaga, Spain: ACM.

Baryun, A. N., & Al-Begain, K. (2008). *A design approach for MANET multicast protocols.* PGNet. ISBN: 978-1-902560-19-9

Basagni, S., Chlamtac, I., Syrotiuk, V. R., & Talebi, R. (2000). On-demand location aware multicast (OLAM) for ad hoc networks. In *Proceedings of Wireless Communications and Networking Conference*, Vol. 3, (pp. 1323-1328).

Basagni, S., Conti, M., Giordano, S., & Stojmenovic, S. (2003). *Mobile ad hoc networks*. IEEE Press, John Wily & Sons, Inc.

Basagni, S., Conti, M., Giordano, S., & Stojmenovic, I. (2004). *Mobile ad hoc networking*. IEEE Press. doi:10.1002/0471656895

Bedworth, M. D., & O'Brien, J. C. (1999). The omnibus model: A new model for data fusion? In *Proceedings of the 2nd International Conference on Information Fusion (FUSION)*, ISIF, Sunnyvale, (pp. 437–444).

Benkler, Y. (2002). Some economics of wireless communications. *Harvard Journal of Law & Technology*, 16.

Bertsekas, D. P., & Gallager, R. G. (1987). Distributed asynchronous Bellman- Ford algorithm . In *Data networks* (pp. 325–333). Englewood Cliffs, NJ: Prentice Hall.

Bettstetter, C. (2002). On the minimum node degree and connectivity of a wireless multi-hop network. *Proceedings ACM MobiHoc, 2002*, 80–91.

Biradar, R., Manvi, S., & Reddy, M. (2010, April). Mesh based multicast routing in MANET: Stable link based approach. *International Journal of Computer and Electrical Engineering, 2*(2), 1793–8163.

Biswas, J., Barai, M., & Nandy, S. K. (2004). Efficient hybrid multicast routing protocol for ad hoc wireless networks. In *Proceedings of 29th Annual IEEE International Conference on Local Computer Networks*, (pp. 180-187).

Blake, S., Black, D., Carlson, M., Davies, E., Wang, Z., & Weiss, W. (1998). *An architecture for differentiated services.* IETF RFC 2475.

Blazevic, L., Boudec, J. Y. L., & Giordano, S. (2005). A location-based routing method for mobile ad hoc networks. *IEEE Transactions on Mobile Computing, 4*(2). doi:10.1109/TMC.2005.16

Blough, D., Leoncini, M., Resta, G., & Santi, P. (2003). The k-neighbors protocol for symmetric topology control in ad hoc networks. *Proceedings of 4th ACM International Symposium on Mobile Ad Hoc Networking & Computing,* Annapolis, MD, (pp. 141-152).

Bonn Motion. (2005). *A mobility scenario generation and analysis tool*. Retrieved from http://web.informatik.unibonn.de/iv/ mitarbeiter/decvaal/bonnmotion

Borbash, S., & Jennings, E. (2002). Distributed topology control algorithm for multihop wireless networks. In *Proceedings of IEEE International Joint Conference on Neural* Networks, Honolulu, HI, (pp. 355-360).

Boudec, J.-Y. L., & Vojnovi'c, M. (2005). Perfect simulation and stationarity of a class of mobility models. *Proceedings - IEEE INFOCOM*, 2005.

Boulgouris, N. V., Thomos, N., & Strintzis, M. G. (2006, January). Optimized transmission of JPEG2000 streams over wireless channels. *IEEE Transactions on Image Processing, 15*(1).

Boulis, A., Ganeriwal, S., & Srivastava, M. B. (2003, September). Aggregation in sensor networks: An energy-accuracy trade-off. *Ad Hoc Networks, 1*, 317–331. doi:10.1016/S1570-8705(03)00009-X

Boyd, J. R. (1987). *A discourse on winning and losing. Unpublished set of briefing slides available at Air University Library*. Alabama: Maxwell AFB.

Bracio, B. R., Horn, W., & Moller, D. P. F. (1997). Sensor fusion in biomedical systems. In *Proceedings of the 19th Annual International Conference of the IEEE Engineering in Medicine and Biology Society*, Vol. 3, (pp. 1387–1390). Chicago, IL: IEEE.

Braden R., Clark, D., & Shenker, S. (1994). *Integrated services in the Internet architecture: An overview*. IETF RFC 1633.

Broch, J., Johnson, D. B., & Maltz, D. A. (1996). *The dynamic source routing in ad hoc wireless networks* (pp. 153–181). Dordrecht, The Netherlands: Kluwer Academic Publishers.

Broch, J., Maltz, D. A., Johnson, D. B., Hu, Y.-C., & Jetcheva, J. (1998). *A performance comparison of multi-hop wireless ad hoc network routing protocols. Mobile Computing and Networking* (pp. 85–97). MobiCom.

Brodersen, R. W., Wolisz, A., Cabric, D., Mishra, S. M., & Willkomm, D. (2004). *CORVUS: A cognitive radio approach for usage of virtual unlicensed spectrum.* University of California, Berkley. Retrieved from http://bwrc.eecs.berkeley.edu/Research/ MCMA/CR_White_chapter_final1.pdf

Brown, C., Durrant-Whyte, H., Leonard, J., Rao, B., & Steer, B. (1992). Distributed data fusion using Kalman □ltering: A robotics application . In Abidi, M. A. (Ed.), *Data fusion in robotics and machine intelligence* (pp. 267–309).

Buchegger, S., & Boudec, J. (2002). Performance analysis of the confidant protocol: Cooperation of nodes - fairness in distributed ad-hoc networks. *Proceedings of the IEEE/ACM Workshop on Mobile Ad Hoc Networking and Computing (MobiHOC)*, (pp. 226–236).

Buchegger, S., & Le Boudec, J.-Y. (2002). *Cooperative routing in mobile ad hoc networks: Current efforts against malice and selfishness.* In Lecture Notes on Informatics, Mobile Internet Workshop, Germany, October 2002. Springer.

Buede, D. M. (1988). Shafer-Dempster and Bayesian reasoning: A response to Shafer-Dempster reasoning with applications to multisensor target identi□cation systems. *IEEE Transactions on Systems, Man, and Cybernetics*, *18*, 1009–1011. doi:10.1109/21.23099

Bui, T. N., & Zrncic, C. M. (2006). An ant-based algorithm for finding degree-constrained minimum spanning tree. In *Proceedings of the 8th Annual Conference on Genetic and Evolutionary Computation*, (pp. 11-18).

Burleigh, S., Hooke, A., Torgerson, L., Fall, K., Cerf, V., & Durst, B. (2003). Delay-tolerant networking–An approach to interplanetary Internet. *IEEE Communications Magazine*, *41*(6), 128–136. doi:10.1109/MCOM.2003.1204759

Burri, N., Rickenbach, P., Watternhofer, R., & Weber, Y. (2006). Topology control made practical: Increasing the performance of source routing. *Lecture Notes in Computer Science*, *4325*, 1–12. doi:10.1007/11943952_1

Butenko, S., Cheng, X., Oliveira, C., & Pardalos, P. M. (2004). A new heuristic for the minimum connected dominating set problem on ad hoc wireless networks . In *Recent developments in cooperative control and optimization* (pp. 61–73). Kluwer Academic Publishers. doi:10.1007/978-1-4613-0219-3_4

Buttayan, L., & Hubaux, J.-P. (2007). *Security and cooperation in wireless networks*. Cambridge University Press.

Buttyan, L., & Vajda, I. (2004). Towards provable security for ad hoc routing protocols. In *Proceedings of the 2nd ACM Workshop on Security of Ad hoc and Sensor Networks*, 2004, (pp. 94–105).

Cabri, G. (2008). The LAICA project: An ad-hoc middleware to support ambient intelligence. *Multiagent and Grid Systems Journal*, *2*(3), 235–247.

Cai, J., & Jia, B. (2005). *Network simulation based on OPNET and application.* 1st International Workshop on Education Technology and Computer Science.

Cai, S., & Yang, X. (2003). The performance of PoolODMRP. In *Proceedings of the 6th IFIP/IEEE International Conference on Management of Multimedia Networks and Service (MMNS2003)*, (pp. 90-101).

Cai, S., Yang, X., & Wang, L. (2004). PDAODMRP: An extended PoolODMRP based on passive data acknowledgement. *Journal of Communications and Networks*, *6*(4), 362–375.

Camp, T., Boleng, J., & Davies, V. (2002). A survey of mobility models for ad hoc network research. *Wireless Communication and Mobile Computing (WCMC), special Issue on Mobile Ad Hoc Networking . Research Trends and Applications, 2*(5), 483–502.

Camtepe, S. A., & Yener, B. (2005). *Key distribution mechanisms for wireless sensor networks: A survey*. Technical Report TR-05-07 Rensselaer Polytechnic Institute, Computer Science Department.

Camtepe, S. A., & Yener, B. (2007). Combinatorial design of key distribution mechanisms for wireless sensor networks. *IEEE/ACM Transactions on Networking, 15*(2). doi:10.1109/TNET.2007.892879

Canourgues, L., Lephay, J., Soyer, L., & Beylot, A. L. (2006). STAMP: Shared-tree ad hoc multicast protocol. *IEEE Military Communications Conference, MILCOM 2006*, (pp. 1-7).

Cardell-Oliver, R. (2005). Rope: A reactive, opportunistic protocol for environment monitoring sensor network. *Proceedings of the 2nd IEEE Workshop on Embedded Networked Sensors.*

Cartigny, J., Simplot, D., & Stojmenovic, I. (2003). Localized minimum energy broadcasting in ad-hoc networks. In *Proceedings of IEEE INFOCOM.*

Challapali, K., Mangold, S., & Zhong, Z. (2004). *Spectrum agile radio: Detecting spectrum opportunities*. 6th Annual International Symposium on Advanced Radio Technologies, Boulder, Colorado, 2 - 4 March.

Chan, H., Perrig, A., & Song, D. (2003). *Random key predistribution scheme for sensor networks*. IEEE Symposium on Research in Security and Privacy.

Chang, X. (1999). Network simulations with OPNET. *Proceedings of the 31st Conference on Winter Simulation.*

Chang, B. J., & Kuo, S. L. (2009). Markov chain trust model for trust-value analysis and key management in distributed multicast MANETs. *IEEE Transactions on Vehicular Technology, 58*(4).

Chao, W., Macker, J. P., & Weston, J. W. (2003). *NRL mobile network emulator*. (RL/FR/5523--03-10,054).

Chaudhari, S. P., Boudec, J. L., & Vojnovic, M. (2005). Perfect simulations for random trip mobility models. In *Proceedings of the 38th Annual Symposium on Simulations*, (pp. 72-79).

Chen, C., Ali, A. M., & Wang, H. (2006). Design and testing of robust acoustic arrays for localization and enhancement of several bird sources. In *Proceedings of the 5th International Conference on Information Processing in Sensor Networks (IPSN)*, (pp. 268–275).

Chen, K., & Nahrstedt, K. (2002). Effective location-guided tree construction algorithms for small group multicast in MANET. In *Proceedings of INFOCOM*, (pp. 1180-1189).

Chen, B., Jamieson, K., Balakrishnan, H., & Morris, R. (2002). Span: An energyefficient coordination algorithm for topology maintenance in ad hoc wireless networks. *ACM Wireless Networks Journal, 8*(5), 481–494. doi:10.1023/A:1016542229220

Cheng, H., & Li, X. (2000). Partial encryption of compressed images and video. *IEEE Transactions on Signal Processing, 48*(8), 2439–2451. doi:10.1109/78.852023

Cheng, X., Huang, X., & Du, D. Z. (2006). *Ad hoc wireless networking* (pp. 319–364). Kluwer Academic Publishers.

Chen, T. W., & Gerla, M. (1997). Global state routing: A new routing scheme for ad-hoc wireless networks. In *. Proceedings of IEEE SICON, 97*, 197–211.

Chialvo, D. R., & Millonas, M. M. (1995). How swarms build cognitive maps. *Autonomous Agents, 144*, 439–450.

Chiang, C. C., Gerla, M., & Zhang, L. (1997a). Shared tree wireless network multicast. In *Proceedings of the 6th International Conference on Computer Communications and Networks* (pp. 28-33). Las Vegas, NV: IEEE.

Chiang, C. C., Wu, H. K., Liu, W., & Gerla, M. (1997b). Routing in clustered multi-hop, mobile wireless networks with fading channel. In *Proceedings of the Singapore International Conference on Networks* (pp. 197-211). Singapore: IEEE.

Chiang, T. C., & Huang, Y. M. (2004). Multicast routing representation in ad hoc networks using fuzzy Petri nets. *in Proceedings of the 18th International Conference on Advanced Information Networking and Application*, (pp. 1-4).

Chiang, C. C., Gerla, M., & Zhang, L. (1999). Forwarding group multicast protocol (FGMP) for multihop, mobile wireless networks. *Journal of Cluster Computing, 1*(2), 187–196. doi:10.1023/A:1019037500012

Chiang, T. C., Liu, C. H., & Huang, Y. M. (2007). A near-optimal multicast scheme for mobile ad hoc networks using a hybrid genetic algorithm. *Expert Systems with Applications, 33*, 734–742. doi:10.1016/j.eswa.2006.06.020

Chlamtac, I., Conti, M., & Liu, J. (2003). Mobile ad hoc networking: Imperatives and challenges. *Journal of Ad hoc . Networks, 1*, 13–64.

Choi, H. S., Byun, H., & Lim, J. (2008). Adaptive power control MAC in wireless ad hoc networks. *IEICE Transaction on Communication . E (Norwalk, Conn.), 91–B*(1), 309–313.

Ci, S., Sharif, H., & Nuli, K. A (2004). UKF-based link adaptation scheme to enhance energy efficiency in wireless sensor networks. In *Proceedings of the 15th IEEE International Symposium on Personal, Indoor and Mobile Radio Communications (PIMRC '04),* Vol. 4, (pp. 2483–2488). IEEE.

Clark, B. N., Colbourn, C. J., & Johnson, D. S. (1990). Unit disk graphs. *Discrete Mathematics, 86*, 165–177. doi:10.1016/0012-365X(90)90358-O

Clausen, T., Hansen, G., Christensen, L., & Behrmann, G. (2001). The optimized link state routing protocol. *Proceedings of the 4th International Symposium on Wireless Personal Multimedia Communications,* Aalborg, Denmark, September 2001.

Cohen, N. H., Purakayastha, A., Turek, J., Wong, L., & Yeh, D. (January 2001). *Challenges in flexible aggregation of pervasive data.* IBM Research Report RC 21942 (98646), NY.

Cordasco, J., & Wetzel, S. (2009). An attacker model for MANET routing security. In *Proceeding of the Second ACM Conference on Wireless Network Security*, Zurich, Switzerland, (pp. 87-94).

Cordeiro, C., Gossain, H., & Agrawal, D. (2003, January). Multicast over wireless mobile ad hoc networks: Present and future directions. *IEEE Network, 17*(1).

Cormen, T. H., Leiserson, C. E., Rivest, R. L., & Stein, C. (2001). *Introduction to algorithms.* Cambridge, MA: MIT Press.

Coue, C., Fraichard, T., Bessiere, P., & Mazer, E. (2002). Multi-sensor data fusion using Bayesian programming: An automotive application. In *IEEE/RSJ International Conference Intelligent Robots and Systems,* Vol. 1, Switzerland, (pp. 141–146).

Crawley, E., Nair, R., Rajagopalan, B., & Sandick, H. (1998, August). *A framework for QoS based routing in the Internet.* RFC 2386.

Crisan, D., & Doucet, A. (2002, Mar). A survey of convergence results on particle filtering methods for practitioners. *IEEE Transactions on Signal Processing, 50*, 736–746. doi:10.1109/78.984773

Crow, B., Widjaja, I., Kim, J. G., & Sakai, P. (1997). Investigation of the IEEE 802.11 medium access control (MAC) sublayer functions. In . *Proceedings of INFOCOM, 1*, 26–33.

Curren, D. (2008). A survey of simulation in sensor networks. *Architecture (Washington, D.C.), 2008*, 867–872. Retrieved from http://citeseerx.ist.psu.edu/viewdoc/ download?doi=10.1.1.94.2626&rep=rep1&type=pdf

Dahill, B., Levine, B. N., Royer, E., & Shields, C. (2002). *ARAN: A secure routing protocol for ad hoc networks.* UMass Tech Report 02-32.

Dai, F., & Wu, J. (2004). An extended localized algorithm for connected dominating set formation in ad hoc wireless networks. *IEEE Transactions on Parallel and Distributed Systems, 15*(10). doi:10.1109/TPDS.2004.48

Das, A., Ghose, A., Razdan, H. S., & Shorey, R. (2001). *Enhancing performance of asynchronous data traffic over the Bluetooth wireless ad-hoc network.* In IEEE INFOCOM'2001.

Das, B. S. M. S. K., & Ram Murthy, C. S. (2002). A dynamic core based multicast routing protocol for ad hoc wireless networks. *Proceedings of the 3rd ACM International Symposium on Mobile Ad Hoc Networking and Computing, Switzerland* (pp. 24-35).

Das, B., Sivakumar, R., & Bharghavan, V. (1997). Routing in ad hoc networks using a virtual backbone. In *Proceedings of IC3N'97*, (pp.1-20).

Das, S., Castaneda, R., Yan, J., & Sengupta, R. (1998). Comparative performance evaluation of routing protocols for mobile ad hoc networks. *Proceedings of IEEE 7th International Conference on Computer Communications and Networks* (I3CN), Lafayette, LA, October, (pp. 153-161).

Das, A. N., Lewis, F. L., & Popa, D. O. (2009). Data-logging and supervisory control in wireless sensor networks. *International Journal of Sensor Networks*, *6*(1). doi:10.1504/IJSNET.2009.028022

Dasarathy, B. V. (1997, Jan). Sensor fusion potential exploitation-innovative architectures and illustrative applications. *Proceedings of the IEEE*, *85*, 24–38. doi:10.1109/5.554206

Davies, V. (2000). *Evaluating mobility models within an ad hoc network*. Master Thesis, Colorado School of Mines.

Dawoud, D. S., Richard, L. G., Suliman, A., & Raja, S. V. K. (2011). *Trust establishment in mobile ad hoc networks: Key management*. Intech Publication.

De Renesse, R., & Aghvami, A. H. (2004). *Formal verification of ad-hoc routing protocols using SPIN model checker*. Centre for Telecommunication Research, King's College London, UK.

Deering, S., Estrin, D., Farinacci, D., Jacobson, V., Helmy, A., & Meyer, D. (1999). *Protocol independent multicast version 2 dense mode specification.* Draft-IETF-PIM-V2-DM-03.txt, 1999.

Deering, S., Estrin, D. L., Farinacci, D., Jacobson, V., Liu, C. G., & Wei, L. (1996). The PIM architecture for wide-area multicast routing. *IEEE/ACM Transactions on Networking*, *4*(2), 153–162. doi:10.1109/90.490743

Deneubourg, J. L., Aron, S., Goss, S., & Pasteels, J. M. (1990). The self-organizing exploratory pattern of the Argentine ant. *Journal of Insect Behavior*, *3*, 159–168. doi:10.1007/BF01417909

Deng, J., & Hass, Z. J. (1998). Dual busy tone multiple access (DBTMA): A new medium access control for packet radio networks. *Proceedings of International Conference on Universal Personal Communication.*

Desai, M., & Manjunath, D. (2002). On the connectivity in finite ad hoc networks. *IEEE Communications Letters*, *6*, 437–439. doi:10.1109/LCOMM.2002.804241

Devroye, N., Mitran, P., & Tarokh, V. (2006). Achievable rates in cognitive radio channels. *IEEE Transactions on Information Theory*, 52.

Diffie, W., & Hellman, M. E. (1976). New directions in cryptography. *IEEE Transactions on Information Theory*, *22*, 644–654. doi:10.1109/TIT.1976.1055638

Djenouri, D., Khelladi, L., & Badache, A. N. (2005). A survey of security issues in mobile ad hoc and sensor networks. *IEEE Communications Surveys & Tutorials*, *7*(4), 2–28. doi:10.1109/COMST.2005.1593277

Dolev, D., & Yao, A. (1983). On the security of public key protocols. *IEEE Transactions on Information Theory*, *29*(2), 198–208. doi:10.1109/TIT.1983.1056650

Doolin, D., & Sitar, N. (2005). *Wireless sensors for wild-fire monitoring*. SPIE Symposium on Smart Structures & Materials.

Dorigo, M. (1992). *Optimization, learning and natural algorithms*. Ph.D. thesis, Dipartimento di Elettronica, Politecnico di Milano, Milan, Italy, 1992.

Dorigo, M., Maniezzo, V., & Colorni, A. (1991). *Positive feedback as a search strategy.* Tech. Report 91-016, Dipartimento di Elettronica, Politecnico di Milano, Italy.

Dorigo, M., & Gambardella, L. M. (1997). Ant colony system: A cooperative learning approach to the traveling salesman problem. *IEEE Transactions on Evolutionary Computation*, *1*(1), 53–66. doi:10.1109/4235.585892

Douceur, J. R. (2002). The Sybil attack. In *Proceedings of 1st International Workshop on Peer-to-Peer Systems, LNCS 2429*, (pp. 251-260). March 2002.

Dousse, O., Thiran, P., & Hasler, M. (2002). Connectivity in ad-hoc and hybrid networks. *Proceedings - IEEE INFOCOM*, (March): 2002.

Dow, C. R., Lin, P. J., Chen, S. C., Lin, J. H., & Hwang, S. F. (2005). A study of recent research trends and experimental guidelines in mobile ad-hoc networks. *Proceedings of the IEEE 19th International Conference on Advanced Information Networking and Applications (AINA'05).*

Downard, I. (2004). *Simulating sensor networks in ns-2*. NRL Formal Report 5522, April.

Du, W., Deng, J., Han, Y. S., & Varshney, P. (2003). A pairwise key pre-distribution scheme for wireless sensor networks. *Proceedings of the 10th ACM Conference on Computer and Communications Security.*

Dube, R., Rais, C. D., Wang, K. Y., & Tripathi, S. K. (1997). *Signal stability adaptive routing (SSA) for ad hoc mobile networks.* IEEE PC, 1997.

Dubey, R., Rais, D., Wang, K. Y., & Tripathi, S. K. (1996). Signal stability based adaptive routing (SSA) for ad-hoc mobile networks. *Computer Science Technical Report Series, 3646*, 22.

Durrant-Whyte, H. F. (1988, December). Sensor models and multisensor integration. *The International Journal of Robotics Research, 7*, 97–113. doi:10.1177/027836498800700608

Elfes, A. (1987, June). Sonar-based real-world mapping and navigation. *IEEE Journal on Robotics and Automation, 3*, 249–265. doi:10.1109/JRA.1987.1087096

Elmenreich, W. (2002). *Sensor fusion in time-triggered systems*. Ph.D. thesis, Institut fur Technische Informatik, Vienna University of Technology, Vienna, Austria.

Ern, V. C. (1985). A thermodynamically approach to the travelling salesman problem: An efficient simulation algorithm. *Journal of Optimization Theory and Applications, 45*, 41–51. doi:10.1007/BF00940812

Eschenauer, V., & Gligor, D. (2002). A key-management scheme for distributed sensor networks. *Proceedings of the 9th ACM Conference on Computer and Communications Security.*

Estrin, D., Farinacci, D., Helmy, A., Thaler, D., Deering, S., & Handley, M. (1998). *Protocol independent multicast-sparse mode (PIM-SM)*. Protocol specification . *RFC, 2362*, 1998.

ETSI, Universal Mobile Telecommunications System (UMTS). (1998). *Selection procedures for choice of radio transmission technologies of the UMTS*, UMTS 30.03 Version 3.2.0. Retrieved from http://www.3gpp.org/ftp/specs/ html-info/3003u.htm

Fang, Y., Zhu, X., & Zhang, Y. (2009). Securing resource-constrained wireless ad hoc networks. *Wireless Communications, 16*(2), 24–29. doi:10.1109/MWC.2009.4907556

Farkas, K., Hossmann, T., Ruf, L., & Plattner, B. (2006). Pattern matching based link quality prediction in wireless mobile ad hoc networks. *International Symposium on Modeling, Analysis and Simulation of Wireless and Mobile Systems,* (pp. 239-246).

Faulhaber, G., & Farber, D. (2003). Spectrum management: Property rights, markets and the commons. In *Proceedings of the Telecommunications Policy Research Conference*, Alexandria, VA.

Feeney, L. M. (2001). An energy-consumption model for performance analysis of routing protocols for mobile ad hoc networks. *Mobile Networks and Applications Journal, 6*(3).

Feeney, M. L. (2005). *Introduction to MANET routing.* Swedish Institute of Computer Science. Retrieved from http://www.nada.kth.se/kurser/kth/ 2D1490/05/lectures/feeney_mobile_adhoc_routing.pdf

Feeney, L. M. (2004). Energy efficient communication in ad hoc networks . In Basagni, S., Conti, M., Giordano, S., & Stojmenovic, I. (Eds.), *Mobile ad hoc networking* (pp. 301–327). IEEE Press and Wiley-Interscience. doi:10.1002/0471656895.ch11

Fiore, M., Harri, J., Filali, F., & Bonnet, C. (2007). Vehicular mobility simulation for VANETs simulation. *ANSS '07, 40th Annual Simulation Symposium*, (pp. 301-309).

Fischer, D. A. (2011). Mobility model for the realistic simulation of social context. *KiVS, 2011*, 215–220.

Fisher, S. (1999, February 5). *Policy management: An introduction.* Gartner Group/Datapro Technical report.

Fishman, G. S. (1978). *Principles of discrete event simulation.* New York, NY: John Wiley & Sons, Inc.

Fogel, L. J., Owens, A. J., & Walsh, M. J. (1966). *Artificial intelligence through simulated evolution.* New York, NY: Wiley.

Foh, C. H., & Lee, B. S. (2004). A closed form network connectivity formula for one-dimensional MANETs. *Proceedings IEEE ICC*, June 2004.

Forouzan, A. B. (2005). *TCP/IP protocol suite* (3rd ed.). McGraw Hill.

Forouzan, B. A. (2003). *TCP/IP protocol suite* (2nd ed.). McGraw-Hill.

Frankel, C. B. (1999). *Such order from confusion sprung: Adaptive competence and affect regulation*. Ph.D. thesis, Pacific Graduate School of Psychology, Palo Alto, CA.

Frankel, C. B., & Bedworth, M. D. (2000). Control, estimation and abstraction in fusion architectures: Lessons from human information processing. In *Proceedings of the 3rd International Conference on Information Fusion (Fusion),* Vol. 1. IEEE.

Friedlander, D. S. (2005). Semantic information extraction . In Iyengar, S. S., & Brooks, R. R. (Eds.), *Distributed sensor networks* (pp. 409–417). Boca Raton, FL: CRC Press.

Friedlander, D. S., & Phoha, S. (2002). Semantic information fusion for coordinated signal processing in mobile sensor networks. *International Journal of High Performance Computing Applications*, *16*, 235–241. do i:10.1177/10943420020160030401

Friginal, J., David, A., Ruiz, J., & Gil, P. (2010). Attack injection to support the evaluation of ad hoc networks. *29th IEEE International Symposium on Reliable Distributed Systems*, (pp. 21-29).

Gambardella, L. M., & Dorigo, M. (1996). Solving symmetric and asymmetric TSPS by ant colonies. *Proceedings of the IEEE International Conference on Evolutionary Computing* (ICEC'96), (pp. 622–627).

Ganeriwal, S., Kumar, R., & Srivastava, M. B. (2003). Timing-sync protocol for sensor networks. In *Proceedings of the 1st International Conference on Embedded Networks Sensor Systems (SenSys '03)*, (pp. 138–149).

Ganesan, G., & Li, Y. (2005). *Agility improvement through cooperative diversity in cognitive radio*. IEEE GLOBECOM Proceedings.

Gao, B., Yang, Y., & Ma, H. (2005). A new distributed approximation algorithm for constructing minimum connected dominating set in wireless ad hoc networks. *International Journal of Communication Systems*, *18*, 743–762. doi:10.1002/dac.726

Garcia-Luna-Aceves, J. J., & Madruga, E. L. (1999). Core-assisted mesh protocol. *IEEE Journal on Selected Areas in Communications*, *17*(8), 784–792. doi:10.1109/49.779921

Garg, N., & Mahapatra, R. P. (2009). MANET security issues. *International Journal of Computer Science and Network Security*, *9*(8).

Gergely, A., Buttyan, L., & Vajda, I. (2006). Provably secure on-demand source routing in mobile ad hoc networks. *IEEE Transactions on Mobile Computing*, *5*(11), 1533–1546. doi:10.1109/TMC.2006.170

Gerharz, M., de Waal, C., Martini, P., & James, P. (2003). Strategies for finding stable paths in mobile wireless ad hoc networks. *IEEE International Conference on Local Computer Networks*, (pp. 130-139).

Gerla, M., Chiang, C. C., & Zhang, L. (1999). Tree multicast strategies in mobile, multi-hop wireless networks. *ACM Baltzer Mobile Networks and Applications*, *4*(3), 193–207. doi:10.1023/A:1019150932356

Gipps, P. G. (1986). A model for the structure of lane-changing decisions. *Transportation Research Board, 20*(5), 403–414.

Gipps, P. G. (1981). A behavioral car following model for computer simulation. *Transportation Research Part B: Methodological*, *15*, 105–111. doi:10.1016/0191-2615(81)90037-0

Glisic, S. G. (2006). *Advanced wireless networks: 4G technologies*. John Wily. doi:10.1002/9780470035405

Glover, F. (1989). Tabu search—Part I. *Computers*, *1*(3), 190–206.

Goff, T., Abu-Ghazaleh, N. B., Phatak, D. S., & Kahvecioglu, R. (2001). *Preemptive routing in ad hoc networks* (pp. 43–52). ACM/IEEE MobiCom.

Goldberg, D. E. (1989). *Genetic algorithms in search, optimization, and machine learning*. Addison-Wesley Pub Co.

Govindaswamy, V., Blackstone, W. L., & Balasekaran, G. (2011). Survey of recent position based routing mobile ad-hoc network protocols. *Proceedings of* 2011 UKSim 13th International Conference on Modelling and Simulation, (pp. 467-471).

Guerrero Zapata, M. (2002). *Secure ad hoc on-demand distance vector (SAODV) routing*. Retrieved from http:// ietfreport.isoc.org/idref/ draft-guerrero-manet-saodv/

Guha, S., & Khuller, S. (1998). Approximation algorithms for connected dominating sets. *Algorithmica, 20*(4), 374–387. doi:10.1007/PL00009201

Gunnarsson, F., & Gustafsson, F. (2003). Positioning using time-difference of arrival measurements. In Proceedings of the *IEEE International Conference on Acoustics, Speech, and Signal Processing. Vol. 6,* IEEE, (pp. 553–556).

Guntsch, M., & Middendorf, M. (2001). Pheromone modification strategies for ant algorithms applied to dynamic TSP. In *Applications of Evolutionary Computing: Proceedings of Evo Workshops 2001, Lecture Notes in Computer Science, Vol. 2037*, (pp. 213–222).

Guo, D., Wang, X., & Chen, R. (2005, Apr.). New sequential Monte Carlo methods for nonlinear dynamic systems. *Statistics and Computing, 15*, 135–147. doi:10.1007/s11222-005-6846-5

Gupta, P., & Kumar, P. (2000). The capacity of wireless networks. *IEEE Transactions on Information Theory, 46*(2), 388–404. doi:10.1109/18.825799

Gupta, S. K. S. (2001). Cored-based tree with forwarding regions (CBT-FR): A protocol for reliable multicasting in mobile ad hoc networks. *Journal of Parallel and Distributed Computing, 61*, 1249–1277. doi:10.1006/jpdc.2001.1749

Gupta, S. K. S., & Srimani, P. K. (2003). Adaptive core selection and migration method for multicast routing in mobile ad hoc networks. *IEEE Transactions on Parallel and Distributed Systems, 14*(1), 27–38. doi:10.1109/TPDS.2003.1167368

Haas, Z. J., Pearlman, M. R., & Samar, P. (2002). The zone routing protocol (ZRP) for ad hoc networks. Draft-ietf-manet-zone-zrp-04.txt, Internet-Draft, IETF, 2002.

Haas, Z. J., Pearlman, M. R., & Samar, P. (2002b). *The intrazone routing protocol (IARP) for ad hoc networks.* Internet Draft: draft-ietfmanet-zone-iarp-02.txt, 2002.

Haas, Z. J., Pearlman, M. R., & Samar, P. (2002c). The bordercast resolution protocol (BRP) for ad hoc networks. Draft-ietf-manet-zone-brp-02.txt, Internet-Draft, IETF, 2002.

Haas, Z. J., Pearlman, M. R., & Samar, P. (2002d). *The interzone routing protocol (IERP) for ad hoc networks.* Draft-ietf-manet-zone-ierp-02.txt, Internet-Draft, IETF, 2002.

Haas, Z. J., & Pearlman, M. R. (2001). The performance of query control schemes for the zone routing protocol. *IEEE/ACM Transactions on Networking, 9*(4), 427–438. doi:10.1109/90.944341

Halfslund, A., Tonnesen, A., Rotvik, R. B., Andersson, J., & Kure, O. (2004). *Secure extension to the OLSR protocol.* OLSR Interop and Workshop, 2004.

Halgamuge, K. S., & Wang, P. L. (2005). *Classification and clustering for knowledge discovery* (1st ed.). Springer.

Hall, D. L., & Garga, A. K. (1999). New perspectives on level four processing in data fusion systems. In *Proceedings of SPIE AeroSense '99 Conference on Digitization of the Battle□eld IV.*

Hall, D. L., & Llinas, J. (1997, January). An introduction to multi-sensor data fusion. *Proceedings of the IEEE, 85*, 6–23. doi:10.1109/5.554205

Hall, D. L., & Llinas, J. (2001). *Handbook of multisensor data fusion.*

Han, B. (2009). Zone-based virtual backbone formation in wireless ad hoc networks. *Ad Hoc Networks, 7*(1), 183–200. doi:10.1016/j.adhoc.2008.01.003

Han, Y., La, R. J., Makowski, A. M., & Lee, S. (2006). Distribution of Path durations in mobile ad hoc networks palms theorem to the rescue. *International Journal of Computer and Telecommunications Networking, 50*, 1887–1900.

Hartung, C., Han, R., Seielstad, C., & Holbrook, S. (2006). FireWxNet: A multi-tiered portable wireless system for monitoring weather conditions in wildland fire environments. *Proceedings of the 4th International Conference on Mobile Systems, Applications and Services.*

Haykin, S. (2005). Cognitive radio: Brain-empowered wireless communications. *IEEE Journal on Selected Areas in Communications, 23*(2), 9–16. doi:10.1109/JSAC.2004.839380

He, C. (2003). *Throughput and delay in wireless ad hoc networks. Final report of EE359 Class project*. Stanford University.

Henderson, T. R., Roy, S., Floyd, S., & Riley, G. F. (2006). ns-3 project goals. *Proceeding of the 2006 Workshop on ns-2: The IP Network Simulator.*

He, T., Blum, B. M., Stankovic, J. A., & Abdelzaher, T. (2004). AIDA: Adaptive application independent data aggregation in wireless sensor network. *ACM Transactions in Embedded Computer Systems*, *3*, 426–457. doi:10.1145/993396.993406

Hofmann-Wellenhof, B., Lichtenegger, H., & Collins, J. (2004). *Global positioning system: Theory and practice*. Berlin, Germany: Springer Publishers.

Holland, J. H. (1975). *Adaption in natural and artificial systems*. Ann Arbor, MI: The University of Michigan Press.

Hong, X., Gerla, M., Pei, G., & Chiang, C.-C. (1999). *A group mobility model for ad hoc wireless networks*. ACM/IEEE International Symposium on Modeling Analysis and Simulation of Wireless and Mobile Systems (MSWiM), Seattle, WA, USA.

Hong, D., & Rappaport, S. S. (1986). Traffic model and performance analysis for cellular mobile radio telephone systems with prioritized and nonprioritized handoff procedures. *IEEE Transactions in OVT*, *35*, 77–92.

Hoover, A., & Olsen, B. D. (2000). Sensor network perception for mobile robotics. In *Proceedings of the IEEE International Conference on Robotics and Automation*, Vol. 1, (pp. 342–347).

Hu, Y. C., & Johnson, D. B. (2002). *Ensuring cache freshness in on-demand ad hoc network routing protocols*. POMC'02, October 30–31, 2002, Toulouse, France. ACM.

Hu, Y. C., Johnson, D. B., & Perrig, A. (2003). SEAD: Secure efficient distance vector routing for mobile wireless ad hoc networks. In *Proceedings of 4th IEEE Workshop on Mobile Computing Systems and Applications.*

Hu, Y. C., Perrig, A., & Johnson, D. B. (2002). Ariadne: A secure on-demand routing protocol for ad hoc networks. In *Proceedings of 8th ACM International Conference on Mobile Computing and Networking (Mobicom'02)*, Atlanta, Georgia, (pp. 12-23).

Hu, Y. C., Perrig, A., & Johnson, D. B. (2002). *Wormhole detection in wireless ad hoc networks*. Tech. Rep. TR01-384, Department of Computer Science, Rice University, June 2002.

Hu, Y. C., Perrig, A., & Johnson, D. B. (2003). Packet leashes: A defense against wormhole attacks in wireless ad hoc network. *Proceedings of the 22nd Annual Joint Conf. IEEE Computer and Communication Societies,* San Francisco, CA, April 2003.

Huang, Y. M., Chiang, T., & Hou, T. W. (2005). An efficient encoding tree for location-aware multicast routing protocol in wireless ad hoc networks. *Journal of Internet Technology*, *6*(1).

Hu, C. C., Wu, E. H. K., & Chen, G. H. (2006). OGHAM: On-demand global hosts for mobile ad hoc multicast services. *Journal of Ad hoc . Networks*, *4*, 709–723.

Hu, Y. C., & Perrig, A. (2004). A survey of secure wireless ad hoc routing. *IEEE Security & Privacy*, *2*(3), 28–39. doi:10.1109/MSP.2004.1

Hu, Y.-C., Johnson, D. B., & Perrig, A. (2003). SEAD: Secure efficient distance vector routing for mobile wireless ad hoc networks. *Ad Hoc Networks*, *1*(1), 175–192. doi:10.1016/S1570-8705(03)00019-2

Hwang, R. H., Do, W. Y., & Yang, S. C. (2000). Multicast routing based on genetic algorithms. *Journal of Information Science and Engineering*, *16*, 885–901.

IEEE 802.11 Working Group. (1999). *Wireless LAN medium access control (MAC) and physical layer (PHY) specification.*

IEEE USA Position. (2003). *Improving spectrum usage through cognitive radio technology.* Retrieved from http://www.ieeeusa.org/policy/ positions/cognitiveradio.asp

IETF MANET Working Group. (n.d.). *Mobile ad hoc networks (MANET).* Working Group. Retrieved from http://www.ietf.org/html.charters/ manet-charter.html

IETF. (2000). *MANET terms*. Retrieved from http://tools.ietf.org/html/ draft-ietf-manet-term-00

IETF. (2004). *MIKEY: Multimedia internet keying*. The Internet Society 2004. Retrieved from http://tools.ietf.org/html/rfc3830 ISO/IEC 14496-1. (May 1998). *Coding of audio-visual objects, part 1: Systems*. Final committee draft.

Iguchi-Cartigny, J., Ruiz, P. M., Simplot-Ryl, D., Stojmenovic, I., & Yago, C. M. (2009). Localized minimum-energy broadcasting for wireless multihop networks with directional antennas. *IEEE Transactions on Computers*, *58*(1), 120–131. doi:10.1109/TC.2008.125

Ilyas, M. (2003). *The handbook of ad hoc wireless networks*. New York, NY: CRC Press.

INET. (n.d.). *Framework*. Retrieved from http://inet.omnetpp.org/

Ingelrest, F., & Simplot-Ryl, D. (2005). Localized broadcast incremental power protocol for wireless ad hoc networks. In *Proceedings of 10th IEEE Symposium on Computer and Communication*, ISCC 200, (pp. 28-33).

Ingelrest, F., Simplot-Ryl, D., & Stojmenovic, I. (2006). Optimal transmission radius for energy efficient broadcasting protocol in ad hoc and sensor networks. *IEEE Transactions on Parallel and Distributed Systems*, *17*(6), 536–547. doi:10.1109/TPDS.2006.74

Intanagonwiwat, C., Estrin, D., Govindan, R., & Heidemann, J. (2002). Impact of network density on data aggregation in wireless sensor networks. In *Proceedings of the 22nd IEEE International Conference on Distributed Computing Systems (ICDCS)*, IEEE, Vienna, Austria, (pp. 457–458).

Intanagonwiwat, C., Govindan, R., & Estrin, D. (2000). Directed diffusion: A scalable and robust communication paradigm for sensor networks. In *Proceedings of the 6th Annual International Conference on Mobile Computing and Networking (MobiCom)*, (pp. 56–67). Boston, MA: ACM Press.

Internet Engineering Task Force (IETF). (n.d.). *Website*. Retrieved from www.ietf.org

INVENT. (n.d.). Retrieved from http://www.njit.edu/~borcea/invent/ invent-vehiculartraffic- generator.tar.gz

Ioannis, C., et al. (2001). An experimental study of basic communication protocols in ad-hoc mobile networks. In *Proceedings of the 5th International Workshop on Algorithm Engineering*. Springer-Verlag.

ISO/IEC 14496-2. (May 1998). *Coding of audio-visual objects, part 2: Visual*. Final committee draft.

ISO/IEC 14496-3. (May 1998). *Coding of audio-visual objects, part 3: Audio*. Final committee draft.

Iyengar, S. S., Chakrabarty, K., & Qi, H. (2001, September). Introduction to special issue on Distributed sensor networks for real-time systems with adaptive con□guration. *Journal of the Franklin Institute*, *338*, 651–653. doi:10.1016/S0016-0032(01)00027-8

Jai, K. E. H., & Kannammal, A. (2009). Privacy and security on anonymous routing protocols in MANET. *Proceedings of IEEE Second International Conference on Computer and Electrical Engineering*, (pp. 431-435).

Jain, J., Fatima, M., Gupta, R., & Bandhopadhyay, K. (2005). Overview and challenges of routing protocol and MAC layer in mobile ad-hoc network. *Journal of Theoretical and Applied Information Technology*, *8*(1).

Jawandhiya, P. M., Ghonge, M. M., Ali, M. S., & Deshpande, J. S. (2010). A survey of mobile ad hoc network attacks. *International Journal of Engineering Science & Technology*, *2*(9), 4063-4071. ISSN 0975-5462

Jeon, J., Lee, K., & Kim, C. (2011). Fast route recovery scheme for mobile ad hoc networks. Proceedings of *ICOIN*, (pp. 419-423).

Jetcheva, J. G., & Johnson, D. B. (2001). Adaptive demand-driven multicast routing in multi-hop wireless ad hoc networks. In *Proceedings of the ACM International Symposium on Mobile Ad hoc Networking and Computing (MobiHoc)*, 2001.

Jiang, S., He, D., & Rao, J. (2005). A prediction-based link availability estimation for routing metrics in MANETs. *IEEE/ACM Transactions on Networking*, *13*, 1302–1312. doi:10.1109/TNET.2005.860094

Jiang, X. (2004). Siren: Context-aware computing for firefighting. [Springer Verlag.]. *IEEE Pervasive Computing / IEEE Computer Society [and] IEEE Communications Society*, *2004*, 87–105.

Ji, L. S., & Corson, M. S. (2001). Differential destination multicast-A MANET multicast routing for multihop ad hoc network. *Proceedings - IEEE INFOCOM, 2*, 1192–1201.

Ji, L., & Scott Corson, M. (2001). *Differential destination multicast–A MANET multicast routing protocol for small groups*. IEEE INFOCOM.

Johnson, D. B. (1994). Routing in ad hoc networks of mobile hosts. In *Proceedings of the Workshop on Mobile Computing Systems and Applications* (pp. 158-163). Santa Cruz, CA: IEEE.

Johnson, D., Hu, Y., & Maltz, D. (2007). *The dynamic source routing protocol (DSR) for mobile ad hoc networks for IPv4.* RFC-4728.

Johnson, D. B., & Maltz, D. A. (1996). Dynamic source routing in ad hoc wireless networks . In Imielinski, T., & Korth, H. F. (Eds.), *Mobile computing* (*Vol. 353*, pp. 152–181). Kluwer Academic Publishers. doi:10.1007/978-0-585-29603-6_5

Julier, S. J., & Uhlmann, J. K. (1997). New extension of the Kalman □lter to nonlinear systems . In *Signal Processing, Sensor Fusion, and Target Recognition VI* (*Vol. 3068*, pp. 182–193). San Diego: SPIE.

Jung, E. S., & Vaidya, N. H. (2002). A power control MAC protocol for ad hoc networks. *Proceedings of ACM International Conference on Mobile Computing and Networking (MOBICOM)*, (pp. 36-47).

Jung, E. S., & Vaidya, N. H. (2002). *An energy efficient MAC protocol for wireless LANs*. IEEE INFOCOM.

Kaddar, L., & Mehaoua, A. (2007). ESTREL: Transmission and reception energy saving model for wireless ad hoc networks. In *Proceedings 32nd IEEE Conference on Local Computer Networks (LCN)* (Short paper), (pp. 784-785).

Kaddar, L., & Mehaoua, A. (2009). *SecVLC: Secure transmission over multimedia wireless ad hoc networks with energy-awareness*. In Global Information Infrastructure Symposium 2009 (GIIS 2009).

Kaddar, L., Hadjadj Aoul, Y., & Mehaoua, A. (2011). *EVAN: Energy-aware SVC video streaming over wireless ad hoc networks*. IEEE 73rd Vehicular Technology Conference: VTC2011-Spring.

Kalia, M., Bansal, D., & Shorey, R. (1999). MAC scheduling and SAR policies for Bluetooth: A master driven TDD pico-cellular wireless system. In *IEEE International Workshop on Mobile Multimedia Communications (MoMuC)*, (pp. 384–388).

Kaliaperumal, B., & Ebenezer, Jeyakumar, A. (2005). Adaptive core based scalable multicasting networks. *INDICON, 2005 Annual IEEE*, (pp. 198-202).

Kallel, L., Naudts, B., & Rogers, A. (Eds.). (2001). *Theoretical aspects of evolutionary computing*. New York, NY: Springer.

Kalman, R. E. (1960). A new approach to linear □ltering and prediction problems. *Transactions of ASME Journal of Basic Engineering*, *82*, 35–45. doi:10.1115/1.3662552

Kang, I., & Poovendran, R. (2003). A comparison of power efficient broadcast routing algorithms. In *Proceedings of IEEE Global Telecommunications Conference (GLOBECOM),* December.

Kant, K., & Awasthi, L. K. (2010). Unicast and multicast routing protocols for MANETs: A comparative survey. *International Journal of IT & Knowledge Management, Special Issue*. ISSN: 0973-4414.

Karimi, M., & Pan, D. (2009). Challenges for quality of service (QoS) in mobile ad-hoc networks (MANETs). *Proceedings of IEEE 10th Annual Wireless and Microwave Technology Conference,* WAMICON '09, (pp. 1-5).

Karp, R. M. (1972). Reducibility among combinatorial problems . In Miller, R. E., & Thatcher, J. W. (Eds.), *Complexity of computer computations* (pp. 85–103). Plenum Press.

Kaufman, S. (1995). *At home in the universe*. Wiley.

Kessler, J., et al. (January 1992). *Functional description of the data fusion process*. Tech. rep., Naval Air Development Center, Warminster, PA, USA.

Khan, A. A., Stojmenovic, I., & Zaguia, N. (2008). Parameterless broadcasting in static to highly mobile wireless ad hoc, sensor and actuator networks. In *Proceedings of* 22nd IEEE *International Conference on Advanced Information Networking and Applications (AINA2008)*, Ginowan, Okinawa, Japan, March 25-28.

Khan, J., & Hayder, S. I. (2010). *A comprehensive performance comparison of on-demand routing protocols in mobile ad-hoc network*. FGCN2010 Jeju Island South Korea 13-15 December 2010.

Khan, J., Shah Khattak, N., & Jan, H. (2011). A step of mobile ad-hoc on-demand routing protocols towards 4G cellular networks. *International Journal of Computers and Applications*, *6*(4). Retrieved from http://www.ijcaonline.org/archives/ volume16/number6/2017-2721

Kim, J., & Bohacek, S. (2005, July). *A survey-based mobility model of people for simulation of urban mesh networks*. In MeshNets'05, Budapest, Hungary.

Kim, J., Lin, X., & Shroff, N. B. (2011). Optimal anycast technique for delay-sensitive energy-constrained asynchronous sensor networks. *IEEE/ACM Transactions on Networking*, *19*(2), 484–497. doi:10.1109/TNET.2010.2072515

Kim, T. H. (Eds.). (2010). *FGCN 2010, Part II, CCIS 120* (pp. 354–369). Heidelberg, Germany: Springer.

Kirkpatrick, S., Gelatt, C. D., & Vecchi, M. P. (1983). Optimization by simulated annealing. *Science*, *220*(4598), 671–680. doi:10.1126/science.220.4598.671

Ko, Y.-B., & Vaidya, N. H. (1998). Location-aided routing (LAR) in mobile ad hoc networks. *Proceedings of 4th ACM International Conference on Mobile Computing and Networking (ACM MobiCom 1998)*, Dallas, Texas, United States, 1998, (pp. 66-75).

Kokar, M. M., Bedworth, M. D., & Frankel, C. B. (2000). A reference model for data fusion systems . In *Sensor Fusion: Architectures, Algorithms and Applications IV* (pp. 191–202). Orlando, FL: SPIE. doi:10.1117/12.381632

Kondylis, G. D., Krishnamurthy, S. V., Dao, S. K., & Pottie, G. J. (2000). Multicasting sustained CBR and VBR traffic in wireless ad hoc networks. In *Proceedings of IEEE International Conference on Communications*, Vol. 1, (pp. 543-549).

Kopke, A., Swigulski, M., Wessel, K., Willkomm, D., Haneveld, P. T. K., Parker, T., et al. (2008). Simulating wireless and mobile networks in OMNeT++ - the MiXiM vision. *Proceedings of the 1st International Workshop on OMNeT++*.

Krenik, W., & Batra, A. (2005). *Cognitive radio techniques for wide area networks*. DAC 2005, Anaheim, California, USA.

Krishnamachari, B., Estrin, D., & Wicker, S. (2002). The impact of data aggregation in wireless sensor networks. *Proceedings of the 22nd International Conference on Distributed Computing Systems*.

Krishnamachari, B., & Iyengar, S. (2004, Mar). Distributed Bayesian algorithms for fault-tolerant event region detection in wireless sensor networks. *IEEE Transactions on Computers*, *53*, 241–250. doi:10.1109/TC.2004.1261832

Kulik, J., Heinzelman, W., & Balakrishnan, H. (2002, March). Negotiation-based protocols for disseminating information in wireless sensor networks. *Wireless Networks*, *8*, 169–185. doi:10.1023/A:1013715909417

Kurkowski, S., Camp, T., & Navidi, W. (2002). *Minimal standards for rigorous MANET routing protocol evaluation. Technical Report MCS 06-02*. Colorado School of Mines.

Labrador, M. A., & Wightman, P. M. (2009). *Topology control in wireless sensor networks*. Springer Science + Business Media B.V. ISBN 978-1-4020-9584-9

Lan, H., & Nguyen, U. T. (2006). Study of different types of attacks on multicast in mobile ad hoc networks. *Proceedings of ICNICONSMCL'06*. ISBN 0-7695-2552-0/06

Laneman, J. N., & Tse, D. N. C. (2004). Cooperative diversity in wireless networks: Efficient protocols and outage behaviour. *IEEE Transactions on Information Theory*, *50*, 3062–3080. doi:10.1109/TIT.2004.838089

Lang, D. (2006). *On the evaluation and classification of routing protocols for mobile ad hoc networks*. Retrieved from http://home.leo.org/~dl/dissertation.pdf

Lee, C. Y., & Cho, H. K. (2001). Multicast routing considering reliability and network load in wireless ad hoc networks. In *Proceedings of the 53rd VTS Spring Vehicular Technology Conference* (pp. 2203-2207). Rhodes, Greece: IEEE.

Lee, H., Cho, C., King, C., Fang, Y., & Lee, B. (2009). *Design and implementation of non-autonomous mobile wireless sensor for debris flow monitoring*. IEEE 6thI-International Conference on Mobile Adhoc and Sensor Systems.

Lee, J., & Stinson, D. R. (2005). Deterministic key predistribution schemes for distributed sensor networks. *Proceedings of ACM Symposium on Applied Computing.*

Lee, M., & Kim, T. (2001). PatchODMRP: An ad hoc multicast routing protocol. In *Proceedings of the 15th International Conference on Information Networking (ICOIN '01)*, (pp. 537-543).

Lee, S. (2000). *Routing and multicasting strategies in wireless mobile ad hoc networks*. Retrieved from http://www.sigmobile.org/ phd/2000/theses/sjlee.pdf

Lee, S. J., Gerla, M., & Chiang, C. C. (1999). *On-demand multicast routing protocol (ODMRP) for ad hoc networks.* Internet draft, draft-retf-manet-odmrp-01.txt, work in progress, 1999.

Lee, S. J., Su, W., Hsu, J., Gerla, M., & Bagrodia, R. (2000). A performance comparison study of ad hoc wireless multicast protocols. In *Proceedings of the 19th Annual Joint Conference on Computer Communications* (pp. 565-574). Tel Aviv, Israel: IEEE.

Lee, S., & Kim, C. (2000). Neighbor supporting ad hoc multicast routing protocol. In *Proceedings of the 1st Annual Workshop on Mobile and Ad hoc Networking and Computing* (pp. 37-44). Boston, MA: ACM.

Lee, S., Gerla, M., & Chaing, C. C. (1999). On-demand multicast routing protocol. In *Proceedings of the Wireless Communications and Networking Conference* (pp. 1298-1302). New Orleans, LA: IEEE.

Lessmann, J., Janacik, P., Lachev, L., & Orfanus, D. (2008). Comparative study of wireless network simulators. *Proceedings of the 7th International Conference of Networking.*

Lewis, T. W., & Powers, D. M. W. (2002). Audio-visual speech recognition using red exclusion and neural networks. In *Proceedings of the 25th Australasian Conference on Computer Science. Australian Computer Society*, Melbourne, Victoria, Australia, (pp. 149–156).

Li, J., & Kato, J. (2008). Future trust management framework for mobile ad hoc networks. *IEEE Communications Magazine, April.*

Li, N., Hou, J., & Sha, L. (2003). Design and analysis of an MST-based topology control algorithm. In *Proceedings of IEEE INFOCOM'03*, San Francisco, CA, (pp. 1702-1712).

Li, Y., Thai, M. T., Wang, F., Yi, C. W., Wang, P. J., & Du, D. Z. (2005). On greedy construction of connected dominating sets in wireless networks. *Special Issue of Wireless Communications and Mobile Computing (WCMC), 5*(8).

Liang, H. B., & Zygmunt, J. (1999). Predictive distance based mobility management for a PCS networks. *Proceedings of the IEEE Infocom.*

Li, L., Halpern, J. Y., Bahl, P., Wang, Y.-M., & Wattenhofer, R. (2005). A cone-based distributed topology-control algorithm for wireless multi-hop networks. *IEEE Transactions on Networking, 13*(1), 147–159. doi:10.1109/TNET.2004.842229

Lim, H., & Kim, C. (2000). Multicast tree construction and flooding in wireless ad hoc networks. In *Proceedings of the 3rd ACM International Workshop on Modeling, Analysis and Simulation of Wireless and Mobile Systems*, (pp. 61-68).

Lima, A., Carvalho, L., Souza, J., & Mota, E. (2007). VoIP management: Speech quality monitoring architecture based on statistical analysis. *International Journal of Network Management, 17*(4).

Lin, C. (n.d.). *AODV routing implementation for scalable wireless ad hoc network simulation (SWANS)*. Retrieved from http://jist.ece.cornell.edu/docs/ 040421-swans-aodv.pdf

Lin, C., & Chao, S. (1999). A multicast routing protocol for multihop wireless networks. *IEEE Global Telecommunication Conference (GlobeCom 1999)*, (pp. 235-239).

Lin, X. H., & Lau, V. K. N. (2003). Power control for IEEE 802.11 ad hoc networks: Issues and a new algorithm. *Proceedings of International Conference on Parallel Processing, ICPP'03*, Kaohsiung, Taiwan.

Lin, C., & Gerla, M. (1997). Adaptive clustering for mobile wireless networks. *IEEE Journal on Selected Areas in Communications, 15*(7), 1265–1275. doi:10.1109/49.622910

Li, T., Ekpenyong, A., & Huang, Y. (2006, Oct). Source localization and tracking using distributed asynchronous sensor. *IEEE Transactions on Signal Processing, 54*, 3991–4003. doi:10.1109/TSP.2006.880213

Liu, J., Zhang, Y., & Zhao, F. (2006). Robust distributed node localization with error management. In *Proceedings of the 7th ACM International Symposium on Mobile Ad Hoc Networking and Computing (MobiHoc '06)*, Florence, Italy, (pp. 250–261).

Liu, B. H., Huang, P. C., & Tsai, M. J. (2008). Distributed reformation of core-based group-shared multicast trees in mobile ad hoc networks. *Journal of Parallel and Distributed Computing, 68*, 582–595. doi:10.1016/j.jpdc.2007.10.002

Liu, W., Chiang, C., Wiu, H., & Gerla, C. (1997). Routing in clustered multihop mobile adhoc networks with fading channel. In . *Proceedings of IEEE SICON, 97*, 197–211.

Lochert, C., et al. (2003). A routing strategy for vehicular ad hoc networks in city environments. *Proceedings IEEE Intelligent Vehicular Symposium*, June 2003, (pp. 156-161).

Lou, W., & Wu, J. (2004). Double-covered broadcast (DCB): A simple reliable broadcast algorithm in MANETs. In *Proceedings of the 23st Annual Joint Conference of the IEEE Computer and Communications Societies (INFOCOM '04),* Hong Kong.

Lou, W., & Wu, J. (2002). On reducing broadcast redundancy in ad hoc wireless networks. *IEEE Transactions on Mobile Computing, 1*(2), 111–122. doi:10.1109/TMC.2002.1038347

Lou, W., & Wu, J. (2007). Toward broadcast reliability in mobile ad hoc networks with double coverage. *IEEE Transactions on Mobile Computing, 6*(2), 148–163. doi:10.1109/TMC.2007.31

Lowdermilk, R., & Harris, F. (1996). Interference mitigation in orthogonal frequency division multiplexing (OFDM). In *Proceedings of IEEE International Conference on Universal Personal Communications,* Vol. 2, Cambridge, MA, (pp. 623–627).

Lundberg, J. (2000). *Routing security in ad-hoc networks*. Retrieved from http://citeseer.nec.com/400961.html

Lundin, R., & Lindskog, S. (2010). Security implications of selective encryption. In *Proceedings of Metric2010*, Bilzano-Bozen, Italy, 2010.

Luo, Q., Chen, G., Huang, C., & Chen, G. (2010). *Study of MANET routing evaluation model based on best-first.* The Telecommunication Engineering Institute, Air Force Engineering University, Xi'an, China. 978-1-4244-5849-3/10

Luo, J., Eugster, P. T., & Hubaux, J. P. (2003). Route driven gossip: Probabilistic reliable multicast in ad hoc networks. *Proceedings - IEEE INFOCOM, 03*, 2229–2239.

Luo, R. C., & Kay, M. G. (1992). Data fusion and sensor integration: State-of-the-art 1990s . In Abidi, M. A., & Gonzalez, R. C. (Eds.), *Data fusion in robotics and machine intelligence* (pp. 7–135). San Diego, CA: Academic Press, Inc.

Luo, R. C., Yih, C.-C., & Su, K. L. (2002, April). Multisensor fusion and integration: Approaches, applications, and future research directions. *IEEE Sensors Journal, 2*, 107–119. doi:10.1109/JSEN.2002.1000251

Luo, X., Dong, M., & Huang, Y. (2006, Jan). On distributed fault-tolerant detection in wireless sensor networks. *IEEE Transactions on Computers*, 55.

Madden, S. R., Franklin, M. J., Hellerstein, J. M., & Hong, W. (2002). TAG: A tiny aggregation service for ad-hoc sensor networks. *ACM SIGOPS Operation Systems Review, 36*, 131–146. doi:10.1145/844128.844142

Madden, S. R., Franklin, M. J., Hellerstein, J. M., & Hong, W. (2005, March). TinyDB: An acqusitional query processing system for sensor networks. *ACM Transactions on Database Systems, 30*, 122–173. doi:10.1145/1061318.1061322

Madhu Viswanatham, V., & Chari, A. A. (2008). An approach for detecting attacks in mobile ad hoc networks. *Journal of Computer Science, 4*(3), 245-251. ISSN 1549-3636

Madruga, E. L., & Garcia, J. J. (2001). Scalable multicasting: The core-assisted mesh protocol. *Kluwer Mobile Network, 6*(2), 151–165. doi:10.1023/A:1011400407314

Mainwaring, A., Polastre, J., Szewczyk, R., Culler, D., & Anderson, J. (2002). *Wireless sensor networks for habitat monitoring.* ACM International Workshop on Wireless Sensor Networks and Applications.

Mamatha, G. S., & Sharma, S. C. (2010). –August). Analyzing the MANET variations, challenges, capacity and protocol issues. *International Journal of Computer Science & Engineering Survey*, *1*(1). doi:10.5121/ijcses.2010.1102

MANET. (n.d.). *The MANET web page*. Retrieved from http://www.ietf.org /html.charters/manet-charter.html

Marathe, M. V., Breu, H., Hunt, H. B. III, Ravi, S. S., & Rosenkrantz, D. J. (1995). Simple heuristics for unit disk graphs. *Networks*, *25*, 59–68. doi:10.1002/net.3230250205

Marcus, M. J. (2005). Unlicensed cognitive sharing of the TV spectrum: The controversy at the Federal Communications Commission. *IEEE Communications Magazine*, 51.

Marti, S., Giuli, T., Lai, K., & Baker, M. (2000). Mitigating routing misbehavior in mobile ad hoc networks. *Proceedings of the Sixth Annual International Conference on Mobile Computing and Networking (MobiCom)*, (pp. 255–265).

Marzullo, K. (1990, November). Tolerating failures of continuous-valued sensors. *ACM Transactions on Computer Systems*, *8*, 284–304. doi:10.1145/128733.128735

Mascolo, C., & Musolesi, M. (2006). SCAR: Context-aware adaptive routing in delay tolerant mobile sensor networks. In *Proceeding of the International Conference on Communications and Mobile Computing (IWCMC)*, ACM, Vancouver, Canada, (pp. 533–538).

Maureira, J., Uribe, P., Dalle, O., Asahi, T., & Amaya, J. (2009). Component based approach using OMNeT++ for train communication modelling. *9th Int. Conf. Intelligent Transport Systems Telecommunications*, (pp. 441-446).

McDonald, A. B., & Znati, T. (1999). A path availability model for wireless ad-hoc networks. *IEEE Wireless Communications and Networking Conference*, (pp. 35-40).

McDonald, A. B., & Znati, T. (2005). *Predicting node proximity in ad-hoc networks: A least overhead adaptive model for selecting stable routes* (pp. 39–43). First Annual Work Shop on Mobile Ad Hoc Networking and Computing.

Meghanathan, N. (2011). Performance comparison study of multicast routing protocols for mobile ad hoc networks under default flooding and density and mobility aware energy-efficient (DMEF) broadcast strategies. *Informatica*, *35*, 165–184.

Merkle, R. (1978). Secure communication over insecure channels. *Communications of the ACM*, *21*(4). doi:10.1145/359460.359473

Meyer, J., & Gadegast, F. (1995). *Security mechanisms for multimedia data with example MPEG-I video.*

Michiardi, P., & Molva, R. (2002). *Game theoretic analysis of security in mobile ad hoc networks.* Institut Eurecom Research Report RR-02-070, April 2002.

Mikki, M. A. (2009). Energy efficient location aided routing protocol for wireless MANETs. *International Journal of Computer Science and Information Security*, *4*(1-2).

Millonas, M. M. (1992). A connectionist-type model of self-organized foraging and emergent behavior in ant swarms. *Journal of Theoretical Biology*, *159*, 529. doi:10.1016/S0022-5193(05)80697-6

Millonas, M. M. (1994). Swarms, phase transitions and collective intelligence . In Langton, C. G. (Ed.), *Artificial Life III, Santa Fe Institute, Studies in the Sciences of Complexity* (*Vol. XVII*, pp. 417–445). Reading, MA: Addison- Wesley.

Misgra, A., & Nadkarni, K. M. (2003). Security in wireless ad hoc networks . In Ilyas, M. (Ed.), *The handbook of ad hoc wireless networks*. CRC Press LLC.

Mitola, J. (1999). Cognitive radio for flexible multimedia communications. *1999 IEEE International Workshop on Mobile Multimedia Communications,* (pp. 3 –10)

Mitola, J. (2000). *Cognitive radio: An integrated agent architecture for software defined radio.* Ph.D. dissertation, Computer Communication System Laboratory, Department of Teleinformatics, Royal Institute of Technology (KTH), Stockholm, Sweden, May 2000.

Mobility Framework. (n.d.). Retrieved from http://mobility-fw.sourceforge.net/

Mohapatra, P., & Krishnamurthy, S. (2005). *Ad hoc networks: Technologies and protocols.* Springer Science, 2005.

Moy, J. (1998). *Open shortest path first (OSPF) version 2.* RFC 2328, April 1998.

Moy, J. (1994). Multicast routing extensions for OSPF. *Communications of the ACM, 37*(8), 61–66. doi:10.1145/179606.179654

Muhammad, R. B. (2006). Distributed Steiner tree algorithm and its application in ad hoc wireless networks. In *Proceedings of The 2006 International Conference on Wireless Networks (ICWN'06)*, USA, (pp. 173-178).

Muhammad, R. B. (2007). A distributed graph algorithm for geometric routing in ad hoc wireless networks. *Journal of Networks, 2*(6), 50–57. doi:10.4304/jnw.2.6.50-57

Mukija, A. (2001). *Reactive routing protocols for mobile ad hoc network.* Retrieved from http://www.ifi.unizh.ch/~mukhija/papers/rrp_thesis.pdf

Mukija, A. (2001). *Reactive routing protocols for mobile ad hoc network.* Retrieved from http://www.ifi.unizh.ch/~mukhija/ papers/rrp_thesis.pdf

Murphy, S. (2002). *Routing protocol threat analysis.* Internet Draft, draft-murphy-threat-00.txt, October 2002.

Murthy, S., & Garcia-Luna-Aceves, J. J. (1996). An efficient routing protocol for wireless networks. *ACM Mobile Networks and Applications Journal, Special Issue on Routing in Mobile Communication Networks*, 183-97.

Murthy, S., & Garcia-Luna-Aceves, J. J. (1996). An efficient routing protocol for wireless networks. *Mobile Networks and Applications, 1*(2), 183–197. doi:10.1007/BF01193336

Musolesi, M., Hailes, S., & Mascolo, C. (2004). An ad hoc mobility model founded on social network theory. In *Proceeding MSWiM '04 Proceedings of the 7th ACM International Symposium on Modeling, Analysis and Simulation of Wireless and Mobile Systems.* New York, NY: ACM.

Nadeem, T., & Parthasarathy, S. (2006). Mobility control for throughput maximization in ad hoc networks. *Wireless Communication and Mobile Computing, 6*, 951–967. doi:10.1002/wcm.431

Nait-Abdesselam, F., Bensaou, B., & Taleb, T. (2003). Detecting and avoiding wormhole attack in wireless ad hoc networks. *IEEE Communications Magazine, 46*(4), 127–133. doi:10.1109/MCOM.2008.4481351

Nakamura, E. F., Nakamura, F. G., Figueiredo, C. M., & Loureiro, A. A. (2005, November). Using information fusion to assist data dissemination in wireless sensor networks. *Telecommunication Systems, 30*, 237–254. doi:10.1007/s11235-005-4327-y

Naoumov, V., & Gross, T. (2003). Simulation of large ad hoc networks. *Proceedings of the 6th ACM International Workshop on Modeling Analysis and Simulation of Wireless and Mobile Systems.* OMNeT++. (n.d.). *Website.* Retrieved from http:/www.omnetpp.org/

National Institute for Standards and Technology (NIST). (2001). *Advanced encryption standard (AES).* FIPS 197. Retrieved from http://csrc.nist.gov/encryption/aes/

Naval Research Laboratory. (n.d.). *Products.* Retrieved from http://cs.itd.nrl.navy.mil/products/

Neel, J., Menon, R., MacKenzie, A., & Reed, J. (2005). *Using game theory to aid the design of physical layer cognitive radio algorithms.* Conference on Economics, Technology and Policy of Unlicensed Spectrum, May 16-17, Lansing, Michigan.

Network Working Group. (2004). *The secure real-time transport protocol (STRP).* Retrieved from http://www.ietf.org/rfc/rfc3711.txt

Neumann, B. C., & Tso, T. (1994). Kerberos: An authentication service for computer networks. *IEEE Communications, 32*(9), 33–39. doi:10.1109/35.312841

Ni, S., Tseng, Y. C., & Sheu, J. (1999) "The broadcast storm problem in a mobile ad hoc network. In *Proceedings of the ACM/IEEE International Conference on Mobile Computing and Networking (MOBICOM)*, (pp. 151–162).

Nie, J., Wen, J., Luo, J., He, X., & Zhou, Z. (2006). An adaptive fuzzy logic based secure routing protocol in mobile ad hoc networks. *Journal of Fuzzy Sets and Systems, 157*, 1704–1712. doi:10.1016/j.fss.2005.12.007

Nordlund, P.-J., Gunnarsson, F., & Gustafsson, F. (2002). Particle □lters for positioning in wireless networks. In *Proceedings of the XI European Signal Processing Conference (EURSIPCO'02). Vol. II*, TeSA, Toulouse, France, (pp. 311–314).

NS-2. (n.d.). *Network Simulator- ns-2*. Retrieved Febuary 15, 2011, from http://www.isi.edu/nsnam/ns

NS-2. (n.d.A). *Documentation*. Retrieved from http://www.isi.edu/nsnam/ ns/ns-documentation.html

NS-2. (n.d.B). *Tutorial*.

Oh, S. J., Kim, Y. J., & Bae, I. H. (2001). A fuzzy multicast method providing linguistic guarantee of disruption free service in mobile wireless networks. *Eighth International IEEE Conference on Parallel and Distributed Systems*, (pp. 135-140).

Oh, S. Y., Park, J. S., & Gerla, M. (2008). E-ODMRP: Enhanced ODMRP with motion adaptive refresh. *Journal of Parallel and Distributed Computing, 68*, 1044–1053. doi:10.1016/j.jpdc.2008.04.004

OPNET. (2007). *Modeler wireless suite for defence*. Retrieved from http://www.opnet.com/ products/modeler/home.html

Ozaki, T., Kim, J. B., & Suda, T. (1999). Bandwidth-efficient multicast routing protocol for ad hoc networks. In *Proceedings of the 8th International Conference on Computer Communications and Networks* (pp. 10-17). Boston, MA: IEEE.

Ozaki, T., Kim, J. B., & Suda, T. (2001). Bandwidth-efficient multicast routing for multi-hop ad hoc networks. *Proceedings - IEEE INFOCOM, 2*, 1182–1191.

Paek, J., Chintalapudi, K., Caffrey, J., Govindan, R., & Masri, S. (2005). A wireless sensor network for structural health monitoring: performance and experience. *Proceedings of the 2nd IEEE Workshop on Embedded Networked Sensors.*

Pagac, D., Nebot, E. M., & Durrant-Whyte, H. (1998, August). An evidential approach to map-building for autonomous vehicles. *IEEE Transactions on Robotics and Automation, 14*, 623–629. doi:10.1109/70.704234

Pakzad, S. N., Kim, S., Fenves, G. L., Glaser, S. D., Culler, D. E., & Demmel, J. W. (2005). Multi-purpose wireless accelerometers for civil infrastructure monitoring. *Proceedings of the 5th International Workshop on Structural Health Monitoring.*

Pal Chaudhuri, S. (2011). *Ns-2 code for random trip mobility model.* Retrieved from http://monarch.s.rice.edu/ ~santa/research /mobility/

Pan, Y. (2007). *Design routing protocol performance comparision in NS2: AODV comparing to DSR as example.* Department of CS, SUNY Binghamton, Vestal NY.

Pan, Y. (2010). *Introduction to OPNET simulator*. Retrieved from http://www.scribd.com/doc/ 37951842/ Introduction to OPNET Simulator

Papadimitratos, P., & Haas, Z. (2002). Secure routing for mobile ad hoc networks. In *Proceedings of SCS Communication Networks and Distributed Systems Modeling and Simulation Conference (CNDS)*, 2002.

Papadimitratos, P., & Haas, Z. J. (2006). Secure data communication in mobile ad hoc networks. *IEEE Journal on Selected Areas in Communication, 24*(2).

Papadimitriou, C. H., & Steiglitz, K. (1982). *Combinatorial optimization—Algorithms and complexity*. New York, NY: Dover Publications Inc.

Park, V. D., & Corson, M. S. (1997). A highly adaptive distributed routing algorithm for mobile wireless networks. In *Proceedings of the 16th Annual International Conference on Computer Communications* (pp. 1405-1413). Las Vegas, NV: IEEE.

Park, V., & Corson, S. (2001). *IETF MANET working group.* Retrieved from http://tools.ietf.org/id/ draft-ietf-manet-tora-spec-04.txt

Patwari, N., Hero, A. O., Perkins, M., Correal, N. S., & O'Dea, R. J. (2003, Aug). Relative location estimation in wireless sensor networks. *IEEE Transactions on Signal Processing, 51*, 2137–2148. doi:10.1109/TSP.2003.814469

Peng, W., & Lu, X. C. (2000). On the reduction of broadcast redundancy in mobile ad hoc networks. In *Proceedings First Annual Workshop on Mobile and Ad Hoc Networking and Computing*, (pp. 129-130).

Perkin, C. E., & Bhagwat, P. (1994). *Highly dynamic destination-sequenced distance-vector routing (DSDV) for mobile computers.* ACM SIGCOMM, 1994.

Perkin, C. E., & Royer, E. M. (2000). The ad hoc on-demand distance vector routing protocol . In Perkin, C. E. (Ed.), *Ad hoc networking* (pp. 173–219). Addison Wesley.

Perkins, C. E., & Bhagwat, P. (1994). Highly dynamic destination –Sequence distant vector routing (DSDV) for mobile compute. *Proceedings of the Conference on Communications Architectures, Protocols and Applications*, England, August 1994, (pp. 234-244).

Perkins, C. E., & Royer, E. M. (1999). Ad hoc on demand distance vector routing. *Proceedings Second IEEE Workshop on Mobile Computing Systems and Applications, WMCSA '99,* (pp. 90-100).

Perkins, C. E., Belding-Royer, E. M., & Das, S. R. (2002). *Ad hoc on-demand distance vector (AODV) routing.* Internet Engineering Task Force, Internet Draft (Work in Progress), June 2002.

Perkins, C. E., Belding-Royer, E. M., & Das, S. R. (2003, February 17). *Ad hoc on-demand distance vector (AODV) routing.* IETF Internet Draft- draft-ietf-manet-aodv-13.txt.

Perkins, C., & Royer, E. M. (1998). *Ad hoc on demand distance vector (AODV) routing.* Internet draft, 1998.

Perkins, C., Belding-Royer, E., & Das, S. (2003). *Ad hoc on-demand distance vector (AODV) routing.* RFC-3561 8.

Perkins, C. E. (2000). *Ad hoc networking.* Addison-Wesley Longman.

Perrig, A., & Canetti, R. (2000). *The TESLA broadcast authentication protocol.* Internet Draft, 2000.

Perrig, A., Canetti, R., Tygar, J. D., & Song, D. (2000). Efficient authentication and signing of multicast streams over lossy channels. In *Proceedings of IEEE Symposium on Security and Privacy,* 2000.

Perrig, A., Hu, Y.-C., & Johnson, D. B. (2001). *Wormhole protection in wireless ad hoc networks.* Technical Report TR01-384, Dept. of Computer Science, Rice University.

Pham, H. N., Pediaditakis, D., & Boulis, A. (2007). From simulation to real deployments in wsn and back. *Proceedings of the 2007 IEEE International Symposium on a World of Wireless, Mobile and Multimedia Networks.*

Pirzada, A. A., Datta, A., & McDonald, C. (2006). Incorporating trust and reputation in the DSR protocol for dependable routing. *Computer Communications* [Elsevier Press.]. *Special Issue on Internet Communications Security, 29,* 2806–2821.

Podesser, M., Schmidt, H.-P., & Uhl, A. (2002). *Selective bitplane encryption for secure transmission of image data in mobile environments.* 5th Nordic Signal Processing Symposium, on board Hurtigruten, Norway, October 4-7.

Polastre, J., Hill, J., & Culler, D. (2004). Versatile low power media access for wireless sensor networks. In *SenSys '04*, (pp. 95–107).

Pommer, A., & Uhl, A. (2003). Selective encryption of wavelet-packet encoded image data. *ACM Multimedia Systems Journal, Special Issue on Multimedia Security, 9*(3).

Prasithsangaree, P., & Krishnamurthy, P. (2004). On a framework for energy-efficient security protocols in wireless networks. *Computer Communications, 27,* 1716–1729. doi:10.1016/j.comcom.2004.05.022

Prizada, A. A., & McDonald, C. (2003). Secure routing protocols for mobile ad hoc wireless networks. In *Proceedings of 2nd Workshop on Internet, Telecommunication and Signal Processing," WITSP 2003*, Australia, (pp. 57-80).

Pushkarev, B., & Zupan, J. M. (1975). *Urban space for pedestrians.* MIT Press.

Qabajeh, L. K., Kiah, L. M., & Qabajeh, M. M. (2009). A qualitative comparison of position-based routing protocols for ad-hoc networks. *International Journal of Computer Science and Network Security, 9*(2).

Qayyum, A., Viennot, L., & Laouiti, A. (2001). Multipoint relaying: An efficient technique for flooding in mobile wireless networks. In *Proceedings of the 35th Annual Hawaii International Conference on System Sciences.*

Qiao, L., & Nahrstedt, K. (1998). Comparison of MPEG encryption algorithms. *International Journal on Computer and Graphics, Special Issue on Data Security in Image Communication and Network, 22*(3).

Rai, A. K., Tewari, R. R., & Upadhyay, S. K. (2010). Different types of attacks on integrated MANET-Internet communication. [IJCSS]. *International Journal of Computer Science and Security*, *4*(3).

Rajaraman, R. (2002). Topology control and routing in ad hoc networks: A survey. *ACM SIGACT News*, *33*(2), 60–73. doi:10.1145/564585.564602

Ramachandran, K. (2006). *Technical report, University of California, Santa Barbara, USA*. Retrieved April 15, 2006, from http://www.cs.ucsb.edu/krishna/aodv-st

Ramos, V., Fernandes, C., & Rosa, A. C. (2005). Social cognitive maps, swarm collective perception and distributed search on dynamic landscapes. *Brains, Minds & Media – Journal of New Media in Neural and Cognitive Science.*

Ram, S. C., & Manog, S. B. (2004). *Ad hoc wireless networks: Architecture and protocol* (1st ed.). Prentice Hall.

Rappaport, T. (1996). *Wireless communications: Principles and practice*. Englewood Cliffs, NJ: Prentice Hall.

Ratnasamy, S., Karp, B., Shenker, S., Estrin, D., Govindan, R., Yin, L., & Yu, F. (2003, Aug). Data-centric storage in sensornets with GHT, a geographic hash table. [MONET]. *Mobile Networks and Applications*, *8*, 427–442. doi:10.1023/A:1024591915518

Raychaudhuri, D., Mandayam, N. B., Evans, J. B., Ewy, B. J., Seshan, S., & Steenkiste, S. (2006). *CogNet – An architecture for experimental cognitive radio networks within the future Internet*. MobiArch'06 - First International Workshop on Mobility in the Evolving Internet Architecture.

Rechenberg, I. (1973). *Evolutionsstrategie: Optimierung technischer Systeme nach Prinzipien der biologischen Evolution*. Frommann-Holzboog.

Rieser, C. (2004). *Biologically inspired cognitive radio engine model utilizing distributed genetic algorithms for secure and robust wireless communications and networking.* PhD Dissertation, Virginia Tech, August.

Rivest, R. L., Shamir, A., & Adleman, L. M. (1978). A method for obtaining digital signatures and public key cryptosystems. *Communications of the ACM*, *21*(2), 120–126. doi:10.1145/359340.359342

Roche, A., Westphall, C. B., Pack, S., & Mecklenburg, G. V. (2002). Quality of service for ad-hoc wireless network. *Proceedings of the 22nd International Conference of the Chilean,* (pp. 100-105).

Rodolakis, G., Laouiti, A., Jacquet, P., & Naimi, A. M. (2008). Multicast overlay spanning trees in ad hoc networks: Capacity bounds protocol design and performance evaluation. *Computer Communications*, *31*, 1400–1412. doi:10.1016/j.comcom.2008.01.019

Rodoplu, V., & Meng, T. (1999). Minimum energy mobile wireless networks. *IEEE Journal on Selected Areas in Communications*, *17*(8), 1333–1344. doi:10.1109/49.779917

Romano, N. C., & Numamaker, J. F. (2001). Meeting analysis: Findings from research and practice. in *Proceedings of the 34th Hawaii International Conference on Systems Science*.

Romer, K., & Mattern, F. (2004). The design space of wireless sensor networks. *IEEE Journals on Wireless Communications*, *11*(6), 54–61. doi:10.1109/MWC.2004.1368897

Rondeau, T. W., Rieser, C. J., Gallagher, T. M., & Bostian, C. W. (2004). Online modeling of wireless channels with hidden Markov models and channel impulse responses for cognitive radio. *IEEE Proceedings of IMS 2004*.

Rouskas, G. N., & Baldine, I. (1997). Multicast routing with end-to-end delay and delay variation constraints. *IEEE Journal on Selected Areas in Communications*, 1–9.

Royer, E. M., & Perkins, C. E. (1999). Multicast operation of the ad hoc on-demand distance vector routing protocol. In *Proceedings of MobiCom*, Seattle, (pp. 207-218).

Royer, E. M., & Toh, C. (1999). A review of current routing protocols for ad hoc mobile wireless networks. *IEEE Personal Communications*, *6*(2), 46–55. doi:10.1109/98.760423

Royer, E. M., & Toh, C. K. (2007). A review of current routing protocols for ad hoc mobile wireless networks. *IEEE Personal Communications*, *2*(6), 46–55.

Sadagopan, N., Bai, F., Krishnamachari, B., & Helmy, A. (2003). Paths: Analysis of PATH duration statistics and their impact on reactive MANET routing protocols. *International Symposium on Mobile Ad Hoc Networking and Computing*, (pp. 245-256).

Samar, P., & Wicker, S. B. (2004). On the behavior of communication links of a node in a multi-hop mobile environment. *ACM International Symposium on Mobile Ad Hoc Networking and Computing*, pp. 145-156.

Sanchez, M. (2002). *Mobility models*. Retrieved from http://www.disca.upv.es/ misan/mobmodel.htm

Sanchez, M., & Manzoni, P. (1999). A Java-based ad hoc networks simulator. In *Proceedings of the SCS Western Multiconference Web-based Simulation Track.*

Sanchez, M., & Manzoni, P. (2001). Anejos: A Java based simulator for ad-hoc networks. *Future Generation Computer Systems*, *17*(5), 573–583. doi:10.1016/S0167-739X(00)00040-6

Santi, P., & Blough, D. M. (2002). An evaluation of connectivity in mobile wireless ad hoc networks. *Proceedings IEEE DSN*, June 2002.

Santini, S., & Romer, K. (2006). An adaptive strategy for quality-based data reduction in wireless sensor networks. In *Proceedings of the 3rd International Conference on Networked Sensing Systems (INSS)*, TRF, Chicago, IL, (pp. 29–36).

Santi, P. (2005). The critical transmitting range for connectivity in mobile ad hoc networks. *IEEE Transactions on Mobile Computing*, *4*, 310–317. doi:10.1109/TMC.2005.45

Sanzgiri, K., Dahill, B., Levine, B. N., Shields, C., & Belding-Royer, E. M. (2002). *A secure routing protocol for ad hoc networks*. In International Conference on Network Protocols (ICNP), 2002.

Sanzgiri, K., Laflamme, D., Dahill, B., Levine, B. N., Shields, C., & Belding-Royer, E. M. (2005). Authenticated routing for ad hoc networks. *Selected Areas in Communications*, *23*(3), 598–610. doi:10.1109/JSAC.2004.842547

Sarela, M. (2004). *Measuring the effects of mobility on reactive ad hoc routing protocols*, Research Report 91, Helsinki University of Technology Laboratory for Theoretical Computer Science, 2004.

Sarkar, S. K., Basavaraju, T. G., & Puttamadappa, C. (2008). *Ad hoc mobile wireless networks* (pp. 261–270). Boca Raton, FL: Auerbach Publications.

Sasson, Y., Cavin, D., & Schiper, A. (2003). Probabilistic broadcast for flooding in wireless mobile ad hoc networks. In *Proceedings IEEE Wireless Communications & Networking Conference (WCNC 2003)*, (pp. 1124-1130). March 2003.

Satapathy, D., & Peha, J. M. (1998). Etiquette modification for unlicensed spectrum: Approach and impact. In *Proceedings of the IEEE Vehicular Technology Conference*, (pp. 272–276).

Savvides, A., Han, C., & Strivastava, M. B. (2003, August). The n-hop multilateration primitive for node localization. *Mobile Networks and Applications*, *8*, 443–451. doi:10.1023/A:1024544032357

Schmid, U., & Schossmaier, K. (2001, April). How to reconcile fault-tolerant interval intersection with the Lipschitz condition. *Distributed Computing*, *14*, 101–111. doi:10.1007/PL00008927

Scott, M., Joseph, P., & Gregory, H. (1999). *Internet based mobile ad hoc networking*. IEEE Internet Computing Magazine.

Sen, J. (2010). An intrusion detection architecture for clustered wireless ad hoc networks. *Proceedings of IEEE Second International Conference on Computational Intelligence, Communication Systems and Networks*, (pp. 202-207).

Sendonaris, A., Erkip, E., & Aazhang, B. (2003). User cooperation in diversity - Part I: System description. *IEEE Transactions on Communications*, *51*, 1927–1938. doi:10.1109/TCOMM.2003.818096

Shanthi, N., Lganesan, & Ramar, K. (2009). Study of different attacks on multicast mobile ad hoc network. *Journal of Theoretical and Applied Information Technology*, *10*(1).

Sheikh, R., Singh, C., & Mishra, D. K. (2010). Security issues in MANET: A review. *Proceedings of IEEE Seventh International Conference on Wireless and Optical Communications Networks* (WOCN), (pp. 1-5).

Sheu, J., Tu, S., & Hsu, C. (2008). Location-free topology control protocol in wireless ad hoc networks. *Elsevier . Computer Communications*, *31*(14), 3410–3419. doi:10.1016/j.comcom.2008.05.039

Shu, H., & Liang, Q. (2005). Fuzzy optimization for distributed sensor deployment. In *IEEE Wireless Communications and Networking Conference (WCNC), Vol. 3*, IEEE, New Orleans, LA, (pp. 1903–1908).

Shulsky, A. N., & Schmitt, G. J. (2002). *Silent warfare: Understanding the world of intelligence* (3rd ed.). New York, NY: Brasseys, Inc.

Sichitiu, M. L., & Ramadurai, V. (2004). Localization of wireless sensor networks with a mobile beacon. In *Proceedings of the 1st IEEE International Conference on Mobile Ad Hoc and Sensor Systems (MASS 2004)*, IEEE, Fort Lauderdale, FL, (pp. 174–183).

Singh, A., Nowak, R., & Ramanathan, P. (2006). Active learning for adaptive mobile sensing networks. In *Proceedings of the 5th International Conference on Information Processing in Sensor Networks (IPSN)*, (pp. 60–68).

Singh, G., & Vellanki, K. (1998). A distributed protocol for constructing multicast trees. In *Proceedings of the International Conference on Principles of Distributed Systems*, France, 1998.

Singh, S., & Raghavendra, C. S. (1998). PAMAS Power aware multi-access protocol with signaling for ad hoc networks. *ACM Computer Communication Review*, *28*(3), 5–26. doi:10.1145/293927.293928

Sinha, P., Sivakumar, R., & Bharghavan, V. (1999). MCEDAR: Multicast core extraction distributed ad hoc routing. *IEEE Wireless Communications and Networking Conference (WCNC)*, (pp. 1313-1317).

Sithitavorn, K., & Qiu, B. (2005). *Mobility prediction with direction tracking on dynamic source routing* (pp. 1–6). IEEE TENCON.

Siva Ram Murthy, C., & Manoj, B. S. (2004). *Ad hoc wireless networks, architecture and protocols*. Prentice Hall PTR.

Sivalingam, K. M., Chen, J., & Agrawal, P. (2000). Design and analysis of low-power access protocols for wireless and mobile ATM networks. *Wireless Networks*, *6*, 73–87. doi:10.1023/A:1019152506607

Slideshare. (n.d.). *Open problems in mobile ad hoc networking*. Retrieved from http://www.slideshare.net/networkingcentral/ open-problems-in-mobile-ad-hoc-networking

So, B., Jeon, H., & Lee, J. (2004). PEODMRP: Performance enhanced on-demand multicast routing protocol. In *Proceedings of the 8th Joint Conference on Communications & Informations (JCCI'04)*, Korea, 2004.

Song, W., Huang, R., Xu, M., Shirazi, B. A., & LaHusen, R. (2010). Design and deployment of sensor network for real-time high-fidelity volcano monitoring. *IEEE Transactions on Parallel and Distributed Systems*, *21*(11), 1658–1674. doi:10.1109/TPDS.2010.37

Spanos, G. A., & Maples, T. B. (1995). Performance study of a selective encryption scheme for security of networked, real-time video. In *Proceedings of the 4th International Conference on Computer Communications and Networks* (ICCCN'95), (pp. 72-78). Las Vegas, Nevada, USA.

Stemm, M., & Katz, R. H. (1997). Measuring and reducing energy consumption of network interface in handheld devices. *IEICE Transaction on Communications . E (Norwalk, Conn.)*, *80-B*(8), 1125–1131.

Stojmenovic, I., & Seddigh, M. (2001). Dominating sets and neighbor elimination based broadcasting algorithms in wireless network. *IEEE Transactions on Parallel and Distributed Systems*, *2*(12), 1–12.

Stützle, T., & Hoos, H. H. (2000). MAX-MIN ant system. *Future Generation Computer Systems*, *16*(8), 889–914.

Su, W., & Gerla, M. (1999). IPv6 flow handoff in ad hoc wireless networks using mobility prediction. In *Proceedings of the Global Telecommunications Conference* (pp. 271-275). New Orleans, LA: IEEE.

Su, B. L., Wang, M. S., & Huang, Y. M. (2008). Fuzzy logic weighted multi-criteria of dynamic route lifetime for reliable multicast routing in ad hoc networks. *Expert Systems with Applications*, *35*, 476–484. doi:10.1016/j.eswa.2007.07.010

Sun, Y., *et al.* (2008). Defense of trust management vulnerabilities in distributed networks. *IEEE Communications Magazine, February*.

Sun, Y. (2006). Information theoretic framework of trust modeling and evaluation for ad hoc networks. *IEEE Journal on Selected Areas in Communications*, *24*(2).

Suresh, A. (2005). *Performance analysis of ad hoc on-demand distance vector routing (AODV) using OPNET simulator*. University of Bremen. Retrieved from http://www.comnets.uni-bremen.de/ ~koo/OPNET-AODV-asn.pdf

Su, W., Lee, S. J., & Gerla, M. (2000). Mobility prediction and routing in ad hoc wireless networks. *International Journal of Network Management*, *11*(1).

Su, W., Lee, S., & Gerla, M. (2001). Mobility prediction and routing in ad hoc wireless networks. *Wireless Communications and Mobile Computing*, *11*, 3–30.

Swarnapriyaa, U. G., Vinodhini, V., Anthoniraj, S., & Anand, R. (2011). Auto configuration in mobile ad hoc networks. *Proceedings of the National Conference on Innovations in Emerging Technology*, (pp. 61-66).

Szewczyk, R., Polastre, J., Mainwaring, A., Anderson, J., & Culler, D. (2004). An analysis of a large scale habitat monitoring application. *Proceedings of the 2nd International Conference on Embedded Networked Sensor Systems*, (pp. 214-226).

Tamilselvan, G. M., & Shanmugam, A. (2010, April). Performance analysis of coexistent heterogeneous networks for various routing protocols using Qualnet simulation. *International Journal of Computer Theory and Engineering*, *2*(2), 1793–1820.

Tang, L. (2007). Methods for encrypting and decrypting MPEG video data efficiently. *Proceedings of the 4th ACM International Multimedia Conference,* Boston, MA, November 18-22, (pp. 219-230).

Tardo, J. J., & Algappan, K. (1991). SPX: Global authentication using public key certificates. In *Proceedings of the 1991 IEEE Symposium on Security and Privacy*, (pp. 232–244). Oakland, CA, May 1991. IEEE Computer Society Press.

Tavel, P. (2007). *Modeling and simulation design*. Natick, MA: AK Peter Ltd.

Thakur, M., Sirisena, H., & McLoughlin, I. V. (2005). *Link layer mitigation in rural UHF- MIMO linking systems*. TENCON 2005 IEEE Region 10.

Thaler, D. (1996, November). *Interoperability rules for multicast routing protocols.* Internet draft draft-thaler-interop.

Thaler, D., Alaettinoglu, C., Estrin, D., & Handley, M. (1998, September). The MASC/BGMP architecture for inter-domain multicast routing. In SIGCOMM, (pp. 93-104). Vancouver, British Columbia.

The CMU Monarch Project. (1998). The CMU Monarch projects wireless and mobility extensions to NS. Retrieved from http://www.monarch.cs.cmu.edu

The Network Simulator – NS-2. (n.d.). Retrieved from http://www.isi.edu/nsam/ns/

Theodorakopoulos, G., & Baras, G. S. (2006). On trust models and trust evaluation metrics for ad hoc networks. *IEEE Journal on Selected Areas in Communications*, *24*(2). doi:10.1109/JSAC.2005.861390

Theraulaz, G., & Bonabeau, E. (1999). A brief history of stigmergy. *Artificial Life*, *5*(2), 97–116. doi:10.1162/106454699568700

Thyagarajan, A. S., & Deering, S. E. (1995). Hierarchical distance vector multicast routing for the MBone. In *ACM SIGCOMM Computer Communication Review*, (pp. 60-66).

Tirthraj, R., & Verma, A. K. (2009). Survey and analysis of secure routing protocols for MANETs. In the *Proceedings of National Conference on Cutting Edge Computer and Electronics Technology* (CECT 2009), Pantnager, February 14-16, (pp. 501-06).

Tiwari, A., Ballal, P., & Lewis, F. L. (2007). Energy-efficient wireless sensor network design and implementation for condition-based maintenance. *ACM Transactions on Sensor Networks, 3*(1).

Todd, R. A., & Alec, Y. (2008). Adaptive threat modeling for secure ad hoc routing protocols. [ENTCS]. *Electronic Notes in Theoretical Computer Science*, *197*(2), 3–14. doi:10.1016/j.entcs.2007.12.013

Toh, C. K., Guichal, G., & Bunchua, S. (2000). ABAM: On-demand associativity-based multicast routing for ad hoc mobile networks. In *Proceedings of the 52nd Fall Vehicular Technology Conference* (pp. 987-993-97). Boston, MA: IEEE.

Toh, C. K. (1997). Associativity-based routing for ad hoc mobile networks. *IEEE Personal Communications*, *4*(2), 103–139.

Toh, C. K. (2002). *Ad hoc mobile wireless networks: Protocols and systems*. Englewood Cliff, NJ: Prentice Hall.

Tolety, V. (1999). *Load reduction in ad hoc networks using mobile servers*. Master's Thesis, Colorado School of Mines.

Tolle, G., Polastre, J., Szewczyk, R., Culler, D., Turner, N., Tu, K., et al. (2005). A macroscope in the redwoods. *Proceedings of the ACM Conference on Embedded Networked Sensor Systems*.

Treiber, M., Hennecke, A., & Helbing, D. (2000). Congested traffic states in empirical observations and microscopic simulations. *Physical Review E: Statistical Physics, Plasmas, Fluids, and Related Interdisciplinary Topics*, *62*, 1805–1824. doi:10.1103/PhysRevE.62.1805

Treiber, M., & Kesting, A. (2009). Models for traffic flow and vehicle motion . In Olariu, S., & Weigle, M. C. (Eds.), *Vehicular networks from theory to practice* (pp. 26–28). Boca Raton, FL: CRC Press. doi:10.1201/9781420085891-c2

Tseng, S. Y., Lin, C. C., & Huang, Y. M. (2008). Ant colony-based algorithm for constructing broadcasting tree with degree and delay constraints. *Expert Systems with Applications*, *35*, 1473–1481. doi:10.1016/j.eswa.2007.08.018

Tseng, Y., Hsu, C., & Hsieh, T. (2003). Power-saving protocols for IEEE 802.11-based multi-hop ad hoc networks. *Computer Networks . The International Journal of Computer and Telecommunications Networking*, *43*(3), 317–337.

Umang, S., Reddy, B. V. R., & Hoda, M. N. (2010). Enhanced intrusion detection system for malicious node detection in ad hoc routing protocols using minimal energy consumption. *IET Communications*, *4*(17), 2084–2094. doi:10.1049/iet-com.2009.0616

Valery, N., et al. (2006). An evaluation of inter-vehicle ad hoc networks based on realistic vehicular traces. In *International Symposium on Mobile Ad Hoc Networking & Computing*, (pp. 108–119).

Van Droogenbroeck, M., & Benedett, R. (2002). Techniques for a selective encryption of uncompressed and compressed images. *Proceedings of Advanced Concepts for Intelligent Vision Systems (ACIVS) 2002*, Ghent, Belgium, September 9-11.

Varga, A., & Hornig, R. (2008). *An overview of the OMNeT++ simulation environment*. 1st International Conference of Simulation tools and Techniques for Communications, Networks and Systems.

Verma, A. K., Dave, M., & Joshi, R. C. (in press). Secure data sharing in mobile adhoc networks. *International Review on Computers and Software (IRECOS), in press*. ISSN 1828-6003

Vidhyasanker, V., Manoj, B. S., & Murthy, C. S. R. (2006). Slot allocation schemes for delay sensitive traffic support in asynchronous wireless mesh networks. *The International Journal of Computer and Telecommunications Networking*, *50*(15), 2595–2613.

Vijayaragavan, S. (2009). A performance study of reactive multicast routing protocols in virtual class room using mobile ad hoc network. *Journal of Computer Science*, *5*(11), 788–793. doi:10.3844/jcssp.2009.788.793

Waitzman, D., Partridge, C., & Deering, S. (1988). Distance vector multicast routing protocol. *Request for Comments, 1075*.

Wald, L. (1999, May). Some terms of reference in data fusion. *IEEE Transactions on Geoscience and Remote Sensing*, *13*, 1190–1193. doi:10.1109/36.763269

Wang, J. C. P., Abolhasan, M., Safaei, F., & Franklin, D. (2007). A survey on control separation techniques in multi-radio multi-channel MAC protocols. In *the 7th IEEE International Symposium on Comm. and Information Technologies* (IEEE ISCIT), Sydney, Australia.

Wang, K., & Chang, C. T. (2001). An intelligent on-demand multicast routing protocol in ad hoc networks. In *Proceedings of the 15th International Conference on Information Networking* (pp. 909-914). Beppu City, Japan: IEEE.

Wang, Y., & Xie, J. (2000). Ant colony optimization for multicast routing. *IEEE APCCAS 2000, The 2000 IEEE Asia-Pacific Conference on Circuits and Systems,* (pp. 54-57).

Wang, B., Sandeep, K., & Gupta, S. (2004). Extending lifetime of multicast trees in wireless ad hoc networks. *Journal of Information Science and Engineering . Special Issue on Mobile Computing*, *20*, 219–243.

Wasef, A., Lin, X., & Shen, X. (2010). *Complementing public key infrastructure to secure vehicular ad hoc network* (pp. 22–28). IEEE Wireless Communication.

Wattenhofer, R., & Zollinger, A. (2004). Xtc: A practical topology control algorithm for ad hoc networks. *Proceedings of Fourth International Workshop on Algorithms for Wireless, Mobile, Ad Hoc and Sensor Networks (WMAN)*, Santa Fe, NM.

Weingrtner, E., Vom Lehn, H., & Wehrle, K. (2009). A performance comparison of recent network simulators. *Proceedings of the 2009 IEEE International Conference on Communications.*

Welch, G., & Bishop, G. (2001). *An introduction to the Kalman □lter*. In SIGGRAPH 2001 Course Notes. *ACM*, Los Angeles, CA. Course 8.

Weniger, K., & Zitterbart, M. (2004). Address autoconfiguration in mobile ad hoc networks: Current approaches and future directions. *IEEE Network, 18*(4), 6–11. doi:10.1109/MNET.2004.1316754

Werner-Allen, G., Johnson, J., Ruiz, M., Lees, J., & Welsh, M. (2005). Monitoring volcanic eruptions with a wireless sensor network. *Proceedings of the 2nd European Workshop on Wireless Sensor Networks*

Werner-Allen, G., Lorincz, K., Johnson, J., Lees, J., & Welsh, M. (2006). Fidelity and yield in a volcano monitoring sensor network. *Proceedings of the 7th Symposium on Operating Systems Design and Implementation.*

Werner-Allen, G., Lorincz, K., Ruiz, M., Marcillo, O., Johnson, J., Lees, J., & Welsh, M. (2006). Deploying a wireless sensor network on an active volcano. *IEEE Internet Computing, Special Issue on Data- Driven Applications in Sensor Networks, 10*(2).

Wieselthier, J. E., Nguyen, G. D., & Ephremides, A. (2002). Energy limited wireless networking with directional antennas: The case of session-based multicasting. In *Proceedings of IEEE INFOCOM.*

Wieselthier, J. E., Nguyen, G. D., & Ephremides, A. (2000). On the construction of energy-efficient broadcast and multicast trees in wireless networks. *Proceedings - IEEE INFOCOM, 00*, 585–594.

Wikipedia. (n.d.). *Mobile ad hoc network*. Retrieved from http://en.wikipedia.org/wiki/ Mobile_ad_hoc_network

Williams, B., & Camp, T. (2002). Comparison of broadcasting techniques for mobile ad hoc networks. *In Proceedings of MOBIHOC'02*, June 2002

Wong, K. J., et al. (2003). Busnet: model and usage of regular traffic patterns in mobile ad hoc networks for inter-vehicular communications. *Proceedings of International Conference on Information and Communication Technologies (ICICT)*, 2003.

Wongngamnit, C., & Angluin, D. (2001, Mar). Robot localization in a grid. *Information Processing Letters, 77*, 261–267. doi:10.1016/S0020-0190(00)00154-X

Wu, C. W., & Tay, Y. C. (1999). AMRIS: A multicast protocol for ad hoc wireless networks. In *Proceedings of the IEEE Military Communications Conference* (pp. 25-29). Atlantic City, NJ: IEEE.

Wu, C. W., Tay, Y. C., & Toh, C. K. (1998). *Ad hoc multicast routing protocol utilizing increasing id-numbers (AMRIS) functional specification.* Internet-Draft, draft-ietf-manet-amris-spec-00.txt, 1998.

Wu, H. D., Siegel, M., Stiefelhagen, R., & Yang, J. (May 2002). *Sensor fusion using Dempster–Shafer theory.* In IEEE Instrumentation and Measurement Technology Conference, Anchorage, AK, USA.

Wu, J., & Li, H. (1999). A dominating set based routing scheme in ad hoc wireless network. In *Proceedings of Third International Workshop Discrete Algorithm and Methods for Mobile Computing and Communication*, (pp. 7-14).

Wu, J., & Li, H. (1999). On calculating connected dominating set for efficient routing in ad hoc wireless networks. In *Proceedings of the Third ACM International Workshop on Discrete Algorithms and Methods for Mobile Computing and Communications (ACM DIALM'1999)*, (pp. 7-14).

Wu, S.-L., Lin, C.-Y., Tseng, Y.-C., & Sheu, J.-P. (2000). *A new multi-channel MAC protocol with on-demand channel assignment for multi-hop mobile ad hoc networks.* International Symposium on Parallel Architectures. Algorithms and Networks.

Wu, Y. F., Xu, Y. L., Chen, C. L., & Wang, K. (2004). On the construction of virtual multicast backbone for wireless ad hoc networks. *IEEE International Conference on Mobile Ad hoc and Sensor Systems*, (pp. 294-303).

Wu, B., Chen, J., Wu, J., & Cardei, M. (2006). A survey on attacks and countermeasures in mobile ad hoc networks . In Xiao, Y., Shen, X., & Du, D.-Z. (Eds.), *Wireless/mobile network security*. doi:10.1007/978-0-387-33112-6_5

Wu, J., & Li, H. (2001). A dominating-set-based routing scheme in ad hoc wireless networks. *Telecommunication Systems Journal, 3*, 63–84.

Wu, S.-L., Tseng, Y.-C., & Sheu, J.-P. (2000). Intelligent medium access for mobile ad hoc networks with busy tones and power control. *IEEE Journal on Selected Areas in Communications, 18*(9), 1647–1657. doi:10.1109/49.872953

Xian, G., Tao, F., Ting, Y. Z., & Feng, M. J. (2010). The hierarchical threat model of routing security for wireless ad hoc networks. *Proceedings of IEEE First International Conference on Networking and Distributed Computing* (pp. 36-40).

Xian, X., Shi, W., & Huang, H. (2005). *Comparison of OMNET++ and other simulator for WSN simulation.* 3rd IEEE Conference of Industrial Electronics and Applications.

Xiao, L., Boyd, S., & Lall, S. (2005) A scheme for robust distributed sensor fusion based on average consensus. In *Proceedings of the 4th International Symposium on Information Processing in Sensor Networks (IPSN'05)*, (pp. 63–70).

Xiao, L., Boyd, S., & Lall, S. (2006). A space-time diffusion scheme for peer-to-peer least-squares estimation. In *Proceedings of the 5th International Conference on Information Processing in Sensor Networks (IPSN'06)*, (pp. 168–176).

Xie, B., Kumar, A., Agrawal, D., & Srinivasan, S. (2006). Secured macro/micro-mobility protocol for multi-hop cellular IP. *Journal of Pervasive and Mobile Computing, 2*(2), 111–136. doi:10.1016/j.pmcj.2005.11.004

Xie, J., Talpade, R. R., Mcauley, A., & Liu, M. (2002). AMRoute: Ad hoc multicast routing protocol. *Mobile Networks and Applications, 7*(6), 429–439. doi:10.1023/A:1020748431138

Xie, J., Talpade, R., McAuley, T., & Liu, M. (2002). AMRoute: Ad hoc multicast routing protocol. *ACM Mobile Networks and Applications (MONET) . Journal, 7*(6), 429–439.

Xie, R., Qi, D., Li, Y., & Wang, J. Z. (2007). A novel distributed MCDS approximation algorithm for wireless sensor networks. *Journal of Wireless Communications and Mobile Computing, 9*(3).

Xing, G., Tan, R., Liu, B., Wang, J., Jia, X., & Yi, C. W. (2009). *Data fusion improves the coverage of wireless sensor networks.* In MobiCom.

Xiong, Z., Liveris, A. D., & Cheng, S. (2004, September). Distributed source coding for sensor networks. *IEEE Signal Processing Magazine, 21*, 80–94. doi:10.1109/MSP.2004.1328091

Xu, N., Rangwala, S., Chintalapudi, K., & Ganesan, D. Broad, A., Govindan, R., & Estrin, D. (2004). A wireless sensor network for structural monitoring. *Proceedings of the 2nd International Conference on Embedded Networked Sensor Systems.*

Xu, Y., & Heidemann, J. & Estrin. D. (2000). *Adaptive energy-conserving routing for multi-hop ad hoc networks.* Technical Report 527, USC/Information Sciences Institute.

Xue, Y., Lee, H. S., Yang, M., Kumarawadu, P., Ghenniwa, H., & Shen, W. (2007). Performance evaluation of ns-2 simulator for wireless sensor networks. *Proceedings of the Canadian Conference on Electrical and Computer Engineering.*

Xu, S., Blackmore, K., & Jones, H. (2007). An analysis framework for mobility metrics in mobile ad hoc networks. *EURASIP Journal on Wireless Communications and Networking, 1*, 26–42.

Yadav, N. S., & Yadav, R. P. (2007). Performance comparison AND Analysis of on demand & table driven routing protocols. *International Journal of Information Technology, 4*(2). ISSN 13052403

Yagan, D., & Tham, C. K. (2005). *Adaptive QoS provisioning in wireless ad hoc networks: A semi-MDP approach. WCNC 2005* (pp. 2238–2244). IEEE Communications Society.

Yang, C.-L., Bagchi, S., & Chappell, W. J. (2005). Location tracking with directional antennas in wireless sensor networks. In *IEEE MTT-S International Microwave Symposium Digest.* Long Beach, CA: IEEE.

Yang, P., & Zheng, S. (2001). Security management in hierachical ad hoc network. *Proceedings of IEEE International Conference on Info-tech and Info-net*, (pp. 642-649).

Yang, H., Luo, H., Ye, F., Lu, S., & Zhang, L. (2004). Security in MANET: Challenges and solutions. *IEEE Wireless Communications*, *11*(1), 38–47. doi:10.1109/MWC.2004.1269716

Yang, H., Luo, H., Ye, F., Lu, S., & Zhang, S. (2004). Security in mobile ad hoc networks: Challenges and solutions. *IEEE Wireless Communications*, *11*(1), 38–47. doi:10.1109/MWC.2004.1269716

Yang, S., Zhang, X. Y. C., & Spyrou, E. (2010). Using social network theory for modeling human mobility. *IEEE Network*, 6–13. doi:10.1109/MNET.2010.5578912

Yen, Y. S., Chan, Y. K., Chao, H. C., & Park, J. H. (2008). A genetic algorithm for energy-efficient based multicast routing on MANETs. *Computer Communications*, *31*, 858–869. doi:10.1016/j.comcom.2007.10.033

Yi, S., Naldurg, P., & Kravets, R. (2002). *Security aware ad hoc routing for wireless networks*. Report No. UIUCDCS-R-2002-2290, UIUC, 2002.

Yih-Chun, H., & Johnson, D. B. (2002). *Ensuring cache freshness in on-demand ad hoc network routing protocol*. POMC'02, October 30–31, 2002, Toulouse, France. ACM.

Yi, P., Wu, Y., Zou, F., & Liu, N. (2010). A survey on security in wireless mesh networks. *IETE Technical Review*, *27*(1). doi:10.4103/0256-4602.58969

Yi, P., Zhong, Y. P., Zhang, S. Y., & Dai, Z. L. (2006). Flooding attack and defence in ad hoc network. *Journal of Systems Engineering and Electronics*, *17*(2), 410–416. doi:10.1016/S1004-4132(06)60070-4

Yoon, J., Liu, M., & Noble, B. (2003). Random waypoint considered harmful. *IEEE International Conference on Computer Communications*, (pp. 1312-1321).

Yoon, J., Liu, M., & Noble, B. (2003, April). Random waypoint considered harmful. In *Proceedings IEEE INFOCOM 2003*, San Francisco, CA.

Yu, B., Sycara, K., Giampapa, J. A., & Owens, S. R. (2004). Uncertain information fusion for force aggregation and classi□cation in airborne sensor networks. In *AAAI-04 Workshop on Sensor Networks*. AAAI Press.

Yuan, Y., & Kam, M. (2004, August). Distributed decision fusion with a random-access channel for sensor network applications. *IEEE Transactions on Instrumentation and Measurement*, *53*, 1339–1344. doi:10.1109/TIM.2004.830598

Zadeh, L. A. (1965). Fuzzy sets. *Information and Control*, *8*(3), 338–353. doi:10.1016/S0019-9958(65)90241-X

Zahrani, M. S., Loomes, M. J., Malcolm, J. A., & Dayem Ullah, A. Z. M., Steinhöfel, K., & Albrecht, A. A. (2008). Genetic local search for multicast routing with pre-processing by logarithmic simulated annealing. *Computers & Operations Research*, *35*, 2049–2070. doi:10.1016/j.cor.2006.10.001

Zapata, M. (2002). *Secure ad hoc on-demand distance vector* (SAODV). Internet Draft, draft-guerrero-manet-saodv-01.txt, 2002.

Zapata, M. G. (2001). *Secure ad-hoc on-demand distance vector (SAODV) routing*. IETF MANET, internet draft (Work in progress), draft-guerreromanet- saodv-00.txt.

Zapata, M. G., & Asokan, N. (2002). Secure ad hoc routing protocols. In *Proceeding of the ACM Workshop on Wireless Security*, Atlanta, GA September, 2002.

Zhang, H., & Jiang, Z. P. (2005). Theoretic analysis of broadcasting schemes for two-dimensional MANETs. In *Proceedings of the Workshop on Performance of Wireless Networks and Communication Systems (WiNCS)*, July.

Zhang, J., Chen, J., Fan, J., Xu, W., & Sun, Y. (2008). *OMNeT++ based simulation for topology control in wireless sensor network: A case study*. International Wireless Communications and Mobile Computing Conference.

Zhang, J., Fang, Z., & Bensaou, B. (2005). Adaptive power control for single channel ad hoc network. *Proceedings of IEEE ICC'05*, Seoul, (pp. 3156-3160).

Zhang, X., & Jacob, L. (2003). Multicast zone routing protocol in mobile ad hoc wireless networks. In *Proceedings of the 28th International Conference on Local Computer Networks* (pp. 150-159). Bonn, Germany: IEEE.

Zhang, Y., Cai, H., Lin, Y., & Xiao, J. (2007). *An ant colony system algorithm for the multicast routing problem*. Third International Conference on Natural Computation, ICNC 2007, 2007.

Zhang, Z., Pang, Z., Chen, J., Chen, Q., Tenhunen, H., Zheng, L., & Yan, X. (2009). *Two- layered wireless sensor networks for warehouses and supermarkets.* 3rd International Conference of Mobile Ubiquitous Computing, Systems, Services and Technologies.

Zhang, G. P. (2000). Neural networks for classification: A survey. *IEEE Transactions on Systems, Man, and Cybernetics*, *30*(4), 451–462. doi:10.1109/5326.897072

Zhang, L., Deering, S., & Estrin, D. (1993). RSVP: A new resource reservation protocol. *IEEE Network*, *7*(5), 8–18. doi:10.1109/65.238150

Zhang, Q., & Leung, Y. W. (1999). An orthogonal genetic algorithm for multimedia multicast routing. *IEEE Transactions on Evolutionary Computation*, *3*(1), 53–62. doi:10.1109/4235.752920

Zhang, W., & Cao, W. (2007). Defending against cache consistency attacks in wireless ad hoc networks. *Journal of Ad Hoc Networks*, *6*(3).

Zhang, X., & Jacob, L. (2004). MZRP: An extension of the zone routing protocol for multicasting in MANETs. *Journal of Information Science and Engineering*, *20*, 535–551.

Zhang, Y., Liu, W., Lou, W., & Fang, Y. (2006). Securing mobile ad hoc networks with certificateless public keys. *IEEE Transactions on Dependable and Secure Computing*, *3*(4), 386–399. doi:10.1109/TDSC.2006.58

Zhang, Y., & Soong, B. H. (2005). The effect of unreliable wireless channel on the call performance in mobile network. *IEEE Transactions on Wireless Communications*, *4*, 653. doi:10.1109/TWC.2004.842947

Zhao, J., Govindan, R., & Estrin, D. (2002). Residual energy scans for monitoring wireless sensor networks. In *Proceedings of the IEEE Wireless Communications and Networking Conference (WCNC'02)*, Vol. 1, IEEE, Orlando, FL, (pp. 356–362).

Zhao, Y., Xu, L., & Shi, M. (2003). On-demand multicast routing protocol with multipoint relay (ODMRP-MPR) in mobile ad hoc network. In *Proceedings of International Conference on Communication Technology (ICCT 2003)*, (pp. 1295-1300).

Zhao, S., Aggarwal, A., Frost, R., & Bai, X. (2011). A survey of applications of identity-based cryptography in mobile ad-hoc network. *IEEE Communication Surveys and Tutorials*, *99*, 1–21.

Zhou, L., & Haas, Z. J. (1999). Securing ad hoc networks. *IEEE Network Magazine*, *13*(6), 24–30. doi:10.1109/65.806983

Zhu, Q., Parsa, M., & Garcia-Luna-Aceves, J. (1995). *A source-based algorithm for delay-constrained minimum-cost multicasting* (pp. 377–385). IEEE INFOCOM.

About the Contributors

Kamaljit I. Lakhtaria is an Assistant Professor of CSE Department at the School of Engineering, Sir Padampat Singhania University, Udaipur, India. He holds position of Editor-in-Chief in many International Journal, He is reviewer in many International Journal including IEEE WSN, IEEE Potential. He leads many Research Groups, working with as team member with other members in Academic and Research Institutes.

Jiwa Abdullah received his B.Eng(Hons) in Electronic Engineering from Liverpool University, Master of Science in Digital Communication System and PhD from Loughborough University. Currently he is an Associate Professor at the Universiti Tun Hussein Onn, Malaysia. His main research interest is in the area of vehicular and mobile ad hoc networks, sensor networks, cognitive radio, broadband networks, and wireless communications. Other research areas are the application of computational intelligence to solve engineering problems. In the capacity as a lecturer he is involved in implementing OBE and Quality Assurance in the teaching and learning aspect within the Faculty of Electrical and Electronic Engineering, University Tun Hussein Onn Malaysia.

Lalit Agrawal is currently a Software Engineer in Amdocs Development Center India Limited. He completed his B.E. in Electronics and Communication Engineering from Purbanchal University (Nepal) in 2009 and M. Tech. with Specialization in Wireless Communication and Computing from IIIT-Allahabad. His areas of interests are wireless sensor networks, data fusion, analog and digital communication, telecommunication, and digital signal processing.

Ana V. Alejos received the M.S. and PhD. degree from University of Vigo, Spain in 2000 and 2006. Her Maste'sr Thesis obtained in 2002 the Ericcson Award by the Spanish Association of Electrical Engineers, as the best Multimedia Wireless Project. She is granted with the Marie Curie International Outgoing Fellowship, carrying out the outgoing phase in the New Mexico State University (NM, USA), at the Klipsch School of ECE, with a research focused on measurement and modeling of propagation through dispersive media, and radar waveform generation. Her research work includes radio propagation, communication electronics, wideband radio channel modeling, multimedia wireless systems, waveform and noise code design, and radar. Dr. Alejos is an IEEE member and also a reviewer of several IEEE and IET journals.

Erwin Anggadjaja gained his Bachelor degree in Computer Engineering (Embedded Systems concentration) from BINUS University, Indonesia in 2001. In 2004 he received his Master's degree from

University of Technology, Sydney (UTS), Australia, majoring in Telecommunications Engineering. He was a full-time lecturer in Electrical Engineering Department of Universitas Pelita Harapan, Indonesia prior to taking a Doctoral degree in 2008. Now he is a PhD. Candidate at Nanyang Technological University (NTU), Singapore, under the School of Computer Engineering, in affiliation with Earth Observatory of Singapore. His research interests include embedded systems, wireless networking & communication systems, and microcontroller design & application. He is currently working on wireless sensor network modelling extending wireless-TCP protocol concept.

Alagan Anpalagan received the B.A.Sc. (H), M.A.Sc. and Ph.D. degrees in Electrical Engineering from the University of Toronto, Canada. He joined the Department of Electrical and Computer Engineering at Ryerson University in 2001 and was promoted to Full Professor in 2010. He served the department as Graduate Program Director (2004-09) and the Interim Electrical Engineering Program Director (2009-10). During his sabbatical (2010-11), he was a Visiting Professor at Asian Institute of Technology and Visiting Researcher at Kyoto University. Dr. Anpalagan's industrial experience includes working at Bell Mobility on 1xRTT system deployment studies (2001), at Nortel Networks on SECORE R&D projects (1997) and at IBM Canada as IIP Intern (1994). Anpalagan directs a research group working on radio resource management (RRM) and radio access & networking (RAN) areas within the WINCORE Lab. His current research interests include cognitive radio resource allocation and management, wireless cross layer design and optimization, collaborative communication, green communications technologies and QoE-aware femtocells. Dr. Anpalagan serves as Associate Editor for the *IEEE Communications Letters* (2010-) and *Springer Wireless Personal Communications* (2009-12), and past Editor for *EURASIP Journal of Wireless Communications and Networking* (2004-2009). He also served as EURASIP Guest Editor for two special issues in Radio Resource Management in 3G+ Systems (2006) and Fairness in Radio Resource Management for Wireless Networks (2008). Dr. Anpalagan served as TPC Co-Chair of: IEEE PIMRC'11 Track on Cognitive Radio and Spectrum Management, IEEE IWCMC'11 Workshop on Cooperative and Cognitive Networks, IEEE CCECE'04/'08 and WirelessCom'05 Symposium on Radio Resource Management. Anpalagan served as IEEE Toronto Section Chair (2006-07), ComSoc Toronto Chapter Chair (2004-05), Chair of IEEE Canada Professional Activities Committee (2009-11). He is the recipient of the Dean's Teaching Award (2011), Faculty Scholastic, Research and Creativity Award (2010), Faculty Service Award (2010) at Ryerson University. Dr. Anpalagan also completed a course on Project Management for Scientist and Engineers at the University of Oxford CPD Center. He is a registered Professional Engineer in the province of Ontario, Canada.

Reshma Channappa's penchant for creative thought process and a single-minded dedication towards knowledge-gathering propelled her to successfully earn a Bachelor's in Computer Science from one of the most well-reputed institutions in the country, the *M. S. Ramaiah Institute of Technology (MSRIT)*. The sequential understanding of the core concepts in Computer Science from the expert faculty lead to the subsequent career in the capacity of a Research & Development Engineer with the telecom major *Nokia Siemens Networks (NSN)* for 3+ years.

Ashraf Darwish received his PhD degree in Computer Science in 2006 from Computer Science Department at Saint Petersburg State University (specialization in Artificial Intelligence), and joined as Lecturer (Assistant Professor) at the Computer Science department, Faculty of Science, Helwan University in 25 June 2006. He is a member of such notable computing associations as the IEEE, ACM, EMS (Egyptian Mathematical Society), QAAP (Quality Assurance and Accreditation Project, Egyptian Supreme Council), Quality Assurance and Accreditation Authority (Egypt), board of the Russian-Egyptian Association for graduates. Dr. Darwish is author of many scientific publications and his publications include papers, abstracts and book chapters concerning computational intelligence. He keeps in touch with his mathematical background through his research. His research, teaching and consulting mainly focuses on artificial intelligence, information security, data and Web mining, intelligent computing, image processing(in particular image retrieval, medical imaging), modeling and simulation, intelligent environment, body sensor networking. This has resulted in publications in journals like *IEEE, Springer, IJCSN,* and *Advances in Computer Science and Engineering*. Dr. Darwish is one of the founders of CYPER SECURITY RESEARCH LAB (CSRL), organized by Professor Aboul Ella Hassanein, Faculty of Computers and Information, Cairo University (Egypt), and he is a member of research projects like Bio-Inspired HCV Analysis (Egypt), Breast Cancer (Egypt), E-Learning (Egypt).

Muhammad Dawood received his BE degree from the NED University of Engineering and Technology, Karachi, Pakistan, in 1985, and his MS and Ph.D. degrees from the University of Nebraska-Lincoln (UNL), in 1998 and 2001. Dr. Dawood is involved in teaching both nationally and internationally, and has been actively performing research in RF/microwave systems and subsystems, and signal processing algorithms since 1995. At present, he is an Associate Professor at the Klipsch School of Electrical and Computer Engineering, the New Mexico State University (NMSU), Las Cruces, New Mexico. Prior to joining the NMSU, Dr. Dawood served as a Research Assistant Professor and as a Research Engineer, respectively, at the University of Kansas, Lawrence, and with Tellabs Research Center, Misshawaka, IN. His research interests lie in the area of electromagnetics. He is a member of IEEE.

Supriya Gupta is currently pursuing her research for PhD in the Department of Computer Sc. and IT, University of Jammu, Jammu, India. She obtained her master's degree in Computer Applications from University of Jammu. Besides, she is also working as Project Associate in the same department in a major research project to trace out vulnerabilities in network applications sanctioned to Dr. Lalitsen Sharma by the University Grants Commission, Ministry of Human Resource Development, Govt. of India. She has published five research articles in national and international journals and delivered conference presentations. Her area of interest is information security, network protocol analysis, and data mining.

Muhammad Jaseemuddin received B.E. from N.E.D. University of Engg. & Tech., Karachi, Pakistan, in 1989, M. S. from The University of Texas at Arlington in 1991, and Ph.D. from The University of Toronto in 1997. He worked as a post-doctoral research fellow in the Network Architecture Lab at The University of Toronto on Network Management using Virtual Network. He worked in Advanced IP group and Wireless Technology Lab (WTL) at Nortel Networks from 1998 to 2001. He worked on Wireless Service Delivery Platform and UMTS VHE prototype for mobile service delivery. He contributed to the development and implementation of Open IP suite of IP protocols. In WTL, he worked on QoS, Routing and Handover issues in mobile wireless IP access network. He has been Associate Professor at Ryerson University since 2002. His research interests include IP mobility, IP traffic engineering, routing protocols, multi-hop mobile wireless networks (ad hoc, mesh, and sensor), vertical system integration, MAC and routing for multi-hop wireless networks with smart beamforming antennas, routing and relay selection in multi-hop cooperative communication networks, and multicast routing, caching, and P2P overlay networks for multimedia delivery and video streaming in mobile wireless ad hoc networks.

Lamia Kaddar, PhD at PRiSM, University of Versailles Saint Quentin en Yvelines, and her main research interests include improvement of transmission power, security, and energy efficiency in both sensor and ad hoc wireless multimedia network.

Lamiaa Khalid received her Bachelor degree (B.Sc.) in Electrical and Electronics Engineering from Ain Shams University, Egypt in 2000 and her Master degree (MASc.) in Electrical and Computer Engineering from Ryerson University, Canada in 2008. Lamiaa is currently a PhD candidate in the Department of Electrical and Computer Engineering at Ryerson University. Her current research interests include spectrum sensing and resource allocation and management in cognitive radio networks.

Jahangir Khan belongs to a small village of Dadawar shamozai Swat, currently serving as Assistant Professor since January 2011, in Department of Computer Science & IT, Sarhad University of Science & Information Technology Peshawar Pakistan. He is PhD (Computer Science) scholar at Graduate school of science & Engineering, PAF-KIET Pakistan. He work as a Research Associate during (Sep 2004-Oct 2010), in college of computing and Information sciences, PAF-KIET Pakistan. He do his MS in (computer & Information Technology), from IICT University of Sindh Jamshoro Pakistan in 2004 and Graduation in (Double math's and physics) from Peshawar University Pakistan (2000). His area of interest includes Future Generation Technology, Handover cellular technology and Mobile Ad-Hoc Networks, Software engineering and intelligent effort estimation, using multiple tools for program development, i.e. OPNET, MATLAB, and Bayesian belief networks.

Shrirang Ambaji Kulkarni is working as Assistant Professor in the Department of Computer Science and Engineering at Gogte Institute of Technology, Belgaum. He received his B.E. in Computer Science and Engineering from Karnataka University Dharwar, in the year 2000 and M.Tech. in Computer Network Engineering from VTU Belgaum, in the year 2004. He is currently pursuing Ph.D. in the area of Ad-Hoc networks at N.I.E Mysore. He has published more than 15 papers in international, national journals and international conferences. His current research interest includes reinforcement learning algorithms and wireless ad hoc networks. He is a life member of CSI and ISTE.

Alok Kumar received the B.E. degree from J.N.Vyas University, Jodhpur, India, in Information Technology, in August, 2005, and the M.Tech. degree from Centre for Development of Advanced Computing, Noida, India, in Computer Science and Engineering, in July, 2007. He submitted his Ph.D. degree in the Information Technology, from Indian Institute of Information Technology, Allahabad, India. His Ph.D. work mainly focuses on the geometrical aspects networks and complexity of quality of service (QoS) routing under both static and dynamic network. Currently, he is working as an Assistant Professor at National Institute of Technology Patna. He is a member of the IEEE and ACM. He published many journals, and conference papers in repute.

Ian McLoughlin completed his PhD at the University of Birmingham, UK in 1997 funded by Philips Telecommunications, Cambridge. During his 25 year career, he has worked for GEC Hirst Research

Centre, Her Majesty's Government Communications Centre, Tait Electronics Group Research (Christchurch, New Zealand), and Nanyang Technological University (Singapore), where he is presently an Associate Professor. Dr McLoughlin has published over 100 peer-reviewed papers, has books in print with Cambridge University Press and McGraw-Hill, and has several worldwide patents. He is a member of the IET, a senior member of the IEEE, a chartered engineer (UK) and an European Ingénieur.

Natarajan Meghanathan is a tenured Associate Professor of Computer Science at Jackson State University, Jackson, MS. He graduated with a Ph.D. in Computer Science from The University of Texas at Dallas in May 2005. Dr. Meghanathan has published more than 120 peer-reviewed articles (more than half of them being journal publications). He has also received federal education and research grants from the U. S. National Science Foundation and the Army Research Lab. Dr. Meghanathan has been serving in the editorial board of several international journals and in the technical program committees and organization committees of several international conferences. His research interests are wireless ad hoc networks and sensor networks, graph theory, network and software security, bioinformatics, and computational biology. For more information, visit http://www.jsums.edu/cms/nmeghanathan.

Mohammad Reza Meybodi received the B.S. and M.S. degrees in Economics from Shahid Beheshti University in Iran, in 1973 and 1977, respectively. He also received the M.S. and Ph.D. degree from Oklahoma University, USA, in 1980 and 1983, respectively in Computer Science. Currently, he is a full professor in Computer Engineering Department, Amirkabir University of Technology, Tehran, Iran. Prior to current position, he worked from 1983 to 1985 as an Assistant Professor at Western Michigan University, and from 1985 to 1991 as an associate professor at Ohio University, USA. His research interests include wireless networks, fault tolerant systems, learning systems, parallel algorithms, soft computing, and software development.

Sudipta Mondal did B.Tech(Hons.) in Computer Science & Engineering from Haldia Institute of Technology, Haldia and did M.Tech in Computer Science & Engineering from National Institute of Technology, Durgapur. He is gold medalist in M.Tech in Computer Science & Engineering in 2010 from National Institute of Technology, Durgapur. He is currently working as an Assistant Professor in Information Technology department at Bengal College of Engineering & Technology, Durgapur. He has 7 years of teaching experience. His area of interests includes data structures & algorithms, formal language & automata theory, computer graphics, visual cryptography, multi-objective optimization, and delay tolerant networks. He has guided several projects at under graduate level.

Jaya Nagori completed B.E. in Computer Engineering from Purbanchal University (Nepal) in 2008 and M. Tech. with Specialization in Wireless Communication and Computing at IIIT-Allahabad in 2011. She also worked as Lecturer in Eastern Engineering College, Biratnagar, Nepal after graduation. She is currently working as a Business Analyst in Groupsoft Technologies. Her area of interests includes wireless networks, mobile database computing, and computer graphics.

Abou Bakar Nauman is Assistant Professor at Sarhad University of Science and Information Technology. He is Doctoral candidate at Comsats Institute of Information technology, Pakistan. He has done

M.Sc. Computer Science in 2000 and after that has professional experience in multinational organizations. His areas of interest are Bayesian statistics, Information Systems, and networks planning.

Bhaskar N. Patel had completed his 12th science with 86% from small village named Ranasan in 1999. After that he has take admission in computer engineering at Sankalchand Patel College of Engineering at Visnagar and complete his Bachelor degree with distinction in 2003. After that He had Work as lecturer at Bhandu engineering college for 5 year and during that period, he was sponsored from the college for Master and completed his Master's from Birla Viswakarma Mahavidhyalaya (BVM) with first class in Aug 2008. Since February 2009, he has been working as a head of Computer and IT Department in B.S.Patel Polytechnic, Kherva, Ganpat University, Mehsana, Gujarat, India.

Paula Gómez Pérez received the M.S. and PhD. degree from University of Vigo, Spain, in 2003 and 2010, respectively. She is responsible of the subjects Transmission Media, and the Laboratory of several subjects related with Radiofrequency technologies, such as Radiocommunication, Radiobroadcast, Radiocommunication Technologies, Optical Communications, Radiodetermination. E.T.S.I. Telecomunicación, University of Vigo (Vigo, Spain). As a Researcher Engineer at University of Vigo, she has performed several projects concerning radar systems, remote sensing and antenna design for several industries.

Satish G. Prajapati had completed his 12th science with 75% from a small village named Jetpur in 2000. After that he has taken admission in Information Technology at AITS at Rajkot and completed his Bachelor degree with First class in 2005. After completed B.E, he joined a Master's at Birla Viswakarma Mahavidhyalaya (BVM). He completed his Master's in Computer Engineering with second class in Aug 2011. During the dissertation period (Aug 2007) he was Lecturer at B.S.Patel Polytechnic, Kherva, Ganpat University, Mehsana, Gujarat, India.

G. Raghavendra Rao received his Master's in Engineering from Indian Institute of Science Bangalore in the year 1987 and his Ph.D. from University of Mysore in the year 1999. He is currently working as Professor in Department of Computer Science and Engineering at NIE Mysore. He has published more than 45 papers in International, National Journals and International Conferences His current research interest includes Genetic Algorithms, Wireless Networks, Cryptography and Network Security. He is a member of IEEE, ACM and ISTE.

T. Rama Rao currently, working as Professor & Head, Department of Telecommunication Engineering, Faculty of Engineering & Technology, SRM University, India. He received his PhD degree on Radio Wave Propagation studies for Fixed and Mobile Communications over Southern India from Sri Venkateswara University, Tirupati, India in the year 2000. He worked with Aalborg University, Denmark as Assistant Research Professor; with Universidad Carlos III de Madrid, Spain and at the University of Sydney, Australia as 'Visiting Professor'. He served as a PostDoc Research Fellow at National Chio Tung Univeristy, Hsinchu, Taiwan. He has a long-standing research history on radiowave propagation studies for wireless communications. His research interests are, radio channel measurements & modeling, broadband wireless communications/networks, and wireless information networks.

Niranjan Kumar Ray received his MTech degree in Computer Science from Utkal University, Bhubaneswar in 2007. He has been a PhD student in the department of Computer Science and Engineering at National Institute of Technology, Rourkela since 2008. His research interest includes wireless ad hoc and sensor networks, and at present, he is working on MAC and routing protocols for energy effciency. He is a student member of IEEE.

Manuel García Sánchez (S'88-M'93) received the Ingeniero de Telecomunicación degree from the Universidad de Santiago de Compostela, Spain, in 1990 and the Doctor Ingeniero de Telecomunicación (Ph.D.) degree from the Universidad de Vigo, Spain, in 1996. Professor Sánchez is currently with the Teoría de la Señal y Comunicaciones Department at Universidad de Vigo, where he teaches "Broadcasting" and "Radiodetermination" to undergraduate students. He is the head of the Radio Systems research group. His research interests include studies of indoor and outdoor radio channel sounding and modelling for narrow and wide-band applications at microwave and millimetre wave frequencies and radio interference detection and mitigation, radio network deployment, radio location, and radio system development.

Lalitsen Sharma received his Doctorate in Computer Science and Engineering form Guru Nanak Dev University Amritsar, Punjab, India. He also holds Master's degree in Mathematics and Computer Applications from the same university. He has been teaching to post graduate students in computer applications of University of Jammu for more than 13years. He is a life member of India Science Congress Association, Institute of Electronics and Communication Engineers and National HRD Network, India. He is specialized in Data Communication and Network, Internet and WWW and Data Structures. He is currently working on a major research project to trace out vulnerabilities in network applications funded by University Grants Commission, Ministry of Human Resource Development, Govt. of India. He is currently supervising three PhD scholars.

Dharmendra K Singh is presently working as Head, Department of Electronics and Communication & Information Technology, BIT Sindri, Dhanbad. He has more than 20 years of teaching experience. He is heading the department of Electronics and Communication & Information technology since 2002. He is instrumental in starting the curriculum on information technology. He has published more than 30 papers in journals and conferences. He has already supervised 01 thesis at doctoral level in computer Science & Engg and 05 research scholars are presently enrolled under him for their doctoral degree. The area of research he works are coding theory, cryptography, optical amplifiers, photonic crystal fibers, e-governance, and technical education planning

Sudha Singh received her M.Tech. degree in Computer Science from Birla Institute of Technology, Mesra, Ranchi, India, in 2002 and was placed First Class. She obtained her Ph.D. degree in Computer Science and Engineering from BIT Sindri,(VBU),India in 2009. She is currently an Associate Professor in PG Department of Computer Science and Engineering. Bengal college of Engineering and Technology, Bidhan Nagar, Durgapur, India. She has more than 25 publications. Her research interest is in recent advances in Data communication, network security, and information security, coding theory, signal processing and combinatorics.

K G Srinivasa received his PhD in Computer Science and Engineering from Bangalore University in 2007. He is now working as a professor in the Department of Computer Science and Engineering, M S Ramaiah Institute of Technology, Bangalore. He is the recipient of All India Council for Technical Education - Career Award for Young Teachers, Indian Society of Technical Education – ISGITS National Award for Best Research Work Done by Young Teachers, Institution of Engineers(India) – IEI Young Engineer Award in Computer Engineering, IMS Singapore – Visiting Scientist Fellowship Award. He has published more than fifty research papers in International Conferences and Journals. He has visited many Universities abroad as a visiting researcher – He has visited University of Oklahoma, USA, Iowa State University, USA, Hong Kong University, Korean University, National University of Singapore are few prominent visits. He has authored two books namely File Structures using C++ by TMH and Soft Computer for Data Mining Applications LNAI Series – Springer. Recently he has been awarded BOYSCAST Fellowship by DST, for conducting collaborative Research in University of Melbourne.

Javad Akbari Torkestani received the B.S. and M.S. degrees in Computer Engineering in Iran, in 2001 and 2004, respectively. He also received the Ph.D. degree in Computer Engineering from Science and Research University, Iran, in 2009. Currently, he is an Assistant Professor in Computer Engineering Department at Arak Azad University, Arak, Iran. Prior to the current position, he joined the faculty of the Computer Engineering Department at Arak Azad University as a Lecturer. His research interests include wireless networks, mobile ad hoc networks, fault tolerant systems, learning systems, parallel algorithms, and soft computing.

Ashok Kumar Turuk received his PhD degree in Computer Science and Engineering from IIT Kharagpur in 2005. He joined National Institute of Technology, Rourkela (Formerly REC Rourkela) as a faculty in the department of Computer Science and Engineering in 1997 and presently held the position of Associates Professor and working as the head of the department. He published more than 50 different papers in journals and conference proceedings. His research interest includes wireles ad hoc, sensor, and optical networks. He was the program chair of of International Conference on Communication Computing and Security (ICCCS-2011). He is the principal investigator (PI) of different projects funded by Government agencys like DST and DIT. He is a member of IEEE Computer Scociety, Indian Society for Technical Education (ISTE), Computer Society of India (CSI), and Orissa Bigyan Academy.

Shirshu Varma graduated in Electronics and Communication Engineering from Allahabad University and post graduated in Communication Engineering from BIT Mesra Ranchi, India. He completed his Ph.D in Optical Communication from University of Lucknow. He has served many organizations like BIT Mesra Ranchi, IET Lucknow, C-DAC Noida in the capacity of Lecturer, Sr. Lecturer, & IT Consultant. Presently he is working Associate Professor in IIIT Allahabad. Dr. Varma has published about 27 papers in international and national journals and conferences of repute. He is a member of IEEE and life member of ISTE. He has been a recipient of many national awards in this area. His areas of interest are intelligent sensor network, wireless sensor network, optical wireless communication, wireless communication, and networks.

Revathi Venkataraman is doing her PhD and working as Assistant Professor in the Department of Computer Science and Engineering, SRM University. Her research interests include wireless networks and security, trust computing and routing in ad hoc networks. She was a visiting research faculty at the Electrical Enggdept, Viterbi School of Engineering, University of Southern California for a duration of six months. Her other research interests are wireless ad hoc and sensor network test bed development swhich are ongoing research activities funded by Indian government.

Archana Venkatesha is working as a Senior Software Executive at *Nokia Siemens Networks*, a global data networking and telecommunications company. I have been working there from the past 3+ years, from 2008. I completed my undergraduate program from *M S Ramaiah Institute of Technology, Bangalore* in the year 2008. Our final year project titled *"Combinatorial Approach to Key Generation using Multiple Key Spaces for Wireless Sensor Networks"* under the guidance of our professor, Dr. K. G. Srinivasa was published in the 16th International Conference on Advanced Computing and Communication (*ADCOM 2008*) held at Chennai, and chosen as one the best papers in the conference.

Poornima Venkatesha is an MS student at the Computer Science Department of State University of New York at Stony Brook, since Fall 2010. She has worked as a Software Engineering Intern at Akamai Technologies during summer, 2011. She was working as a Software Engineer at the Research & Development division of Avaya (previous Nortel Networks), Bangalore, India between July 2008 and July 2010. She finished undergraduate studies in Computer Science at M S Ramaiah Institute of Technology, Bangalore in the year 2008.

Index

A

B

C

D

E

F

G

H

I

J

L

M

N

O

P

Q

R

S

www.ingramcontent.com/pod-product-compliance
Lightning Source LLC
LaVergne TN
LVHW062137200926
848484LV00007B/1823
* 9 7 8 1 4 6 6 6 0 3 2 1 9 *